Fundamentals and Basic Optical Instruments
Volume 1

OPTICAL SCIENCE AND ENGINEERING

Founding Editor
Brian J. Thompson
University of Rochester
Rochester, New York

RECENTLY PUBLISHED

*Please visit our website **www.crcpress.com** for a full list of titles*

Fundamentals and Basic Optical Instruments
Volume 1

Edited by
Daniel Malacara-Hernández
Brian J. Thompson

CRC Press
Taylor & Francis Group
Boca Raton London New York

CRC Press is an imprint of the
Taylor & Francis Group, an **informa** business

CRC Press
Taylor & Francis Group
6000 Broken Sound Parkway NW, Suite 300
Boca Raton, FL 33487-2742

First issued in paperback 2019

ISBN-13: 978-1-4987-2074-8 (hbk)
ISBN-13: 978-0-367-87296-0 (pbk)

Library of Congress Cataloging-in-Publication Data

Names: Malacara, Daniel, 1937- editor. | Thompson, Brian J., editor.
Title: Fundamentals and basic optical Instruments ; Advanced optical instruments and techniques / [edited by] Daniel Malacara Hernandez and Brian J. Thompson.
Other titles: Handbook of optical engineering. | Advanced optical instruments and techniques
Description: Second edition. | Boca Raton : CRC Press, 2017-2018. | Series: Optical science and engineering | "The second edition of the Handbook of Optical Engineering has been enlarged and it is now divided in two volumes. The first volume contains thirteen chapters and the second volume contains twenty one, making a total of 34 chapters"--Volume 1, preface. | Includes bibliographical references and index. Contents: volume 1. Fundamentals and basic optical instruments -- volume 2. Advanced optical instruments and techniques.
Identifiers: LCCN 2017021177| ISBN 9781498720748 (hardback : v. 1) | ISBN 9781315119984 (ebook : v. 1) | ISBN 9781498720670 (hardback : v. 2) | ISBN 9781315119977 (ebook : v. 2)
Subjects: LCSH: Optics--Handbooks, manuals, etc. | Optical instruments--Handbooks, manuals, etc.
Classification: LCC TA1520 .H368 2018 | DDC 621.36--dc23
LC record available at https://lccn.loc.gov/2017021177

Contents

Preface

The Second Edition of the Handbook of Optical Engineering has been enlarged and is now divided into two volumes. The first volume contains thirteen chapters and the second volume contains twenty-one, making a total of thirty-four chapters. In the first volume, Chapter 4, "Ray Tracing," by Ricardo Florez-Hernández and Armando Gómez-Vieyra; Chapter 5, "Optical Design and Aberrations," by Armando Gómez-Vieyra and Daniel Malacara-Hernández; Chapter 9, "Polarization and Polarizing Optical Devices," by Rafael Espinosa-Luna and Qiwen Zhan; and Chapter 12, "Microscopes," by Daniel Malacara-Doblado and Alejandro Téllez-Quiñones, have been added. In Volume 2, Chapter 1, "Optics of Biomedical Instrumentation," by Shaun Pacheco, Zhenyue Chen and Rongguang Liang; Chapter 7, "Active and Adaptive Optics," by Daniel Malacara-Hernández and Pablo Artal; Chapter 13, "Color and Colorimetry," by Daniel Malacara-Doblado; and Chapter 14, "The Human Eye and Its Aberrations," by Jim Schwiegerling, are new.

Most of the remaining chapters were also updated, some with many changes and others with slight improvements. The contributing authors of some chapters have been modified to include a new collaborator.

The general philosophy has been preserved to include many practical concepts and useful data.

This work would have been impossible without the collaboration of all authors of the different chapters and many other persons. The first editor is grateful to his assistant, Marissa Vásquez, whose continuous and great help was of fundamental importance. Also, the first editor is especially grateful to his family, mainly his wife, Isabel, who has always encouraged and helped him in many ways.

Daniel Malacara-Hernández
Brian J. Thompson

Contributors

Sofia E. Acosta-Ortiz
Laser Tech, S.A. de C.V.
Aguascalientes, Ags., México

Maximino Avendaño-Alejo
CCADET
Universidad Nacional Autónoma de México
CDMX, México

Glenn D. Boreman
Department of Physics and Optical Science
University of North Carolina, Charlotte
Charlotte, North Carolina, USA

Rafael Espinosa-Luna
Centro de Investigaciones en Óptica, AC
León, Gto., México

Ricardo Flores-Hernández
Centro de Investigaciones en Óptica, AC
León, Gto., México

Armando Gómez-Vieyra
Universidad Autónoma Metropolitana
CDMX, México

Maureen S. Kirk
Department of Veterinary Medicine
Texas A&M University
College Station, Texas, USA

Daniel Malacara-Doblado
Centro de Investigaciones en Óptica, AC
León, Gto., México

Daniel Malacara-Hernández
Centro de Investigaciones en Óptica, AC
León, Gto., México

Duncan T. Moore
University of Rochester
Rochester, New York, USA

Cristina Solano
Centro de Investigaciones en Óptica, AC
León, Gto., México

Orestes Stavroudis
Centro de Investigaciones en Óptica, AC
León, Gto., México

Marija Strojnik
Centro de Investigaciones en Óptica, AC
León, Gto., México

Alejandro Téllez-Quiñones
Centro de Investigaciones en Geografía y
    Geomática "Ing. Jorge L. Tamay", AC
Mérida, Yuc., Mexico

William Wolfe
College of Optical Sciences
University of Arizona
Tucson, Arizona, USA

Qiwen Zhan
Department of Electro-Optics and Photonics
University of Dayton
Dayton, Ohio, USA

1

Basic Ray Optics

Orestes Stavroudis
and Maximino
Avendaño-Alejo

1.1 Introduction

Geometrical optics is a peculiar science. It consists of the physics of the seventeenth and eighteenth centuries thinly disguised by the mathematics of the nineteenth and twentieth centuries. Its contemporary applications are almost entirely in optical design which, like all good engineering, remains more of an art even after the advent of the modern computer. This brief chapter is intended to convey the basic formulas as well as the flavor of geometrical optics and optical design in a concise and compact form. We have attempted to arrange the subject matter logically, although not necessarily in historical order.

The basic elements of geometrical optics are *rays* and *wavefronts*: neither exist, except as mathematical abstractions. A ray can be thought of as a beam of light with an infinitesimal diameter. However, to make a ray experimentally by passing light through a very small aperture causes diffraction to rear its ugly head and the light spreads out over a large solid angle. The result is not a physical approximation to a ray but a distribution of light in which the small aperture is a point source. A wavefront is defined as a surface of constant phase to which definite properties can be attributed, such as principal directions, principal curvatures, cusps, and other singularities. But, like the ray, the wavefront cannot be observed. Its existence can only be inferred circumstantially with interferometric methods.

However, there is in geometrical optics an object that is observable and measurable: the *caustic surface* [1]. It can be defined in distinct but equivalent ways:

- As the envelope of an *orthotomic* system of rays; i.e., rays ultimately from a single object point.
- As the cusp locus of a wavefront train, or, equivalently, the locus of points where the element of area of the wavefront vanishes.

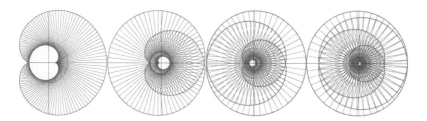

FIGURE 1.1 The first four caustics and wavefronts formed considering multiple reflections when the point source is placed at the vertex of the surface. The caustics are represented by the inner curves and their respective wavefronts are represented by outer curves. For the caustic in the first reflection, a curve called cardioid is formed. We can clearly see that the straight lines are the locus of the principal centers of curvature for each of the wavefronts, respectively.

We think the most useful definition is that the caustic is the locus of principal centers of curvature of a wavefront. In general, every surface has two principal curvatures at each of its points. This definition then shows clearly that the caustic is a two-sheeted surface. On the other hand, a system of rays originating from a common object is called a normal congruence.

A normal congruency of rays incident on either a refracting or a reflecting surface gives rise to an aggregate of refracted or reflected rays, which also constitutes a normal congruence. A second property of a normal congruence is that associated with it is a family of orthogonal surfaces, which are commonly regarded as wavefronts. The theorem of Malus and Dupin states that: A normal congruence will remain a normal congruence after any number of reflections or refractions [2]. In Figure 1.1, four caustics and wavefronts considering a circular reflecting surface when a point source is placed at the vertex of circular surface are shown.

1.2 Gaussian Optics à la Maxwell

Usually the formulas of Gaussian optics are derived from paraxial optics, a system based on approximations to the equations for ray tracing. We will encounter these in a subsequent section. Maxwell, on the other hand, took a global approach. He used a model of a perfect optical instrument and from that model, in a very elegant but straightforward way, deduced its properties, defined its parameters, and derived the various equations associated with Gaussian optics. Gauss actually found the equations for paraxial optics from the first-order terms of two power series expansions. While this is not a forum appropriate for a detailed discussion of the method Maxwell used, we will present an outline of his argument.

Maxwell [3] began by assuming that a perfect lens maps each point in object space into one and only one point in image space. Since a lens turned around is still a lens, the inverse of this mapping has to have exactly the same mathematical structure. Included in this mapping and its inverse are points at infinity whose images are the focal points of the instrument. The mapping that Maxwell chose is the linear fractional transformation,

$$x' = \frac{a_1 x + b_1 y + c_1 z + d_1}{a x + b y + c z + d}$$

$$y' = \frac{a_2 x + b_2 y + c_2 z + d_2}{a x + b y + c z + d},$$

$$z' = \frac{a_3 x + b_3 y + c_3 z + d_3}{a x + b y + c z + d}$$

(1.1)

where (x, y, z) represents a point in object space and where (x', y', z') is its image. The inverse transform has an identical structure,

$$x = \frac{A_1 x' + B_1 y' + C_1 z' + D_1}{Ax' + By' + Cz' + D}$$

$$y = \frac{A_2 x' + B_2 y' + C_2 z' + D_2}{Ax' + By' + Cz' + D}, \qquad (1.2)$$

$$z = \frac{A_3 x' + B_3 y' + C_3 z' + D_3}{Ax' + By' + Cz' + D}$$

another linear fractional transform. Here, the coefficients, denoted by capital letters, are determinants whose elements are the coefficients that appear in Equation (1.1).

The fractional-linear transformation maps planes into planes. This can be seen in the following way. Suppose a plane in object space is given by the equation,

$$px + qy + rz + s = 0, \qquad (1.3)$$

into which we substitute (x, y, z) from Equation (1.2). The result is

$$(pA_1 + qA_2 + rA_3 + sA)x' + (pB_1 + qB_2 + rB_3 + sB)y' +$$
$$(pC_1 + qC_2 + rC_3 + sC)z' + (pD_1 + qD_2 + rD_3 + sD) = 0, \qquad (1.4)$$

clearly the equation of a plane in image space that is evidently the image of the plane in object space.

This transformation, therefore, maps planes into planes. Since a straight line can be represented as the intersection of two planes, it follows that this transform maps straight lines into straight lines.

From Equation (1.1), we can see that the plane in object space, $ax + by + cz + d = 0$ is imaged at infinity in object space; from Equation (1.2), infinity in object space is imaged into the plane $Ax' + By' + Cz' + D = 0$ in image space.

We have established coordinate systems in both object and image space. Now we impose conditions on the coefficients that bring the coordinate axes into correspondence. First, we look at a plane through the coordinate origin of object space perpendicular to the z-axis, Equation (1.3), with $r = s = 0$, as its equation. From this, and Equation (1.4), we obtain the equation of its image,

$$(pA_1 + qA_2)x' + (pB_1 + qB_2)y' + (pC_1 + qC_2)'z + pD_1 + qD_2 = 0.$$

For this plane to pass through the image space coordinate origin and be perpendicular to the z'-axis, the coefficient of z' and the constant term must vanish identically, yielding

$$C_1 = C_2 = D_1 = D_2 = 0, \qquad (1.5)$$

Again using Equation (1.3), by setting $q = 0$, we get the equation of a plane perpendicular to the y-axis whose image, by substituting the values given in Equation (1.5) into Equation (1.4), we get

$$(pA_1 + rA_3 + sA)x' + (pB_1 + rB_3 + sB)y' + (pC_1 + rC_3 + sC)z' + pD_1 + rD_3 + sD = 0.$$

For this to be perpendicular to the y'-axis, the coefficient of y' must equal zero, yielding

$$B_1 = B_3 = B = 0. \qquad (1.6)$$

The final step in this argument involves a plane perpendicular to the x-axis, obtained by setting $p = 0$ in Equation (1.3). Its image, by substituting those values provided in Equations (1.5) and (1.6) into Equation (1.4), is

$$(qA_2 + rA_3 + sA)x' + (qB_2)y' + (rC_3 + sC)z' + rD_3 + sD = 0.$$

Now, the coefficient of x' must vanish, yielding the last of these conditions,

$$A_2 = A_3 = A = 0. \tag{1.7}$$

These conditions assure that the coordinate axes in image space are the images of those in object space. Nothing has been done to change any of the optical properties of this ideal instrument.

Substituting these, from Equations (1.5) through (1.7), into Equation (1.2) yields

$$x = \frac{A_1 x'}{Cz' + D}, \quad y = \frac{B_2 y'}{Cz' + D}, \quad z = \frac{C_3 z' + D_3}{Cz' + D}. \tag{1.8}$$

It is a simple matter to invert this transformation to obtain

$$A_1 = \frac{cd_3 - c_3 d}{a_1}, \quad B_2 = \frac{cd_3 - c_3 d}{b_2}$$
$$C_3 = -d, \quad C = c, \quad D_3 = d_3 \quad D = -c_3, \tag{1.9}$$

so that Equations (1.1) and (1.2) now read

$$x' = \frac{a_1 x}{cz + d}, \quad y' = \frac{b_2 y}{cz + d}, \quad z' = \frac{c_3 z + d_3}{cz + d}, \tag{1.10}$$

and

$$x = \frac{x'(cd_3 - c_3 d)}{a_1(cz' - c_3)},$$
$$y = \frac{y'(cd_3 - c_3 d)}{b_2(cz' - c_3)}, \tag{1.11}$$
$$z = \frac{dz' - d_3}{cz' - c_3}.$$

Now, we impose a restriction on the instrument itself by assuming that it is rotationally symmetric with the z-axis, and its image, the z'-axis as its axis of symmetry. Then, in Equations (1.10) and (1.11), $b_2 = a_1$, so that we need only the y- and z-coordinates. They then degenerate into

$$y' = \frac{a_1 y}{cz + d}, \quad z' = \frac{c_3 z + d_3}{cz + d}, \tag{1.12}$$

and

$$y = \frac{y'(cd_3 - c_3 d)}{a_1(cz' - c_3)}, \quad z = \frac{dz' - d_3}{cz' - c_3}. \tag{1.13}$$

To recapitulate, the image of a point (y, z) in object space is the point (y', z') as determined by Equation (1.12). Conversely, the image of (y', z') in image space is the point (y, z) in object space obtained from Equation (1.13).

The plane perpendicular to the z-axis, given by the equation $cz + d = 0$, has, as its image, the plane at infinity, as can be seen from Equation (1.12). Therefore, $z_f = -d/c$ is the z-coordinate of the focal point of the instrument in object space. In exactly the same way, we can find the z'-coordinate of the focal point in image space is $z'_f = c_3/c$, from Equation (1.13). To summarize, we have shown that

$$
\begin{aligned}
zf &= -d/c, \\
z'f &= c_3/c.
\end{aligned}
\tag{1.14}
$$

At this point, we make a change of variables, shifting both coordinate origins to the two focal points; thus,

$$
\begin{aligned}
z &= \bar{z} + zf, \\
z' &= \bar{z}' + z'f.
\end{aligned}
$$

so that, from the second equation of Equation (1.12), we obtain

$$
\bar{z}' = \frac{cd_3 - c_3 d}{c^2 \bar{z}}.
\tag{1.15}
$$

From the second equation of Equation (1.13), using the same transformation, we obtain an identical result. When the transformation is applied to the first equation of Equation (1.12), we obtain

$$
\bar{y}' = \frac{a_1 y}{c\bar{z}'},
\tag{1.16}
$$

while the first equation of Equation (1.13) yields

$$
y = \frac{(cd_3 - c_3 d)y'}{a_1 c\bar{z}'}.
\tag{1.17}
$$

Now, define *lateral magnification* as $m = y'/y$. Then, from Equations (1.16) and (1.17), it follows that

$$
m = \frac{a_1}{c\bar{z}} = \frac{a_1 c\bar{z}'}{cd_3 - c_3 d},
\tag{1.18}
$$

from which we can see that the conjugate planes of unit magnification are given by

$$
\begin{aligned}
\bar{z}_p &= \frac{a_1}{c}, \\
\bar{z}'_p &= \frac{cd_3 - c_3 d}{a_1 c}.
\end{aligned}
$$

These are called the *principal planes* of the instrument. Now $\bar{z}_p$ and $\bar{z}'_p$ are the distances, along the axis of symmetry, between the foci and the principal points. These distances are called the *front* and *rear focal lengths* of the instrument and are denoted by f and f', respectively; thus,

$$
\begin{aligned}
f &= a_1/c, \\
f' &= (cd_3 - c_3 d) / (a_1 c).
\end{aligned}
\tag{1.19}
$$

Next, we substitute these relations into Equation (1.15) and get Newton's formula

$$\bar{z}\,\bar{z}' = f\,f',\qquad\qquad(1.20)$$

while from Equations (1.16) and (1.17), it follows that

$$y' = fy\,/\,\bar{z},$$
$$y = f'y'\,/\,\bar{z}'.\qquad\qquad(1.21)$$

Suppose now that $\bar{y}$ and $\bar{z}$ define a right triangle with a corner at the focus in object space and let θ be the angle subtended by the z-axis and the hypotenuse. Then, the first equation of Equation (1.21) becomes the familiar

$$y' = f\tan\theta.\qquad\qquad(1.22)$$

Finally, let e equal the distance of an axial point in object space to the first principal point and let e' be the distance between its conjugate and the second principal point. Then, it follows that

$$e = \bar{z} + f,$$
$$e' = \bar{z}' + f'.\qquad\qquad(1.23)$$

Substituting these relations into Newton's formula, Equation (1.20), results in the familiar

$$\frac{f}{e} + \frac{f'}{e'} = 1.\qquad\qquad(1.24)$$

We have seen that straight lines are mapped into straight lines. Now, we complete the argument and assume that such a line and its image constitute a single ray that is traced through the instrument.

From these results, we can find object–image relationships using a graphic method. In Figure 1.2, the points z_f and z_f' are the instrument's foci and z_p and z_p' are its principal planes. Let O be any object

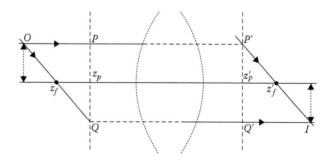

FIGURE 1.2 Graphical construction of an object–image relationship. The points z_f and z_f' are the instruments foci and z_p and z_p' its principal planes. From object point O, ray $\overline{OP}$ is parallel to the axis. Since P is on the object principal plane, its image P' must be at the same height. Image ray must therefore pass through $\overline{P'z_f'}$. A second ray Oz_fQ, passes through the object focus and therefore must emerge in image space parallel to the axis. It must also pass through Q', the image of Q. The two rays cross at I, the image point.

point. Let $\overline{OP}$ be a ray parallel to the axis, passing through P. Let its extension pass through P'. Since $\overline{\frac{P}{Oz_f Q}}$ and P' lie on the conjugate principal planes, the ray in image space must pass through P'. Since this ray is parallel to the axis in object space, its image must pass through z_f'. These two points completely determine this ray in image space. Now take a second ray through the object point O. Since it passes through z_f it must emerge in image space parallel to the axis. Since it passes through Q on the principal plane, it must also pass through its image Q'. These two points determine this ray in image space. Where the two rays cross is I, the image of O.

With this concept, we can find a most important third pair of conjugates for which the instrument's *angular magnification* is unity. Then, a ray passing through one of these points will emerge from the instrument and pass undeviated through the other. These are the *nodal* points.

Refer now to Figure 1.3. Suppose a ray passes through the axis at z_0, at an angle θ, and intersects the principal plane at y_p. After passing through this ideal instrument, it intersects the axis in image space at z_0', at an angle θ', and passes through the principal plane at y_p'. Newton's formula, Equation (1.20), provides a relationship between z_0 and z_0'.

$$z_0 z_0' = f f'. \tag{1.25}$$

Moreover, y_1 and y_1' are equal, since they represent the heights of conjugate points on the principal planes. From Figure 1.3, we can see that

$$\begin{aligned} y_p &= -(z_0 - f)\tan\theta, \\ y_p' &= -(f' - z_0')\tan\theta'. \end{aligned} \tag{1.26}$$

so that the angular magnification is given by

$$M = \frac{\tan\theta'}{\tan\theta} = \frac{z_0 - f}{f' - z_0'}.$$

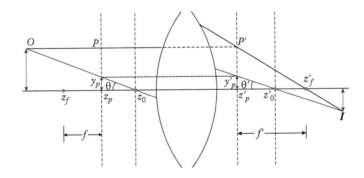

FIGURE 1.3 Graphical construction of the nodal points. The points z_f and z_f' are the two focal points and z_p and z_p' are its two principal planes; z_0 and z_0' are the nodal points, where $\theta = \theta'$ and f and f' are the front and rear focal lengths.

With the aid of Equation (1.25), this becomes

$$M = \frac{z_0}{f'} = \frac{f}{z_0'}.$$

For z_p and z_p' to represent points where the angular magnification is unity, it must be that

$$z_0 = f',$$
$$z_0' = f. \tag{1.27}$$

These nodal points are important for two reasons. In optical testing, there is an instrument called the nodal slide, which is based on the properties of the nodal points and is used to find, quickly and accurately, the focal length of a lens. A more subtle property is that images in image space bear the same perspective relationship to the second nodal point, as do the corresponding objects in object space to the first nodal point.

With this in mind, we make another change of variables: a translation of the z-axes to place the origins at the two nodal points. The new z-coordinates will be g and g'. The change is realized by

$$g = \bar{z} - \bar{z}_n = \bar{z} - f',$$
$$g' = \bar{z}' - \bar{z}_n' = \bar{z}' - f'.$$

Again, using Newton's formula, Equation (1.20), we obtain

$$gg' + gf + g'f' = 0,$$

from which comes

$$\frac{f}{g'} + \frac{f'}{g} + 1 = 0. \tag{1.28}$$

This concludes the study of the ideal optical instrument. We have found the six *cardinal points*, the foci, the principal points and the nodal points, solely from considering the Maxwell model of an ideal instrument [4]. The model is a static one; there is no mention of wavefronts, velocities, or refractive indices. These characteristics will be introduced in subsequent sections of this chapter.

1.3 The Eikonal Function and Its Antecedents

This subject has a rather odd pedigree. It was first discovered by Hamilton, who called it the *characteristic function*. Then, it was rediscovered by Bruns who dubbed it the *eikonal* [5]. Its origins are much earlier. The law of refraction, discovered by Willebrord Snell using empirical methods, after his death, came into the hands of Descartes who derived for it what he claimed to be an analytic proof. Fermat disagreed. In his opinion, Snell's law was only an approximation and Descartes' proof was erroneous. He then set out to find the exact formula for refraction. But, to his surprise, Snell's law was indeed exact.

The approach to the derivation that he used has come down to us as Fermat's principle: light consists of a flow of particles, termed *corpuscles*, the trajectories of which are such that their time of transit from point to point is an extremum, either a maximum or a minimum. These trajectories are what we now call *rays*. Fermat's justification for this principle goes back to observations by Heron of Alexandria, but that is the subject of an entirely different story. What does concern us here is the interpretation of his principle in mathematical terms: its representation in terms of the variational calculus, which deals specifically with the determination of extremal of functions.

To set the stage, let us consider an optical medium in which a point is represented by a vector $\mathbf{P} = (x, y, z)$ and in which the refractive index is given by a vector function of position: $n = n(\mathbf{P})$. We will represent a curve in this medium by the vector function $\mathbf{P}(s)$, where the parameter "s" is the geometric distance along the curve. It follows that $d\mathbf{P}/ds = \mathbf{P}'$ is a tangent vector to the curve. It can be shown that $\mathbf{P}'$ is a *unit* tangent vector.

Note that

$$d\mathbf{P} = \mathbf{P}'ds, \tag{1.29}$$

so that

$$d\mathbf{P} = \mathbf{P}'ds, \quad \mathbf{P} \cdot d\mathbf{P} = \mathbf{P}'^2 ds = ds. \tag{1.30}$$

The velocity of light in this medium is c/n where c is its velocity *in vacuo*. The time of transit between any two points on a ray is therefore given by

$$\int (n/c)ds,$$

which is proportional to the *optical path length*

$$I = \int nds. \tag{1.31}$$

For $\mathbf{P}(s)$ to be a ray, according to Fermat's principle, I must be an *extremum*, a term from the variational calculus. A necessary condition for the existence of an extremum is that a set of differential equations, the *Euler equations*, must be satisfied [6]. The Euler equations that guarantee that I be an extremum, for geometric optics, take the form [7],

$$\frac{d}{ds}\left(n\frac{d\mathbf{P}}{ds}\right) = \nabla n, \tag{1.32}$$

which, when expanded becomes

$$n\mathbf{P}'' + (\nabla n \cdot \mathbf{P}')\mathbf{P}' = \nabla n. \tag{1.33}$$

What we have here is the general case, a ray in an inhomogeneous (but isotropic) medium, which has come to be called a *gradient index medium* and is treated more broadly in Chapter 6, Volume 1.

It is useful to look at this differential equation from the point of view of the differential geometry of space curves. We define the unit tangent vector to the ray $\mathbf{t}$, the unit normal vector, $\mathbf{n}$; and the unit binormal, $\mathbf{b}$, as follows [8]:

$$\begin{aligned} \mathbf{t} &= \mathbf{P}', \\ \mathbf{n} &= \rho\mathbf{P}'', \\ \mathbf{b} &= \mathbf{t} \times \mathbf{n}. \end{aligned} \tag{1.34}$$

With these, Equation (1.31) can be rewritten as

$$\frac{n}{\rho}\mathbf{n}+(\nabla n\cdot\mathbf{t})\mathbf{t}=\nabla n, \tag{1.35}$$

which shows that the tangent vector, the normal vector, and the gradient of the refractive index must be collinear. It follows that the binormal vector must be perpendicular to the index gradient,

$$\nabla n\cdot\mathbf{b}=0, \tag{1.36}$$

or, stated differently, that

$$\frac{n}{\rho}\mathbf{b}=\mathbf{t}\times\nabla n. \tag{1.37}$$

At any point on a space curve the vectors **t**, **n**, and **b** can be regarded as a set of orthonormal axes for a local coordinate system. As the parameter s varies, this point, along with the three associated vectors, slides along the curve (Figure 1.4). Their motion is governed by the Frenet–Serret equations [9],

$$t'=\frac{1}{\rho}\mathbf{n},$$

$$n'=-\frac{1}{\rho}\mathbf{t}+\frac{1}{\tau}\mathbf{b}, \tag{1.38}$$

$$b'=-\frac{1}{\tau}\mathbf{n},$$

$$\frac{1}{\rho^2}=\frac{(\mathbf{t}\times\nabla n)^2}{n^2}. \tag{1.39}$$

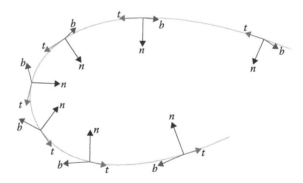

FIGURE 1.4   The sliding trihedron. As a point moves along the space curve, the vectors *t*, *n*, and *b* slide along with it. The Frenet-Serret equations, Equation (1.34), describe their rates of change, where $1/\rho$, as before, is the curve's curvature at the point in question and $1/\tau$ is its torsion. These formulas show that the curvature is the rate of change of **t** and that torsion is the rate of change of the **b** vector. Both these motions are in the direction of **n**. By squaring the expression in Equation (1.37), we obtain a formula for the ray's curvature.

The torsion refers to the third of the Frenet–Serret equations in Equation (1.38) and obtains

$$\frac{1}{\tau}\mathbf{n}\cdot\mathbf{b}=\rho^2(\mathbf{P}'\times\mathbf{P}'')\cdot\mathbf{P}''',\tag{1.40}$$

where we have used Equation (1.34). By taking the derivative of Equation (1.33) and multiplying the result by $\mathbf{P}'\times\mathbf{P}''$, we obtain

$$\frac{1}{\tau}=\frac{\rho^2}{n}=(\mathbf{P}'\times\mathbf{P}'')\cdot(\Delta n)',\tag{1.41}$$

where $(\Delta n)'$ represents the derivative of the gradient with respect to s. To show how this works, consider Maxwell's fish eye [10] in which the refractive index function is given, in idealized form, by

$$n(\mathbf{P})=\frac{1}{1+\mathbf{P}^2},$$

so that its gradient is

$$\nabla n=\frac{-2\mathbf{P}}{(1+\mathbf{P}^2)^2}.$$

Substituting these into the ray equation, Equation (1.33), yields

$$(1+\mathbf{P}^2)\mathbf{P}''-2(\mathbf{P}\cdot\mathbf{P}'')\mathbf{P}'+2\mathbf{P}=0.$$

When this is differentiated, one obtains

$$(1+\mathbf{P}^2)\mathbf{P}''-2(\mathbf{P}\cdot\mathbf{P}'')\mathbf{P}'=0.\tag{1.42}$$

Now multiply by $\mathbf{P}''$ to get

$$\mathbf{P}''\cdot\mathbf{P}''=(1/2)(\mathbf{P}''^2)'=(1/2)(1/\rho^2)'=0,$$

which shows that the curvature is constant. From Equations (1.40) and (1.42), we get

$$\frac{1}{\tau}=0.\tag{1.43}$$

The torsion is everywhere zero so that the ray is a plane curve. It follows that the ray path is the arc of a circle, exactly as it should be.

Let us return for the moment to Equation (1.32), the ray equation. Its vector product with $\mathbf{P}$ results in

$$\mathbf{P}\times\frac{\mathrm{d}}{\mathrm{d}s}\left(n\frac{\mathrm{d}\mathbf{P}}{\mathrm{d}s}\right)=\frac{\mathrm{d}}{\mathrm{d}s}\left[\mathbf{P}\times\left(n\frac{\mathrm{d}\mathbf{P}}{\mathrm{d}s}\right)\right]=\mathbf{P}\times\nabla n.\tag{1.44}$$

Now assume that the refractive index function is symmetric with respect to the z-axis, so that $n(\mathbf{P}) = n(\rho, z)$, where $\rho^2 = x^2 + y^2$. Then its gradient is

$$\nabla n = \left(\frac{x}{\rho} \frac{\partial n}{\partial \rho}, \frac{y}{n} \frac{\partial n}{\partial \rho}, \frac{\partial n}{\partial z} \right), \tag{1.45}$$

and

$$\mathbf{Z} \cdot \left[\mathbf{P} \times \frac{d}{ds} \left(n \frac{d\mathbf{P}}{ds} \right) \right] = \frac{d}{ds} \left\{ \mathbf{Z} \cdot \left[\mathbf{P} \times \left(n \frac{d\mathbf{P}}{ds} \right) \right] \right\}$$
$$= \mathbf{Z} \cdot (\mathbf{P} \times \nabla n) = \mathbf{P} \cdot (\nabla n \times \mathbf{Z}). \tag{1.46}$$

But

$$\nabla n \times \mathbf{Z} = \left(\frac{x}{\rho} \frac{\partial n}{\partial \rho}, \frac{y}{n} \frac{\partial n}{\partial \rho}, \frac{\partial n}{\partial z} \right) \times (0,0,1) = \frac{1}{\rho} \frac{\partial n}{\partial \rho} (y,-x,0), \tag{1.47}$$

so that

$$\mathbf{P} \cdot (\nabla n \times \mathbf{Z}) = 0, \tag{1.48}$$

showing that, from Equation (1.38),

$$\frac{d}{ds} \left\{ \mathbf{Z} \cdot \left[\mathbf{P} \times \left(n \frac{d\mathbf{P}}{ds} \right) \right] \right\} = 0. \tag{1.49}$$

Thus,

$$\mathbf{Z} \cdot \left[\mathbf{P} \times \left(n \frac{d\mathbf{P}}{ds} \right) \right] = \text{constant}. \tag{1.50}$$

This is therefore independent of s and is known as the *skewness invariant* or, more simply, the *skewness* [11].

Now, we return to Equation (1.31), which, when we apply Equation (1.30), becomes the line integral

$$I = \int n\mathbf{P}' \cdot d\mathbf{P}, \tag{1.51}$$

where $\mathbf{P}(s)$ is a solution of the ray equation, Equation (1.32) or Equation (1.33). Now, let $\mathbf{P}_0$ be a starting point of a ray in this medium and let $\mathbf{P}_1$ be its endpoint. Then, the line integral is

$$I(\mathbf{P}_0, \mathbf{P}_1) = \int_{\mathbf{P}_0}^{\mathbf{P}_1} n\mathbf{P}' \cdot d\mathbf{P}. \tag{1.52}$$

Define two nabla operators,

$$\nabla_0 = \left(\frac{\partial}{\partial x_0}, \frac{\partial}{\partial y_0}, \frac{\partial}{\partial z_0} \right); \; \nabla_1 = \left(\frac{\partial}{\partial x_1}, \frac{\partial}{\partial y_1}, \frac{\partial}{\partial z_1} \right).$$

Then,

$$\nabla_0 I = n_0 \mathbf{P}_0'; \quad \nabla_1 I = -n_1 \mathbf{P}_1', \tag{1.53}$$

where $n_0 = n(\mathbf{P}_0)$, $n_1 = n(\mathbf{P}_1)$, $\mathbf{P}_0' = \mathbf{P}'|_{\mathrm{P0}}$, and $\mathbf{P}_1' = \mathbf{P}'|_{\mathrm{P1}}$. The function I, given by Equation (1.52), is known as *Hamilton's characteristic function* or, more simply, the *eikonal,* while the equations in Equation (1.53) are *Hamilton's characteristic equations* [12]. By squaring either of the expressions in Equation (1.53), we obtain the *eikonal equation,*

$$(\nabla I)^2 = n^2. \tag{1.54}$$

The eikonal equation can be derived from the Maxwell equations in several different ways. By assuming that a scalar wave equation represents light propagation, along with an application of Huygens' principle, Kirchhoff obtained a harmonic solution for light intensity at a point. He showed that the eikonal equation was obtained as a limit as wavelength approached zero. Kline and Kay [13] give a critique as well as a detailed account of this method.

Luneburg [14], on the other hand, took a radically different approach. He started with an integral version of the Maxwell equations, regarded a wavefront as a singularity in the solution of these equations and that radiation transfer consisted of the propagation of these singularities. Then, he used Huygens' principle to obtain what we have called the eikonal equation. This has led to speculation that geometric optics is a limiting case of physical optics as frequency becomes small. Perhaps. Suffice it to say that the eikonal equation, since it can be derived from sundry starting points, remains a crucial point in optical theory.

It is important to state that recently a general solution to the eikonal equation for propagation in a homogeneous medium [15] has been presented. This solution was presented in Cartesian coordinates and expressed in terms of an arbitrary function that was denominated as the *k-function*. A solution of the eikonal equation gives the equations for the wavefront and caustic surfaces either after reflection or refraction. As was pointed out, the *k-function* contains all information about the aberrations introduced to the wavefront by the refracting or reflecting surface.

Now suppose the optical medium is discontinuous: that there is a surface S where the index of refraction function has a jump discontinuity. To fix ideas, assume that light travels from left to right. Choose a point $\mathbf{P}_0$ to the left of S and a second, $\mathbf{P}_1$, to its right. Unless $\mathbf{P}_0$ and $\mathbf{P}_1$ are conjugates, they will be connected by a unique ray path determined by Fermat's principle and, moreover, the segments of the ray path will be solutions of Equations (1.32) or (1.33). Let $\overline{\mathbf{P}}$ be the point where this ray path crosses the discontinuity S and let n_- and n_+ be the left- and right-hand limits of the refractive index function along the ray path at $\overline{\mathbf{P}}$. For convenience let $\mathbf{S} = \mathrm{d}\mathbf{P}/\mathrm{d}s$ represent a ray vector and let $\mathbf{S}_-$ and $\mathbf{S}_+$ represent the left and right limits of $\mathbf{S}$ at the point $\overline{\mathbf{P}}$.

To best describe the consequences of a discontinuity with mathematical rigor and vigor, one should apply the Hilbert integral of the calculus of variations [16]. For our purposes, a schoolboy explanation is more appropriate. Joos [17] uses the definition of the gradient in terms of limits of surface integrals to define a surface gradient as a gradient that straddles a surface of discontinuity so that

$$\nabla\phi = (\phi_+ - \phi_-)\mathbf{N}, \tag{1.55}$$

where $\mathbf{N}$ is a unit normal vector to the surface. Suppose we have Equation (1.32) straddle $\mathbf{S}$ and replace the derivative by a difference quotient; then we obtain

$$\frac{\mathrm{d}}{\mathrm{d}s}(n\mathbf{S}) = \frac{1}{\Delta s}(n_+\mathbf{S}_+ - n_-\mathbf{S}_-) = (n_+ - n_-)\mathbf{N}. \tag{1.56}$$

When we take the vector product of this with **N**, we obtain

$$n_+(\mathbf{S}_+ \times \mathbf{N}) = n_-(\mathbf{S}_- \times \mathbf{N}). \tag{1.57}$$

As we shall see in the next section, this leads to the vector form of Snell's law for homogeneous media.

There is another use for the eikonal concept: the Cartesian oval, a refracting surface that images an object point perfectly onto an image point [18]. Let us assume that such a surface passes through the coordinate origin and that the two conjugate points be located at distances t and t' from that origin on the z-axis, so that their coordinates are $\mathbf{P} = (0, 0, -t)$ and $\mathbf{P}' = (0, 0, t')$. And let any point on the presumed refracting surface have the coordinates $(\bar{x}, \bar{y}, \bar{z})$. As usual, let n and n' be the refractive indices.

It turns out that for **P** and **P'** to be perfect conjugates, the optical path length *along any ray* must be constant [19]. In other words,

$$n\sqrt{\bar{x}^2 + \bar{y}^2 + (\bar{z}+t)^2} + n'\sqrt{\bar{x}^2 + \bar{y}^2 + (\bar{z}-t')^2} = \text{constant}. \tag{1.58}$$

Now take a ray through the coordinate origin so that $\bar{x} = \bar{y} = \bar{z} = 0$. From this, we see that the constant is exactly equal to $nt = n't'$, so that the equation for the surface is

$$n\sqrt{\bar{x}^2 + \bar{y}^2 + (\bar{z}+t)^2} + n'\sqrt{\bar{x}^2 + \bar{y}^2 + (\bar{z}-t')^2} = nt + nt'. \tag{1.59}$$

By eliminating the square roots, we get the rather formidable polynomial

$$\left[(n^2 - n'^2)(\bar{x}^2 + \bar{y}^2 + \bar{z}^2) + 2\bar{z}(n^2 t + n'^2 t')\right]^2 \\ - 4nn'(nt + n't')\left[(n't + nt')(\bar{x}^2 + \bar{y}^2 + \bar{z}^2) + 2tt'(n - n')\bar{z}\right] = 0, \tag{1.60}$$

a quartic surface in the shape of an oval. This is shown in Figure 1.5. Kepler found this numerically early in the seventeenth century. Descartes, a generation later, found the mathematical formula. And it has been rediscovered over and over again, even by Maxwell, ever since.

An interesting (indeed, fascinating) consequence obtains when the object point approaches infinity. Divide this formula by $t^2 t'^2$ and then let t become large. The result is

$$n^2\bar{z}^2 - n'^2(\bar{x}^2 + \bar{y}^2 + \bar{z}^2) - 2n't'(n - n')\bar{z} = 0, \tag{1.61}$$

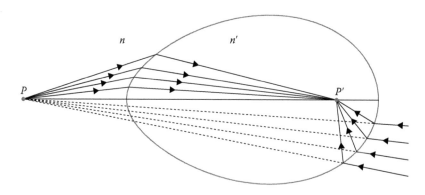

FIGURE 1.5 The Cartesian oval. Point P is imaged perfectly on point P'. Rays coming from the left are from the real object P. For rays coming from the right, P is a virtual object.

which can be rearranged in the form

$$\left(\frac{n+n'}{n't'}\right)^2\left(\bar{z}-\frac{n't'}{n+n'}\right)^2+\left[\frac{n+n'}{(n'-n)t'2}\right](\bar{x}^2+\bar{y}^2)=1, \tag{1.62}$$

the equation of a conic section of revolution whose eccentricity is $\varepsilon=n/n'$, whose center is at $z_c=n't'/(n+n')$, whose z intercepts are $z_-=0$ and $z_+=2n't'/(n+n')$ and whose vertex curvature is $c=1/(1-\varepsilon)t'$. Clearly the surface is an ellipsoid when $n'>n$ and a hyperboloid when $n'<n$. This can be seen in Figure 1.6.

Finally, we come to the aplanatic surfaces of a sphere [20]. Let

$$t=-k(1+n'/n),\ t'=k(1+n/n'), \tag{1.63}$$

and substitute it into Equation (1.60). Since $nt+n't'=0$, this degenerates into the equation of a sphere that passes through the origin and has a radius of k,

$$\bar{x}^2+\bar{y}^2+\bar{z}^2+2k\bar{z}=0. \tag{1.64}$$

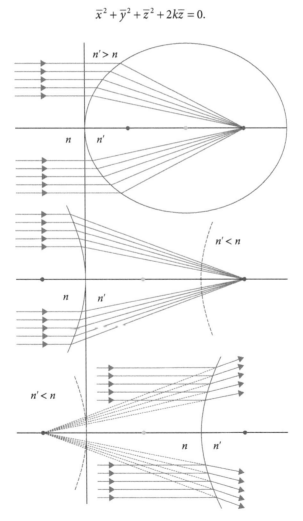

FIGURE 1.6 The Cartesian oval when the object is at infinity. In this case, the quartic surface degenerates into a quadratic surface: an ellipse when $n'>n$; a hyperbola when $n'<n$. "·" indicates the location of a focus or the center of symmetry, respectively.

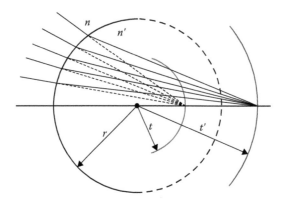

FIGURE 1.7 The aplanatic surfaces of a sphere. Here, k is the radius of the refracting sphere; $t = k(1 + n'/n)$ is the radius of the object surface; $t' = k(1 + n/n')$, that of the image surface.

Since the refracting sphere has central symmetry, t and t' can be taken as the radii of two other spheres, which are perfect conjugates. These are the aplanatic surfaces and they are shown in Figure 1.7. The aplanatic surfaces were the basis for a system for tracing meridional rays before the advent of computers.

1.4 Ray Tracing and Its Generalization

Now, we take up the very practical problem of tracing rays in a homogeneous, isotropic medium, a medium in which the refractive index, n, is constant. It follows that its gradient, Δn, is zero so that Equation (1.32) becomes

$$\frac{d^2 \mathbf{P}}{ds^2} = 0, \tag{1.65}$$

a second-order differential equation whose solution is a linear function of s; therefore, a ray in this medium must be a straight line. Note here that in media of constant refractive index,

$$\frac{d\mathbf{P}}{ds} = \mathbf{S} = (\xi, \eta, \zeta). \tag{1.66}$$

Snell's law, from Equation (1.57), now takes the form

$$n'(\mathbf{S'} \times \mathbf{N}) = n(\mathbf{S} \times \mathbf{N}), \tag{1.67}$$

where $\mathbf{S}$ and $\mathbf{S'}$ are the direction cosine vectors of a ray before and after refraction, respectively; where $\mathbf{N}$ is the unit normal vector to the refracting surface at the point of incidence; and where n and n' are the refractive indices of the media before and after the refracting surface. Note that in the preceding section, we used the *prime* symbol (') to denote differentiation with respect to the parameter s; here, we use it to signal refraction or (subsequently) transfer.

By taking the absolute value of Equation (1.67), we get the more familiar form of Snell's law

$$n' \sin i' = n \sin i, \tag{1.68}$$

where i and i' are the angles of incidence and refraction, respectively. This statement, unlike its vector form, does not tell the whole story. Equation (1.67) provides the additional information that the vectors $\mathbf{S'}$, $\mathbf{S}$, and $\mathbf{N}$ are coplanar and determines the *plane of incidence*.

In what follows, we will develop the equations for ray tracing. The form in general use today, developed by T. Smith over a period of several decades [21], will be cast in vector form here [22]. An earlier scalar version, designed particularly for computer use, is by Feder [23]. If we rearrange the terms of Equation (1.67) to get

$$n'(\mathbf{S}' - n\mathbf{S}) \times \mathbf{N} = 0, \tag{1.69}$$

we can see that the vector n'**S**' − n**S** is parallel to the unit normal vector **N** so that, for some γ,

$$n'\mathbf{S}' - n\mathbf{S} = \gamma\mathbf{N} \tag{1.70}$$

from which we can get the *refraction equation*,

$$n'\mathbf{S}' = n\mathbf{S} + \gamma\mathbf{N}. \tag{1.71}$$

Note that cos $i = \mathbf{S} \cdot \mathbf{N}$ and cos $i' = \mathbf{S}' \cdot \mathbf{N}$ so that by taking the scalar product of Equation (1.71) with **N**, we find that γ is given by

$$\gamma = n'(\mathbf{S}' \cdot \mathbf{N}) - n(\mathbf{S} \cdot \mathbf{N}). \tag{1.72}$$

The convention for reflecting surfaces is only a convention, quite divorced from any physical reality. One sets $n' = -n$ in Equations (1.71) and (1.72). Since $i' = i$, it follows that

$$\mathbf{S}' = -\mathbf{S} + 2\mathbf{N}\cos i. \tag{1.73}$$

This takes care of the *refraction* or *reflection operation*. It involves only local properties of the refracting surface: the location of the point of incidence and the unit normal vector **N**, at that point. On the other hand, the *transfer operation,* by means of which the point of incidence and the surface normal are found, involves the global properties of the surface.

Suppose the surface is given by a vector function of position,

$$\mathcal{F}(\mathbf{P}) = 0. \tag{1.74}$$

A ray can be represented in parametric form by

$$\mathbf{P} = \mathbf{P}_0 + \lambda\mathbf{S}, \tag{1.75}$$

where λ represents the distance along the ray from $\mathbf{P}_0$ to **P**. If **P** is the *point of incidence,* then λ must be a solution for the equation

$$\mathcal{F}(\mathbf{P}_0 + \lambda\mathbf{S}) = 0. \tag{1.76}$$

With the value of λ so obtained, Equation (1.75) provides the point of incidence. The normal to a surface is best given by the gradient of its equation, $\Delta\mathcal{F}$, so that the unit normal vector is found from

$$N = \frac{\nabla\mathcal{F}}{\sqrt{(\nabla\mathcal{F})^2}}, \tag{1.77}$$

calculated, of course, at the point of incidence. But there is a problem here. Equation (1.76) may have multiple roots. For the sphere, or for that matter, for any other conic section, $\mathcal{F}$ is a quadratic and will have two solutions: either two real roots, indicating that the ray intersects the surface at two points;

or two complex roots, in which case the ray misses the surface completely. More complicated surfaces produce more complicated solutions. A torus, a quartic surface, may have up to four real roots, corresponding to four points where ray and surface intersect. Deciding which is which is a daunting problem.

A particularly useful method is to identify that region of the surface that is of interest, then do a translation of coordinates to a point in that region, and then solve the equation for the reformulated $\mathcal{F}$ function and choose that solution that lies within that region or is nearest to the chosen point. Thus, the transfer operation becomes a two-step process.

To illustrate this, consider a rotationally symmetric optical system consisting of spherical refracting surfaces. Let each surface have a local coordinate system with the z-axis as the axis of symmetry and the x- and y-axes are tangent to the sphere where it is intersected by the z-axis. This x, y-plane is called the *vertex plane*. Suppose $\mathbf{P_0}$ is the point of incidence of a ray with a refracting surface whose coordinates are relative to the local coordinates associated with that surface; suppose, further, that the distance along the z-axis, between this surface and the next succeeding surface, is t; then $\overline{\mathbf{P}}$, the point of intersection of the ray with the next vertex plane, is given by Equation (1.75), which in scalar form is

$$\overline{x} = x_0 + \overline{\lambda}\xi,$$

$$\overline{y} = y_0 + \overline{\lambda}\eta, \tag{1.78}$$

$$\overline{z} = z_0 + \overline{\lambda}\zeta = t,$$

so that

$$\overline{\lambda} = -(z_0 - t)/\zeta. \tag{1.79}$$

This next sphere passes through the origin at its vertex and has the formula

$$\mathcal{F}(\mathbf{P}) = x^2 + y^2 + (z - r)^2 - r^2 = 0, \tag{1.80}$$

relative to its own coordinate system. Substituting from Equation (1.75) yields the equation

$$\left[(\overline{\mathbf{P}} - r\mathbf{Z}) + \lambda\mathbf{S} \right]^2 = r^2, \tag{1.81}$$

a quadratic equation in λ,

$$\lambda^2 + 2\lambda(\overline{\mathbf{P}} - r\mathbf{Z})\cdot\mathbf{S} + (\overline{\mathbf{P}} - r\mathbf{Z})^2 - r^2 = 0, \tag{1.82}$$

whose solution is

$$\lambda = -(\overline{\mathbf{P}} - r\mathbf{Z})\cdot\mathbf{S} \pm \sigma, \tag{1.83}$$

where

$$\sigma^2 = r^2 - \left[(\overline{\mathbf{P}} - r\mathbf{Z}) \times \mathbf{S} \right]. \tag{1.84}$$

This constitutes the second part of this two-part transfer operation. The ambiguity in sign in Equation (1.80) has an easy explanation. In general, a ray will intercept a sphere at two points. We are almost always interested in the point of incidence nearest the vertex plane and therefore choose the appropriate branch of the solution.

The unit normal vector is easily obtained from the expression for the gradient in Equation (1.77) and that for the sphere in Equation (1.80), and is

$$\mathbf{N} = \frac{1}{r}(\overline{\mathbf{P}} - r\mathbf{Z}) = \frac{1}{r}(x, y, z - r) = (cx, cy, cz - 1),$$ (1.85)

where $c = 1/r$ is the sphere's curvature. For surfaces more complicated than the sphere, we need only substitute their formulas into Equation (1.75) and proceed.

The *skewness invariant,* shown in Equation (1.50), takes a slightly different form. In media of constant refractive index,

$$\frac{d\mathbf{P}}{d\mathbf{S}} = \mathbf{S} = (\xi, \eta, \zeta),$$

so that

$$n\mathbf{Z} \cdot (\mathbf{P} \times \mathbf{S}) = \text{constant}.$$ (1.86)

Since $\mathbf{P} = (x, y, z)$, the skewness invariant becomes

$$n'(x'\eta' - y'\xi') = n(x\eta - y\xi),$$ (1.87)

valid for both the refraction and transfer operations. Geometric wavefronts (we exclude wavefronts that arise from diffraction) are defined in several equivalent ways [24]. As a *surface of constant phase,* it is the locus of points that have the same optical path length from some object point. A system of rays originating from some common object is termed an *orthotomic system* or a *normal congruence* as was explained above. In these terms, a wavefront can be thought of as a *transversal surface* orthogonal to each of the rays in the system. A third definition is based on Huygens' principle, in which the wavefront is taken to be the envelope of a family of spherical wavelets centered on a preceding wavefront.

However they are defined, wavefronts have structures and properties that are best described using the language of the differential geometry of surfaces [25]. As we shall see, wavefronts are smooth surfaces that may possess cusps but which have continuous gradients.

In general, a smooth surface, at almost every point, possesses two unique directions, called the *principal directions,* which may be indicated by a pair of orthogonal vectors tangent to the surface. They have the property that curvatures of arcs embedded in the surface in these directions have curvatures that are extrema; the arc curvature in one principal direction is a maximum relative to that of all other arcs through the same point. The arc curvature in the other principal direction is a minimum. These two maximum and minimum curvatures are called the *principal curvatures.* Obvious exceptions are the plane, in which curvature is everywhere zero, and the sphere, where it is everywhere constant. In both cases, principal directions cannot be defined. Another exception is the *umbilical point,* a point on a surface at which the two principal curvatures are equal. There the surface is best fit by a sphere.

What follows is a method for determining the changes in the principal directions and principal curvatures of a wavefront in the neighborhood of a traced ray. These calculations depend on and are appended to the usual methods of tracing rays. We have called them *generalized ray tracing* [26].

Consider now a ray traced through an optical system. Through each of its points passes a wavefront that has two orthogonal principal directions and two principal curvatures. As before, let $\mathbf{S}$ be a unit vector in the direction of ray propagation and therefore normal to the wavefront. Suppose one of these principal directions is given by the unit vector $\mathbf{T}$ so that the other principal direction is $\mathbf{T} \times \mathbf{S}$. Let the two principal curvatures be $1/\rho_1$ and $1/\rho_2$. The quantities $1/\rho_1$, $1/\rho_2$, and $\mathbf{T}$ are found using general methods of differential geometry. These will not be treated here.

Suppose this ray is intercepted by a refracting surface that has, at the point of incidence, a unit normal vector $\mathbf{N}$, a principal direction $\bar{\mathbf{T}}$ and as principal curvatures, $1/\bar{\rho}_1$, and $1/\bar{\rho}_2$. Through this point passes one of the incident wavefronts, with parameters defined as above. The equations for refraction, Equations (1.71) and (1.72), define the plane of incidence. The unit normal vector $\mathbf{P}$ to this plane is defined by

$$\mathbf{P} = \frac{\mathbf{N} \times \mathbf{S}}{\sin i} = \frac{\mathbf{N} \times \mathbf{S'}}{\sin i'}, \tag{1.88}$$

where we have used Equations (1.67) and (1.68). Note that $\mathbf{P}$ is invariant with respect to refraction. From this, we may define three unit vectors lying in the plane of incidence:

$$\mathbf{Q} = \mathbf{P} \times \mathbf{S}; \quad \bar{\mathbf{Q}} = \mathbf{P} \times \mathbf{N}; \quad \mathbf{Q'} = \mathbf{P} \times \mathbf{S'}. \tag{1.89}$$

Taking the vector product of $\mathbf{P}$ and Equation (1.71) gives us the refraction equations for the $\mathbf{Q}$ vectors,

$$n'\mathbf{Q'} = n\mathbf{Q} + \gamma\bar{\mathbf{Q}}, \tag{1.90}$$

where γ is given in Equation (1.72). The next step in these calculations is to find the curvatures of sections of the wavefront lying in and normal to the plane of incidence. Let θ be the angle between the wavefront principal direction $\mathbf{T}$ and the normal to the plane of incidence $\mathbf{P}$ so that

$$\cos\theta = \mathbf{T} \cdot \mathbf{P}. \tag{1.91}$$

Then, we find the desired curvatures, $1/\rho_p$ and $1/\rho_q$, as well as a third quantity, $1/\sigma$, related to the torsion of these curves. The equations are

$$\frac{1}{\rho_p} = \frac{\cos^2\theta}{\rho_1} + \frac{\sin^2\theta}{\rho_2},$$
$$\frac{1}{\rho_q} = \frac{\sin^2\theta}{\rho_1} + \frac{\cos^2\theta}{\rho_2}, \tag{1.92}$$
$$\frac{1}{\sigma} = \left[\frac{1}{\rho_1} - \frac{1}{\rho_2}\right] + \frac{\sin 2\theta}{2}.$$

We do exactly the same thing for the refracting surface:

$$\cos\bar{\theta} = \bar{\mathbf{T}} \cdot \mathbf{P}. \tag{1.93}$$

$$\frac{1}{\bar{\rho}_p} = \frac{\cos^2\bar{\theta}}{\bar{\rho}_1} + \frac{\sin^2\bar{\theta}}{\bar{\rho}_2},$$

$$\frac{1}{\bar{\rho}_q} = \frac{\sin^2\bar{\theta}}{\bar{\rho}_1} + \frac{\cos^2\bar{\theta}}{\bar{\rho}_2}, \tag{1.94}$$

$$\frac{1}{\bar{\sigma}} = \left[\frac{1}{\bar{\rho}_1} - \frac{1}{\bar{\rho}_2}\right] + \frac{\sin 2\bar{\theta}}{2}.$$

Note that Equations (1.91) and (1.93) and Equations (1.92) and (1.94) are quite similar. In a computer program, both can be calculated with the same subroutine. The refraction equations we use to relate all these are

$$
\frac{n'}{\rho'_P} = \frac{n}{\rho_P} + \frac{\gamma}{\overline{\rho}_P},
$$

$$
\frac{n'\cos i'}{\sigma'} = \frac{n\cos i}{\sigma} + \frac{\gamma}{\overline{\sigma}}, \tag{1.95}
$$

$$
\frac{n'\cos^2 i'}{\rho'_q} = \frac{n\cos^2 i}{\rho_q} + \frac{\gamma}{\overline{\rho}_q},
$$

where γ is given in Equation (1.72). The next step is to find θ', the angle between $\mathbf{P}$ and one of the principal directions of the wavefront after refraction,

$$
\tan 2\theta' = \frac{2}{\sigma'\left[\dfrac{1}{\rho'_q} - \dfrac{1}{\rho'_q}\right]}. \tag{1.96}
$$

The penultimate step is to find $\mathbf{T}'$, the vector in a principal direction,

$$
\mathbf{T}' = \mathbf{P}\cos\theta' + \mathbf{Q}'\sin\theta', \tag{1.97}
$$

where $\mathbf{Q}'$ is found using Equation (1.89), and the two principal curvatures,

$$
\frac{1}{\rho'_1} = \frac{\cos^2\theta'}{\rho'_P} + \frac{\sin^2\theta'}{\rho'_q} + \frac{\sin 2\theta'}{\sigma'},
$$

$$
\frac{1}{\rho'_2} = \frac{\sin^2\theta'}{\rho'_P} + \frac{\cos^2\theta'}{\rho'_q} + \frac{\sin 2\theta'}{\sigma'}. \tag{1.98}
$$

This takes care of refraction at a surface. The transfer operation is far simpler. The principal directions $\mathbf{T}$ and $\mathbf{T}\times\mathbf{S}$ are unchanged and the principal curvatures vary in an intuitively obvious way,

$$
\rho'_1 = \rho_1 - \lambda; \quad \rho'_2 = \rho_2 - \lambda, \tag{1.99}
$$

where λ is obtained by adding Equations (1.80) and (1.83). Note that the first and third equations in Equation (1.95) are exactly the Coddington equations [27]. If a principal direction of the wavefront lies in the plane of incidence, then θ and therefore $1/\sigma$ are zero. The same is true for the principal directions of the refracting surface at the point of incidence: $\overline{\theta}$ and $1/\overline{\sigma}$ both vanish. If both occur, if both the wavefront and the refracting surface have a principal direction lying in the plane of incidence, then Equations (1.92), (1.94) and (1.98) become ephemeral, in which case Equation (1.95) reduces to the two Coddington equations. However, generally speaking, the surface principal directions will not lie in the plane of incidence. Indeed, this will happen only if the refracting surface is rotationally symmetric and the plane of incidence includes the axis of symmetry.

In a rotationally symmetric system, a plane containing the axis of symmetry is called a *meridional plane*; a ray lying entirely in that plane is a *meridional ray*. A ray that is not a meridional ray is a *skew ray*.

What we have shown here is that the Coddington equations are valid only for meridional rays in rotationally symmetric optical systems where a principal direction of the incident wavefront lies in the meridional plane. No such restriction applies to the equations of generalized ray tracing.

This concludes the discussion on generalized ray tracing except for a few observations. If the incident wavefront is a plane or a sphere, or if the traced ray is at an umbilical point of the wavefronts, then the principal curvatures are equal and the principal directions are undefined. In that case, as a *modus operandi,* the incident **T** vector may be chosen arbitrarily as long as it is perpendicular to **S**.

It is no secret that rays are not real and that the existence of wavefronts can be inferred only by interferometry. The only artifact in geometric optics that can be observed directly is the caustic surface [28]. Unlike waves and wavefronts, the caustic can be seen, photographed, and measured. It also can be calculated using the methods of generalized ray tracing.

Like the wavefront, the *caustic surface* can be defined in several different but equivalent ways, each revealing one of its properties [29]. As the envelope of an orthotomic system of rays, it is clear that light is concentrated on its locus. The caustic is also where the differential element of area of a wavefront vanishes, showing that it is a cusp locus of the wavefront train. Where the wavefront and caustic touch, the wavefront fold back on itself. Our final definition is that the caustic is the locus of the principal centers of curvature of a wavefront. Since there are two principal curvatures for each ray, it is clear that the caustic is, in general, a complicated, two-sheeted surface.

It is the last of these definitions that is relevant to generalized ray tracing and that provides a means for calculating the caustic. Suppose that **P** represents a point on a ray where it intersects the final surface or the exit pupil of an optical system and that **S** is the ray's direction vector, both found using ordinary ray tracing. At that point, ρ_1 and ρ_2 are found using generalized ray tracing. Then, the point of contact of the ray with each of the two sheets of the caustic is given by

$$\mathbf{C}_i = \mathbf{P} + \rho_i \mathbf{S}, \quad (i = 1, 2). \tag{1.100}$$

The importance of the caustic cannot be underestimated. For perfect image formation, it degenerates into a single image point [30]. Its departure from the ideal point then is a measure of the extent of the geometric aberrations associated with that image point; its location is an indicator of the optical system's distortion and curvature of field.

1.5 The Paraxial Approximation and Its Uses

In the conclusion of his two-volume opus in which he derived his equations for the local properties of a wavefront, Coddington [31] remarked that his formulas were OK (here we paraphrase) but were far too complicated to have any practical value and that he intended to continue to use his familiar, tried-and-true methods to design lenses. Such was the case until the advent of the modern computer. Tracing a single ray was a long tedious process. Tracing enough rays to evaluate an optical system was tedium raised to an impossible power. Indeed, most traced rays were meridional rays; skew rays were far too difficult for routine use. Folklore has it that only paraxial rays were used in the design process, which, when concluded, was followed by the tracing of fans of meridian rays to determine whether the design was good enough to make a prototype. If it was and if the prototype proved satisfactory, the lens went into production; otherwise, it was destroyed or relegated to a museum and the designer tried again. It was cheaper to make and destroy unsatisfactory prototypes than to trace the skew rays required to make a decision as to the quality of a design.

The paraxial approximation is simplicity itself [32]. The quantities in the finite ray-tracing equations are expanded into a power series and then all but the linear turns are dropped. Thus, the sine of an angle is replaced by the angle itself (in radians, of course) while the cosines become unity. Rays calculated using these approximations are termed *paraxial* rays because their domain of validity is an ϵ region that hugs the axis of symmetry of the optical system.

However, it needs to be mentioned that Gauss, whose name is often attached to these equations, experimented with a power series solution to the eikonal equation, shown in Equation (1.54) [33]. He got no further than terms of the first order but found that these enabled him to obtain the paraxial equations that we are about to derive. For that reason, these equations are frequently referred to as the *first-order equations*.

So, we make the following assumptions and approximations. First, we set $x = \xi = 0$, so that we are confined to meridional rays and we can write η and ζ as trigonometric functions of u, the angle between the ray and the axis. Here, we need to set $\eta = -\sin u$ and $\zeta = \cos u$; the minus sign is needed to conform with the sign convention used in paraxial ray tracing. Next, we assume the u is so small that $\eta^2 \sim 0$, so that $\eta \sim u$ and $\zeta \sim 1$. The ray vector now takes the form

$$S \approx (0, -u, 1). \tag{1.101}$$

The point of incidence can be represented by the vector

$$\mathbf{P} = (0, y, z), \tag{1.102}$$

where we further assume that $y^2 \sim 0$ and $z^2 \sim 0$. From Equation (1.85), the unit normal vector is

$$\mathbf{N} = \frac{1}{r}(\mathbf{P} - r\mathbf{Z}) = (0, cy, cz - 1), \tag{1.103}$$

where $\mathbf{Z}$ is the unit vector along the axis of rotation. It follows that

$$\mathbf{N}^2 = c^2 y^2 + (c^2 z^2 - 2cz + 1) \approx 1 - 2cz. \tag{1.104}$$

But $\mathbf{N}$ is a unit vector, so that $cz \sim 0$ and

$$\mathbf{N} = (0, cy, -1). \tag{1.105}$$

Finally, the sine of the angle of incidence is given by

$$\sin i = |\mathbf{N} \times \mathbf{S}| = cy - u, \tag{1.106}$$

with a similar expression valid for $\sin i'$. But these, too, are paraxial quantities and

$$\begin{aligned} i &\approx cy - u, \\ i' &\approx cy - u', \end{aligned} \tag{1.107}$$

When we apply the scalar form of Snell's law, Equation (1.68), we get its paraxial equivalent,

$$n'i' = ni. \tag{1.108}$$

$$n'u' = nu + cy(n' - n), \tag{1.109}$$

the formula for paraxial refraction. Next, we take up the problem of paraxial transfer. Recall that transfer consists of two parts. The generic formula is given by Equation (1.75)

$$\mathbf{P}' = \mathbf{P} + \lambda \mathbf{S}. \tag{1.110}$$

The first part, transfer from vertex plane to vertex plane, is represented by Equation (1.79),

$$\bar{\lambda} = -(z_0 - t)/\zeta,$$ (1.111)

and the second part, transfer from vertex plane to sphere, by Equations (1.83) and (1.84),

$$\lambda = -(\bar{\mathbf{P}} - r\mathbf{Z}) \cdot \mathbf{S} - \sigma$$
$$\sigma^2 = r^2 - \left[(\bar{\mathbf{P}} - r\mathbf{Z}) \times \mathbf{S} \right]^2.$$ (1.112)

Consider the first transfer. To the paraxial approximation, $z_0 \sim 0$ and $\zeta \sim 1$, so that

$$\bar{\lambda} = t.$$ (1.113)

The second transfer is a little less straightforward. The second term in the expression for σ^2 in Equation (1.112) is the square of a paraxial quantity and is therefore equal to zero so that $\sigma = r$. The distance λ then becomes

$$\lambda = -(\bar{\mathbf{P}} - r\mathbf{Z}) \cdot \mathbf{S} - r,$$
$$= -(0, \bar{y}, -r) \cdot (0, -u, 1) - r$$ (1.114)
$$= \bar{y}u \approx 0,$$

which follows from the paraxial assumptions. Paraxial transfer is therefore given by

$$y' = y + tu.$$ (1.115)

Strictly speaking, these formulas are valid only for rays exceedingly close to the optical axis. (Indeed, because of the underlying assumptions, they are not valid for systems lacking rotational symmetry.) Yet they are of immense value in defining properties of lenses from a theoretical point of view and at the same time are indispensable to the optical designer and optical engineer.

We can use these paraxial formulas for refraction to derive the Smith–Helmholtz invariant [34]. Figure 1.8 shows a single refracting surface $\bar{\mathbf{P}}$ and two planes, located at $\mathbf{P}$ and $\mathbf{P}'$, that are conjugates. Let t and t' be the distance of each of these planes from the surface. Because the two planes are conjugates, a ray from $\mathbf{P}$ to a point $\bar{\mathbf{P}}$ on the surface must then pass through $\mathbf{P}'$. Let the height $\mathbf{P}'$ be h and let the angles that these two rays make with the axis be u and u', respectively. Then, from the figure, $u = h/t$ and $u' = h/t'$, so that

$$h = tu = t'u'.$$ (1.116)

Now take a point on the object plane at a height y and its image whose height is y' and consider a ray connecting these two points that intersects the refracting surface at the axis. The angle subtended by this ray and the axis is its angle of incidence i, so that $i = y/t$. A similar expression holds for the refracted ray, $i' = y'/t'$ with i' being the angle of refraction. This yields

$$y = it; \quad y' = i't'.$$ (1.117)

Finally, we invoke the paraxial form of Snell's law, Equation (1.108), $n'i' = ni$. We make the following cascade of calculations,

$$n'y'u' = n'i't'u' = nit'u' = nyu;$$ (1.118)

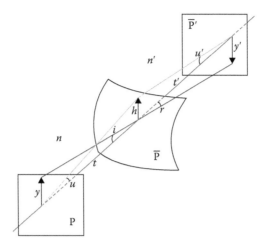

FIGURE 1.8   In the Smith-Helmholtz invariant, $\overline{P}$ is a refracting surface separating two media of refractive index n and n', respectively. P and $\overline{P}'$ are a pair of conjugate planes.

in other words,

$$n'y'u' = nyu. \tag{1.119}$$

Suppose we are dealing with an optical system consisting of n surfaces. Then, this relation can be iterated at each surface in the following way:

$$nyu = n_1 y_1 u_1 = \cdots = n_k - 1 y_{k-1} u_{k-1} = n_k y_k u_k = \cdots n'y'u'. \tag{1.120}$$

the Smith–Helmholtz invariant. This derivation is based on the paraxial ray tracing equations. An identical result can be obtained by using the general equations for a perfect optical system, as given in Section 1.1, applied to a single surface.

We say that paraxial rays are defined by Equations (1.109) and (1.115). We also speak of the paraxial image of a point. Consider two paraxial rays emanating from some point in object space. Where they intersect in image space is the location of its image. All other paraxial rays from that point will also intersect at the same paraxial image point. We can conclude from this that any distinct pair of paraxial rays provides a basis for all other paraxial rays; any other paraxial ray can be written as a linear combination of these two. We can generalize this in an obvious way to paraxial images of surfaces.

Now consider an optical system that consists of a sequence of glass elements, all rotationally symmetric, all aligned on a common axis. Associated with each is a surface; each surface is limited by an edge that defines its boundary. By tracing paraxial rays through the surfaces preceding it, an image of that surface is formed in object space together with an image of its boundary. Any ray in object space that passes through this image and that lies inside the image of the boundary will pass through the surface itself; a ray that fails to clear the boundary of this image will not pass. All refracting surfaces in the lens will have images in object space; any ray that clears the boundary of each of these images will pass through the lens. Those that fail to do so will exceed the boundary of some surface within the lens and be blocked [35].

In many cases, particularly for camera lenses, where the regulation of the light passing through the lens is important for the control of film exposure, an additional surface is introduced whose diameter is controlled externally. The paraxial image in object space of this *diaphragm* or *stop* is called the *entrance pupil*: in image space, the *exit pupil*. To be effective as a light regulator, the stop must be the dominant aperture. This is assured by locating the entrance pupil and setting its diameter so that it is smaller than the other aperture images in object space. (Its location is often dictated by a need to adjust values of the third-order aberrations.) Indeed, once its location and diameter have been established, the diameters of the other elements in the lens can be adjusted so that their images match that of the entrance pupil.

The entrance pupil defines the boundary of the aggregate of rays from any object point that passes through the lens; that ray that passes through its axial point is, ideally, the barycenter of the aggregate. Such a paraxial ray is termed a *chief ray* or a *principal ray*. The *marginal ray* is a ray from the object point that just clears the boundary of the entrance pupil. Almost always these two paraxial rays are taken to be the *basis* rays; every other paraxial ray can be represented as a linear combination of these two.

From Equations (1.109) and (1.115), the two sets of formulas for marginal rays and chief rays, the latter indicated by a superior "bar" [36],

$$n'u' = nu + cy(n' - n), \quad y' = y + tu,$$
$$n'\bar{u} = n\bar{u} + c\bar{y}(n' - n), \quad \bar{y}' = \bar{y} + t\bar{u}. \tag{1.121}$$

It is easy to see that

$$\mathcal{L} = n(y\bar{u} - \bar{y}u) = n'(y'\bar{u}' - \bar{y}'u'), \tag{1.122}$$

is a paraxial invariant for the entire optical system and is called the Lagrange invariant. Note that if $\bar{y}$ is equal to zero, the Lagrange invariant reduces to the Smith–Helmholtz invariant, which makes it clear that they are versions of each other.

The stop or diaphragm controls the amount of radiation that passes through the lens. The entrance pupil and the exit pupil can be likened to holes through which the light pours. Apart from this, there is another important property associated with these pupils. The larger the hole, for an ideal lens, the better its resolving power. But lenses are rarely ideal and certain aberrations grow with the stop aperture.

Several parameters are used to quantify the light transmission properties of a lens [37]. The most familiar of these is the *f/number*, defined as the ratio of the system's focal length to the diameter of the entrance pupil. It can be used directly in the formula for calculating the diameter of the Airy disk, for an equivalent ideal lens, which provides an estimate of the upper limit of the lens' resolving power. If the geometric aberrations of such a lens are of the same magnitude as the Airy disk, then it is said to be *diffraction limited*. The details of this are treated in following chapters.

A second parameter, applied exclusively to microscope objectives is the *numerical aperature,* defined as the sine of one-half the angle subtended by the entrance pupil at the axial object point multiplied by the refractive index.

1.6 The Huygens' Principle and Developments

Christian Huygens (1629–1695) seemed to be a far different man than his renowned contemporaries René Descartes and Pierre de Fermat. They were scholastic types content to dream away in their ivory towers: the one constructed magnificent philosophies out of his reveries; the other, after his magisterial and political duties, retreated to his library where he entered notes in the margins of books. Huygens was not like that. He made things; among his creations were lenses and clocks. He was an observer. The legend is that he watched waves in a pond of water and noticed that when these waves encountered an obstruction, they gave rise to circular ripples, centered where wave and obstruction met, that spread out and interfered with each other and the original waves.

The principle that he expounded was that each point on a water wave was the center of a circular wavelet and that the subsequent position and shape of that wave was the envelope of all those wavelets with equal radii [38].

The extension of Huygens' observations on interfering wavelets to optical phenomena is the basis of what we have come to know as optical interference, interferometry, and diffraction, subjects that go far beyond the scope of this chapter [39]. However, his principle on the formation and propagation of waves, when applied to geometrical optics, provides an alternative route to the law of refraction.

Figure 1.9 shows the evolution of a train of wavefronts constructed as envelopes of wavelets developed from an arbitrary starting wavefront. The medium is assumed to be homogeneous and isotropic. There are two things to notice. As mentioned previously, in the neighborhood of a center of curvature of the initial wavefront, cusps are formed. The locus of these cusps constitutes the caustic surface associated with the wavefront train. In the same region, wavefronts in the train tend to intersect each other, giving rise to the interference patterns sometimes observed in the neighborhood of the caustic surface [40].

A ray in this construction can be realized as the straight line connecting the center of a wavelet with the point where it touches the envelope. It is not possible to see but it turns out that the envelope of these rays is exactly and precisely the caustic and therefore coincides with the cusp locus of the wavefront.

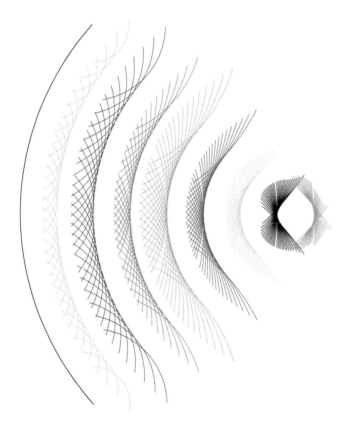

FIGURE 1.9 Demonstrating Huygens' principle. The curve on the left side of the figure represents the initial state of a wavefront. To its right are six wavefronts generated by families of wavelets centered on the initial wavefront. Note that after the six wavefronts, the second and third of these have cusps and intersect themselves. The cusp locus is the caustic.

Now for the law of refraction. Suppose a plane refracting surface separating media of refractive index n and n' lies on the x, y-plane of a coordinate system so that the surface unit normal vector $\mathbf{Z}$ is the z-axis. Take the incident ray vector to be $\mathbf{S} = (\xi, \eta, \zeta)$. If $\mathbf{W} = (p, q, r)$ represents a point on a wavefront normal to $\mathbf{S}$, then an equation for this wavefront is

$$\mathbf{S} \cdot \mathbf{W} = p\xi + q\eta + r\zeta = s, \tag{1.123}$$

where s is a parameter for the distance from the wavefront to the coordinate origin. The wavefront intersects the refracting surface on a line obtained from Equation (1.123) by setting, $p = \bar{x}$, $q = \bar{y}$ and $r = 0$, where $\bar{x}$ and $\bar{y}$ are coordinates on the refracting plane; thus,

$$\bar{x}\xi + \bar{y}\eta = s. \tag{1.124}$$

The point $(\bar{x}, \bar{y}, 0)$ is to be the center of a spherical wavelet with radius s'. The total optical path length to the wavefront will be $ns + n's' = $ const. Then $s' = \mu (s + k)$, where $\mu = n/n'$ and k is that constant.

The equation of the wavelet is then

$$\mathcal{F} \equiv (x - \bar{x})^2 + (y - \bar{y})^2 + z^2 = \mu^2(s + k)^2 = \mu^2(\bar{x}\xi + \bar{y}\eta + k)^2, \tag{1.125}$$

an equation for a two-parameter family of spherical wavelets all centered on the refracting surface and all with radii arising from an incident wavefront. To find the envelope of this family we take the partial derivatives of $\mathcal{F}$ with respect to $\bar{x}$ and $\bar{y}$ and solve for those two parameters and then substitute their values back into the original equation in Equation (1.125). These derivatives are

$$\begin{aligned}\mathcal{F}_{\bar{x}} &\equiv x - \bar{x} + \mu^2(\bar{x}\xi + \bar{y}\eta + k)\xi = 0, \\ \mathcal{F}_{\bar{y}} &\equiv y - \bar{y} + \mu^2(\bar{x}\xi + \bar{y}\eta + k)\eta = 0,\end{aligned} \tag{1.126}$$

which rearranges itself into

$$\begin{aligned}(1 - \mu^2\xi^2)\bar{x} - \mu^2\xi\eta\bar{y} &= x - \mu^2 k\xi, \\ -\mu^2\xi\eta\bar{x} + (1 - \mu^2\eta^2)\bar{y} &= y - \mu^2 k\eta,\end{aligned} \tag{1.127}$$

a simultaneous pair whose determinant of coefficients is

$$\Delta = 1 - \mu^2(\xi^2 + \eta^2) = 1 - \mu^2(1 - \zeta^2). \tag{1.128}$$

The solution is then

$$\begin{aligned}\Delta\bar{x} &= (1 - \mu^2\eta^2)x + \mu^2\xi[\eta y - k], \\ \Delta\bar{y} &= (1 - \mu^2\xi^2)y + \mu^2\eta[\xi x - k],\end{aligned} \tag{1.129}$$

From this, we can show that

$$\bar{x}\xi + \bar{y}\eta - k = \frac{1}{\Delta}(x\xi + y\eta - k), \tag{1.130}$$

and with this and Equation (1.126) we obtain

$$x - \bar{x} = \frac{\mu^2}{\Delta}(x\xi + y\eta - k)\xi,$$

$$y - \bar{y} = \frac{\mu^2}{\Delta}(x\xi + y\eta - k)\eta. \tag{1.131}$$

When this is substituted into Equation (1.125), we obtain

$$\Delta^2 z^2 - \mu^2(x\xi + y\eta - k)^2[1 - \mu^2(\xi^2 + \eta^2)] = 0, \tag{1.132}$$

which reduces to

$$\Delta z^2 - \mu^2(x\xi + y\eta - k)^2 = 0. \tag{1.133}$$

Here, we have used Equation (1.128). So, the three components of the vector that goes from the center of the wavelet to its point of contact with the envelope are

$$x - \bar{x} = \frac{\mu^2}{\Delta}(x\xi + y\eta - k)\xi,$$

$$y - \bar{y} = \frac{\mu^2}{\Delta}(x\xi + y\eta - k)\eta. \tag{1.134}$$

$$z = \frac{\mu}{\sqrt{\Delta}}(x\xi + y\eta - k).$$

Normalizing this vector yields **S**, the direction cosine vector for the refracted ray,

$$\mathbf{S}' = (\mu\xi, \mu\eta, \sqrt{\Delta}) = \mu(\xi, \eta, \zeta) + (\sqrt{\Delta} - \mu\zeta)(0,0,1). \tag{1.135}$$

But $\mathbf{S} = (\xi, \eta, \zeta)$, the unit normal vector to the refracting surface is $\mathbf{N} = (0, 0, 1)$ and $\sqrt{\Delta} - \mu\zeta$ is exactly y/n', this from Equation (1.73). Thus, Equation (1.135) is identical to the refraction equation, Equation (1.71).

This concludes the section of Huygens' principle. We have shown how this system of elementary wavelets and their envelope can be used to derive the refraction equation. The same technique can be applied in media in which the wavelets are not spheres; for example, in birefringent media, where the wavelets are two-sheeted surfaces [41].

1.7 The Aberrations, Seidel, and Otherwise

Aberration theory and optical design are so profoundly intertwined as to be almost indistinguishable. The most obvious difference is that mathematicians do one and optical engineers do the other with some physicists playing an intermediate role. Perhaps another distinction is that what aberration theorists consider publishable, optical engineers guard jealously as trade secrets. In this day and age, computers and their programmers yield wide control over both areas and to them accrue much of the glory and much of the funding.

The etiology of the concept of an aberration makes an interesting, and sometimes droll, history. Kepler, nearly 400 years ago, identified spherical aberration. Subsequently, it was realized that it was independent of object position, and that other aberrations, namely coma and astigmatism, were not. In this century, we deal with a power series that may not converge, whose terms are identified with the physical appearance of the aberrations. Much of this history is contained in a brief article by R. B. Johnson [42].

The power series on which modern aberration theory is based on a solution of the eikonal equation, Equation (1.54), in which Hamilton's equations, given in Equation (1.53), play an important part. Under the assumptions of rotational symmetry, the power series solution possesses only even terms; therefore, only odd terms appear in its derivatives. Gauss experimented with these odd series but his studies did not go beyond terms of the first order. These terms comprise the *first-order* equations, which are identical to the equations for paraxial optics. For this reason, they are often referred to as *Gaussian optics*. Seidel extended the work of Gauss to terms of the third order and correlated their coefficients with observable image errors [42]. The *third-order* or *Seidel* aberrations (in Europe frequently called the *primary* aberrations) are terms of the third order in this expansion and relate quantities obtained from the paraxial equations with other components of an optical system to observable phenomena associated with defects in optical image formation. Recently, by using the caustic formulas and a paraxial approximation, analytic expressions have been derived to evaluate the spherical aberration to the third order, and also a formula to reduce this aberration has been provided [43].

1.7.1 The Classical Third-Order Aberration Equations

As in Section 1.5, we distinguish two types of paraxial rays. The chief ray passes through the center of the entrance pupil and is indicated by a superior bar. The marginal ray is a ray parallel to the axis of the lens that just clears the edge of the aperture of the entrance pupil.

Further, we adapt the notation introduced by Feder [44]. A symbol with no subscript denotes a quantity at a particular surface or in the following medium. One with a subscript −1 lies on the next preceding surface or in the next preceding medium. Finally, a symbol with the subscript 1 is on the next following surface.

Finally, we modify the equations in Equations (1.125), (1.127), and (1.129) to conform with the notation of T. Smith [45],

$$u = cy(1-\mu)+\mu u_{-1} \qquad \bar{u} = c\bar{y}(1-\mu)+\mu\bar{u}_{-1},$$

$$y_1 = y - tu \qquad \bar{y}_1 = \bar{y} - t\bar{u}, \tag{1.136}$$

$$i = cy - u_{-1} \qquad \bar{i} = c\bar{y} - \bar{u}_{-1},$$

where c is the curvature of the refracting surface; t is the separation between two adjacent surfaces; and $\mu = n/n_{-1}$, with n_{-1} being the refractive index of the medium next preceding the refracting surface and n that of the next following. The Lagrange invariant is, of course, from Equation (1.122):

$$\mathcal{L} = n(y\bar{u}-\bar{y}u) = n_{-1}(y_{-1}u_{-1}-\bar{y}_{-1}u_{-1}) = n_1(y_1\bar{u}_1-\bar{y}_1u_1). \tag{1.137}$$

In order to calculate the aberration contributions at each surface, we need several auxiliary quantities. The first of these is the Petzval contribution

$$P = \frac{(u-1)c}{n_{-1}}. \tag{1.138}$$

Two additional auxiliary quantities are

$$S = \frac{n_{-1}y(u-i)(1-\mu)}{2\mathcal{L}},$$
$$\bar{S} = \frac{n_{-1}\bar{y}(\bar{u}-\bar{i})(1-\mu)}{2\mathcal{L}},$$

(1.139)

that are used, along with P, in calculating the aberration contributions from the individual surfaces. It is remarkable that P is the only quantity that does not depend on paraxial quantities but only on surface curvature and the surrounding refractive indices.

The surface contributions are

	Image Contributions	Pupil Contributions	
Spherical Aberration	$B = Si^2$	$\bar{B} = -\bar{S}\bar{i}^2$	
Coma	$F = Si\bar{i}$	$\bar{F} = -\bar{S}i\bar{i}$	(1.140)
Astigmatism	$C = S\bar{i}^2$	$\bar{C} = -\bar{S}i^2$	
Curvature	$D = C + P\mathcal{L}/2$	$\bar{D} = \bar{C} - P\mathcal{L}/2$	
Distorsion	$E = -\bar{F} + (\bar{u}^2 - \bar{u}_{-1}^2)/2$	$\bar{E} = -F + (u^2 - u_{-1}^2)/2$	

These are the third-order contributions from a single surface; the aberrations for the entire system are obtained by summing the individual terms; thus,

Image errors	Field errors	
$\mathcal{B} = \Sigma B$	$\mathcal{D} = \Sigma D$	(1.141)
$\mathcal{F} = \Sigma F$	$\varepsilon = \Sigma E$	
$c = \Sigma C$		

for the aberrations associated with image formation and similar "bared" expressions for the aberrations associated with the pupils. It is remarkable that these third-order aberrations add algebraically. It is not the case for aberrations of a degree greater than third.

Now, take a ray in object space with coordinates on the object that are y and z and that pass through the point $\bar{y}$ and $\bar{z}$ on the entrance pupil plane. Let the nominal format radius on the object plane be h and let the radius of the entrance pupil aperture be $\bar{h}$. Finally, let

$$H = \varepsilon \frac{y^2 + z^2}{h^2} + 2C \frac{y\bar{y} + z\bar{z}}{h\bar{h}} + F \frac{\bar{y}^2 + \bar{z}^2}{\bar{h}},$$
$$K = D \frac{y^2 + z^2}{h^2} + 2F \frac{y\bar{y} + z\bar{z}}{h\bar{h}} + B \frac{\bar{y}^2 - \bar{z}^2}{\bar{h}},$$

(1.142)

then

$$y' = h'\left[(1+H)(y/h) + K(\bar{y}/\bar{h})\right],$$
$$z' = h'\left[(1+H)(z/h) + K(\bar{z}/\bar{h})\right],$$

(1.143)

where y' and z' are the coordinates of this ray on the image plane. When the ray is a chief ray, then $\bar{y} = \bar{z} = 0$ and $y'/h' = (1 + H)(y/h)$ and $z'/h' = (1 + H)(z/h)$. In the absence of aberrations, $H = 0$, so that h' must be the radius of the image field.

As the object plane approaches infinity, the ratios y/h and z/h remain constant and less than unity and are replaced by appropriate trigonometric functions.

The third-order chromatic aberrations are treated in the same way. If ΔN is the partial dispersion of the medium after the refracting surface and if ΔN_{-1} is that for the next preceding medium, then the chromatic aberrations are given by

$$a = y(\Delta N_{-1} - \mu \Delta N)i,$$
$$b = y(\Delta N_{-1} - \mu \Delta N)\bar{i}. \tag{1.144}$$

These represent the blurring of the image due to the dispersion of the component glasses and can be thought of as the displacement of an image point as the wavelength changes. A lateral displacement of the image of a point, called *lateral color* or *chromatic difference of magnification*, is given by Σb; *longitudinal color* is the shift of the image plane along the axis and is given by Σa.

1.7.2 Aberrations of the Fifth Order

The completion of Seidel's work represented an enormous step in the transformation of optical design from an ill-understood art to the beginnings of an engineering science. That it was only an improvement over paraxial optics was clear. Moreover, it suggested that it was only the second of an infinite number of steps in the power series expansion on which it was based.

The progression to the third step is an interesting story. In 1904, Schwarzschild [46] derived what he called the *Seidel eikonal,* a function of five variables whose first partial derivatives were, exactly, the five Seidel aberration coefficients. Four years later, Kohlschütter [47] calculated the second partial derivatives. These he believed, by analogy, were the 15 fifth-order aberrations. Rightly or not, the formulas were well beyond the capability of routine computation at the time. Forty years later, Wachendorf would describe them as *schrechlich*. Both Wachendorf [48] and Herzberger [49] attempted the derivation of these complicated expressions; both succeeded, but both failed to provide a significant improvement in accuracy that warranted their over-complexity. It was Buchdahl who realized that these fifth-order terms were not complete; that, at each surface, functions of third-order terms from it and all preceding surfaces needed to be added. We have come to identify the fifth-order contributions at a surface as the *intrinsic contributions* and the functions of the third-order contributions of the preceding surfaces as the *extrinsic contributions*.

And here arises a paradox. Within the context of geometric optics, it is well known that rays are reversible; a ray traced from object space to image space is identical to the ray traced, with the same parameters, from image space to object space. It is also acknowledged that the aberration terms are approximations to real ray tracing. But in the case of the fifth order, this is not so. The intrinsic contributions are the same in both cases but the extrinsic contributions are not. The following formulas are from Buchdahl [50] by way of Rimmer [51] and Stavroudis [52]. Certain changes have been introduced to make them compatible with Equations (1.136) through (1.140).

Auxiliary Quantities

$$x_{73} = 3ii' + 2u^2 - 3u_{-1}^2$$

$$x_{74} = 3i\bar{i}' + 2u\bar{u} - 3u_{-1}\bar{u}_{-1}$$

$$x_{75} = 3\bar{i}\bar{i}' + 2\bar{u}^2 - 3\bar{u}_{-1}^2$$

$$x_{76} = -i(3u_1 - u)$$

$$x_{42} = \bar{y}^2 ci - y\bar{i}(\bar{u} + \bar{u}_{-1})$$

$$x_{82} = -\bar{y}^2 cu_{-1} + y\bar{i}'(\bar{u} - \bar{u}_{-1})$$

$$\bar{x}_{42} = y^2 c\bar{i} + \bar{y}i(u + u_{-1})$$

$$\bar{x}_{82} = -y^2 c\bar{u}_{-1} + \bar{y}i'(u + u_{-1})$$

$$x_{77} = -\bar{i}\,(2u_{-1} - u) - i\bar{u}_{-1}$$

$$x_{78} = -\bar{i}\,(3u_{-1} - \bar{u})$$

$$\omega = (i^2 + i'^2 + u^2 - 3u_{-1}^2)/8$$

$$\hat{S}_{1p} = 3\omega Si/2$$

$$\hat{S}_{2p} = S(\bar{i}x_{73} + ix_{74} - \bar{u}x_{76} - ux_{77})/4$$

$$\hat{S}_{3p} = n_{-1}(\mu-1)[x_{42}x_{73} + x_{76}x_{82} + y(i+u)(ix_{75} - ux_{78})/8$$

$$\hat{S}_{4p} = S(\bar{i}x_{74} - \bar{u}x_{77})/2$$

$$\hat{S}_{5p} = n_{-1}(\mu-1)[x_{42}x_{72} + x_{77}x_{82} + y(i+u)(\bar{i}x_{75} - \bar{u}x_{78})]/4$$

$$\hat{S}_{6p} = n_{-1}(\mu-1)(x_{42}x_{75} + x_{78}x_{82})/8$$

$$\hat{S}_{1q} = n_{-1}(\mu-1)(\bar{x}_{42}x_{75} + x_{78}x_{82})/8$$

Intrinsic Fifth Order

Spherical aberration	$\breve{B}_5 = 2i\breve{S}_{2p}$
Coma	$\vec{F}_1 = 2\,\vec{i}\breve{S}_{1p} + i\breve{S}_{2p}$
	$\vec{F}_2 = \vec{i}\breve{S}_{2p}$
Oblique Spherical aberration	$\bar{M}_1 = 2\vec{i}\breve{S}_{2p}$
	$\bar{M}_2 = 2\vec{i}\breve{S}_{3p}$
	$\bar{M}_3 = 2\vec{i}\breve{S}_{4p}$
Elliptical coma	$\bar{N}_1 = 2\vec{i}\hat{S}_{3p}$
	$\bar{N}_2 = 2\vec{i}\hat{S}_{4p} + 2i\hat{S}_{5p}$
	$\bar{N}_3 = 2\vec{i}\hat{S}_{5p}$
Astigmatism	$\bar{C}_5 = \vec{i}\hat{S}_{5p}/2$
"Petzval"	$\bar{\pi}_5 = 2i\hat{S}_{6p} - \vec{i}\hat{S}_{5p}/2$
Image distortion	$\bar{E}_5 = 2\vec{i}\hat{S}_{6p}$
Pupil distortion	$\bar{\bar{E}} = 2i\hat{S}_{1p}$

In what follows, Σ' denotes the sum of the third-order aberrations calculated on all the refracting surfaces up to but not including the surface on which the fifth-order intrinsic aberrations are calculated.

Extrinsic Fifth Order

Spherical aberration	$B^0 = 3[F\Sigma'B - B\Sigma'F]/2\mathcal{L}$
Coma	$F_1^0 = [(P+4c)\Sigma'B + (5\Sigma'F - 4\Sigma'E)F$
	$\quad - (2\Sigma'P + 5\Sigma'\bar{C})B]/2\mathcal{L}$

$$F_2^0 = [(P + 2c)\sum{}' B + 2(2\sum{}' F - \sum{}' E)F$$

$$- (\sum{}' P + 4\sum{}' \bar{C})B]/2\mathcal{L}$$

Oblique Spherical

$$M_1^0 = [E\sum{}' B + (4\sum{}' F - \sum{}' \bar{E})C$$

$$+ (\sum{}' C - 4\sum{}' \bar{C} - 2\sum{}' P)F - B\sum{}' \bar{F}]/\mathcal{L}$$

$$M_2^0 = [E\sum{}' B + (P + C)(2\sum{}' F - \sum{}' \bar{E})$$

$$+ (\sum{}' P + 3\sum{}' \bar{C} - 2\sum{}' \bar{C})F - 3B\sum{}' \bar{F}]/2\mathcal{L}$$

$$M_3^0 = 2[(2C + P)\sum{}' F + (\sum{}' C - 2\sum{}' \bar{C})\bar{F}$$

$$- B\sum{}' \bar{F}]/\mathcal{L}$$

Astigmatism

$$C^0 = [E(4\sum{}' C + \sum{}' P) - P\sum{}' \bar{F}]$$

$$+ 2C(\sum{}' E - 2\sum{}' \bar{F}) - 2F\sum{}' \bar{B}]/4\mathcal{L}$$

"Petzval"

$$P^0 = [E(\sum{}' P - 2\sum{}' C) + P(4\sum{}' E - \sum{}' \bar{F})$$

$$+ 2C(\sum{}' E + \sum{}' \bar{F}) - 2F\sum{}' \bar{B}]/4\mathcal{L}$$

Elliptical coma

$$N_1^0 = 3[E\sum{}' F - (P + C)(\sum{}' P + \sum{}' \bar{C})$$

$$+ 2C(\sum{}' P - \sum{}' \bar{C}) + F(\sum{}' E - 2\sum{}' \bar{F})$$

$$- B\sum{}' B]/2\mathcal{L}$$

$$N_2^0 = 3[E\sum{}' F - (P + 3C)(3\sum{}' C - \sum{}' \bar{C} + \sum{}' P)$$

$$- C(\sum{}' P + \sum{}' \bar{C}) + F(\sum{}' E - 8\sum{}' \bar{F})$$

$$- B\sum{}' B']/\mathcal{L}$$

$$N_3^0 = [E\sum{}' F + (P + C)(3\sum{}' C - \sum{}' \bar{C} + \sum{}' P)$$

$$+ C(\sum{}' C + \sum{}' P) + F(\sum{}' E - 4\sum{}' \bar{F})$$

$$- B\sum{}' \bar{B}]/2\mathcal{L}$$

Image distortion

$$E^0 = [3E\sum{}' E - (P + 3C)\sum{}' \bar{B}]/2\mathcal{L}$$

$$+ 2C(\sum{}' E + \sum{}' \bar{F}) - 2F\sum{}' \bar{B}]/4L$$

Pupil distortion

$$E^0 = [-3E\sum{}' \bar{E} + (P + 3\bar{C})\sum{}' B]/2\mathcal{L}$$

References

Gaussian Optics à la Maxwell

1. Cagnet M, Françon M, Thrier JC. *The Atlas of Optical Phenomena*. Englewood Cliffs, NJ: Prentice Hall, 1962.
2. Cornbleet S. *Microwave and Optical Ray Geometry*. London: Wiley, 1984, pp. 11–35; Avendaño-Alejo M, Moreno I, Castañeda L. Caustics caused by multiple reflections on a circular surface. *Am. J. Phys.*, **78**, 1195–1198; 2010.

3. Maxwell JC. *Scientific Papers I*. Cambridge University Press, 1890, p. 271. Cited in Born M, Wolf E. *Principles of Optics*, 4th edn. London: Pergamon Press, 1970, pp. 150–157.

4. Stavroudis ON. *The Mathematics of Geometrical and Physical Optics, The K-Function and its Ramifications*. Germany: Wiley-VCH Verlag GmbH & Co. KGaA, 2006, Chapter 14, pp. 197–204.

The Eikonal Equation and Its Antecedents

5. Herzberger M. On the characteristic function of Hamilton, the Eikonal of Bruns and their uses in optics. *J. Opt. Soc. Am.*, **26**, 177; 1936; Herzberger M. Hamilton's characteristic function and Bruns' Eikonal. *J. Opt. Soc. Am.*, **27**, 133; 1937; Synge JL. Hamilton's characteristic function and Bruns' Eikonal. *J. Opt. Soc. Am.*, **27**, 138; 1937.

6. A concise account is in *Encyclopedic Dictionary of Mathematics*, 2nd edn, Cambridge, MA: MIT Press, 1993, Section 46. A more detailed reference is, for example, Clegg JC. *Calculus of Variations*. New York, NY: John Wiley, 1968.

7. Stavroudis ON. *The Optics of Rays, Wavefronts and Caustics*, New York, NY: Academic Press, 1972, Chapter II; Klein M, Kaye IW. *Electromagnetic Theory and Geometrical Optics*. New York, NY: Interscience Publishers, 1965, pp. 72–74. Herzberger M. *Modern Geometrical Optics*. New York, NY: Interscience Publishers, 1958, Chapter 35.

8. See *Encyclopedic Dictionary of Mathematics*, 1958, 2nd edn, Cambridge, MA: MIT Press, 1993, Section 111F; App. A, Table 4.1. See also Struik DJ. *Lectures on Classical Differential Geometry*, 2nd edn, Reading, MA: Addison-Wesley, 1961, Chapter 1.

9. *Encyclopedic Dictionary of Mathematics*, 2nd edn, Cambridge, MA: MIT Press, 1993, Section 111D; 111H; App. A, Table 4.1. See also Struik DJ. *Lectures on Classical Differential Geometry*, 2nd edn, Reading, MA: Addison-Wesley, 1961, Chapter 2.

10. Luneburg RK. *Mathematical Theory of Optics*, Berkeley: University of California Press, 1966, pp. 172–182; Stavroudis ON. *The Optics of Rays, Wavefronts and Caustics*. New York, NY: Academic Press, 1972, Chapters 4 and 11; Born M, Wolf E. *Principles of Optics*, 4th edn. Oxford: Pergamon Press, 1970, pp. 147–149.

11. Welford WT. *Aberrations of the Symmetrical Optical System*. Oxford: Academic Press, 1974, pp. 66–69; Stavroudis ON. *The Optics of Rays, Wavefronts and Caustics*, New York: Academic Press, 1972, Chapter 12.

12. Conway AW, Synge JL (eds). *The Mathematical Papers of Sir William Roan Hamilton, Vol. 1, Geometrical Optics*. London: Cambridge University Press, 1931; Rund H. *The Hamilton-Jacobi Theory in The Calculus of Variations*. Princeton, NJ: Van Nostrand-Reinhold, 1966.

13. Klein M, Kay IW. *Electromagnetic Theory and Geometrical Optics*. New York, NY: Interscience Publishers, 1965, Chapter III.

14. Luneburg RK. *Mathematical Theory of Optics*. Berkeley: University of California Press, 1966, Chapters I and II.

15. Stavroudis ON. *The Mathematics of Geometrical and Physical Optics, The K-function and its Ramifications*. Germany: Wiley-VCH Verlag GmbH & Co. KGaA, 2006, Chapter 6, pp. 85–96; Shealy DL, Hoffnagle JA. Wavefront and caustics of a plane wave refracted by an arbitrary surface. *J. Opt. Soc. Am. A*, **25**, 2370–2382 (2008); Hoffnagle JA, Shealy DL. Refracting the k-function: Stavroudis's solution to the eikonal equation for multielement optical systems. *J. Opt. Soc. Am. A*, **28**, 1312–1321; 2011.

16. Bliss GA. *Lectures on the Calculus of Variations*. University of Chicago Press, 1946, Chapter 1; Stavroudis ON. *The Optics of Rays, Wavefronts and Caustics*, New York, NY: Academic Press, 1972, pp. 71–73.

17. Joos G. *Theoretical Physics*, Tr. Freeman IM, London: Blackie and Son, 1934, pp. 40–42; Stavroudis ON. *The Optics of Rays, Wavefronts and Caustics*. New York, NY: Academic Press, 1972, Chapter V.

18. Luneburg RK. *Mathematical Theory of Optics*, Berkeley: University of California Press, 1966, pp. 129–138; Herzberger M. *Modern Geometrical Optics*. New York, NY: Interscience Publishers, 1958, Chapter 17.

19. Born M, Wolf E. *Principles of Optics*, 4th edn. Oxford: Pergamon Press, 1970, pp. 130–132; Stavroudis ON. *The Mathematics of Geometrical and Physical Optics, The K-function and its Ramifications*. Germany: Wiley-VCH Verlag GmbH & Co. KGaA, 2006, Chapter 12, pp. 180–186; Maca-García S, Avendaño-Alejo M, Castañeda L. Caustics in a meridional plane produced by concave conic mirrors. *J. Opt. Soc. Am. A*, **29**, 1977–1985; 2012.
20. Born M, Wolf E. *Principles of Optics*, 4th edn. Oxford: Pergamon Press, 1970, pp. 149–150; Luneburg, R. K. *Mathematical Theory of Optics*. Berkeley: University of California Press, 1966, pp. 136–138; Herzberger, M. *Modern Geometrical Optics*, New York, NY: Interscience Publishers, 1958, p. 190; Welford WT. *Aberrations of the Symmetric Optical System*. London: Academic Press, 1974, pp. 139–142.

Ray Tracing and Its Generalization

21. Smith T. articles in Glazebrook, R. *Dictionary of Applied Physics*. London: MacMillan, 1923.
22. Stavroudis ON. *The Optics of Rays, Wavefronts and Caustics*. New York, NY: Academic Press, 1972, Chapter VI.
23. Feder DP. Optical calculations with automatic computing machinery. *J. Opt. Soc. Am.*, **41**, 630–635; 1951.
24. Welford WT. *Aberrations of the Symmetric Optical System*. London: Academic Press, 1974, pp. 9–11; Herzberger M. *Modern Geometrical Optics*. New York, NY: Interscience Publishers, 1958, pp. 152–153, 269–271; Born M, Wolf E. *Principles of Optics*, 4th edn. London: Pergamon Press, 1970, Chapter 3; Stavroudis ON. *The Optics of Rays, Wavefronts and Caustics*, New York, NY: Academic Press, 1972, Chapter VIII; Klein M, Kay IW. *Electromagnetic Theory and Geometrical Optics*. New York, NY: Interscience Publishers, 1965, Chapter V.
25. *Encyclopedic Dictionary of Mathematics*, 2nd edn. Cambridge, MA: MIT Press, 1993, Section 111. Also Struik DJ. *Lectures on Classical Differential Geometry*, 2nd edn. Reading, MA: Addison-Wesley, 1961, Chapter 2.
26. Stavroudis ON. *The Optics of Rays, Wavefronts and Caustics*. New York, NY: Academic Press, 1972, Chapter X; Stavroudis ON. A simpler derivation of the formulas for generalized ray tracing. *J. Opt. Soc. Am.*, 66, 1330–1333; 1976; Kneisley JA. Local curvatures of wavefronts. *J. Opt. Soc. Am.*, **54**, 229–235; 1964.
27. Coddington H. *A System of Optics. In two parts*. London: Simkin and Marshal, 1829, pp. 1830–1839; Gullstrand A. Die reelle optische abbildung. *Sven. Vetensk. Handl.*, **41**, 1–119 (1906); Altrichter O, Schaffer G. Herleitung der gullstrandschen Grundgleichungen fur schiefe Strahlbuschel aus den Hauptkremmungen der Wellenfla sche. *Optik*, **13**, 241–253; 1956.
28. Cagnet M, Françon M, Thrier JC. *The Atlas of Optical Phenomena*. Englewood Cliffs, NJ: Prentice Hall, 1962.
29. Born M, Wolf E. *Principles of Optics*, 4th edn. London: Pergamon Press, 1970, pp. 127–130, 170; Herzberger M. *Modern Geometrical Optics*. New York, NY: Interscience Publishers, 1958, pp. 156, 196, 260–263; Stavroudis ON. *The Optics of Rays, Wavefronts and Caustics*, New York, NY: Academic Press, 1972, pp. 78–80, 157–160, 173–175; Klein M, Kay IW. *Electromagnetic Theory and Geometrical Optics*, New York, NY: Interscience Publishers, 1965, extend the idea of the caustic to include diffraction, pp. 498–499.
30. Castillo-Santiago G, Avendaño-Alejo M, Díaz-Uribe R, Castañeda L. Analytic aspheric coefficients to reduce the spherical aberration of lens elements used in collimated light. *Appl. Opt.*, **53**, 22, 4939–4946; 2014.

The Paraxial Approximation and Its Uses

31. Coddington H. *A System of Optics. In two parts*. London: Simkin and Marshal, 1829, 1839.
32. Welford WT. *Useful Optics*. Chicago, IL: University of Chicago Press, 1991, Chapter 3; Welford WT. *Aberrations of the Symmetrical Optical System*. London: Academic Press, 1974, Chapter 3;

Herzberger M. *Modern Geometrical Optics*, New York, NY: Interscience Publishers, 1958, Chapter 8; Luneburg RK. *Mathematical Theory of Optics*. Berkeley: University of California Press, 1966, Chapter IV.

33. Johnson RB. An Historical Perspective on Understanding Optical Aberrations, in Lens Design. *Crit. Rev. Opt. Sci. Technol.*, **CR41**, 24; 1992.

34. Born M, Wolf E. *Principles of Optics*, 4th edn. London: Pergamon Press, 1970, pp. 164–166; Welford WT. *Aberrations of the Symmetrical Optical System*. London: Academic Press, 1974, pp. 23.

35. Herzberger M. *Modern Geometrical Optics*. New York, NY: Interscience Publishers, 1958, pp. 101–105, 406; Born M, Wolf E. *Principles of Optics*, 4th edn. London: Pergamon Press, 1970, pp. 186–188; Welford WT. *Aberrations of the Symmetrical Optical System*. London: Academic Press, 1974, pp. 27–29; Smith WJ. *Modern Lens Design. A Resource Manual*. New York, NY: McGraw-Hill, 1992, pp. 33–34.

36. Born M, Wolf E. *Principles of Optics*, 4th edn. London: Pergamon Press, 1970, pp. 193–194; Feder DP. Optical calculations with automatic computing machinery. *J. Opt. Soc. Am.*, **41**, 630–635; 1951.

37. Definitions of f/# and numerical aperture are to be found in Born M, Wolf E, *Principles of Optics*, 4th edn. London: Pergamon Press, 1970, p. 187. The relationship between these quantities and resolution, as defined by Rayleigh, are in Ditchburn RW. *Light*, 3rd edn, Vol. 1. London: Academic Press, 1976, p. 277.

The Huygens' Principle and Developments

38. Herzberger M. *Modern Geometrical Optics*. New York, NY: Interscience Publishers, 1958, p. 482; Avendaño-Alejo M, et al. Huygens principle: Exact wavefronts produced by aspheric lenses. *Opt. Exp.*, **21**, 24, 29874–29884; 2013.

39. Born M, Wolf E. *Principles of Optics*, 4th edn. London: Pergamon Press, 1970, pp. 370–375. Klein M, Kay IW. *Electromagnetic Theory and Geometrical Optics*. New York, NY: Interscience Publishers, 1965, pp. 328, 342–346.

40. Cagnet M, Francon M, Thrier JC. *The Atlas of Optical Phenomena*. Englewood Cliffs, NJ: Prentice Hall, 1962.

41. Stavroudis ON. Ray tracing formulas for uniaxial crystals. *J. Opt. Soc. Am.*, **52**, 187–191; 1962; Avendaño-Alejo M, Stavroudis ON. Huygens's principle and rays in uniaxial anisotropic media II. Crystal axis with arbitrary orientation. *J. Opt. Soc. Am. A*, **19**, 1674–1679; 2002; Avendaño-Alejo M, Díaz-Uribe R, Moreno I. Caustics caused by refraction in the interface between an isotropic medium and a uniaxial crystal. *J. Opt. Soc. Am. A*, **25**, 1586–1593; 2008.

The Aberrations, Seidel and Otherwise

42. Johnson RB. Historical perspective on understanding optical aberrations, in *Lens Design*, Smith WJ, ed., SPIE *Crit. Rev. Opt. Sci. Technol.*, **CR24**, 18–29; 1992.

43. Avendaño-Alejo M. Caustics in a meridional plane produced by plano-convex aspheric lenses. *J. Opt. Soc. Am. A*, **30**, 501–508; 2013; Avendaño-Alejo M, González-Utrera D, Castañeda L. Caustics in a meridional plane produced by plano-convex conic lenses. *J. Opt. Soc. Am. A*, **28**, 2619–2628; 2011; Avendaño-Alejo M, Castañeda L, Moreno I. Properties of caustics produced by a positive lens: Meridional rays. *J. Opt. Soc. Am. A*, **27**, 2252–2260; 2010.

44. Feder DP. Optical calculations with automatic computing machinery. *J. Opt. Soc. Am.*, **41**, 630–635; 1951.

45. Smith T. Sundry articles on optics in Glazebrook, R. *Dictionary of Applied Physics*. London: MacMillan, 1923.

46. Born M, Wolf E. *Principles of Optics*, 4th edn. London: Pergamon Press, 1970, pp. 207–211; Schwarzschild K. *AstronomBeobachtungen mit elementaren Hilfmitteln*. Leipzig: Teubner, 1904.

47. Kohlschutter A. Die Bildfehler funfter Ordnung optischer Systeme. Dissertation, Universitat Gottingen, 1908.

48. Wachendorf F. Bestimmung der Bildfehler 5, Ordnung in zentrierten optischer Systemce. *Optik*, **5**, 80–122; 1949.
49. Herzberger M. *Modern Geometrical Optics*. New York, NY: Interscience Publishers, 1958, Part VII.
50. Buchdahl H. *Optical Aberration Coefficients*. New York, NY: Dover Publications, 1968. This is a reissue of the 1954 edition with a series of journal articles appended.
51. Rimmer M. Optical Aberration Coefficients, in *Ordeals II Program Manual*, Gold MB (ed). Fairport, NY: Tropel, 1965, Appendix IV.
52. Stavroudis ON. *Modular Optical Design*. Berlin: Springer-Verlag, 1982, pp. 112–116.

2

Basic Wave Optics

Glenn D. Boreman

2.1 Diffraction

If one looks closely at the edges of shadows—the transition regions between bright and darkness—the transition is not abrupt as predicted by geometrical optics. There are variations in irradiance—interference fringes—seen at the boundary. As the size of the aperture is reduced, we expect that the size of the illuminated region in the observation plane will also decrease. This is true, but only to a certain point, where decreasing the dimensions of the aperture will produce spreading of the observed irradiance distribution.

Huygens' principle can be used to visualize this situation qualitatively. Each point on the wavefront transmitted by the aperture can be considered a source of secondary spherical waves. The diffracted wave is the summation of the secondary sources, and the envelope of these spherical waves is the new diffracted wavefront. For an aperture whose width is large compared with the wavelength of the radiation, the transmitted wavefront has very nearly the direction of the original wavefront, and the spreading caused by truncation of the spherical waves at the edges of the aperture is negligible. However, when the aperture is sufficiently small that it contains only a few Huygens' sources, the transmitted wavefront will exhibit a large divergence angle. A single Huygens' source would emit uniformly in all directions as a spherical wave.

We can put this phenomenon on a quantitative basis as follows. Let a scalar field V represent an optical disturbance (like the magnitude of the electric field, but a scalar rather than a vector), where V^2 is proportional to irradiance. This field V satisfies the scalar wave equation

$$\nabla^2 V(\bar{x},t) = \frac{1}{c^2} \frac{\partial^2 V(\bar{x},t)}{\partial t^2}. \tag{2.1}$$

If we assume that V is monochromatic, the scalar disturbance takes the form

$$V(\bar{x}, t) = \psi(\bar{x}) e^{-j2\pi \upsilon t}, \tag{2.2}$$

where $\psi(\bar{x})$ accounts for the spatial variation of the amplitude and phase. The spatial variation of the optical disturbance also satisfies a wave equation

$$\nabla^2\psi + \left(\frac{2\pi\upsilon}{c}\right)^2\psi = \nabla^2\psi + k^2\psi = 0. \tag{2.3}$$

An important point to note is that this wave equation is linear in ψ, which ensures the validity of the superposition principle.

We consider two coordinate systems, with subscript s for the source plane and subscript o for the observation plane. Any general point in the source plane $P(x_s, y_s)$ gives rise to a spherical wave that emits equally in all directions. For a general source distribution, the optical disturbance in the observation plane is just a weighted sum of spherical waves that originate at the various source points that comprise the source distribution. The proper expression for such a Huygens' wavelet is

$$\psi = \frac{1}{j\lambda}\frac{e^{jkr(\bar{x}_s,\bar{x}_o)}}{r(\bar{x}_s,\bar{x}_o)}, \tag{2.4}$$

where it is explicit that r is a function of position in both the source and observation planes. Thus the source distribution $\psi_s(x_s, x_s)$ produces a field distribution in the observation plane:

$$\psi(x_o, y_o) = \int_{\text{aperture}} \psi_s(x_s, y_s)\frac{e^{jkr(\bar{x}_s,\bar{x}_o)}}{j\lambda r(\bar{x}_s,\bar{x}_o)}dx_s dy_s. \tag{2.5}$$

In simplifying Equation 2.5, we can approximate r in the denominator of the integrand by z, the axial separation between source and observation planes. In the exponential, we express r using the first two terms of the binomial expansion

$$\sqrt{1+a} \approx 1 + \frac{a}{2} - \frac{a^2}{8} + \cdots, \tag{2.6}$$

yielding

$$r(\bar{x}_s,\bar{x}_o) = \sqrt{(x_o - x_s)^2 + (y_o - y_s)^2 + z^2} = z\sqrt{1 + \left(\frac{x_o - x_s}{z}\right)^2 + \left(\frac{y_o - y_s}{z}\right)^2} \tag{2.7}$$

$$r(\bar{x}_s,\bar{x}_o) \approx z\left(1 + \frac{1}{2}\left(\frac{x_o - x_s}{z}\right)^2 + \frac{1}{2}\left(\frac{y_o - y_s}{z}\right)^2\right) \tag{2.8}$$

$$r(\bar{x}_s,\bar{x}_o) \approx z + \frac{x_o^2 + y_o^2}{2z} + \frac{x_s^2 + y_s^2}{2z} - 2\left(\frac{x_s x_o + y_s y_o}{2z}\right). \tag{2.9}$$

With these substitutions, Equation 2.5 becomes

$$\psi(x_o, y_o) = \frac{1}{j\lambda z} \int_{\text{aperture}} \psi_s(x_s, y_s) \exp\left[jkr(\bar{x}_s, \bar{x}_o)\right] dx_s dy_s \tag{2.10}$$

$$\psi(x_o, y_o) = \frac{1}{j\lambda z} \int_{\text{aperture}} \psi_s(x_s, y_s)$$

$$\exp\left[jk\left(z + \frac{x_o^2 + y_o^2}{2z} + \frac{x_s^2 + y_s^2}{2z} - 2\left(\frac{x_s x_o + y_s y_o}{2z}\right)\right)\right] dx_s dy_s \tag{2.11}$$

$$\psi(x_o, y_o) = \frac{e^{jkz}}{j\lambda z} \exp\left[jk\frac{x_o^2 + y_o^2}{2z}\right] \int_{\text{aperture}} \psi_s(x_s, y_s)$$

$$\exp\left[jk\left(\frac{x_s^2 + y_s^2}{2z} - 2\left(\frac{x_s x_o + y_s y_o}{2z}\right)\right)\right] dx_s dy_s. \tag{2.12}$$

Diffraction problems separate into two classes, based on Equation 2.12. For Fresnel diffraction, the term $(k/2z)(x_s^2 + y_s^2)$ is large enough that it cannot be neglected in the exponent. In Fresnel diffraction, the spherical waves arising from point sources in the aperture are approximated as quadratic-phase surfaces. Fresnel diffraction patterns change functional form continuously with distance z. The validity of the Fresnel approximation is ensured provided the third term in the binomial expansion above is negligible. This condition can be written as

$$z\frac{2\pi}{\lambda}\frac{1}{8}\left[\frac{(x_o - x_s)^2 + (y_o - y_s)^2}{z^2}\right]^2 \ll 1, \tag{2.13}$$

or equivalently

$$z^3 \ll \frac{\pi}{4\lambda}\left[(x_o - x_s)^2 + (y_o - y_s)^2\right]^2. \tag{2.14}$$

Fraunhofer (far-field) diffraction conditions are obtained when the term $(k/2z)(x_s^2 + y_s^2)$ is small. Under this approximation, Equation 2.12 for the diffracted amplitude becomes

$$\psi(x_o, y_o) = \frac{e^{jkz}}{j\lambda z} \exp\left[jk\frac{x_o^2 + y_o^2}{2z}\right] \int_{\text{aperture}} \psi_s(x_s, y_s)$$

$$\exp\left[-j2\pi\left(x_s\left(\frac{x_o}{\lambda z}\right) + y_s\left(\frac{y_o}{\lambda z}\right)\right)\right] dx_s dy_s, \tag{2.15}$$

which aside from an amplitude factor $1/\lambda z$ and a multiplicative phase factor (which is of no importance when irradiance $|\psi(x_o, y_o)|^2$ is calculated) is just a Fourier transform of the amplitude distribution across the aperture, with identification of the Fourier-transform variables

$$\xi = \frac{x_o}{\lambda z} \text{ and } \eta = \frac{y_o}{\lambda z}. \tag{2.16}$$

Thus, Fraunhofer diffraction patterns will scale in size with increasing distance z, but keep the same functional form at all distances consistent with the far-field approximation. Because the distance from the aperture to the observation plane is so large, the Fraunhofer expression approximates the spherical waves arising from point sources in the aperture as plane waves.

2.1.1 Fresnel Diffraction

We must modify Equation 2.9 for $r(\bar{x}_s, \bar{x}_o)$ in the situation of point-source illumination of the aperture from a point at a distance r_{10} away, as seen in Figure 2.1. The aperture-to-observation-plane axial distance is r_{20}, and the distance from the source point to a general point $Q(x, y)$ in the aperture is r_1 and the distance from Q to the observation point P is r_2. We analyze the special case of a rectangular aperture that is separable in x- and y-coordinates. The diffraction integral, Equation 2.10,

$$\psi(x_o, y_o) = \frac{1}{j\lambda z} \int_{\text{aperture}} \psi_s(x_s, y_s) \exp\left[jkr(\bar{x}_s, \bar{x}_o) \right] dx_s dy_s \qquad (2.17)$$

becomes, in terms of on-axis amplitude in the observation plane at a point P,

$$\psi(P) = \frac{1}{j\lambda(r_{10} + r_{20})} \int_{y_1}^{y_2} \int_{x_1}^{x_2} \exp\left[jk(r_1 + r_2) \right] dx dy, \qquad (2.18)$$

where x_1, x_2, y_1, and y_2 are the edge locations of the diffracting aperture. The typical procedure for computation of Fresnel diffraction patterns of separable apertures involves calculation of irradiance at point P for a particular aperture position, then moving the aperture incrementally in x or y and calculating a new on-axis irradiance. For aperture sizes that are small compared with $r_{10} + r_{20}$, this procedure is equivalent to a calculation of irradiance as a function of position in the observation plane, but is simpler mathematically.

The distance terms in the exponent of Equation 2.18 can be expressed as

$$r_1 = \sqrt{r_{10}^2 + x^2 + y^2} \approx r_{10} + \frac{x^2 + y^2}{2r_{10}} \qquad (2.19)$$

$$r_2 = \sqrt{r_{20}^2 + x^2 + y^2} \approx r_{20} + \frac{x^2 + y^2}{2r_{20}}. \qquad (2.20)$$

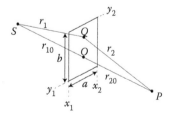

FIGURE 2.1   Geometry for Fresnel diffraction.

We can thus write, for Equation 2.18, the amplitude in the observation plane as

$$\psi(P) = 1 \frac{1}{j\lambda(r_{10} + r_{20})} \int_{y_1}^{y_2} \int_{x_1}^{x_2}$$

$$\exp\left[jk\left\{ (r_{10} + r_{20}) + (x^2 + y^2)\left(\frac{r_{10} + r_{20}}{2r_{10}r_{20}} \right) \right\} \right] dxdy. \tag{2.21}$$

With a change of variables

$$\xi = x\sqrt{\frac{2(r_{10} + r_{20})}{\lambda r_{10}r_{20}}} \tag{2.22}$$

and

$$\eta = y\sqrt{\frac{2(r_{10} + r_{20})}{\lambda r_{10}r_{20}}}, \tag{2.23}$$

we find, for Equation 2.21, the diffracted wave amplitude

$$\psi(P) = \frac{1}{2j} \frac{e^{jk(r_{10}+r_{20})}}{(r_{10} + r_{20})} \int_{\xi_1}^{\xi_2} e^{y\pi\xi^2/2} d\xi \int_{\eta_1}^{\eta_2} e^{j\pi\eta^2/2} d\eta. \tag{2.24}$$

An important special case is that of plane-wave illumination, $r_{10} \to \infty$. Under those conditions, Equations 2.22 and 2.23 become

$$\xi = x\sqrt{\frac{2}{\lambda r_{20}}} \tag{2.25}$$

$$\eta = y\sqrt{\frac{2}{\lambda r_{20}}}. \tag{2.26}$$

In general, we identify the term

$$\psi_0(P) = \frac{e^{jk(r_{10}+r_{20})}}{(r_{10} + r_{20})} \tag{2.27}$$

as the unobstructed wave amplitude at P in the absence of a diffracting screen. Hence, we can write for Equation 2.24,

$$\psi(P) = \frac{\psi_0(P)}{2j} \int_{\xi_1}^{\xi_2} e^{j\pi\xi^2/2} d\xi \int_{\eta_1}^{\eta_2} e^{j\pi\eta^2/2} d\eta. \tag{2.28}$$

We identify the Fresnel cosine and sine integrals

$$C(w) = \int_0^w \cos(\pi\xi^2/2) d\xi \tag{2.29}$$

and

$$S(w) = \int_0^w \sin\left(\pi\xi^2/2\right)d\xi \tag{2.30}$$

seen in Figure 2.2. Using Equations 2.29 and 2.30, we can write for Equation 2.28, the diffracted on-axis amplitude:

$$\psi(P) = \frac{\psi_o(P)}{2j}\left\{\left[C(\xi_2)-C(\xi_1)\right]+j\left[S(\xi_2)-S(\xi_1)\right]\right\}$$
$$\left\{\left[C(\eta_2)-C(\eta_1)\right]+j\left[S(\eta_2)-S(\eta_1)\right]\right\} \tag{2.31}$$

and for the diffracted on-axis irradiance:

$$E(P) = \frac{E_o(P)}{2j}\left\{\left[C(\xi_2)-C(\xi_1)\right]^2+\left[S(\xi_2)-S(\xi_1)\right]^2\right\}$$
$$\left\{\left[C(\eta_2)-C(\eta_1)\right]^2+\left[S(\eta_2)-S(\eta_1)\right]^2\right\}, \tag{2.32}$$

where $E_0(P)$ is the unobstructed irradiance. If the wavelength and geometry are specified, the above expression can be evaluated numerically for separable apertures.

As an example of the calculation method, let us compute the Fresnel diffraction pattern from an edge, illuminated from a source at infinity. Assuming that the edge is oriented vertically, $\eta_1 = -\infty$ and $\eta_2 = \infty$. If we let the left edge of the open aperture be at $-\infty$, we have $\xi_1 = -\infty$ and $\xi_2 = x_2(2/\lambda r_{20})^{1/2}$. Using the asymptotic forms $C(-\infty) = S(-\infty) = -0.5$ and $C(\infty) = S(\infty) = 0.5$, we can write for Equation 2.32:

$$E(P) = \frac{E_o(P)}{4}\left\{\left[C(\xi_2)-0.5\right]^2+\left[S(\xi_2)-0.5\right]^2\right\}\left\{\left[0.5+0.5\right]^2+\left[0.5+0.5\right]^2\right\}. \tag{2.33}$$

The value of irradiance at the edge of the geometrical shadow region $\left(\xi_2 = 0\right)$ is found to be one-quarter of the unobstructed irradiance. This result is intuitive, because the presence of the edge blocks

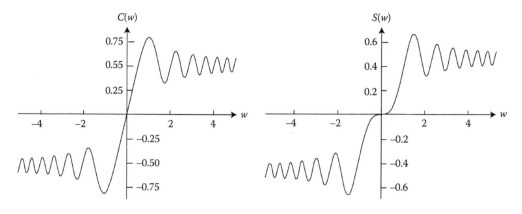

FIGURE 2.2 Fresnel cosine and sine integrals.

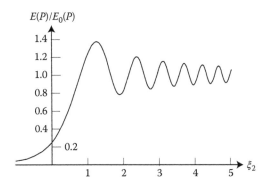

FIGURE 2.3 Fresnel diffraction pattern (irradiance) at an edge.

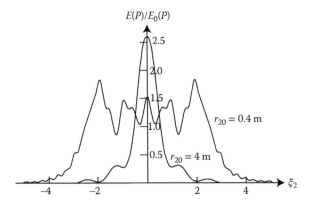

FIGURE 2.4 Fresnel diffraction patterns (irradiance) for an aperture at two distances.

half of the amplitude on-axis, resulting in one-quarter of the irradiance that would be present without the edge. The complete irradiance distribution is shown in Figure 2.3.

Examples of other irradiance distribution calculations are shown in Figure 2.4, for an aperture of dimensions 2 mm × 2 mm, with plane-wave illumination at 0.5 μm wavelength. The distance from the diffracting aperture to the observation screen, r_{20}, is the variable. In the first plot, cases are shown for $r_{20} = 0.4$ m and $r_{20} = 4$ m. The changes in the form of the irradiance pattern with changes in r_{20} are evident. There is not a simple scaling of size as one finds with Fraunhofer diffraction, but a change of functional form depending on the value of r_{20}. For the situation where $r_{20} = 0.4$ m, the geometrical shadow region is within $|\xi_2| < 3.16$; and for the case of $r_{20} = 4$ m, the shadow region is within $|\xi_2| < 1$. The shadow region is just the size of the original aperture, expressed in terms of ξ_2, for a given distance.

Figure 2.5 shows the irradiance distribution at a longer distance, for $r_{20} = 40$ m. This is calculated by means of the Fresnel-integral technique above, but the dimensions involved are such that the observation plane is essentially in the Fraunhofer region of the aperture. Thus, the irradiance distribution approximates that of the Fourier transform squared of the aperture distribution. Note that the peak irradiance has decreased dramatically, consistent with the spreading of flux far beyond the boundaries of the geometrical shadow at $|\xi_2| = 0.32$.

Fresnel diffraction from a nonseparable aperture, even one as simple as a circular disk, is more difficult to describe analytically. The most effective calculation method is to compute the squared modulus of a two-dimensional Fourier transform, but with a multiplicative quadratic phase factor in the aperture. This phase factor accounts for the optical path difference between the axial distance ($r_{10} + r_{20}$) between source and observation points and the distance ($r_1 + r_2$) that traverses the maximum extent of the aperture. The Fresnel pattern can then be calculated by the same means as a Fraunhofer pattern.

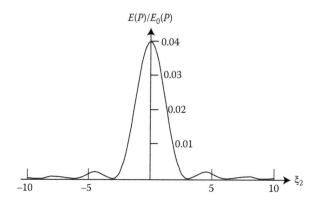

FIGURE 2.5   Fresnel diffraction pattern (irradiance) for a square aperture at a long distance.

2.1.2 Fraunhofer Diffraction

For situations in which the distance from the aperture to the observation plane is sufficiently large, or for which the aperture is small, we can neglect the term $(k/2z)(x_s^2 + y_s^2)$. According to Equation 2.15, the diffracted amplitude can be expressed in terms of the Fourier transform of the amplitude transmittance function of the aperture $c_s(x_s, y_s)$:

$$\psi(x_o, y_o) = \frac{e^{jkz}}{j\lambda z} \exp\left[jk\frac{x_o^2 + y_o^2}{2z} \right] \int\limits_{\text{aperture}} \psi_s(x_s, y_s)$$

$$\exp\left[-j2\pi\left(x_s\left(\frac{x_o}{\lambda z}\right) + y_s\left(\frac{y_o}{\lambda z}\right) \right) \right] dx_s dy_s \tag{2.34}$$

$$\psi(x_o, y_o) = \frac{1}{\lambda z}\exp\left[j\left\{ k\left(z + \frac{x_o^2 + y_o^2}{2z} \right) - \frac{\pi}{2} \right\} \right] \wedge\left\{ \psi_s(x_s, y_s) \right\}\Big|_{\xi=\frac{x_o}{\lambda z}, \eta=\frac{y_o}{\lambda z}}. \tag{2.35}$$

The complex exponential does not affect the value of the diffracted irradiance, $E(x_o, y_o) = |\psi(x_o, y_o)|^2$, so that we find

$$E(x_o, y_o) = \frac{1}{\lambda^2 z^2}\left| \wedge\left\{ \psi_s(x_s, y_s) \right\} \right|_{\xi=\frac{x_o}{\lambda z}, \eta=\frac{y_o}{\lambda z}}. \tag{2.36}$$

Note that the Fourier transform of a spatial-domain aperture function returns a function of spatial frequency, and thus a change of variables seen in Equation 2.16, $\xi = x_o/\lambda z$ and $\eta = y_o/\lambda z$, is required for evaluation of the result, Equation 2.36, in terms of position in the observation plane. The change of variables takes the slightly different form

$$\xi = \frac{x_o}{\lambda f} \quad\text{and}\quad \eta = \frac{y_o}{\lambda f}, \tag{2.37}$$

when the diffraction pattern is observed in the focal plane of a positive lens of focal length f, allowing convenient scaling of the pattern to any desired dimensions by the choice of lens focal length.

Our examples of calculation of Fraunhofer patterns begin with the one-dimensional rectangular aperture of full-width b. Using the standard notation of Gaskill [1], we take the aperture distribution as $\psi_s(x_s) = \text{rect}(x_s/b)$, and for the usual expression of a diffraction pattern normalized to an on-axis value of unity,

$$\frac{E(x_o)}{E(x_o = 0)} = \text{sinc}^2\left(\frac{\pi b x_o}{\lambda z}\right) = \left(\frac{\sin\left(\dfrac{\pi b x_o}{\lambda z}\right)}{\left(\dfrac{\pi b x_o}{\lambda z}\right)}\right)^2. \tag{2.38}$$

This function, seen in Figure 2.6, has a maximum value of 1 and a first zero at $x_o = \lambda z/b$.

Figure 2.7 illustrates the Fraunhofer diffraction patterns for four different aperture shapes.

Another useful aperture distribution is that representing two point apertures separated by a distance l:

$$\psi_s(x_s) = \frac{1}{2}\left[\delta(x_s - l/2) + \delta(x_s + l/2)\right], \tag{2.39}$$

for which the corresponding irradiance distribution in the diffraction pattern is

$$\frac{E(x_o)}{E(x_o = 0)} = \cos^2\left(\frac{\pi l x_o}{\lambda z}\right). \tag{2.40}$$

A circular aperture of full width D, $c_s(x_s) = \text{circ}(r_s/D)$, is another important case. The normalized diffracted-irradiance distribution is

$$\frac{E(r_o)}{E(r_o = 0)} = \left[\frac{2J_1\left(\dfrac{\pi D r_o}{\lambda z}\right)}{\left(\dfrac{\pi D r_o}{\lambda z}\right)}\right], \tag{2.41}$$

which, as seen in Figure 2.8, has its first zero when

$$\frac{\pi D r_o}{\lambda z} = 1.22\pi = 3.83, \tag{2.42}$$

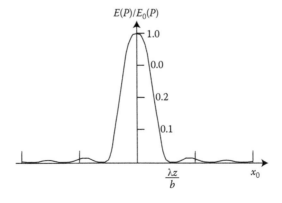

FIGURE 2.6 Sinc-squared function describing diffracted irradiance from a slit.

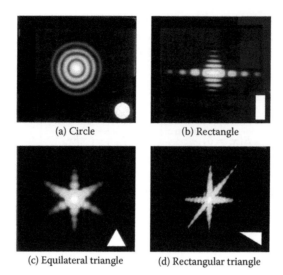

(a) Circle   (b) Rectangle

(c) Equilateral triangle   (d) Rectangular triangle

FIGURE 2.7 Fraunhofer diffraction patterns for four different aperture shapes. (a) A circular aperture, (b) a rectangular aperture, (c) an equilateral triangular aperture and, (d) a right angle triangular aperture.

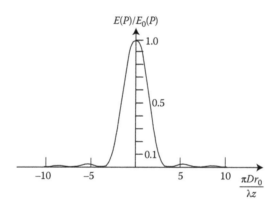

FIGURE 2.8 Bessel-squared function for diffracted irradiance from a circular aperture.

or in terms of radius at

$$r_0 = \frac{1.22\lambda z}{D}. \tag{2.43}$$

Integrating the irradiance distribution, we find that 84% of the power contained in the pattern is concentrated into a diameter equal to $2.44\,\lambda z/D$. When the diffracting aperture is a lens aperture itself, the full width of the diffracted-irradiance distribution in the focal plane is then $2.44\,\lambda F/\#$, where the lens focal ratio is defined as $F/\# \equiv f/D$. This relationship determines the ability of an imaging system to resolve objects of a certain size. The angular resolution β, expressed as the full width of the diffraction spot divided by the focal length of the lens, is

$$\beta = 2.4\frac{\lambda f/D}{f} = 2.4\frac{\lambda}{D}. \tag{2.44}$$

When the angle in Equation 2.45 is multiplied by p, the distance from the lens to the object plane, the resulting dimension is the diffraction-limited resolution in the object, as seen in Figure 2.9.

For example, a diffraction-limited system with an aperture size of 10 cm can resolve a 12 mm target at 1 km if the system operates at a wavelength of 0.5 µm, while a system of the same diameter operating at $\lambda = 10$ µm can resolve targets of 240 mm lateral dimension.

Figure 2.10 illustrates the Fraunhofer diffraction patterns of two close point light sources (stars) at different separations. The left hand column images were given enough exposure to detect the first ring. The images in the right hand column were given a shorter exposure time.

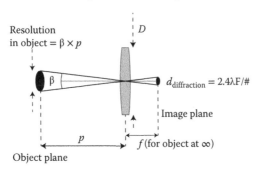

FIGURE 2.9 Diffraction-limited resolution in an optical system.

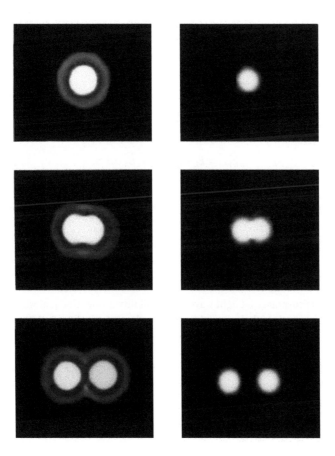

FIGURE 2.10 Fraunhofer diffraction patterns of two close point light sources (stars) at different separations.

2.2 Interference

The phenomenon of the interference depends on the principle of superposition of waves. Suppose that the following sinusoid corresponds to the electric-field magnitude of a light wave:

$$\varepsilon(x,t) = A\cos(kx - \omega t + \phi) \tag{2.45}$$

where $k = 2\pi/\lambda$ is the free-space wavenumber and $\omega = 2\pi/T$ is the radian frequency. Any number of waves can be added together coherently, and the magnitude of the resultant wave will depend on whether the interference is constructive or destructive. As an example, suppose that two waves are added together (with time dependence suppressed):

$$\varepsilon(x) = A_1\cos(kx + f_1) + A_2\cos(kx + \phi_2) \tag{2.46}$$

Constructive interference occurs when the amplitude of the resultant sum wave is larger than the magnitude of either component. This occurs when the relative phase difference between the two waves $\delta = \phi_2 - \phi_1$ is close to zero or an even multiple of 2π. Destructive interference occurs when the amplitude of the resultant is less than the magnitude of either component. This condition occurs when δ is close to an odd multiple of π. The relative phase difference between two waves derived from the same source depends on the optical path difference (OPD) between the two paths. In terms of the refractive index n and the physical path length d for the two paths, the phase difference is defined as

$$\delta = \frac{2\pi}{\lambda} \times \text{OPD} = \frac{2\pi}{\lambda}\left[\sum_{\text{path 2}} nd - \sum_{\text{path 1}} nd\right]. \tag{2.47}$$

Any detector of optical or infrared radiation responds to the long-term time average of the square of the resultant electric field, a quantity proportional to irradiance, $E(\text{W/cm}^2)$. Assuming that the ε field vectors of the two waves are co-polarized, we can write an expression for the total irradiance:

$$E_{\text{total}} = E_1 + E_2 + 2\sqrt{E_1 E_2}\cos\delta. \tag{2.48}$$

If the interfering waves are of equal magnitude, the maximum value of $E_{\text{total}} = 4E$, and the minimum value is zero. We characterize the contrast of the fringes by the visibility V, a number between zero and one, defined as

$$V = \frac{E_{\max} - E_{\min}}{E_{\max} + E_{\min}}. \tag{2.49}$$

For equal-amplitude co-polarized light waves, the visibility is one. Unequal wave amplitudes and any cross-polarized component (which does not interfere) will reduce the fringe visibility. We have assumed the phase difference between the two waves is stable with time. If δ changes rapidly over a measurement time, then the cos δ term averages out, and an incoherent addition of irradiances is obtained.

Some special cases of interference are of particular interest for applications. If we interfere two plane waves of the same frequency and parallel polarization:

$$\bar{\varepsilon}_1 = \hat{z}\varepsilon_1 e^{i(\bar{k}_1\cdot\bar{r}+\phi_1)} \tag{2.50}$$

$$\bar{\varepsilon}_2 = \hat{z}\varepsilon_2 e^{i(\bar{k}_1\cdot\bar{r}+\phi_2)}, \tag{2.51}$$

where r is a position vector in the x–y plane, $\bar{r} = \hat{x}x + \hat{y}y$. Letting the initial phases both be equal to zero, we can write an expression for the constructive-interference condition

$$k[x(\cos\theta_2 - \cos\theta_1 + y(\sin\theta_2 - \sin\theta_1)] = 2\pi m(\min teger). \tag{2.52}$$

This fringe pattern consists of straight fringes of equal spacing, parallel to the line bisecting the ray directions. The fringe spacings along the x- and y-axes are

$$\Delta y = \frac{\lambda}{(\sin\theta_2 - \sin\theta_1)} \tag{2.53}$$

and

$$\Delta x = \frac{\lambda}{(\cos\theta_2 - \cos\theta_1)}. \tag{2.54}$$

As seen from Equations 2.53 and 2.54, the fringe spacing increases as the angle between the interfering beams gets smaller.

Interference of two spherical waves is another important special case, with the two components given by

$$\bar{\varepsilon}_1 = \hat{z}\varepsilon_1 \frac{e^{i(k|\bar{r}-\bar{r}_1|+\phi_1)}}{|\bar{r} - \bar{r}_1|} \tag{2.55}$$

and

$$\bar{\varepsilon}_2 = \hat{z}\varepsilon_2 \frac{e^{i(k|\bar{r}-\bar{r}_2|+\phi_2)}}{|\bar{r} - \bar{r}_2|}. \tag{2.56}$$

Setting the phase offsets to zero in Equations 2.55 and 2.56, we find, for a given fringe, that

$$k(|\bar{r} - \bar{r}_1| - |\bar{r} - \bar{r}_2|) = \text{constant}. \tag{2.57}$$

Because the distance between two points is a constant, this is the equation of a family of hyperboloids of revolution. On an observation screen perpendicular to the line of centers of the sources, one obtains circular fringes, and on an observation screen parallel to the line of centers, one obtains (at least near the center) equally spaced sinusoidal fringes. Taking the circular-fringe case (we will analyze the other case in the next section when dealing with Young's interference experiment), we begin with the condition for a bright fringe:

$$m\lambda = |\bar{r} - \bar{r}_1| - |\bar{r} - \bar{r}_2|. \tag{2.58}$$

Let source #1 be located on the x-axis at $x = -\upsilon$ and source #2 located at $x = -u$. Thus, we can write

$$|\bar{r} - \bar{r}_1| = \sqrt{y^2 + \upsilon^2} = \upsilon\sqrt{1 + \frac{y^2}{\upsilon^2}} \cong \upsilon\left[1 + \frac{y^2}{2\upsilon^2}\right] = \upsilon + \frac{y^2}{2\upsilon} \tag{2.59}$$

and

$$\left|\vec{r} - \vec{r_2}\right| = \sqrt{y^2 + u^2} = u\sqrt{1 + \frac{y^2}{u^2}} \cong u\left[1 + \frac{y^2}{2u^2}\right] = u + \frac{y^2}{2u}. \tag{2.60}$$

The bright-fringe condition, Equation 2.58, becomes

$$m\lambda = \left|\vec{r} - \vec{r_1}\right| - \left|\vec{r} - \vec{r_2}\right| = \upsilon - u + \frac{y^2}{2}\left[\frac{1}{\upsilon} - \frac{1}{u}\right], \tag{2.61}$$

and the resulting fringe pattern corresponds to the case where the radii are proportional to the square root of the order number m.

The final special case to consider is interference of a plane wave and a spherical wave. The components are

$$\bar{\mathcal{E}}_1 = \hat{z}\mathcal{E}_1 e^{i(\vec{k}\cdot\vec{r} + \phi_1)} \tag{2.62}$$

and

$$\bar{\mathcal{E}}_2 = \hat{z}\mathcal{E}_2 \frac{e^{i(k|\vec{r} - \vec{r_2}| + \phi_2)}}{\left|\vec{r} - \vec{r_2}\right|}. \tag{2.63}$$

Assuming that the spherical wave arises from a point source located on the x-axis, at a point $x = -x_0$, the exponent in Equation 2.63 can be written

$$k\left|\vec{r} - \vec{r_2}\right| = k\sqrt{(x - x_0)^2 + y^2 + z^2}. \tag{2.64}$$

The condition for a bright fringe is

$$k\left[x\cos\theta + y\sin\theta - \sqrt{(x - x_0)^2 + y^2 + z^2}\right] = 2\pi m. \tag{2.65}$$

Observing in the y–z ($x = 0$) plane, Equation 2.65 reduces to

$$m\lambda = \left[y\sin\theta - \sqrt{x_0^2 + y^2 + z^2}\right]. \tag{2.66}$$

Assuming that we observe the fringe pattern in a small region of the y–z plane, $|x_0| \gg y$ and $|x_0| \gg z$, we can expand Equation 2.66 as a binomial series:

$$m\lambda = y\sin\theta - x_0\sqrt{1 + \frac{y^2 + z^2}{x_0^2}} \cong y\sin\theta - x_0\left[\frac{1 + y^2 + z^2}{2x_0^2}\right]. \tag{2.67}$$

Considering the case where the plane wave propagates parallel to the x-axis, we find the condition for bright fringes, Equation 2.67, reduces to

$$m\lambda = x_0 + \frac{r^2}{2x_0}.$$ (2.68)

Thus, the fringes are concentric circles where the radius of the circle is proportional to the square root of the order number m. The fringe spacing Δr thus decreases with radius, according to $\Delta r = \lambda x_0/r$.

2.2.1 Wavefront Division

Division-of-wavefront interferometers are configurations where the optical beam is divided by passage through apertures placed side by side. The most important example of this type of interferometer is Young's double-slit configuration. Typically, a point source illuminates two point apertures in an opaque screen. We will consider later the effects of finite source size (spatial coherence) and finite slit width (envelope function). These points are sources of spherical waves (Huygens' wavelets) that will interfere in their region of overlap. The usual situation is that the fringes are observed on a screen that is oriented parallel to the line of centers of the pinholes. The basic setup is seen in Figure 2.11. The bright-fringe condition is

$$m\lambda = |\vec{r} - \vec{r_1}| - |\vec{r} - \vec{r_2}|.$$ (2.69)

If the two pinholes are illuminated with the same phase (source equidistant from each), then there will be a bright fringe on the axis, along the line perpendicular to the line of centers. The expressions for the distances from each pinhole to a general point in the observation plane are

$$|\vec{r} - \vec{r_1}| = \sqrt{(y - l/2)^2 + x_0^2}$$ (2.70)

and

$$|\vec{r} - \vec{r_2}| = \sqrt{(y + l/2)^2 + x_0^2}.$$ (2.71)

Using Equations 2.70 and 2.71, the bright-fringe condition, Equation 2.69 becomes

$$m\lambda = \sqrt{(y - l/2)^2 + x_0^2} - \sqrt{(y + l/2)^2 + x_0^2}$$

$$= x_0\sqrt{1 + \frac{(y - l/2)^2}{x_0^2}} - x_0\sqrt{1 + \frac{(y + l/2)^2}{x_0^2}}.$$ (2.72)

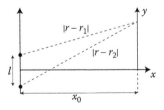

FIGURE 2.11 Young's double-slit configuration.

Under the binomial approximation of Equation 2.6 (equivalent to a small-angle condition), Equation 2.72 becomes

$$m\lambda \cong x_0\left[1 + \frac{(y-l/2)^2}{2x_0^2}\right] - x_0\left[1 + \frac{(y+l/2)^2}{2x_0^2}\right] = \frac{ly}{x_0}. \tag{2.73}$$

A simple mnemonic construction that yields this condition is seen in Figure 2.12. We assume that the rays drawn are essentially parallel and will thus interfere at infinity. The OPD between the two rays is the distance identified as $l \sin\theta$. When the OPD is equal to an even number of wavelengths, there will be constructive interference

$$m\lambda = l\sin\theta\tan\theta = \frac{ly}{x_0}. \tag{2.74}$$

The expression for center-to-center fringe separation is

$$\Delta y = \frac{x_0\lambda}{l}, \tag{2.75}$$

which means that as $\Delta y \propto x_0$, the fringe pattern scales with distance; as $\Delta y \propto \lambda$, the fringe pattern scales with wavelength; and as $\Delta y \propto 1/l$, the fringe pattern scales as the pinhole separation decreases.

It should also be noted from a practical point of view that it is not necessary to make x_0 a very long distance to achieve the overlap of these essentially parallel rays. As seen in Figure 2.13, a positive lens will cause parallel rays incident at an angle of θ to overlap at a distance f behind the lens, at a y height of θf. This configuration is often used to shorten the distance involved in a variety of experimental setups for interference and diffraction.

It can be verified that the fringes obtained in a Young's double-slit configuration have a cosine-squared functional form. Taking a phasor model for the addition of light from the two slits, and remembering that the irradiance is the square of the field, we find

$$E \propto \left|e^{j0} + e^{j\delta}\right|^2 = 2 + \cos\delta = 4\cos^2(\delta/2), \tag{2.76}$$

as seen in Figure 2.14. The phase shift δ can be written as

$$\delta = \frac{2\pi}{\lambda} \times \text{OPD} = \frac{2\pi}{\lambda}[l\sin\theta] \cong \frac{2\pi}{\lambda}l\frac{y}{x_0}. \tag{2.77}$$

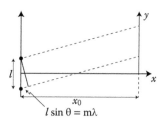

FIGURE 2.12   Simple construction for Young's fringes.

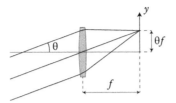

FIGURE 2.13 Use of a lens to produce the far-field condition for Fraunhofer diffraction.

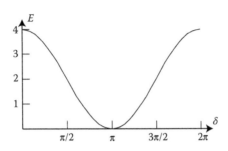

FIGURE 2.14 Irradiance as a function of phase shift δ for two-beam interference.

Thus, for small angles, Equation 2.76 describing the irradiance becomes

$$E = 4\cos^2\left[\dfrac{\dfrac{2\pi}{\lambda}l\dfrac{y}{x_0}}{2}\right] = 4\cos^2\left[\dfrac{\pi l y}{\lambda x_0}\right]. \tag{2.78}$$

The first zero of $\cos^2(\beta)$ is at $\beta = \pi/2$, *so* the first zero of the double-slit irradiance pattern of Equation 2.78 is at

$$y_{\text{first-zero}} = \frac{\lambda x_0}{2l}, \tag{2.79}$$

and the fringe spacing near the center of the pattern is

$$\Delta y = \frac{\lambda x_0}{l}. \tag{2.80}$$

It is useful to develop expressions relating the phase difference δ, the diffraction angle θ, and the position on the observation screen y. First, for small angles, where $\sin \theta \approx \tan \theta \approx \theta$, we have, in terms of diffraction angle θ,

$$l\theta \approx m\lambda. \tag{2.81}$$

Taking the derivative with respect to order number m:

$$\frac{\partial \theta}{\partial m} = \frac{\lambda}{l}. \tag{2.82}$$

Letting Equation 2.82 be a finite difference

$$\Delta\theta = \Delta m \frac{\lambda}{l}, \tag{2.83}$$

we have, for the angular spacing between fringes, $\Delta\theta$ for $\Delta m = 1$:

$$\Delta\theta \mid_{\Delta m=1} = \frac{\lambda}{l}. \tag{2.84}$$

A similar development can be obtained in terms of position in the observation plane. Assuming a small angle,

$$y \approx x_0\theta. \tag{2.85}$$

Taking the derivative with respect to angle, we have

$$\frac{\partial y}{\partial \theta} = x_0. \tag{2.86}$$

Combining Equations 2.85 and 2.86,

$$\frac{\partial y}{\partial m} = \frac{\partial y}{\partial \theta}\frac{\partial \theta}{\partial m} = x_0 \left[\lambda/l\right]. \tag{2.87}$$

Finding the y-position spacing between fringes as the increment in y for which the increment in order number $m = 1$

$$\Delta y = \Delta m \times x_0 \left[\lambda/l\right] \tag{2.88}$$

$$\Delta y \mid_{\Delta m=1} = x_0 \left[\lambda/l\right]. \tag{2.89}$$

Now, leaving the domain of small angles ($l \ll x$ but y not $\ll x$), we begin the development with

$$l \sin\theta = m\lambda. \tag{2.90}$$

Taking the derivative

$$\frac{\partial\theta}{\partial m} = \frac{\lambda}{l\cos\theta}. \tag{2.91}$$

The angular fringe spacing is then

$$\Delta\theta \mid_{\Delta m=1} = \frac{\lambda}{l\cos\theta}. \tag{2.92}$$

We note that the fringe spacing increases as the angle increases.

Now, in terms of y position, we begin with

$$y = x_0 \tan\theta.\tag{2.93}$$

Taking the derivative

$$\frac{\partial y}{\partial \theta} = x_0 \frac{1}{\cos^2\theta}\tag{2.94}$$

$$\frac{\partial y}{\partial m} = \frac{\partial y}{\partial \theta}\frac{\partial \theta}{\partial m} = \frac{x_0}{\cos^2\theta}\frac{\lambda}{l\cos\theta} = \frac{x_0\lambda}{l\cos^3\theta}.\tag{2.95}$$

For the fringe spacing in the observation plane (setting $\Delta m = 1$)

$$\Delta y = \frac{x_0\lambda}{l\cos^3\theta}.\tag{2.96}$$

Again, the fringes have a larger spacing away from the center of the pattern, but with a different functional dependence than seen in Equation 2.92 for the spacing in terms of angle.

Young's interference experiment can be performed with any number of slits, and the result can be easily obtained using phasor addition:

$$E \propto \left| e^{j0} + e^{j\delta} + e^{j2\delta} \right|^2,\tag{2.97}$$

where δ is the phase difference between successive slits. The three phasors are all in phase for $\delta = m \times 2\pi$, where m is an integer (including zero). The value of the irradiance for this condition is nine times the irradiance of a single slit. There are two values of δ, for a three-slit pattern, for which the irradiance equals zero. Using phasor diagrams, we can see that this condition is satisfied for $\delta = 2\pi/3$ and $\delta = 4\pi/3$. There is a subsidiary maximum (with irradiance equal to that from one slit alone) for the condition where $\delta = \pi$. The irradiance as a function of δ is shown in Figure 2.15, for the three-slit and four-slit cases.

2.2.2 Amplitude Division

Division of amplitude interferometers is typically implemented when one surface of the system is a partial mirror—in other words, a beamsplitter—that allows for an optical wave to be split into two

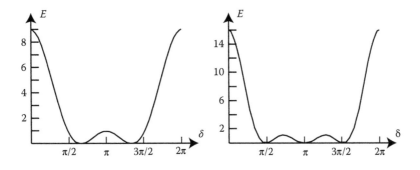

FIGURE 2.15 Irradiance as a function of phase shift S for three- and four-beam interference.

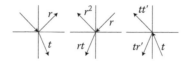

FIGURE 2.16   Stokes' reversibility principle.

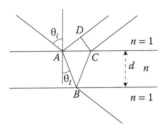

FIGURE 2.17   Interference in a plane-parallel plate.

components that will travel different paths and recombine afterwards. These interferometers are useful for determining the flatness of optical surfaces or the thickness of transparent films and are widely used for measurement purposes.

To properly account for the phase shift on reflection, it is necessary to note that a "rare-to-dense" reflection will differ in phase by π radians from the ray undergoing a "dense-to-rare" reflection. That this is true for either polarization can be shown easily by application of the Stokes' reversibility principle, which states that all light rays are reversible if absorption is negligible. Using Figure 2.16, we let the reflection and transmission coefficients for a ray incident from medium #1 be r and t, respectively. For a ray incident from the second medium, the reflection and transmission coefficients are r' and t'.

From Figure 2.16, we have

$$rt + r't = 0 \tag{2.98}$$

and

$$tt' + r^2 = 1. \tag{2.99}$$

From Equation 2.98, we can see that $r = -r'$, which verifies the required phase shift.

The first interferometer configuration we consider is that of fringes of equal thickness. There is a thin transparent film for which the thickness is a function of position. For example, the film can be an air film, contained between two pieces of glass, an oil film on the surface of water, or a freestanding soap film. We begin our analysis by considering interference in a plane-parallel plate of index n, immersed in air. We assume illumination by a single point source, and follow the ray trajectory as seen in Figure 2.17.

We consider interference between the rays reflected from the top and bottom surfaces of the plate. If the plate is parallel, these two rays will be parallel, and will interfere at infinity, or at the focus of a lens. For this analysis, we ignore other rays that undergo multiple reflections in the plate. Taking the plate as having a refractive index of glass (around 1.5), the two rays shown each carry about 4% of the original power in the incident ray, and all other rays carry a much smaller power. We assume that the thickness d of the plate is small enough to be within the coherence length of the light source used. Typically for films of a few micrometers or less thickness, interference effects are seen even with white light, such as sunlight. As the thickness of the plate increases to several millimeters or larger, interference effects are seen only with coherent sources, such as lasers.

Taking into account the difference in phase shifts on reflection seen in Figure 2.16, the condition for a bright fringe can be written as

$$\frac{2\pi}{\lambda}\left[n\left(\overline{AB}+\overline{BC}\right)-\overline{AD}\right]=\left(m+1/2\right)2\pi \tag{2.100}$$

$$\frac{2\pi}{\lambda}\left[n\frac{2d}{\cos\theta_t}-2nd\tan\theta_t\sin\theta_t\right]=\left(m+1/2\right)2\pi \tag{2.101}$$

$$\frac{2\pi}{\lambda}\frac{2nd}{\cos\theta_t}\left[1-\sin^2\theta_t\right]=\left(m+1/2\right)2\pi \tag{2.102}$$

$$2nd\cos\theta_t=\left(m+1/2\right)\lambda. \tag{2.103}$$

It should be noted that as θ increases, $\cos\theta_t$ decreases, and hence the resonance wavelength decreases. This effect can be important for the "off-angle" response for optical interference filters. In general, d, $\cos\theta$, and λ can all be variables in a given situation. If, by the configuration chosen, we arrange to keep θ_t constant, then we get interference maxima at certain wavelengths (if white light is used), depending on the local thickness of the film. If monochromatic light is used, we will see bright and dark Fizeau fringes across the film that correspond to thickness variations. The fringes are the loci of all points for which the film has a constant optical thickness.

Practically speaking, θ_t varies somewhat except when a collimated point source is used to provide the input flux. However, the variation in $\cos\theta_t$ is typically small enough to neglect this effect. This is especially true if the aperture of the collection lens is small, as is usually the case when viewing fringes directly by eye. Angular variations can also be neglected for the cases where the collection lens is far from the film being measured, or if the flux enters and leaves the film at nearly normal incidence, where the change of $\cos\theta_t$ with angle is minimum.

These conditions are satisfied in the Fizeau interferometer, often called a Newton interferometer, when the surface to be tested is very nearly in contact with a reference flat. As seen in Figure 2.18, a slight tilt (α) between two planar glass surfaces forms a wedge-shaped air film ($n=1$) between For light incident at approximately $\theta=0$, we can write the condition for a bright fringe (taking into account phase shifts on reflection as before) as

$$\frac{2\pi}{\lambda}\left[2h\right]=\left(m+1/2\right)2\pi, \tag{2.104}$$

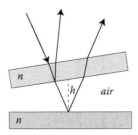

FIGURE 2.18 Interference in a wedge-shaped air gap between two flat surfaces.

where $2h$ is the round-trip path length inside the air film of local height h. Thus, bright fringes represent half-wave differences in film air-gap thickness, and are formed whenever

$$h = \left(m + 1/2\right)\frac{\lambda}{2}. \tag{2.105}$$

Referring to Figure 2.19, if we look down from the top, we will see straight fringes, assuming that both surfaces are flat. From the relationship

$$\tan\alpha \approx \alpha = \frac{\lambda/2}{\Delta x}, \tag{2.106}$$

we find that the fringe separation is

$$\Delta x = \lambda/2\alpha. \tag{2.107}$$

These straight-line fringes do not indicate which way the wedge is oriented. However, by physically pushing down on the low end of the wedge, more fringes are added to the interferogram because the tilt angle is increased. Pushing on the high end of the wedge decreases the tilt angle, and hence fewer fringes are seen.

Fringes that are not straight indicate a departure from flatness in the surface being tested. Referring to Figure 2.20, for a given peak departure from flatness Δ, the peak height error in the piece is

$$\text{height error} = \frac{\Delta}{\Delta x}\frac{\lambda}{2}. \tag{2.108}$$

The interferogram of Figure 2.20 does not show whether the peak error is a high spot or a low spot. Remembering that a fringe is a locus of points of constant gap dimension, a knowledge of which end of the air wedge is open will allow determination of the sign of the height error. Compared with a situation where equally spaced fringes exist between two tilted flat surfaces, a high spot will displace the fringes toward the

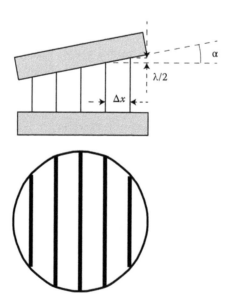

FIGURE 2.19 Relationship of wedge-shaped air gap to interferogram.

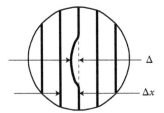

FIGURE 2.20   Interferogram of a surface with a departure from flatness.

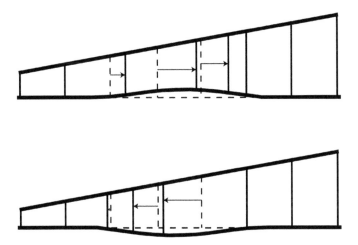

FIGURE 2.21   Fringe displacements for high and low areas.

open end of the wedge, and a low spot will displace the fringes toward the closed end of the wedge. As seen in Figure 2.21, this fringe displacement will keep the local gap height constant along a fringe.

Newton's rings are fringes of equal thickness seen between a flat and a curved surface. If the sphere and the flat are in contact at the center of the curved surface, there will be a dark spot at the center when viewed in reflected light, consistent with the π phase difference between dense-to-rare and rare-to-dense reflections. Referring to Figure 2.22, we can develop the expression for the fringe radii. By the Pythagorean theorem, we can write a relationship between the air gap height h, the fringe radius r, and the radius of curvature of the surface R:

$$(R-h)^2 + r^2 = R^2. \tag{2.109}$$

Under the approximation $h^2 \approx r^2$, we find that

$$h \approx \frac{r^2}{2R}. \tag{2.110}$$

We can thus write the condition for a bright fringe (in reflection) as

$$2h = (m+1/2)\lambda. \tag{2.111}$$

Substituting, we find the fringe radii are proportional to the square root of integers:

$$r_{\text{bright fringe}} = \sqrt{(m+1/2)R\lambda}. \tag{2.112}$$

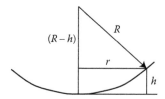

FIGURE 2.22   Construction for Newton's rings.

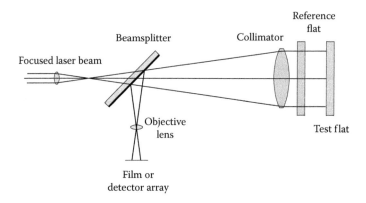

FIGURE 2.23   Laser-based Fizeau interferometer.

For surfaces that are not spherical, the shape of the resulting Fizeau fringes can provide a topographic characterization of the surface profile, because the fringes are loci of points of constant thickness of the air gap. Addition of a small tilt between the reference surface and the surface under test can allow quantitative measurement of departure from flatness on the order of $\lambda/20$ or less, which would otherwise be impossible to measure.

Using filtered arc sources, the Fizeau fringes must be observed in an air gap that is within the coherence length of the source, a few micrometers at most. This requires contact between the reference surface and the test surface. For situations where this is undesirable, a laser-based Fizeau interferometer allows substantial separation between the two surfaces, as seen in Figure 2.23.

A wide variety of configurations for the laser-based Fizeau are possible, allowing characterization of surface figure of both concave and convex surfaces, as well as homogeneity of an optical part in transmission. Schematic diagrams of this instrumentation can be found in Malacara [2]. A Twyman–Green Interferometer, seen in Figure 2.24, allows for a distinct separation of the optical paths of the reference arm and the test arm, and typically uses a collimated point source such as a focused laser beam for illumination.

The interpretation of Twyman–Green interferograms is essentially the same as for Fizeau instruments, because in both cases light travels through the test piece twice. This double-pass configuration yields a fringe-to-fringe spacing that represents $\lambda/2$ of OPD. In a Mach–Zehnder interferometer, seen in Figure 2.25, the test piece is only traversed once, so that fringes represent one wavelength of OPD.

2.2.3 Multiple-Beam Interference

In a typical fringes-of-equal-thickness configuration, where the refractive index difference is between air ($n = 1.0$) and uncoated glass ($n = 1.5$), only the first two reflected rays are significant, having powers referenced to the incoming beam of 4% and 3.69%. The next reflected beam has a power of 6×10^{-3}%. Only the two primary beams interfere, producing cosine-squared fringes. The multiple reflected beams

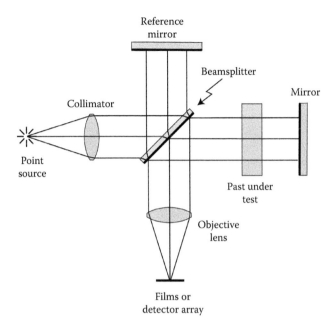

FIGURE 2.24 Twyman-Green interferometer.

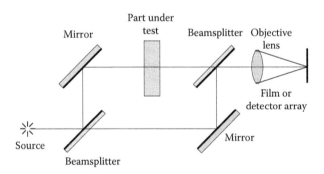

FIGURE 2.25 Mach-Zehnder interferometer.

do not affect the interference in any appreciable way. The situation changes when the reflectivity of the interfaces is high. The example shown in Figure 2.26 is for surfaces having a power reflectivity of 90%. The falloff of power in the multiple reflected beams is slower, both in reflection and in transmission. More rays will add to produce the interference fringes. The fringe shape will be seen to be sharper than sinusoidal, with narrower maxima in reflection and narrower minima in transmission. In Figure 2.27, we show the multiple-beam case for a general surface reflectivity.

By our previous analysis of interference in a plane-parallel plate, the difference in OPD between adjacent rays is

$$\text{OPD} = 2nd\cos\theta_t, \tag{2.113}$$

and the resulting round-trip phase difference between adjacent rays is

$$\delta = \frac{2\pi}{\lambda}2nd\cos\theta_t. \tag{2.114}$$

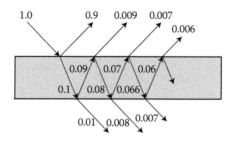

FIGURE 2.26 Multiple reflected beams for a surface power reflectivity of 90%.

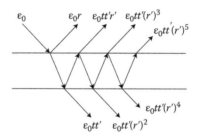

FIGURE 2.27 Multiple reflected beams for a general surface reflectivity.

These rays are parallel if the plate is parallel and can be brought to interfere at infinity, or at the focus of a lens. The symbols t and r in Figure 2.27 are the transmittance and reflectance for the case of incidence from a low-index medium. The symbols r' and t' apply for the case of incidence from a high-index medium. We can write expressions for the complex amplitude of the total reflected ε_r and transmitted ε_t beams, in terms of the amplitude of the incident beam ε_0 and the incremental phase shift δ:

$$\varepsilon_r = \varepsilon_0\left(r + tt'r'e^{i\delta} + tt'\left(r'\right)^3 e^{i2\delta} + tt'\left(r'\right)^5 e^{i3\delta} + \ \dots \ + tt'\left(r'\right)^{2N-3} e^{i(N-1)\delta}\right), \tag{2.115}$$

$$\varepsilon_r = \varepsilon_0\left(r + tt'r'e^{i\delta}\left(1 + \left(r'\right)^2 e^{i\delta} + \left(\left(r'\right)^2 e^{i\delta}\right)^2 + \dots + \left(\left(r'\right)^2 e^{i\delta}\right)^{N-2}\right)\right), \tag{2.116}$$

$$\%_t = \%_0 tt'\left(1 + \left(r'\right)^2 e^{i\delta} + \left(r'\right)^4 e^{i2\delta} + \dots + \left(r'\right)^{2(N-1)} e^{i(N-1)\delta}\right). \tag{2.117}$$

Letting $N \to \infty$, we can sum Equation 2.117 for the reflected amplitude using a geometric series

$$\lim_{n \to \infty}\left(1 + a + a^2 + \dots + a^n\right) \to \frac{1}{1-a} \tag{2.118}$$

to yield

$$\varepsilon_r = \varepsilon_0\left(r + tt'r'e^{i\delta}\left[\frac{1}{1 - \left(r'\right)^2 e^{i\delta}}\right]\right). \tag{2.119}$$

Using the Stokes' relations from Figure 2.14, and Equations 2.98 and 2.99

$$r = -r'$$

(2.120)

and

$$tt' = 1 - r^2$$

(2.121)

we find, for the reflected amplitude

$$\varepsilon_r = \varepsilon_0 \left(\frac{r(1 - e^{i\delta})}{1 - r^2 e^{i\delta}} \right).$$

(2.122)

We now express the ratio of reflected irradiance to incident irradiance:

$$\frac{E_r}{E_0} = \frac{\varepsilon_r \varepsilon_r^*}{\varepsilon_0 \varepsilon_0^*} = \frac{r(1 - e^{i\delta})}{1 - r^2 e^{i\delta}} \frac{r(1 - e^{-i\delta})}{1 - r^2 e^{-i\delta}} = \frac{2r^2(1 - \cos\delta)}{(1 + r^4) - 2r^2 \cos\delta}$$

$$= \frac{\left(\frac{2r}{1 - r^2}\right)^2 \sin^2(\delta/2)}{1 + \left(\frac{2r}{1 - r^2}\right) \sin^2(\delta/2)}$$

(2.123)

Using the power reflectance $R = r^2$, and with identification of the coefficient of finesse F,

$$F = \left(\frac{2r}{1 - r^2}\right)^2 = \frac{4R}{(1 - R)^2},$$

(2.124)

we can write

$$\frac{E_r}{E_0} = \frac{F \sin^2(\delta/2)}{1 + F \sin^2(\delta/2)}.$$

(2.125)

If there is no absorption in the plate, we can write an expression for the transmittance

$$\frac{E_t}{E_0} = 1 - \frac{E_r}{E_0}$$

(2.126)

$$\frac{E_t}{E_0} = \frac{1}{1 + F \sin^2(\delta/2)}.$$

(2.127)

Equation 2.127 for transmittance is called the Airy function. For small surface reflectivity, the reflection and transmission functions are nearly sinusoidal in nature. For higher values of reflectivity (leading to larger values of F) the functions become sharply peaked, as shown in the Figure 2.28. It can be shown that the full width of the Airy function at the 50% level of transmittance is equal to $4/\sqrt{F}$.

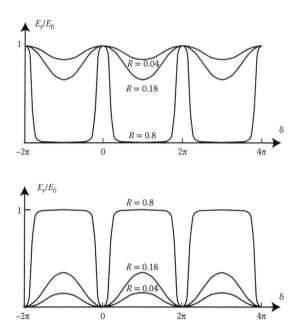

FIGURE 2.28 Multiple-beam transmission and reflection curves as a function of the power reflectance *R*.

2.3 Coherence

Issues of coherence are important in practical optical systems, because ideal sources of light are never actually obtained. No source is truly monochromatic; there is always a finite bandwidth, so we consider temporal coherence effects. Also, any source of interest has a finite spatial extent; there are no true point sources. Thus, we need to consider spatial coherence.

2.3.1 Temporal Coherence

The coherence time t_c is the average time interval over which we can predict the phase of a light wave. If the light wave were truly monochromatic (an infinite sinusoid), one could predict the phase at all times, given the phase at one time. Over time, intervals that are short compared with t_c, the light waves behave as a coherent wavetrain, and interference fringes of good visibility can be formed. For time intervals on the same order as t_c, the fringe visibility begins to decrease. For longer time intervals, interference is not possible because the wavetrain does not have a stable phase relationship.

Beginning with the basic relationship between wavelength and frequency and speed of light c

$$\lambda v = c, \tag{2.128}$$

we take the derivative with respect to λ and find the following relationship between the bandwidths in terms of frequency and wavelength:

$$\Delta v = \Delta \lambda c / \lambda^2. \tag{2.129}$$

The coherence time can be expressed as the inverse of the frequency bandwidth:

$$t_c = 1/\Delta v. \tag{2.130}$$

Some typical magnitudes of these quantities are given in the following examples. An He–Ne laser with mean wavelength $\lambda = 0.63$ μm and spectral bandwidth $\Delta\lambda = 5 \times 10^{-15}$ m will have a frequency bandwidth $\Delta\nu \approx 4$ MHZ and a corresponding coherence time $t_c \approx 0.25$ μs. For a filtered Hg-arc lamp with $\lambda = 0.546$ μm and $\Delta\lambda = 10^{-8} m$, the frequency bandwidth will be — 10^{13}Hz, corresponding to a coherence time of $\approx 10^{-13}$ s.

Coherence length l_c is the distance that light travels during the coherence time:

$$l_c = ct_c = \lambda^2 / \Delta\lambda. \tag{2.131}$$

The coherence length is the maximum OPD allowed in an interference configuration if fringes of high visibility are to be obtained. For the examples above, the He–Ne laser had a coherence length of 75 m, while the filtered Hg-arc lamp had a coherence length of 30 μm. Another interesting example is visible light, such as sunlight, with a mean wavelength of 0.55 μm and a wavelength range from 0.4 to 0.7 μm. We find a coherence length for visible sunlight on the order of 1 mm, which is just sufficient for the operation of thin-film interference filters, for which the round-trip path in the film is on the order of a micrometer.

2.3.2 Spatial Coherence

Spatial coherence is defined by the ability of two points on a source to form interference fringes. Using a Young's interference configuration, seen in Figure 2.27, we ask what is the visibility of the fringes that are formed in the observation plane, as a function of the size of the source, the separation of the pinholes, and the distance from the source to the pinhole plane.

Even for an extended incoherent source, where all points have an independent random-phase relationship, we find interference fringes of good visibility are formed in the observation plane for certain geometries. The key to understanding this phenomenon is that, while independent point sources are incoherent and are incapable of exhibiting mutual interference, the position of the fringes formed in a Young's interference setup will depend on the location of the point source that illuminates the pinholes. For an extended source, a number of separate source locations correspond to various point sources that comprise the extended source. If the radiation from each of two such points gives rise to interference fringes that have maxima and minima that are registered in the same location, the overall fringe pattern, while just an incoherent superposition of the two individual fringe patterns, will still have good fringe visibility. Because the mutual coherence of these two sources is measured in terms of the visibility of the fringes that are formed, these two independent sources are said to exhibit correlation. It is not that the sources themselves are made more correlated by the process of propagation but that the corresponding fringe patterns become co-registered. If the fringes formed by one source had maxima that were located at the position of the minima of the other source, the resulting fringe pattern would be of low visibility and the sources would be said to have small mutual coherence.

We can put these ideas on an analytical footing using the VanCittert–Zernike theorem, which states that the complex degree of coherence γ is just the Fourier transform of the spatial irradiance distribution of the source. We begin the development by considering a particular off-axis source point at location y_s in the source plane, as shown in Figure 2.29. The distances from the source to each of the two pinholes are r_1 and r_2, respectively. The fact that these two distances are different gives rise to a phase shift in the fringes, so that the central maximum will not be located on axis. We can write an expression for the irradiance as a function of y_o in the observation plane:

$$E(y_o) = E_o + Re\left\{ E_o e^{j\Delta\phi} e^{j2\pi l y_o / \lambda x_o} \right\}, \tag{2.132}$$

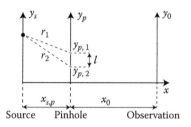

FIGURE 2.29 Configuration for definition of complex degree of coherence.

where the phase shift is

$$\Delta\phi = \frac{2\pi}{\lambda}(r_1 - r_2).$$ (2.133)

Using the Pythagorean theorem, we have expressions for r_1 and r_2 in terms of the axial distance x_{sp} from the source plane to the pinhole plane, and the y position of the two pinholes y_{p1} and y_{p2}:

$$r_1 = \sqrt{x_{sp}^2 + (y_s - y_{p1})^2}$$ (2.134)

and

$$r_2 = \sqrt{x_{sp}^2 + (y_s - y_{p2})^2}.$$ (2.135)

We form an expression for $r_1 - r_2$:

$$r_1 - r_2 = x_{sp}^2 \sqrt{1 + \frac{(y_s - y_{p1})^2}{x_{sp}^2}} - x_{sp}^2 \sqrt{1 + \frac{(y_s - y_{p2})^2}{x_{sp}^2}}.$$ (2.136)

Using the binomial expansion of Equation 2.6, we find

$$r_1 - r_2 \approx x_{sp}\left(1 + \frac{(y_s - y_{p1})^2}{2x_{sp}^2}\right) - x_{sp}\left(1 + \frac{(y_s - y_{p2})^2}{2x_{sp}^2}\right)$$ (2.137)

$$r_1 - r_2 \approx \frac{(y_s - y_{p1})^2 - (y_s - y_{p2})^2}{2x_{sp}} \approx \frac{y_{p1}^2 - y_{p2}^2 - 2y_s(y_{p1} - y_{p2})}{2x_{sp}}.$$ (2.138)

So for the phase shift, Equation 2.133, we find

$$\Delta\phi = \frac{2\pi}{\lambda x_{xp}}(r_1 - r_2) \approx \frac{2\pi}{\lambda x_{xp}}\left[\frac{1}{2}(y_{p1}^2 - y_{p2}^2) - ys(y_{p1} - y_{p2})\right].$$ (2.139)

If we now allow for an extended source, many such point sources will exist in the source plane. Each source will independently form a Young's fringe pattern in the observation plane. Because these fringe patterns are each derived from a point source that is incoherent with all the others, the resulting irradiance distribution in the observation plane is simply the incoherent addition of the fringe patterns, each with some value of spatial shift corresponding to the phase shift $\Delta\phi$:

$$E(y_o) = \int_{\text{source}} E_o(y_s)dy_s + Re\left\{ \int_{\text{source}} E_o(y_s)e^{j\Delta\phi(y_s)}e^{j2\pi ly_o/\lambda x_o}dy_s \right\}. \tag{2.140}$$

Because the periodic phase term is independent of y_s,

$$E(y_o) = \int_{\text{source}} E_o(y_s)dy_s + Re\left\{ e^{j2\pi ly_o/\lambda x_o} \int_{\text{source}} E_o(y_s)e^{j\Delta\phi(y_s)}dy_s \right\}. \tag{2.141}$$

Let us introduce two new parameters, the total source exitance

$$E_{\text{total}} \equiv \int_{\text{source}} E_o(y_s)dy_s \tag{2.142}$$

and the normalized source irradiance distribution

$$\hat{E}_o(y_s) \equiv \frac{E_o(y_s)}{\displaystyle\int_{\text{source}} E_o(y_s)dy_s}. \tag{2.143}$$

Then, noting that $\Delta\phi$ is a function of the pinhole coordinates, the complex degree of coherence is defined as

$$\gamma(y_{p1}, y_{p2}) \equiv \int_{\text{source}} \check{E}_o(y_s)e^{j\Delta\phi(y_s)}dy_s. \tag{2.144}$$

Substituting in the expression for $\Delta\phi$ from Equation 2.139,

$$\gamma(y_{p1}, y_{p2}) \equiv \int_{\text{source}} \check{E}_o(y_s)\exp\left[j\frac{2\pi}{\lambda x_{xp}}\left[\frac{1}{2}(y_{p1}^2 - y_{p2}^2) - y_s(y_{p1} - y_{p2}) \right] \right]dy_s \tag{2.145}$$

and

$$\gamma(y_{p1}, y_{p2}) \equiv \exp\left[\frac{j\pi}{\lambda x_{xp}}(y_{p1}^2 - y_{p2}^2) \right]\int_{\text{source}} \hat{E}_o(y_s)\exp\left[\frac{-j2\pi}{\lambda x_{xp}}y_s(y_{p1} - y_{p2}) \right]dy_s. \tag{2.146}$$

The first term in Equation 2.146 is a purely imaginary phase factor; the integral following is the Fourier transform of the normalized source brightness distribution function. The change of variables

$$\xi = \frac{y_{p1} - y_{p2}}{\lambda x_{sp}} = \frac{l/x_{sp}}{\lambda} = \frac{\theta_{p,s}}{\lambda} \tag{2.147}$$

is used, where $\theta_{p,s}$ is the angular separation of the pinholes (with an in-plane separation l), as viewed from the source plane. With these substitutions, Equation 2.141 becomes

$$E(y_o) = E_{total}\left[1 + \mathrm{Re}\left\{\gamma(y_{p1}, y_{p2})e^{j2\pi l y_o/\lambda x_o}\right\}\right]. \tag{2.148}$$

Comparing this to the expression for double-slit interference, from Equations 2.76 and 2.78

$$E(y_o) = 2E_o\left[1 + \mathrm{Re}\left\{e^{j2\pi l y_o/\lambda x_o}\right\}\right], \tag{2.149}$$

we find the interpretation of the complex degree of coherence, $y(y_{p1}, y_{p2})$. The magnitude of γ determines the contrast of the fringes produced in the observation plane, and the phase portion of γ determines the amount of phase shift relative to an on-axis point source. Given the source size in the y_s plane, γ will be in pinhole (y_p) coordinates, and will depend on the specific values for seven and x_{sp}. However, γ will be independent of x_o, because that distance scales the sizes of the fringe pattern and changing that distance does not affect the fringe visibility.

As the first example, we consider a one-dimensional slit source of total length L, which can be expressed in source-plane coordinates as

$$\hat{E}_o(y_s) = \mathrm{rect}(y_s/L). \tag{2.150}$$

Taking the Fourier transform and making the required change of variables from Equation 2.147, we find that

$$\gamma = \frac{\sin\left(\pi L \dfrac{1}{\lambda x_{sp}}\right)}{\left(\pi L \dfrac{1}{\lambda x_{sp}}\right)}, \tag{2.151}$$

which exhibits a first zero of visibility at a pinhole separation $l = \lambda x_{sp}/L$.

The normalized source function for a two-dimensional uniform source can be written as

$$\hat{E}_o(r_s) = \frac{4}{\pi D^2}\mathrm{cyl}(r_s/D), \tag{2.152}$$

which is Fourier transformed to yield

$$\gamma = \frac{2J_1\left(\dfrac{\pi D l}{\lambda x_{sp}}\right)}{\left(\dfrac{\pi D l}{\lambda x_{sp}}\right)}, \tag{2.153}$$

which has its first zero of visibility at

$$\frac{\pi D l}{\lambda x_{sp}} = \pi \times 1.22, \tag{2.154}$$

corresponding to a pinhole spacing $l = 1.22 x_{sp} \lambda / D$.

Two point sources separated by a distance b can be described in normalized form as

$$\hat{E}_0(y_s) = \frac{1}{2}\left[\delta\left(y_s + b/2\right) + \delta\left(y_s - b/2\right)\right], \tag{2.155}$$

which, upon performing the transformation and the change of variables from Equation 2.147, can then be written as

$$\gamma = \cos\left(2\pi \frac{b/2}{\lambda x_{sp}} l\right). \tag{2.156}$$

In general, for a given size of an extended source, the farther away the source is, we find that γ will approach one, for a given pinhole separation. This means that the light from the extended source, if allowed to interfere after passing through pinholes with that separation, would tend to form fringes with good visibility. This does not mean that the extended source itself acquires a coherence by propagation, but that the fringe patterns formed from each portion of the source would tend to become more nearly registered as the source-to-pinhole distance x_{sp} increases. Because the fringe patterns add on an irradiance basis, an increasing registration of the fringe maxima and minima will produce an overall distribution with higher visibility fringes. Similar arguments apply when the pinhole separation decreases, for a given source size and distance x_{sp}.

2.4 Polarization

Polarization effects on optical systems arise from the vectorial nature of the electric field ε, which is orthogonal to the direction of propagation of the light. Because of the polarization dependence of the Fresnel equations, differences in the power distribution between reflected and transmitted rays occur when the electric field makes a nonzero angle with the boundary between two media. Other effects arise in anisotropic crystals from the fact that the refractive index of a material may depend upon the polarization of the incident light. In this case, the trajectory of rays may depend upon the orientation of the electric field to one particular preferred direction.

2.4.1 Polarizing Elements

Linear polarizers produce light polarized in a single direction from natural light, which has light polarized in all directions. The amount of power transmitted through a linear polarizer is at most 50% when the input light is composed of all polarizations. Light that is already linearly polarized will be transmitted through a linear polarizer with a power transmittance that depends on the angle 6 between the electric field of the input light, and the pass direction of the polarizer, according to Malus' law:

$$\phi_t = \phi_i \cos^2\theta. \tag{2.157}$$

The functions of a linear polarizer may be accomplished by long molecular chains or by metallic wires with spacing less than the wavelength. These act by allowing electric currents to flow in the long direction of the structure. Light with an incident electric field oriented in this direction will be preferentially absorbed because of joule heating arising in the currents, or may be reflected if the absorption losses are small. Light having a polarization orthogonal to the long direction will pass through the structure relatively unattenuated.

Another common means of producing linearly polarized light uses a stack of dielectric plates, all oriented at Brewster's angle to the original propagation direction. Each plate will transmit the entire *p* polarization component and some of the *s* component. By arranging to have the incident light traverse many such plates of the same orientation, the transmitted beam can be made to be strictly *p* polarized.

2.4.2 Anisotropic Media

Usually, optical materials are isotropic in that they have the same properties for any orientation of electric field. However, anisotropic materials have a different refractive index for light polarized in different directions. The most useful case is a so-called birefringent material that has two different refractive indices for light polarized in two orthogonal directions. These refractive indices are denoted n_o, the ordinary index, and n_e, the extraordinary index. By convention, the ordinary index is less than the extraordinary index. By properly orienting the polished faces of a slab of birefringent material, it is possible to produce a retarder, which introduces a phase delay φ between the two orthogonal polarizations of a beam

$$\varphi = \frac{2\pi}{\lambda_0} d\left(n_e - n_o\right).$$

(2.158)

By proper choice of *d*, a phase delay of 90° can be produced. If linearly polarized light is incident with its electric field at angle of 45° to the fast axis of such a slab, a phase delay is introduced between the component of light polarized along the fast axis and that polarized along the slow axis, producing circularly polarized light.

2.4.3 Optical Activity

Certain materials produce a continuous rotation of the electric field of polarized light as the light traverses the bulk of the material. The most common substances having this property are quartz and sugar dissolved in water, but the most important technologically are liquid crystals, which have the advantage that their optical activity can be controlled by an external voltage. By placing such a material between orthogonal linear polarizers, a variable amount of light can be made to pass through the structure, depending upon the amount of rotation of the polarization.

References

1. Gaskill J. *Linear Systems, Fourier Transforms, and Optics*, New York, NY: Wiley, 1978.
2. Malacara D. *Optical Shop Testing*, New York, NY: Wiley, 1978.

Bibliography

1. Born M, Wolf E. *Principles of Optics*, New York, NY: Pergamon, 1975.
2. Stone J. *Radiation and Optics*, New York, NY: McGraw-Hill, 1963.
3. Reynolds G, DeVelis J, Parrent G, Thompson B. *Physical Optics Notebook*, Bellingham, WA: SPIE Press, 1989.

3

Basic Photon and Quantum Optics

Sofia E.
Acosta-Ortiz

3.1 Nature of Quantum Optics

The first indication of the quantum nature of light came in 1900 when M. Planck discovered he could account for the spectral distribution of thermal light by postulating that the energy of a harmonic oscillator is quantized. In 1905, Albert Einstein showed that the photoelectric effect could be explained by the hypothesis that the energy of a light beam was distributed in discrete bundles, later known as *photons*.

With the development of his phenomenological theory in 1917, Einstein also contributed to the understanding of the absorption and emission of light from atoms. Later it was shown that this theory is a natural consequence of the quantum theory of electromagnetic radiation.

A remarkable feature of the theory of light is the large measure of agreement between the predictions of classical and quantum theory, despite the fundamental differences between the two approaches. For example, classical and quantum theories predict identical interference effects and associated degrees of coherence for experiments that use coherent or chaotic light, or light with coherence properties intermediate between the two.

The vast majority of physical-optics experiments can be explained adequately using the classical theory of electromagnetic radiation based on Maxwell's equations. Interference experiments of Young's type do not distinguish between the predictions of classical theory and quantum theory. It is only in higher-order interference experiments involving the interference of intensities that differences between the predictions of both theories appear. Whereas classical theory treats the interference of intensities, in quantum theory the interference is still at the level of probability amplitudes. This is one of the most important differences between quantum theory and classical theory.

The main tool in the quantum description of a beam of light is its photon probability distribution, or more generally, its density operator. Photon-counting experiments provide a fairly direct measurement of the photon probability distribution for all kinds of light embraced by the quantum theory. Such experiments form the observational basis of quantum optics and play a leading role in the study of quantum phenomena in light beams.

The field of quantum optics occupies a central position involving the interaction of atoms with the electromagnetic field. It covers a wide range of topics, ranging from fundamental tests of quantum theory to the development of new laser light sources.

3.1.1 Young's Experiment

Young's experiment is one example of a phenomenon that can be explained by both classical and quantum theory. In this experiment (Figure 3.1), monochromatic light is passed through a pinhole S so as to illuminate a screen containing two further identical pinholes or narrow slits placed close together. The presence of the single pinhole S provides the necessary mutual coherence between the light beams emerging from the slits S_1 and S_2. The wavefronts from S intersect S_1 and S_2 simultaneously in such a way that the light contributions emerging from S_1 and S_2 are derived from the same original wavefront and are therefore coherent. These contributions spread out from S_1 and S_2 as "cylindrical" wavefronts and interfere in the region beyond the screen. If a second screen is placed as shown in the figure, then an interference pattern consisting of straight-line fringes parallel to the slits is observed on it.

The phase difference between the two sets of waves arriving at P from S_1 and S_2 depends on the path difference $(D_2 - D_1)$ as, in general, phase difference = $(2\pi/\lambda)$ (optical phase difference); then

$$\phi = \phi_2 - \phi_1 = (2\pi/\lambda)(D_2 - D_1), \tag{3.1}$$

where D_1 and D_2 are the distances from S_1 and S_2 to P, respectively.

Bright fringes occur when the phase difference is zero or $\pm 2n\pi$, where n is an integer; that is, when

$$(2\pi/\lambda)(D_2 - D_1) = \pm 2n\pi,$$

which is equivalent to $D_2 - D_1 = \pi\lambda$. Therefore, bright fringes will occur whenever the path difference is an integral number of wavelengths. Similarly, dark fringes will occur when $\phi = \pm(2n + 1)\pi$; that is, when the path difference is an odd integral number of half-wavelengths.

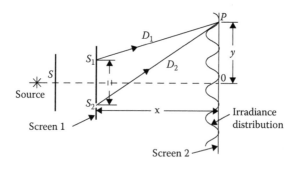

FIGURE 3.1 Schematic diagram showing Young's double-slit interference experiment.

Bright fringes occur at point P, a distance y from O such that

$$y = \pm(n\lambda x)/H, \tag{3.2}$$

provided that both y and H are small compared with x, where H is the slit separation and x is the distance from the screen containing the slits to the observing screen [1].

The quantum theory of the experiment parallels the classical theory. The Heisenberg electric-field operator at position $\mathbf{r}$ on the screen is given by

$$\hat{E}^+(\mathbf{r}t) = u_1\hat{E}^+(\mathbf{r}_1t_1) + u_2\hat{E}^+(\mathbf{r}_2t_2), \tag{3.3}$$

where we have assumed for simplicity a single polarization direction for the light.

The superposition theory proceeds in a similar way to the classical treatment, except that the classical intensity, proportional to an ensemble average of E^*E, must be replaced by the quantum-mechanical detector response, proportional to an average of the intensity operator $\hat{E}^-\hat{E}^+$.

The intensity distribution is given by Loudon [2]:

$$\left\langle \hat{I}(\mathbf{r}t) \right\rangle = 2\varepsilon_0 c \left\{ \begin{array}{l} u_1^2 \left\langle \hat{E}^-(\mathbf{r}_1t_1)\hat{E}^+(\mathbf{r}_1t_1) \right\rangle + u_2^2 \left\langle \hat{E}^-(\mathbf{r}_2t_2)\hat{E}^+(\mathbf{r}_2t_2) \right\rangle \\ + u_1^*u_2 \left\langle \hat{E}^-(\mathbf{r}_1t_1)\hat{E}^+(\mathbf{r}_2t_2) \right\rangle + u_1u_2^* \left\langle \hat{E}^-(\mathbf{r}_2t_2)\hat{E}^+(\mathbf{r}_1t_1) \right\rangle \end{array} \right\}, \tag{3.4}$$

where the angle-bracket expressions are evaluated as

$$\left\langle \hat{E}^-(\mathbf{r}_1t_1)\hat{E}^+(\mathbf{r}_2t_2) \right\rangle = \mathrm{Tr}\left(\rho\hat{E}^-(\mathbf{r}_1t_1)\hat{E}^+(\mathbf{r}_2t_2) \right). \tag{3.5}$$

The two final expectation values in Equation 3.4 are complex conjugates, and bearing in mind the purely imaginary character of u_1 and u_2, the intensity reduces to

$$\left\langle \hat{I}(\mathbf{r}t) \right\rangle = 2\varepsilon_0 c \left\{ u_1^2 \left\langle \hat{E}^-(\mathbf{r}_1t_1)\hat{E}^+(\mathbf{r}_1t_1) \right\rangle + u_2^2 \left\langle \hat{E}^-(\mathbf{r}_2t_2)\hat{E}^+(\mathbf{r}_2t_2) \right\rangle + 2u_1^*u_2\mathrm{Re}\left\langle \hat{E}^-(\mathbf{r}_1t_1)\hat{E}^+(\mathbf{r}_2t_2) \right\rangle \right\}. \tag{3.6}$$

The electric-field operators in Equation 3.6 are related to photon creation and destruction operators by

$$\hat{E}^+(\mathbf{R}t) = -i\sum_k (\hbar\omega_k/2\varepsilon_0 V)^{1/2} \varepsilon_k\hat{a}_k \exp(-i\omega_k t + i\mathbf{k}\cdot\mathbf{R}) \tag{3.7}$$

and

$$\hat{E}^-(\mathbf{R}t) = -i\sum_k (\hbar\omega_k/2\varepsilon_0 V)^{1/2} \varepsilon k\hat{a}_5^\dagger \exp(i\omega_k t - i\mathbf{k}\cdot\mathbf{R}), \tag{3.8}$$

where $\mathbf{R}$ is the position of the nucleus.

The first two terms in Equation 3.6 give the intensities on the second screen, which result from each pinhole in the absence of the other. The interference fringes result from the third term, whose normalized form determines the quantum degree of first-order coherence of the light.

Finer details of the quantum picture of the interference experiment are found in Walls [3] and Loudon [2].

In Young's experiment, each photon must be capable of interfering with itself in such a way that its probability of striking the second screen at a particular point is proportional to the calculated intensity at that point. This is possible only if each photon passes partly through both pinholes, so that it can have a knowledge of the entire pinhole geometry as it strikes the screen. Indeed, there is no way in which a photon can simultaneously be assigned to a particular pinhole and contribute to the interference effects. If a phototube is placed behind one of the pinholes to detect photons passing through, then it is not possible to avoid obscuring that pinhole, with consequent destruction of the interference pattern [2]. These remarks are in agreement with the principles of quantum mechanics.

In the 1960s, improvements in photon counting techniques proceeded along with the development of new laser light sources. Then, light from incoherent (thermal) and coherent (laser) sources could be distinguished by their photon counting properties.

It was not until 1975, when H. J. Carmichael and D. F. Walls predicted that light generated in resonance fluorescence from a two-level atom would exhibit photon antibunching, that a physically accessible system exhibiting non-classical behavior was identified.

3.1.2 Squeezed States

Nine years after the observation of photon antibunching, another prediction of the quantum theory of light was observed: the squeezing of quantum fluctuations. To understand what squeezing of quantum fluctuations means remember that the electric field for a nearly monochromatic plane wave may be decomposed into two quadrature components with a time dependence cos ωt and sin ωt, respectively. In a coherent state, the closest quantum counterpart to a classical field, the fluctuations in the two quadratures are equal and minimize the uncertainty product given by Heisenberg's uncertainty relation. The quantum fluctuations in a coherent state are equal to the zero-point vacuum fluctuation and are randomly distributed in phase. In a squeezed state, the quantum fluctuations are no longer independent of phase. One quadrature phase may have reduced quantum fluctuations at the expense of increased quantum fluctuations in the other quadrature phase such that the product of the fluctuations still obeys Heisenberg's uncertainty relation.

Then, squeezed states have been considered as a general class of minimum-uncertainty states. A squeezed state may, in general, have less noise in one quadrature than a coherent state. To satisfy the requirements of a minimum-uncertainty state, the noise in the other quadrature must be greater than that of a coherent state. Coherent states are a particular member of this more general class of minimum uncertainty states with equal noise in both quadratures.

Some applications of squeezed light are interferometric detection of gravitational radiation and sub-shot-noise phase measurements [4].

3.1.3 The Hanbury-Brown and Twiss Experiment

The experiment carried on by Hanbury-Brown and Twiss in 1957 let us see more clearly the differences between classical and quantum theories. The schematic diagram of the experiment is shown in Figure 3.2. In the experiment, each photon that strikes the semitransparent mirror is either reflected or transmitted, and can only be registered in one of the phototubes. Therefore, the two beams that arrive at the detectors are not identical. The incident photons have equal probabilities of transmission or reflection and their creation and destruction operators can be written as

$$\hat{a}^{\dagger} = \left(\hat{a}_1^{\dagger} + \hat{a}_2^{\dagger} \right) / 2^{1/2}$$

$$\hat{a} = \left(\hat{a}_1 + \hat{a}_2 \right) / 2^{1/2},$$

(3.9)

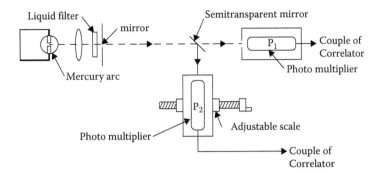

FIGURE 3.2 Experimental arrangement for an intensity interference experiment. (Reprinted by permission from Macmillan Publishers Ltd., *Nature*, Hanbury-Brown and Twiss, 1957, copyright 2017.)

where the $2^{1/2}$ factors ensure that all like pairs of creation and destruction operators have the same commutation rule:

$$\left[\hat{a},\hat{a}^{\dagger}\right]=\hat{a}\hat{a}^{\dagger}-\hat{a}^{\dagger}\hat{a}=1.$$

A state with n incident photons has the form

$$n>=\left(n!\right)^{-1/2}\left(\hat{a}^{\dagger}\right)^{n}0>,$$

and the probability $P_{n1,n2}$ that n_1 photons are transmitted through the mirror to detector 1 while n_2 are reflected to detector 2. $P_{n1,n2}$ is given by

$$P_{n1,n2}=\left\langle n1,n2,n^{2}\right\rangle=n!/\,n_1!n_2!2^{n}.$$

We can use the probability distribution to obtain average properties of the photon counts in the two arms of the experimental system. The mean numbers of counts n_1 and n_2 are both equal to (1/2) [2].

$$\bar{n}_1=\left(1/2\right)n \tag{3.10}$$

$$\bar{n}_2=\left\langle n\hat{a}_2^{\dagger}\,\hat{a}_2\,n\right\rangle=\left(1/2\right)n, \tag{3.11}$$

and the correlation is

$$\left\langle \hat{a}_1^{\dagger}\hat{a}_2^{\dagger}\hat{a}_2\hat{a}_1\right\rangle\equiv\left\langle n_1 n_2\right\rangle=\left(1/4\right)n(n-1); \tag{3.12}$$

$\langle n_1 n_2\rangle$ indicates the average of the product of the photon counts at the two photon photomultipliers.

Table 3.1 shows the photon-count distributions and the degrees of second-order coherence $\left(g^{(2)}(\tau)\right)$, for small numbers n of incident photons.

The Hanbury-Brown and Twiss experiment was designed to produce the average

$$\frac{\left\langle (n_1-\bar{n}_1)(n_2-\bar{n}_2)\right\rangle}{\bar{n}_1\bar{n}_2}=g^{(2)}(\tau)-1, \tag{3.13}$$

TABLE 3.1   Photon Distribution in a Hanbury-Brown and Twiss Experiment

n	n^1	n^2	$\bar{n}_1 = \bar{n}_2$	$\langle n_1 n_2 \rangle$	$g^{(2)}(\tau)$
1	1	0	1/2	0	0
	0	1			
	2	0	1	1/2	1/2
2	1	1			
	1	1			
	0	2			
3	—	—	3/2	3/2	2/3
4	—	—	2	3	3/4

where $g^{(2)}(\tau)$ is given by

$$g^{(2)}(\tau) = \langle n_1 n_2 \rangle / \bar{n}_1 \bar{n}_2. \tag{3.14}$$

The normalized correlation gives [2]:

$$\frac{\langle (n_1 - \bar{n}_1)(n_2 - \bar{n}_2) \rangle}{\bar{n}_1 \bar{n}_2} = -1/n. \tag{3.15}$$

From the above analysis, it is seen that non-classical light with a degree of second order smaller than unity occurs in the Hanbury-Brown and Twiss experiment because the photons must make a choice between reflection and transmission at the mirror. This is an essential quantum requirement. The quantum analysis of the Hanbury-Brown and Twiss experiment is readily extended to incident beams that have some superposition of statistical mixture of the number states of a single mode.

Later in this chapter, some of the most actual quantum effects and their applications will be discussed.

3.2 Nonlinear Optics

Nonlinear optics is the study of the phenomena that occur as a consequence of the modification of the optical properties of matter due to the presence of light. In principle, only laser light has the intensity sufficient enough to modify the optical properties of matter. In fact, the start of the nonlinear optics field is frequently considered as the start of second-harmonic generation by Franken et al. in 1961, soon after the demonstration of the first laser by Maiman in 1960. Nonlinearity is intended in the sense that nonlinear phenomena occur when the response of the material to an applied electric field depends in a nonlinear manner on the magnitude of the optical field.

As an example, second-harmonic generation occurs as a result of the atomic response that depends quadratically on the magnitude of the applied optical field. As a consequence, the intensity of the generated light at the frequency of the second harmonic tends to increase as the square of the intensity of the applied laser light.

To better understand the significance of nonlinear optics, let us consider the dependence of the dipolar moment by unitary volume: that is, the polarization $P(t)$ on the magnitude $E(t)$ of the applied optical field. In the linear case, the relation is given by

$$P(t) = \chi^{(1)} E(t), \tag{3.16}$$

where χ is the constant of proportionality known as "linear susceptibility." In nonlinear optics, the Equation 3.16 can be generalized by expressing $P(t)$ as a power series on the magnitude of the field $E(t)$:

$$P(t) = \chi^{(1)}E(t) + \chi^{(2)}E^2(t) + \chi^{(3)}E^3(t) + \ldots$$

$$P(t) = P^1(t) + P^2(t) + P^3(t) + \ldots,$$

(3.17)

where $\chi^{(2)}$ and $\chi^{(3)}$ are known as the second- and third-order nonlinear optical susceptibilities, respectively.

In general, $P(t)$ and $E(t)$ are nonscalar quantities but vectorial ones. In such cases, $\chi^{(1)}$ becomes a second-rank tensor, $\chi^{(2)}$ a third-rank tensor, $\chi^{(3)}$ a fourth-rank tensor, and so on.

We refer then to $P^2(t) = \chi^{(2)}E^2(t)$ as second-rank nonlinear polarization and to $P^3(t) = \chi^{(3)}E^3(t)$ as third-rank nonlinear polarization, each of which gives rise to different physical phenomena.

Furthermore, second-order nonlinear optical interactions can only occur in noncentrosymmetric crystals—that is, in crystals that do not possess inversion symmetry—while third-order nonlinear phenomena can occur in any medium regardless of whether it possesses inversion symmetry or not.

Optical nonlinearity manifests by changes in the optical properties of a medium as the intensity of the incident light wave increases or when two or more light waves are introduced into a medium.

Optical nonlinearity can be classified into two general categories: extrinsic and intrinsic [5].

- **Extrinsic Optical Nonlinearity.** Extrinsic optical nonlinearity is the change in the optical properties of the medium that is directly related to a change in the composition of the medium as a result of the absorption or emission of light. This change can be a change in the relative population of the base and excited states or in the number of optically effective electrons. The laser medium itself, certain dyes used for laser Q-switching, and laser mirrors with semiconductor covers have this property.
- **Intrinsic Optical Nonlinearity.** The optical phenomena with intrinsic nonlinearity are violations to the superposition principle that arises from the nonlinear response of the individual molecules or unitary cells to the fields of two or more light waves. This category includes a nonlinear response to a single light beam, as it is possible to consider any light beam as the sum of two or more similar light waves, identical in polarization, frequency, and direction.

In any type of nonlinearity, the optical properties of the medium depend on the intensity of the light, and therefore, it is useful to classify them according to the intensity of the light involved. For example, the intensity of the second-harmonic light at $\lambda = 0.53$ µm generated in lithium niobatium by the laser radiation of neo-dymium is observed to be proportional to the square of the intensity of the fundamental 1.06 µm and therefore is classified as a second-order nonlinear process.

Every advance in the development of optics has involved the development of advanced technology. The quantum theory, for example, rests to a high degree in highly sensitive and precise spectroscopic techniques. The key that opened the doors to nonlinear optics was the development of the *maser*. This device uses stimulated emission to generate radiation of narrow bandwidth in the range of microwaves from an adequate prepared medium. Later, the range was extended to optical frequencies by the development of the *laser*, which allows the generation of highly monochromatic light beams that can be concentrated at very high intensities.

A peculiar property of the laser, essential to nonlinear optics, is its high degree of coherence; that is, the several monochromatic light waves are emitted in a synchronized way, being in phase both in time and space. This property allows us to concentrate the radiation from a laser into a small area, whose size is only limited by diffraction and by the optical quality of the laser and the focusing system. In this way, it is possible to obtain fields of local radiation that are extremely intense but in small volumes.

Coherence also allows us to combine small contributions of nonlinear interactions due to very separated parts of an extended medium to produce an appreciable result. For this reason, it is necessary to use laser light as the light power, to be able to observe optical nonlinear phenomena.

Nonlinear optics is a powerful tool to generate coherent sources. There are many types of lasers commercially available with different wavelengths; however, the wavelengths available for materials processing is limited, and nonlinear optics can help us to improve this. For example, there are very few solid-state laser sources in the visible region: He-Ne laser at 632.8 nm, Ruby laser at 649.3 nm.

As examples of second- and third-order nonlinear processes, we have (the number in parenthesis indicates the order of the optical nonlinearity):

- Second-harmonic generation   (2)
- Rectification   (2)
- Frequency generation of sum and   (2)
  difference
- Third-harmonic generation   (3)
- Parametric oscillation   (2)
- Raman scattering   (2)
- Inverse Raman effect   (2)
- Brillouin scattering   (2)
- Rayleigh scattering   (2)
- Inverse Faraday effect   (3)
- Two photons absorption   (2)
- Parametric amplification   (2)
- Induced reflectivity   (2)
- Intensity-dependent refraction index   (3)
- Induced opacity   (2)
- Optical Kerr effect   (3)
- Four-wave mixing   (3)
- Electro-optic effect   (2)
- Birefringence   (2)

We now describe some of these nonlinear phenomena.

3.2.1 Second-Harmonic Generation

The essential feature of second-harmonic generation (SHG) is the existence of a perceptible dependence of the electronic polarization on the square of the electric field intensity at optical frequencies, in addition to the usual direct dependence.

Practically all SHG materials are birefringent crystals. When its SHG is under consideration, the ith component of the total charge polarization **P** in a birefringent medium contains two contributions:

$$P_i = P_i^{(\omega)} + P_i^{(2\omega)} \quad i = 1, 2, 3, \tag{3.18}$$

where

$$P_i^{(\omega)} = \varepsilon_0 \chi_{ij}^{(\omega)} E_j^{(\omega)} \quad i, j = 1, 2, 3 \tag{3.19}$$

and

$$P_i^{(2\omega)} = \varepsilon_0 \chi_{ij}^{(2\omega)} E_j^{(\omega)} E_k^{(\omega)} \quad i, j, k = 1, 2, 3. \tag{3.20}$$

The first term, $P_i^{(\omega)}$, accounts for the linear part of the response of the medium. The second one, $P_i^{(2\omega)}$, is quadratic in the electric field, and introduces the third-rank tensor $\chi_{ijk}^{(2\omega)}$. The superscripts (ω) and (2ω) are now necessary to distinguish the frequencies at which the respective quantities must be evaluated. Thus, two sinusoidal electric field components at frequency ω acting in combination exert a resultant containing the double frequency 2ω (and a dc term). The susceptibility factor χ_{ijk} must be evaluated at the combination frequency.

The components of the nonlinear optical coefficients for three SHG crystals are given in Table 3.2. The nonlinear optical coefficient is defined as $d_{ijk}^{(2\omega)} = (\varepsilon_0/2)\chi_{ijk}^{(2\omega)}$.

Only noncentrosymmetric crystals can possess a nonvanishing d_{ijk} tensor. This follows from the requirement that in a centrosymmetric crystal, a reversal of the signs of $E_j^{(\omega)}$ and $E_k^{(\omega)}$, must cause a reversal in the sign of $P_1^{(2\omega)}$ and not affect the amplitude [6].

It follows that since no physical significance is attached to the order of the electric field components, all the coefficients that are related by a rearrangement of the order of the subscripts are equal. This statement is known as the Kleinman's conjecture [7]. The Kleinman conjecture applies only to lossless media, but since most nonlinear experiments are carried out in the lossless regime, it is a powerful practical relationship.

Table 3.3 gives a list of the nonzero components for the nonlinear optical coefficients of a number of crystals.

The first experiment carried out to demonstrate the optical second-harmonic generation was performed by Franken et al. in 1961. A sketch of the original experiment is shown in Figure 3.3a. In this experiment, a ruby laser beam at 694.3 nm was focused on the front surface of a crystalline quartz plate. The emergent radiation was examined with a spectrometer and was found to contain radiation at twice the input frequency, that is at $\lambda = 347.15$ nm. The conversion efficiency in this experiment was very poor, only about 10^{-8}. In the last few years, the availability of more efficient materials, higher intensity lasers, and index-matching techniques have resulted in conversion efficiencies approaching unity.

Second-harmonic generation is a special case of frequency mixing with the initial waves having a common frequency, that is, $\omega = \omega_1 = \omega_2$ and $\omega_3 = 2\omega$ in Equation 3.20. The generation of a second harmonic can easily be shown if we describe the applied field by $E = E_0 \sin(\omega t)$ in

$$P = \varepsilon_0 \chi^{(1)}E + \varepsilon_0\chi^{(2)}EE + \varepsilon_0 \chi^{(3)}EEE + \ldots, \qquad (3.21)$$

where ε_0 is the permittivity of free space and $\chi^{(1)}$ is the linear susceptibility representing the linear response of the material [8]. The two lowest-order nonlinear responses are accounted for by the

TABLE 3.2 Components of the Nonlinear Optical Coefficient for Several Crystals

	Components j, k					
Component i	x, x	y, y	z, z	y, z	x, z	x, y
Barium titanate						
X	0	0	0	0	d_{15}	0
Y	0	0	0	d_{15}	0	0
Z	d_{31}	d_{31}	d_{33}	0	0	0
Potassium dihydrogen phosphate (KDP)						
X	0	0	0	d_{14}	0	0
Y	0	0	0	0	d_{14}	0
Z	0	0	0	0	0	d_{36}
Quartz						
X	d_{11}	$-d_{11}$	0	d_{14}	0	0
Y	0	0	0	0	$-d_{14}$	$-2d_{11}$
Z	0	0	0	0	0	0

TABLE 3.3   The Nonlinear Optical Coefficients of a Number of Crystals

Crystal	$d_{ijk}^{(2\omega)}$ in units of $(1/9) \times 10^{-22}$ MKS
$LiIO_3$	$d_{15} = 4.4$
$NH_4H_2PO_4$	$d_{36} = 0.45$
(ADP)	$d_{14} = 0.50 \pm 0.02$
KH_2PO_4	$d_{36} = 0.45 \pm 0.03$
(KDP)	$d_{14} = 0.35$
KD_2PO_4	$d_{36} = 0.42 \pm 0.02$
	$d_{14} = 0.42 \pm 0.02$
KH_2ASO_4	$d_{36} = 0.48 \pm 0.03$
	$d_{14} = 0.51 \pm 0.03$
Quartz	$d_{11} = 0.37 \pm 0.02$
$AlPO_4$	$d_{11} = 0.38 \pm 0.03$
ZnO	$d_{33} = 6.5 \pm 0.2$
	$d_{31} = 1.95 \pm 0.2$
	$d_{15} = 2.1 \pm 0.2$
CdS	$d_{33} = 28.6 \pm 2$
	$d_{31} = 30 \pm 10$
	$d_{36} = 33$
GaP	$d_{14} = 80 \pm 14$
GaAs	$d_{14} = 72$
$BaTiO_4$	$d_{33} = 6.4 \pm 0.5$
	$d_{31} = 18 \pm 2$
	$d_{15} = 17 \pm 2$
$LiNbO_3$	$d_{15} = 4.4$
	$d_{22} = 2.3 \pm 1$
Te	$d_{11} = 517$
Se	$d_{11} = 130 \pm 30$
$Ba_2NaNb_5O_{15}$	$d_{33} = 10.4 \pm 0.7$
	$d_{32} = 7.4 \pm 0.7$
Ag_3AsS_3	$d_{22} = 22.5$
(proustite)	$d_{36} = 13.5$
CdSe	$d_{31} = 22.5 \pm 3$
$CdGeAs_2$	$d_{36} = 363 \pm 70$
$AgGaSe_2$	$d_{36} = 27 \pm 3$
$AgSbS_3$	$d_{36} = 9.5$
ZnS	$d_{36} = 13$

Source: Yariv A: *Quantum Electronics.* 1989. Copyright Wiley-VCH Verlag GmbH & Co. KgaA. With permission of John Wiley & Sons.

second- and third-order nonlinear susceptibilities $\chi^{(2)}$ and $\chi^{(3)}$. The induced polarization, according to Equation 3.21 is given by

$$P = \varepsilon_0 \, \chi^{(1)} \, E \sin(\omega t) + \chi^{(2)} \, E^2 \sin^2(\omega t) + \ldots, \tag{3.22}$$

which can be written as

$$P = \varepsilon_0 \left[\chi^{(1)} E_i \sin(\omega t) + \tfrac{1}{2} \chi^{(2)} E_i^2 (1 - \cos 2\omega t) + \ldots, \right] \tag{3.23}$$

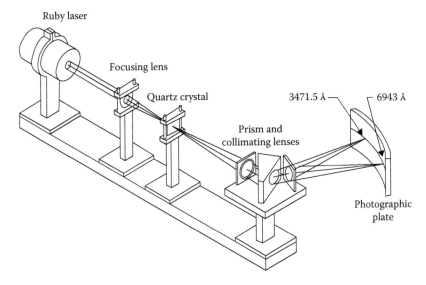

FIGURE 3.3(a)   Arrangement used for the first experimental demonstration of optical second-harmonic generation. (Reprinted with permission from Franken PA, Hill AE, Peters CW, Weinreich G. *Phys. Rev. Lett.,* 7, 118; 1961. Copyright 2017 by the American Physical Society.)

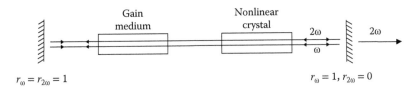

FIGURE 3.3(b)   Intracavity second-harmonic generation. (Reproduced with permission from Milonni P-W, Eberly JH. *Laser Physics,* John Wiley & Sons, 2010. Copyright [2010], **American Institute of Physics.**)

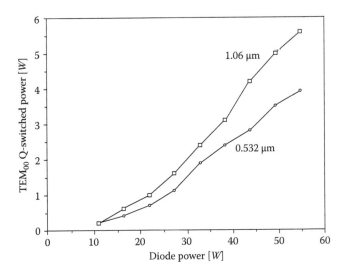

FIGURE 3.3(c)   Fundamental and second harmonic average output power from a repetitively Q-switched cw-pumped Nd:YAG laser. (With kind permission from **Springer Science+Business Media**: *Solid State Laser Engineering,* 2006, p. 630, Koechner W. With permission of Springer.)

In this expression, the presence of the second term indicates that a wave having twice the frequency of the fundamental can be formed in a nonlinear medium.

The power conversion efficiency for second-harmonic generation is proportional to the pump intensity [9]. Therefore, in most cases, a high peak intensity available only from pulsed lasers is required, but for some purposes, it is desirable to generate a cw second-harmonic field. It can be obtained by intracavity second-harmonic generation, that is, by inserting a nonlinear crystal inside the cavity of a cw laser, as shown in Figure 3.3b. The output mirror of the laser is replaced by one that is perfectly reflecting at the laser frequency ω, but perfectly transmitting at the second-harmonic frequency 2ω. Of course, there are losses due to the insertion of the nonlinear crystal because of scattering; however, commercial lasers, such as the Nd-YAG laser with an insertable lithium iodate ($LiIO_3$) crystal uses this technique to provide a second-harmonic output with efficiencies of 10% or more.

Figure 3.3c shows the performance of an acousto-optically Q-switched laser, which exhibits the average power obtained at 1.06 μm and at 0.532 μm when the output coupler is replaced by a total reflector at 1.06 and 0.532 μm.

Mathematical details to obtain the conversion efficiency as well as examples of second-harmonic generation can be found in Yariv [6]; see also Zyss et al. [10].

3.2.2 Raman Scattering

The Raman effect is one of the first discovered and best-known nonlinear optical processes. It is used as a tool in spectroscopic studies, and also in tunable laser development, high-energy pulse compression, and so on. Several review articles exist that summarize the earlier work on the Raman effect [11,12].

Raman scattering involves the inelastic scattering of light from a crystal. The Raman effect belongs to a class of nonlinear optical processes that can be called quasi-resonant [13]. Although none of the fields are in resonance with the atomic or molecular transitions, the sum or difference between two optical frequencies equals a transition frequency.

Raman scattering is one of the physical processes that can lead to spontaneous light scattering. By spontaneous light scattering, we mean light scattering under conditions such that the optical properties of the material system are unmodified by the presence of the incident light beam. Figure 3.4 shows a diagram for an incident beam on a scattering medium (a) and the typical observed spectrum (b) in which Raman, Brillouin, Rayleigh, and Rayleigh-wing features are present. By definition, those components that are shifted to lower frequencies are known as Stokes' components, and those that are shifted to higher frequencies are known as anti-Stokes' components. Typically, the Stokes' lines are orders of magnitude more intense than the anti-Stokes' lines. In Table 3.4, a list of the typical values of the parameter describing these light-scattering processes is given.

Raman scattering results from the interaction of light with the vibrational modes of the molecules constituting the scattering medium and can be equivalently described as the scattering of light from optical phonons.

Light scattering occurs as a consequence of fluctuations in the optical properties of a material medium. A completely homogeneous material can scatter light only in the forward direction [14] .

Spontaneous Raman scattering was discovered in 1928 by Raman. To observe the effect, a beam of light illuminates a material sample and the scattered light is observed spectroscopically. In general, the scattered light contains frequencies different from those of the excitation source; that is, it contains Stokes' and anti-Stokes' lines. Raman-Stokes' scattering consists of a transition from the ground state i to a virtual level associated with the excited state f' followed by a transition from the virtual level to the final state f. Raman-anti-Stokes' scattering entails a transition from level f to level f' serving as the intermediate level (Figure 3.5).The anti-Stokes' lines are typically much weaker than the Stokes' lines because, in thermal equilibrium, the population of level f is smaller than the population of level i by the Boltzmann factor: $\exp(-h\omega_{n_\ell}/kT)$.

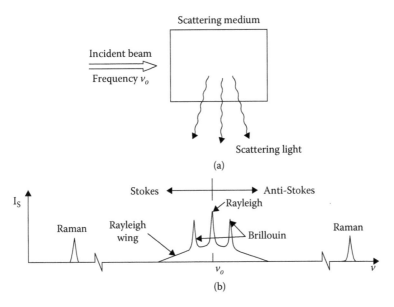

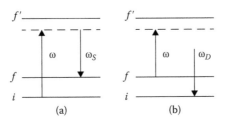

FIGURE 3.4 (a) Diagram for an incident beam on a scattering medium; (b) typical observed Raman, Brillouin, Rayleigh, and Rayleigh-wing spectra.

FIGURE 3.5 Energy level diagrams for (a) Raman-Stokes' and (b) Raman-anti-Stokes' scattering.

TABLE 3.4 Typical Values of the Parameters Describing Several Light-Scattering Processes

Process	Shift (cm⁻¹)	Linewidth (cm⁻¹)	Relaxation time (sec)	Gain[a] (cm/MW)
Raman	1000	5	10^{-12}	5×10^{-3}
Brillouin	0.1	5×10^{-3}	10^{-9}	10^{-2}
Rayleigh	0	5×10^{-4}	10^{-8}	10^{-4}
Rayleigh-wing	0	5	10^{-12}	10^{-3}

Source: Reprinted from Academic Press, Boyd RW. *Nonlinear Optics*, Copyright 1992, with permission from Elsevier.
[a] Gain of the stimulated version of the process.

The Raman effect has important spectroscopic applications because transitions that are one-photon forbidden can often be studied using Raman scattering. Figure 3.6 shows a spectral plot of the spontaneous Raman emission from liquid N2 obtained by Clements and Stoicheff in 1968. In Table 3.5, the Raman scattering cross-sections per molecule of some liquids are given.

Typically, the spontaneous Raman scattering process is a rather weak process. The scattering cross-section per unit volume for Raman–Stokes' scattering in condensed matter is only about 10^{-6} cm⁻¹.

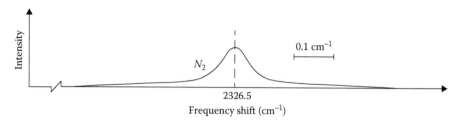

FIGURE 3.6 Spontaneous Raman emission from liquid N_2. (Reproduced with permission from Clements WRL, Stoicheff BP. *Appl. Phys. Lett.*, 12, 246, 1968. Copyright 1968, American Institute of Physics.)

TABLE 3.5 Raman Scattering Cross-Sections per Molecule of Some Liquids

Raman Lines	Wavelength of the Exciting Light (nm)	Raman Scattering Cross-Section ($d\sigma/d\Omega$) ($10^{-29} cm^2 molecule^{-1} sr^{-1}$)
C_6H_6	632.8	0.800 ± 0.029
992 cm^{-1}	514.5	2.57 ± 0.08
Benzene	488.0	3.25 ± 0.10
$C_6H_5CH_3$	632.8	0.353 ± 0.013
1002 cm^{-1}	514.5	1.39 ± 0.05
Chlorobenzene	488.0	1.83 ± 0.06
$C_6H_5NO_2$	632.8	1.57 ± 0.06
1345 cm^{-1}	514.5	9.00 ± 0.29
Nitrobenzene	488.0	10.3 ± 0.4
	694.3	0.755
CS_2	632.8	0.950 ± 0.034
656 cm^{-1}	514.5	3.27 ± 0.10
	488.0	4.35 ± 0.13
CCl_4	632.8	0.628 ± 0.023
459 cm^{-1}	514.5	1.78 ± 0.06
	488.0	2.25 ± 0.07

Source: Reproduced with permission from Kato Y, Takuma H. *J. Chem. Phys.*, **54**, 5398, Copyright 1971, American Institute of Physics.

Hence, in propagating through 1 cm of the scattering medium, only approximately one part in 10^6 of the incident radiation will be scattered into the Stokes' frequency. However, under excitation by an intense laser beam, highly efficient scattering can occur as a result of the stimulated version of the Raman scattering process.

The ability of lasers to produce light of extremely high intensity makes them especially attractive sources for Raman spectroscopy of molecules by increasing the intensity of the anti-Stokes' components in the Raman effect. Each resulting pair of lines, equally displaced with respect to the laser line, reveals a characteristic vibrational frequency of the molecule. Compared to the spontaneous Raman scattering, stimulated Raman scattering is a very strong scattering process: 10% or more of the energy of the incident laser beam is often converted into the Stokes' frequency.

Stimulated Raman scattering was discovered by Woodbury and Ng in 1962 and described by Eckhardt et al. [15] in the same year. Later some authors made a review of the properties of stimulated Raman scattering; see, for example, Bloembergen [11], Kaiser and Maier [16], Penzkofer et al. [12], and Raymer and Walmsley [17].

One of the earliest methods of accomplishing Raman laser action employed a repetitively pulsed Kerr cell as an electro-optical shutter, or "Q-switch," enclosed within the laser cavity, together with

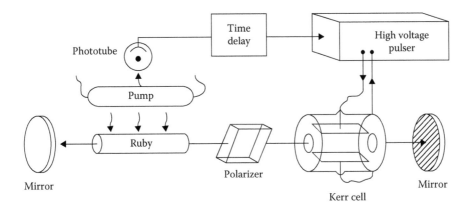

FIGURE 3.7 Experimental setup for pulsed operation of a laser employing a Kerr cell as an electro-optical shutter (With kind permission from **Springer Science+Business Media**: *An Introduction to Nonlinear Optics*, 1969, Baldwin GC. With permission of Springer.)

a polarizing prism, as shown in Figure 3.7 [5]. This allows laser action to occur only during the brief time intervals when the Kerr cell is transmitting; the laser avalanche then discharges energy stored over the much longer time interval since the preceding pulse on the Kerr cell. It was observed that for sufficiently high laser-pulse intensity, the 694.3 nm ruby laser line is accompanied by a satellite line at 767 nm, which originates in the nitrobenzene of the Kerr cell. The satellite line increases markedly in intensity as the laser output is increased above a threshold level of the order of 1 MW/cm², persists only while the laser output is above this threshold, shares the direction of the laser radiation, and becomes spectrally narrower at higher intensities. Its wavelength agrees with that of the known Raman–Stokes' shift in nitrobenzene. The conclusion was therefore reached that the phenomenon is a stimulated Raman scattering process, pumped by the laser beam and resonated by the end-reflectors of the laser.

Raman media are now widely used in conjunction with pulsed lasers to generate coherent radiation at frequencies other than those currently accessible to direct laser action. This is one of the powerful advantages of the Raman scattering process as a spectroscopic technique. Table 3.6 gives some properties of stimulated Raman scattering for several materials.

One application of Raman Scattering is in the Raman lidars. Lidar (Light detection and ranging) dates back to the 1930s but it has become one of the primary tools in atmospheric and environment research. There are several types of lidar, but all involve a transmitter of laser radiation and a receiver for the detection and analysis of backscattered light. The scattered radiation is shifted in frequency from the incident laser radiation by the change in the vibrational-rotational energy and since the energy levels are distributed according to a Boltzmann distribution at the ambient temperature, Raman Lidars are used for temperature profiling of the atmosphere [9]. Unlike other kind of lidars, Raman lidars have the advantage that they do not require specific laser wavelengths that match an absorption line of an atmospheric constituent.

Surface-enhanced Raman scattering (SERS) allows a significant increase in sensitivity compared to normal Raman scattering. It was initially observed by Fleischman et al. in 1974. SERS allows an enhancement of the Raman signal by a factor of up to 1014, provided the sample is in close proximity to a nanostructured metal surface (gold, silver or copper) [18]. SERS has been applied in many fields, such as forensic science, homeland security, biochemistry and medicine [19]. The spectrum obtained by SERS is the result of an analyte's molecular structure, which is useful for real-time detection of certain compounds in biofluids at sub-nanomolar concentrations, so it allows in vivo-measurements. Murphy et al. [20] and Stuart et al. [21] have developed a number of biosensors for environmental as well as in-vivo measurements.

The design of a practical device based on Stimulated Raman scattering in crystals is a very challenging task because high power density is required to obtain efficient conversion in a Raman active material but

TABLE 3.6 Frequency Shift v_v, Scattering Cross-section $N(d\sigma/d\Omega)$ of Spontaneous Raman Scattering (N is the number of molecules per cm³) and Steady-State Gain Factor g_s/I_1 of Stimulated Raman Scattering in Different Substances (λ_1 = 694.3 nm)

Substance	Frequency Shift v_v (cm⁻¹)	Linewidth Δv (cm⁻¹)	Cross-Section $N(d\sigma/d\Omega) \times 10^8$ (cm⁻¹ ster⁻¹)	Gain Factor[a] g_s/I_1 in Units of 10^{-3} (cm/MW)	Temperature (K)
Liquid O₂	1552	0.117	0.48 ± 0.14	14.5 ± 4	
				16 ± 5	
Liquid N₂	2326.5	0.067	0.29 ± 0.09	17 ± 5	
				16 ± 5	
Benzene	992	2.15	3.06	2.8	300
		2.3	3.3	3.0	
			4.1	3.8	
CS₂	655.6	0.50	7.55	24.0	300
Nitrobenzene	1345	6.6	6.4	2.1	300
			7.9	2.6	
Bromobenzene	1000	1.9	1.5	1.5	300
Chlorobenzene	1002	1.6	1.5	1.9	300
Toluene	1003	1.94	1.1	1.2	300
LiNbO₃	256	23.0	381.0	8.9	300
	258	7.0	262.0	28.7	80
	637	20.0	231.0	9.4	300
	643	16.0	231.0	12.6	80
Li₆NbO₃	256			17.8	300
	266			35.6	80
	637			9.4	300
	643			12.6	80
Ba₂NaNb₅O₁₅	650			6.7	300
	655			18.9	80
LiTaO₃	201	22.0	238.0	4.4	300
	215	12.0	167.0	10.0	80
Li₆TaO₃	600			4.3	300
	608			7.9	80
SiO₂	467			0.8	300
				0.6	300
H₂ gas	4155			1.5 (P > 10 atm)	300

Source: Reprinted from North Holland, Kaiser W, Maier M. *Stimulated Rayleigh, Brillouin and Raman Spectroscopy Laser Handbook*, Copyright 1972, with permission from Elsevier.

Note: Linewidths and Temperatures are also indicated.

[a] To obtain the gain constant g_s(cm⁻¹) at v_s, multiply by v_s/v_1 and by the intensity in MW/cm².

at the same time, damage to the crystal must be avoided. In contrast to the harmonic generation, which is a lossless process, Raman conversion deposits heat into the crystal, which leads to strong thermal lensing and birefringence. However, despite these problems, several diode-pumped solid-state Raman lasers have been successfully operated at the 1 W level and above [8].

As an example, an externally pumped Raman laser containing a Ba(NO₃)₂ crystal produced up to 1.3 W at the first Stokes wavelength of 1197 nm, where the pump laser was a diode-pumped Nd:YAG slab laser [22].

Raman scattering is currently used in several medical applications, from cancer detection to osteoporosis detection (scanning directly over the skin in the arms or legs of the patient) [23,24].

Raman spectroscopy is an inelastic light scattering process, generally implemented with a laser in the near ultraviolet, visible or near infrared range [25]. Light from the laser interacts with vibrational excitations in the system producing characteristic shifts in the energy of the scattered light [18]. Raman spectroscopy allows us to identify the molecular composition of a sample, giving a chemical/structural fingerprint as well as information concerning conformation and bond structure [23].

One of the main advantages of Raman spectroscopy is the fact that it is relatively insensitive to water and requires no special sample preparation. For this reason, it has been used extensively in a variety of biomedical testing applications, such as the quantification of blood components (glucose, cholesterol, and urea), estimation of protein structure, evaluation of skin composition as well as cancer and pre-cancer diagnosis [23,24].

Raman spectroscopy is increasingly used to identify single bacteria as it can help to avoid production downtime in pharmaceutical clean rooms and reduce health hazards in clinical situations and food processing [26–29].

For detailed theoretical discussions of the Raman process, see Baldwin [5], Boyd [30], Yariv [6], Mostowsky and Raymer [13]. For experimental techniques, see Demtröder [31]; see also Ederer and McGuire [32].

Recent medical applications include the identification of microcalcifications and underlying breast lesions at stereotactic core needle biopsy [33] and the measurement of abnormal bone compositions in vivo using noninvasive Raman spectroscopy [34].

3.2.3 Rayleigh Scattering

The spectrum of scattered light generally contains an *elastic* contribution, where the scattered frequency ω_s equals the incident frequency ω, together with several *inelastic* components, where ω_s differs from ω. The *elastic* component is known as Rayleigh scattering.

Rayleigh scattering is the scattering of light from nonpropagating density fluctuations. It is known as elastic or quasielastic scattering because it induces no frequency shift and can be described as scattering from entropy fluctuations [30].

Rayleigh-wing scattering (i.e., scattering in the wing of the Rayleigh line) is scattering from fluctuations in the orientation of anisotropic molecules. Since the molecular reorientation process is very rapid, this component is spectrally very broad. This type of scattering does not occur for molecules with an isotropic polarizability tensor.

Imagine a gaseous medium subdivided into volume elements in the form of small cubes of $\lambda = 2$ on a side, as shown in Figure 3.8 [5]. The radiation scattered by two adjacent elements will be in phase in the forward direction but oppositely phased at 90°; if they contain exactly equal numbers of molecules, there will be no 90° scattering. However, for green light, $\lambda = 500$ nm, the volume of each element will be $\lambda^3/8 = 1.6 \times 10^{-20}$ m^3, and at atmospheric pressure, it will contain approximately

$$N\lambda^3 / 8 = \left[2.7 \times 10^{25} \, \text{m}^{-3}\right] \times \left[1.6 \times 10^{-20} \, \text{m}^3\right] = 4.3 \times 10^5 \, \text{molecules},$$

where N is the density of molecules. The root-mean-square fluctuation, approximately 600 molecules, is 0.15% of the total number. At shorter wavelengths, these numbers are less, but the fluctuations become more pronounced. These produce the net scattering. This phenomenon is responsible for the blue color of the sky. The scattered power is proportional to the total number of scattering centers and to the fourth power of the frequency. The scattered light is incoherent.

The Rayleigh scattering reduces the intensity of a light beam without actually absorbing the radiation. Despite the regular arrangement of atoms in a crystalline media, which reduces the fluctuations, light can also be scattered by crystals, as thermal motion causes slight random fluctuations on density. Then, we can observe Rayleigh scattering not only in gases and liquids, but also in crystalline media.

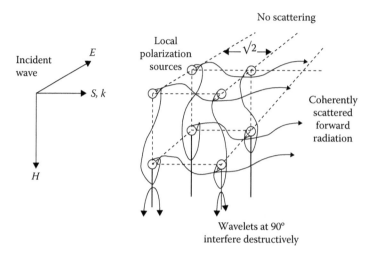

FIGURE 3.8 Coherent scattering of light by elementary dipoles induced in a medium by an electromagnetic wave incident from the left. The wavelets reradiated by the dipoles interfere constructively in the forward direction while those at 90° interfere destructively. (With kind permission from **Springer Science+Business Media**: *An Introduction to Nonlinear Optics*, 1969, Baldwin GC.)

Rayleigh scattering can be used for astronomical purposes. The Rayleigh-backscattered light from a laser beam propagating up into the atmosphere can serve as a guide star [9].

Details of how to obtain the total average cross sections can be found in Loudon [2] and Chu [35]; other recommended texts are Bloembergen [36], Khoo et al. [37], and Keller [38].

3.3 Multiphoton Processes

Multiphoton processes refer to the processes during which transitions between discrete levels of atoms are produced by the simultaneous absorption of n photons.

Multiphoton transitions are found whenever electrons bound in atoms or molecules interact with sufficiently intense electromagnetic radiation. Multiphoton ionization is the ultimate outcome of multiphoton transitions provided the radiation is reasonably intense, which is usually taken to mean about 10^6 W/cm^2, or more, at optical frequencies. With intense sources and interaction times of the order of a nanosecond or more, ionization may be expected for atomic gases, while dissociation, with varying degrees of ionization, for molecular gases.

Multiphoton processes have features that are different from the traditional interaction of radiation with atoms and molecules. First, the interactions are nonlinear; secondly, a multiphoton transition probes atomic and molecular structure in a more involved and detailed way than can be imagined in single-photon spectroscopy; thirdly, owing to the high intensities available, extremely high levels of excitation can be reached, thus introducing qualitatively new physics; and fourthly, new applications, such as isotope separation, generation of coherent radiation at wavelengths shorter than ultraviolet, and plasma diagnostics, can be realized.

Multiphoton processes are one of the main sources of nonlinearity in the interaction of intense laser fields with atoms and molecules. Resonant multiphoton processes are of special interest because the multiphoton transition probabilities are enhanced under resonance conditions, and can be observed in fields of moderate intensity. The resonance is intended in the sense that the sum of the energies of the n photons is equal to the energy difference between the two concerned levels. It excludes the case where an intermediate level is resonantly excited during the process. However, some authors have to deal with the problem of processes involving both types of resonance.

The first detailed theoretical treatment of two-photon processes was reported more than 60 years ago, by Göppert-Mayer in 1931. However, experiments could not be realized until 30 years later, when lasers were available [39,40].

3.3.1 Two-Photon Absorption

Two-photon absorption can be described by a two-step process from initial level $|i>$ via a "virtual level" $|v>$ to the final level $|f>$ as shown in Figure 3.9. The virtual level is represented by a linear combination of the wave functions of all real molecular levels $|k_n)$, which combine with $|i>$ by allowed one-photon transitions. The excitation of $|v>$ is equivalent to the off-resonance excitation of all these real levels $|k_n>$. The probability amplitude for a transition $|i> \rightarrow |v>$ is represented by the sum of the amplitudes of all allowed transitions $|i> \rightarrow |k>$ with off-resonance detuning $(\omega - \omega_{ik})$. The same arguments apply to the second step $|v> \rightarrow |f>$.

The probability of two-photon absorption at low intensity is proportional to the square of the intensity of the light field I^2 (in the case of a single laser beam), the probability of absorption of each photon being proportional to I. In the case of n-photon transitions, the probability of excitation is proportional to I^n. For this reason, pulsed lasers are generally used, in order to have sufficiently large peak powers. The spectral linewidth of these lasers is often comparable to or even larger than the Doppler width.

For a molecule moving with a velocity $\mathbf{v}$, the probability A_{if} for a two-photon transition between the ground state E_i and an excited state E_f, induced by the photons $\hbar\omega_1$ and $\hbar\omega_2$ from two light waves with wave vectors $\mathbf{k}_1$ and $\mathbf{k}_2$, polarization unit vectors $\hat{\mathbf{e}}_1$, $\hat{\mathbf{e}}_2$, and intensities I_1, I_2, respectively, can be written as [31]:

$$A_{ij} \alpha \frac{\gamma_{if} I_1 I_2}{\left[\omega_{if} - \omega_1 - \omega_2 - \mathbf{v} \cdot (\mathbf{k}_1 + \mathbf{k}_2)\right]^2 + \left(\gamma_{if}/2\right)^2}$$
$$\times \left| \sum_k \frac{\mathbf{R}_{ik} \cdot \hat{\mathbf{e}}_1 \cdot \mathbf{R}_{kf} \cdot \hat{\mathbf{e}}_2}{\omega_{if} - \omega_1 - \omega_2 - \mathbf{v} \cdot \mathbf{k}_1} + \frac{\mathbf{R}_{ik} \cdot \hat{\mathbf{e}}_2 \cdot \mathbf{R}_{kf} \cdot \hat{\mathbf{e}}_1}{\omega_{if} - \omega_1 - \omega_2 - \mathbf{v} \cdot \mathbf{k}_2} \right|$$

(3.24)

The first factor gives the spectral line profile of the two-photon transition of a single molecule. It corresponds exactly to that of a single-photon transition of a moving molecule at the center frequency $\omega_{if} = \omega_1 + \omega_2 + \mathbf{v}(\mathbf{k}_1 + \mathbf{k}_2)$ with a homogeneous linewidth γ_{if}.

The second factor describes the transition probability for the two-photon transition and can be derived quantum mechanically by second-order perturbation theory [41,42]. It contains a sum of products of

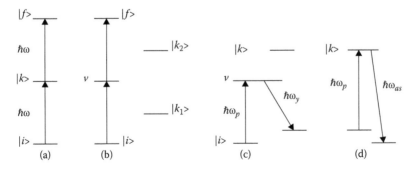

FIGURE 3.9 Energy level diagrams for several two-photon transitions: (a) resonant two-photon absorption with a real intermediate level $|k>$; (b) nonresonant two-photon absorption; (c) Raman transition; and (d) resonant and anti-Stokes' Raman scattering.

matrix elements $\mathbf{R}_{ik}\,\mathbf{R}_{kf}$ for transitions between the initial state i and intermediate molecular levels k or between these levels k and the final state f.

The summation extends over all molecular levels k that are accessible by allowed one-photon transitions from the initial state $|i>$. However, only those levels k, which are not too far off resonance with one of the Doppler-shifted laser frequencies $\omega'_1 = \omega_1 - \mathbf{v}\cdot\mathbf{k}_1$, $\omega'_2 = \omega_2 - \mathbf{v}\cdot\mathbf{k}_2$, will mainly contribute, as can be seen from the denominator.

The frequencies ω_1 and ω_2 can be selected in such a way that the virtual level is close to a real molecular eigenstate. This greatly enhances the transition probability and then is generally advantageous to excite the final level E_f by two different photons with $\omega_1 + \omega_2 = (E_f - E_i)/h$ rather than by two photons out of the same laser with $2\omega = (E_f - E_i)/h$.

Multiphoton processes have several applications in laser spectroscopy. For example, two-quantum resonant transitions in a standing wavefield are an important method for eliminating Doppler broadening. One of the techniques of high-resolution laser spectroscopy is based on this approach [43,44].

3.3.2 Doppler-Free Multiphoton Transitions

Doppler broadening is due to the thermal velocities of the atoms in the vapor. If $\mathbf{v}$ is the velocity of the atom and $\mathbf{k}$ is the wave vector of the light beam, the first-order Doppler shift is $\mathbf{k}\,\mathbf{v}$. If the sense of the propagation of light is reversed $(\mathbf{k} \rightarrow -\mathbf{k})$, the first-order Doppler shift is reversed in sign.

Suppose that a two-photon transition can occur between the levels E_i and E_f of an atom in a standing electromagnetic wave of angular frequency ω (produced for example by reflecting a laser beam onto itself using a mirror). In its rest frame, the atom interacts with two oppositely traveling waves of angular frequencies $\omega + \mathbf{k}\,\mathbf{v}$ and $\omega - \mathbf{k}\,\mathbf{v}$. If the atom absorbs one photon from each traveling wave, the energy conservation implies that

$$E_f - E_i = E_{fi} = h(\omega + \mathbf{k}\cdot\mathbf{v}) + h(\omega - \mathbf{k}\cdot\mathbf{v}) = 2h\omega. \tag{3.25}$$

The term, depending on the velocity of the atom, disappears, indicating that at resonance, all the atoms, irrespective of their velocities, can absorb two photons [45,46].

In theory, the Doppler-free two-photon absorption resonance must have a Lorentzian shape, the width of the resonance being the natural one [47]. However, the wings of the resonance generally differ from the Lorentzian curve because, if the frequency ω of the laser does not fulfill the resonant condition, Equation 3.25, but is still close to it, the atoms cannot absorb two photons propagating in opposite directions, although some atoms of definite velocity can absorb two photons propagating in the same direction, provided that the energy defect $(E_{fi} - 2h\omega)$ is equal to the Doppler shift $\pm 2\mathbf{k}\,\mathbf{v}$. For each value of ω, there is only one group of velocities that contribute to this signal, whereas at resonance (due to the absorption photons propagating in opposite directions), all the atoms contribute.

The two-photon line shape appears as the superposition of two curves, as shown in Figure 3.10. A Lorentzian curve of large intensity and narrow width (natural width) corresponds to the absorption of photons from the oppositely traveling waves, while a Gaussian curve of small intensity and broad width (Doppler width) corresponds to the absorption of photons from the same traveling wave. Typically, the Doppler width of the Gaussian curve is 100 or 1000 times larger than the natural width of the Lorentzian curve and the Gaussian curve appears as a very small background. In some cases, the choice of different polarizations permits one to suppress the Doppler background completely, using the different selection rules corresponding to different polarization [48].

The first experimental demonstrations of Doppler-free two-photon transitions were performed with pulsed dye lasers on the 3S–5S transition in sodium [48–50]. The precision of the measurements has been increased by the use of cw dye lasers in single-mode operation. A typical setup using a cw laser is shown in Figure 3.11.

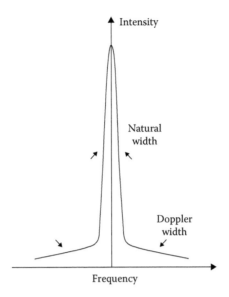

FIGURE 3.10 Theoretical Doppler-free two-photon absorption resonance. The two-photon line shape appears as the superposition of two curves. (With kind permission from **Springer Science+Business Media**: *Coherent Nonlinear Optics: Recent Advances*, Multiphoton resonant processes in atoms, 1980, Grynberg G, Cagnac B, Biraben F. With permission of Springer.)

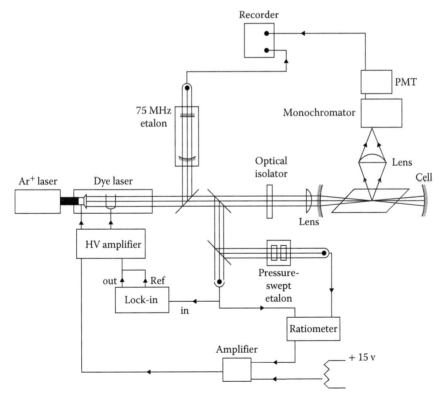

FIGURE 3.11 Typical setup for Doppler-free two-photon experiments using cw dye laser. (With kind permission from **Springer Science+Business Media**: *Coherent Nonlinear Optics, Recent Advances*, 1980, Feld MS, Letokhov VS (eds). With permission of Springer.)

The cw laser is pumped by an argon ion laser. In order to obtain good control of the laser frequency, two servo loops are used. The purpose of the first one is to maintain the single-frequency oscillation of the dye laser. The Fabry-Perot etalon inside the laser cavity selects one particular longitudinal mode of the cavity. The second servo loop is used to control the frequency of the laser cavity and does not include any modulation. For details see Grynberg et al. [47].

The light coming from the laser is focused into the experimental cell in order to increase the energy density. The transmitted light is refocused from the other side into the cell using a concave mirror whose center coincides with the focus of the lens. In some experiments, the energy is increased by placing the experimental cell in a spherical concentric Fabry–Perot cavity [51]. The lens is chosen to match the radius of curvature of the wavefront to the radius of the first mirror. In order to reduce the losses in the cavity, the windows of the experimental cell are tilted to the Brewster angle. The length of the cavity is locked to the laser frequency to obtain the maximum transmitted signal. The two-photon resonance is detected by collecting photons emitted from the excited state at a wavelength λ_{vf} (see Figure 3.9) different from the exciting wavelength λ. Sometimes it is still more convenient to detect the resonance on another wavelength λ_{ab} emitted by the atom in a cascade. The characteristic λ_{vf} is selected with an interference filter or a monochromator. The difference between λ_{vf} and λ allows the complete elimination of the stray light of the laser, despite its high intensity, and the observation of very small signals on a black background.

A simpler experimental arrangement for Doppler-free two-photon spectroscopy is shown in Figure 3.12 [31]. The two oppositely traveling waves are formed by reflection of the output beam from a single-mode tunable dye laser. The Faraday rotator prevents feedback into the laser. The two-photon absorption is monitored by the fluorescence emitted from the final state E_f into other states E_m. The two beams are focused into the sample cell by the lens L and the spherical mirror M.

More examples of Doppler-free two-photon spectroscopy can be found in Demtröder [31].

3.3.3 Multiphoton Spectroscopy

If the incident intensity is sufficiently large, a molecule may absorb several photons simultaneously. Equation 3.24 can be generalized in order to obtain the probability for absorption of a photon $h\omega_k$ on the transition $|i >\rightarrow |f >$ with $E_f - E_i; = \Sigma h\omega_k$. In this case, the first factor in Equation 3.24 contains the

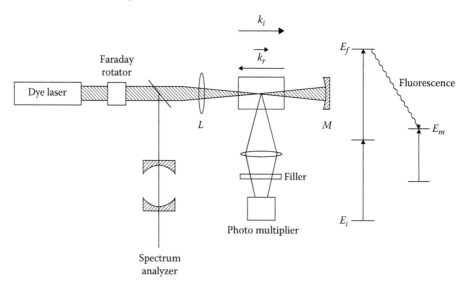

FIGURE 3.12 A simpler experimental arrangement for Doppler-free two-photon experiments. (With kind permission from **Springer Science+Business Media**; *Laser Spectroscopy: Basic Concepts and Instrumentation*, 1996, Demtröder W. With permission of Springer.)

product $\Pi_k I_k$ of the intensities I_k of the different beams. In the case of n-photon absorption of a single laser beam, this product becomes I^n. The second factor in Equation 3.24 includes the sum over products of n one-photon matrix elements.

In the case of Doppler-free multiphoton absorption besides the energy conservation $\Sigma\hbar\omega_k = E_f - E_i$, the momentum conservation

$$\sum_k \mathbf{p}_k = \hbar \sum_k \mathbf{k}_k = 0 \tag{3.26}$$

has also to be fulfilled. Each of the absorbed photons transfers the momentum $\hbar\mathbf{k}_k$ to the molecule.

If Equation 3.26 holds, the total transfer of momentum is zero, which implies that the velocity of the absorbing molecule has not changed. This means that the photon energy is completely converted into excitation energy of the molecule without changing its kinetic energy. As this is independent of the initial velocity of the molecule, the transition is Doppler-free.

Figure 3.13 shows a possible experimental arrangement for Doppler-free three-photon absorption spectroscopy, while Figure 3.14 shows the three-photon excited resonance fluorescence in Xe at $\lambda = 147$ nm, excited with a pulsed dye laser at $\lambda = 441$ nm with 80 kW peak power.

3.3.4 Multiphoton Ionization Using an Intermediate Resonant Step

Resonant multiphoton excitation often occurs as an intermediate step in other processes. In the case of multiphoton ionization, the number of ions increases by a huge factor when the wavelength of the exciting laser is adjusted to obtain a resonant multiphoton with an intermediate level, as shown in Figure 3.15.

The figure shows the four-photon ionization of cesium with an intermediate resonant level. The variation of the number of atomic cesium ions as a function of the laser frequency (Nd glass laser) is also shown. The curves have a big enhancement in the neighborhood of the three-photon transition 6S → 6F. The center of the resonance is shifted by an amount that is proportional to the intensity. For an intensity $I = 1$ GW $=$ cm², the wavelength of excitation is reduced by an amount larger than 0.1 nm.

This light shift explains the strange behavior of the order of nonlinearity of a multiphoton ionization process near an n-photon resonance [52].

The order of nonlinearity K is defined by [47]:

$$K = \frac{\partial \log N_t}{\partial \log I}. \tag{3.27}$$

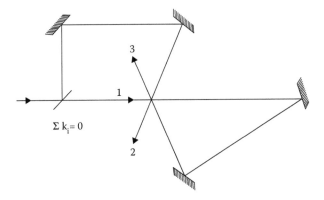

FIGURE 3.13 Possible experimental arrangement for Doppler-free three-photon absorption spectroscopy.

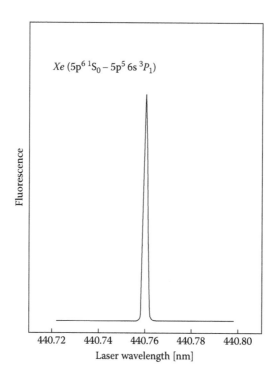

FIGURE 3.14   Three-photon excited resonance fluorescence in Xe at $\lambda = 147$ nm, excited with a pulsed dye laser at $\lambda = 441$ nm with 80 kW peak power.

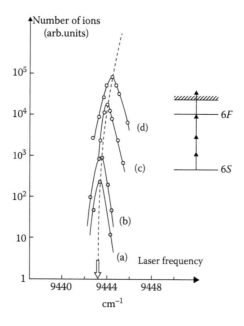

FIGURE 3.15   Four-photon ionization of cesium with an intermediate resonant level. The variation of the number of atomic cesium ions as a function of the laser frequency in the neighborhood of the three-photon transition $6S \rightarrow 6F$ is also shown. (With kind permission from **Springer Science+Business Media**; *Coherent Nonlinear Optics, Recent Advances*, 1980, Feld MS, Letokhov VS [eds]. With permission of Springer.)

where N_i is the number of ions obtained in the multiphoton ionization and I is the light intensity. Far away from any intermediate resonance, K is equal to the number of photons K_0, which is needed for photoionization, but close to a resonance, this is not true.

Because of the light shift, the effective detuning from the resonance increases or decreases, depending on the sign of the laser detuning ($E_f - nh\omega$). This explains why, on one side of the resonance, K is much larger than K_0 while, on the other side, K is smaller.

The resonant multistep processes permit the selective photoionization of atoms. This approach is fundamental for laser methods of single-atom detection [53].

3.4 Phase Conjugate Optics

Phase conjugate optics refers in its most elemental basics to the conversion in *real time* of a monochromatic optical field $\mathbf{E}_1$ to a new field $\mathbf{E}_2$ such that

$$E_1(\mathbf{r},t) = \mathrm{Re}[\psi(\mathbf{r})e^{i(\omega t - kz)}]$$

$$E_2(\mathbf{r},t) = \mathrm{Re}[\psi^*(\mathbf{r})e^{i(\omega t + kz)}].$$

We refer to E_2 as the phase conjugate replica of E_1.

Suppose that a monochromatic optical beam E_1 propagates to the right and then incides into a lossless medium that distorts it. If the distorted beam is reflected by a mirror, then we will have a reflected beam traveling to the left that, when it incides into the lossless medium, is distorted in exactly the opposite way such that the emerging beam E_2 is the replica of the initial beam E_1 everywhere.

To show that it is possible that both beams E_1 and E_2 coincide everywhere, we refer and demonstrate the Distortion Correction Theorem [54], which may be stated as follows: "If a (scalar) wave $E_1(r)$ propagates from left to right through an arbitrary dielectric (but lossless) medium, then if we generate in some region of space (say, near $z = 0$), its phase conjugate replica $E_2(r)$ then E_2 will propagate backward from right to left through the dielectric medium remaining everywhere the phase conjugate of E_1."

Consider the (scalar) right-going wave E_1:

$$E_1 = \psi_1(r)e^{i(\omega t - kz)},$$

where k is a real constant. E_1 obeys in the paraxial limit the wave equation:

$$^2E_1 + \omega^2\mu\varepsilon(\mathbf{r})E_1 = 0. \tag{3.28}$$

If we substitute E_1 into the wave Equation 3.28, we obtain

$$^2\psi_1 + \left[\omega^2\mu\varepsilon(\mathbf{r}) - k\right]\psi_1(r) - 2ik\partial\psi_1/\partial z = 0. \tag{3.29}$$

The complex conjugate of Equation 3.29 leads to:

$$\nabla^2\psi_1^* + \left[\omega^2\mu\varepsilon^*(\mathbf{r}) - k\right]\psi_1^*(r) + 2ik\partial\psi_1^*/\partial z = 0. \tag{3.30}$$

Now take the wave E_2 propagating to the left, into the wave Equation 3.26. We obtain

$$^2\psi_2 + \left[\omega^2\mu\varepsilon(\mathbf{r}) - k\right]\psi_2(r) + 2ik\partial\psi_2/\partial z = 0. \tag{3.31}$$

We see from Equations 3.30 and 3.31 that ψ_1^* and ψ_2 obey the same differential equation if $\varepsilon(\mathbf{r}) = \varepsilon^*(\mathbf{r})$; that is, if we have a lossless and gainless medium.

If $\psi_2 = a\psi_1^*$, where a is an arbitrary constant over any plane, say $z = 0$, Equation 3.29 is still valid. Then due to the uniqueness property of the solutions to second-order linear differential equations:

$$\psi_2(x, y, z) = a\psi_1^*(x, y, z) \text{ for all } x, y, z < 0. \tag{3.32}$$

This completes the proof of the distortion correction theorem.

Phase conjugate waves can be generated by means of nonlinear optical techniques. As second-order nonlinear optics gives rise to phenomena such as second-harmonic generation and parametric amplification, third-order nonlinear optics, which involves the third power of the electric field $P \sim E$, gives rise to phenomena, such as third-harmonic generation and to the phenomenon of *four-wave mixing* (FWM) where, if three waves of different frequencies $\omega_1, \omega_2, \omega_3$, incide into a medium material, this radiates with a frequency $\omega 4$ such that $\omega_4 = \omega_1 + \omega_2 - \omega_3$.

We will show that if three waves A_1, A_2, A_3, are mixed into a medium as shown in Figure 3.16, this medium generates and radiates a new wave A_4, which is the phase conjugate of A_1.

The nonlinear medium is crossed simultaneously by four waves at the same frequency:

$$E_1(\mathbf{r},t) = (1/2)A_1'(z)e^{i(\omega t - kz)} + c.c.$$

$$E_2(\mathbf{r},t) = (1/2)A_2'(z)e^{i(\omega t - kz)} + c.c.$$

$$E_3(\mathbf{r},t) = (1/2)A_3'(z)e^{i(\omega t - k_3 \cdot z)} + c.c.$$

$$E_4(\mathbf{r},t) = (1/2)A_4'(z)e^{i(\omega t - k_2)} + c.c.,$$

where $k^2 = \omega^2 \mu \varepsilon$.

The output power of beams 1 and 4 increases as they cross the nonlinear medium to the expense of beams 2 and 3. The quantum mechanics description [55] of the processes shows that in atomic scale, two photons—one from beam 2 and another one from beam 3—are simultaneously annihilated while two photons are created. One of these photons is added to beam 1 and the other one to beam 4.

The mathematical analysis gives us the expression [6,56]:

$$A_4(x, y, z < 0) = -i\left[\kappa^* / |\kappa|\right]\tan|\kappa|A_1^*(x, y, z < 0)$$

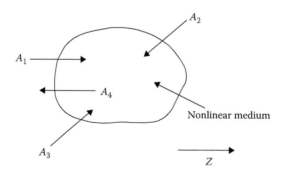

FIGURE 3.16 Four-wave mixing into a nonlinear medium.

This is the phase conjugation basic result for the four-wave mixing. This expression shows that the reflected beam $A_4(\mathbf{r})$ to the left of the nonlinear medium ($z < 0$) is the phase conjugate of the input beam $A_1(\mathbf{r})$. Here $\kappa^* = (\omega/2)\sqrt{\left[(\mu/c)\chi A_2 A_3\right]}$ and χ is a fourth-rank tensor.

3.4.1 Optical Resonators with Phase Conjugate Reflectors

In the optical resonators with phase conjugate optics, one of the mirrors of the resonator is substituted by a phase conjugate mirror (PCM), as shown in Figure 3.17.

Let us consider that the two mirrors are separated a distance l and that a Gaussian beam with quantum numbers m, n, is reflected by both mirrors. Call $\Phi_1(m, n)$ the phase shift suffered by the beam due to its propagation between the two mirrors; let Φ_R be the phase shift after reflection from the conventional mirror and α the phase shift after reflection from the PCM.

If f_1 is the initial phase of the beam in, for example, the plane P, then the several phases of the beam after each reflection will be

$$f_2 = -f_1 + \alpha$$

$$f_3 = f_2 + \Phi_1(m,n) = -f_1 + \alpha + \Phi_1(m,n)$$

$$f_4 = f_3 + \Phi_R = -f_1 + \alpha + \Phi_1(m,n) + \Phi_R$$

$$f_5 = f_4 + \Phi_1(m,n) = -f_1 + \alpha + 2\Phi_1(m,n) + \Phi_R$$

$$f_6 = -f_5 + \alpha = f_1 - 2\Phi_1(m,n) - \Phi_R$$

$$f_7 = f_6 + \Phi_1(m,n) = f_1 - \Phi_1(m,n) - \Phi_R$$

$$f_8 = f_7 + \Phi_R = f_1 - \Phi_1(m,n)$$

$$f_9 = f_8 + \Phi_1(m,n) = f_1.$$

We have shown that the phase of the beam reflected into the resonator reproduces itself after two round trips. The phase conjugate resonator has a resonance at the frequency of the pump beams. The resonance condition is satisfied independently of the length l of the resonator and the transversal order (m, n) of the Gaussian beam. The phase conjugate resonator is stable independently of the radius of curvature R of the mirror and the distance l between both mirrors.

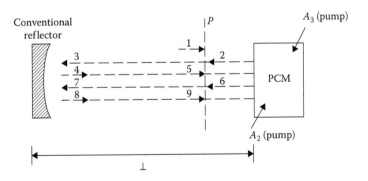

FIGURE 3.17 Optical resonator formed by a conventional mirror and a phase conjugate mirror (PCM).

3.4.2 Applications

We will consider only some of the most common applications of phase conjugation.

1. **Dynamic Correction of Distortion into a Laser Resonator**

   One important application of phase conjugate optics is the real-time dynamic correction of distortion in optical resonators. Let us consider the optical resonator shown in Figure 3.17, but now with an amplifier and a distortion (it could be the gain medium itself or even a "bad" optics), as shown in Figure 3.18. If a Gaussian beam is distorted when it passes through the distortion, it recovers its form when it is reflected by the PCM and passes the distortion again now in the opposite direction, according to the Distortion Correction Theorem. Then, at the output of the resonator, we will have a Gaussian beam again.

   The experimental arrangement for a laser oscillator with phase conjugate dynamic distortion correction is shown in Figure 3.19.

2. **Aberration Compensation**

   During the generation, transmission, processing, or imaging of coherent light, we can have aberrations due to turbulence, vibrations, thermal heating, and/or imperfections of optical elements. Most of these aberrations can be compensated by the use of a PCM if we let the wavefront follow the path shown in Figure 3.20.

   As stated before, the complex amplitudes of forward- and backward-going waves are complex conjugates of each other and the function of the PCM is to generate a conjugate replica of a general incident field, having an arbitrary spatial phase and amplitude distribution.

   One of the many experiments carried out to show aberration compensation is that of Jain and Lind [57]. In this experiment, the input light was that of a focused, pulsed ruby laser beam operating in a TEM_{00} mode. The phase aberrator was an etched sheet of glass, and the PCM was realized via degenerate four-wave mixing (DFWM), by using a semiconductor-doped glass as the nonlinear optical medium, pumped by a ruby laser.

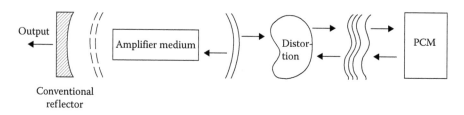

FIGURE 3.18 Laser oscillator with distortion.

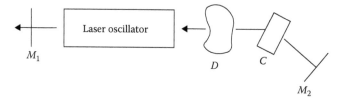

FIGURE 3.19 Experimental system for a laser oscillator with a distortion D, a phase conjugate crystal C, and two mirrors M_1 and M_2.

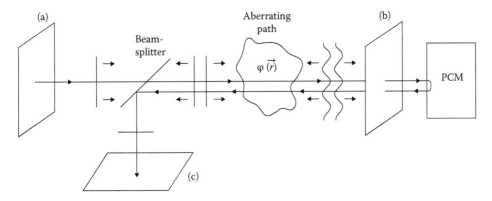

FIGURE 3.20 Experimental arrangement for aberration compensation. An input field at plane (a) becomes aberrated at plane (b) after propagation through a distortion path $\varphi(\mathbf{r})$. As a consequence of the conjugation process, the conjugate of the initial field is recovered at plane (a). The beamsplitter allows one to view the same compensator field at the plane (c). PCM, phase conjugate mirror.

This experiment shows the near-perfect recovery of a diffraction-limited focal spot using nonlinear phase conjugate techniques.

Another example of aberration compensation was carried out by Bloom and Bjorklund [58]. In this experiment, a resolution chart is illuminated by a planar probe wave and placed at plane (a) of Figure 3.20. The PCM is realized using a DFWM process, with carbon disulfide as the nonlinear medium and a doubled Nd:YAG laser as the light source.

This example shows both the compensation ability of the system and its lensless imaging capabilities. Also, it points out the importance of generating a true phase conjugate replica, both in terms of its wavefront contours and in terms of the propagation direction.

3. **Lensless Imaging: Applications to High-Resolution Photolithography Using Four-Wave Mixing**

Nonlinear phase conjugation (NLPC) allows us not only to compensate for defects in optical elements, such as lenses, but permits the elimination of the lenses altogether. Further, this lensless approach permits a major reduction in the effective F number, thereby improving the spatial resolution of the system.

An example of an NLPC application on photolithography was carried out by Levenson in 1980 and by Levenson et al. in 1981. A typical photolithography requirement is diffraction-limited resolution over a wafer as large as 3–4 inches in diameter. Levenson *et al.* obtained a resolution of 800 lines/mm over a 6.8 mm^2 field and a features size of 0.75 mm. This is equivalent to a numerical aperture (NA) of 0.48, which far exceeds the NA of conventional imaging systems. A crystal of LiNbO$_3$ was used as the FWM phase conjugate mirror, pumped by a 413 nm Kr ion laser. As in all phase conjugators, the accuracy of the conjugated wave produced by the PCM is critically dependent upon the quality of the nonlinear medium and of the pump waves.

The speckle typical problem in laser imaging systems is also eliminated, and according to Levenson [50], the elimination is a consequence of using the PCM in conjunction with planewave illumination.

4. **Interferometry**

Phase conjugate mirrors can also be used in several conventional and unconventional interferometers. For example, in the Mach–Zender type (Figure 3.21), the object wavefront interferes with

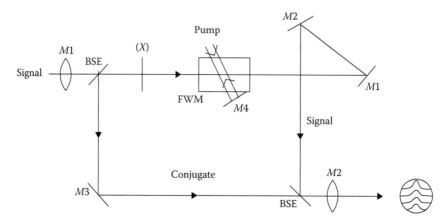

FIGURE 3.21 Mach-Zender interferometer using phase conjugate mirror. (Hopf FA, Interferometry using conjugate-wave generation, *J. Opt. Soc. Am.*, **70**, 1320; 1980. With permission of Optical Society of America.)

its phase conjugate instead of a plane-wave reference [59]. In Figure 3.21, the input signal $I(x; y)$ evaluated at plane x is given by

$$I(x,y) = A(x,y)\exp\left[-if(x,y)\right].$$

This field is imaged to the observation plane with an arbitrary transmission T, yielding an amplitude $I_T = TI(x, y)$. The conjugate of the input is generated by a four-wave mixing process in reflection from a phase conjugate mirror and arrives at the observation plane with an amplitude

$$I_R = RA(x,y)\exp[-if(x,y)],$$

where R and T depend on the reflectivity and transmittance of the PCM and other components within the interferometer. The time-averaged intensity at the observation plane will be proportional to

$$I = |I_T + I_R|^2 = \left[A(x,y)\right]^2\left\{T^2 + R^2 + 2RT\cos[2f(x,y)]\right\}.$$

An advantage of this system is that we can adjust the reflectivity of the components, including the PCM, so that $T = R$, yielding a fringe visibility of unity, independent of intensity variations in or across the sample beam.

5. **Nonlinear Laser Spectroscopy**

   Some spectroscopic studies have been carried out to obtain information about line-widths, atomic motion, excitation diffusion, and so on, by measuring the four-wave mirror-frequency response, angular sensitivity, or polarization dependence. Table 3.7 gives some examples of these spectroscopic studies.

   In addition, the spatial structure of the nonlinear susceptibilities can be studied using nonlinear microscopy by three-wave mixing (TWM), as reported by Hellwarth and Christensen [60] or four-wave mixing, as has been done by Pepper et al. [61].

   Some other interesting applications include optical computing, communications, laser fusion, image processing, temporal signal processing, and low-noise detection schemes. Also, the extension of these applications to other regions of the electromagnetic spectrum can provide new classes of quantum electronic processors.

TABLE 3.7   Laser Spectroscopy Using Four-Wave Mixing

Phase Conjugate Mirror Reflectivity Versus	Physical Mechanism	Reference
Angle	Atomic-motional effects	Wandzura (1979); Steel et al. (1979); Nilsen et al. (1981); Nilsen and Yariv (1979, 1981); Humphrey et al. (1980); Fu and Sargent (1980); Saikan and Wakata (1981); Bloch et al. (1981)
Buffer gas pressure	Collisional excitation	Liao et al. (1977); Fujita et al. (1979); Bodgan et al. (1981)
	Pressure-broadening mechanisms	Raj et al. (1980a, b); Woerdman and Schuurmans (1981); Bloch et al. (1981)
Pump probe polarization	Coherent-state phenomena	Lam et al. (1981)
	Multiphoton transitions	Steel and Lam (1979); Bloch et al. (1981)
	Quadrupole optical transitions	[a]
	Electronic and nuclear contributions to nonlinear optical susceptibility	Hellwarth (1977)
Magnetic fields	Optical pumping	Economou and Liao (1978); Steel et al. (1981a)
	Zeeman state coupling	Economou and Liao (1978); Yamada et al. (1981); Steel et al. (1981a)
	Liquid-crystal phase transitions	Khoo (1981)
Electric fields	Stark effects	[a]
	Liquid-crystal phase transitions	Jain and Lind (1983)
RF and microwave fields	Hyperfine state coupling	[a]
Pump-probe detuning	Atomic populations relaxation rates	Bloch (1980); Bloch et al. (1981); Steel et al. (1981a)
	Atomic linewidth effects	Bloch (1980); Bloch et al. (1981); Steel et al. (1981a)
	Optical pumping effects	Bloch (1980); Bloch et al. (1981); Steel et al. (1981a)
	Doppler-free one- and two-photon spectroscopy	Bloch (1980); Bloch et al. (1981); Steel et al. (1981a)
Frequency ($\omega_{pump} = \omega_{probe}$)	Natural linewidth measurements	Bloom et al. (1978); Lam et al. (1981)
	Atomic motionally-induced nonlinear coherences	Bloom et al. (1978); Lam et al. (1981)
Pump frequency scanning	Laser-induced cooling of vapors	Palmer (1979)
	Sub-Doppler spectroscopy	[a]
Transient regime (temporal effects)	Atomic coherence times	Liao et al. (1977)
	Population relaxation rates	Fujita et al. (1979); Steel et al. (1981a)
	Inter- and intramolecular relaxation	Steel and Lam (1979); Steel et al. (1981b)
	Carrier diffusion coefficients	Eichler (1977); Hamilton et al. (1979); Moss et al. (1981)
Pump-wave intensity	Saturation effects	Fu and Sargent (1979); Harter and Boyd (1980); Raj et al. (1980a, b); Woerdman and Schuurmans (1981); Steel and Lind (1981)
	Inter- and intramolecular population coupling	Dunning and Lind (1982)
	Optically induced level shifts and splitting	Bloom et al. (1978); Nilsen et al. (1981); Nilsen and Yariv (1979, 1981); Fu and Sargent (1979); Saikan and Wakata (1981); Steel and Lind (1981)

[a] No evidence regarding this mechanism has been reported to date.

3.5 Ultrashort Optical Pulses

It has been more than three decades since the era of ultrashort optical pulse generation was ushered in with the report of passive model-locking of the ruby laser by Mocker and Collins in 1965. One year later, the first optical pulses in the picosecond range were generated with an Nd: glass laser by Demaria et al. [62]. Since then, the width of optical pulses has been reduced at an exponential rate, as shown in Figure 3.22, where the logarithm of the shortest reported optical pulse width versus year is graphed.

Each reduction in pulse width has been accompanied by an advance in technology. As an example, a pulse width of about 10^{-14} s was possible thanks to the development of pulse compression. Optical pulse widths as short as 6 fs have been generated, approaching the fundamental limits of what is possible in the visible region of the spectrum.

Progress in generating intense ultrashort laser pulses has made systematic studies of coherent interactions between picosecond laser pulses and molecular vibrations possible [63]. The first method used for this purpose [64,65] has proved to be the most efficient. In this method, the sample is simultaneously irradiated by two coherent collimated ultra-shortlight pulses, whose frequency difference is equal to the molecular vibrational frequency. This induces an excitation, of the Raman-type, of the N molecules contained in a coherent interaction volume. An ultrashort pulse of variable delay then probes the state of the system as it decays. Both the intensity and direction of the scattered probe pulse can then be studied.

Because of the coherent nature of the interaction between the excitation and probe pulses, the interaction efficiency for short delay is proportional to N^2, and depends on the relative orientation of the wave vectors of the exciting and probe fields. However, as the molecular vibrations dephase, the interaction becomes incoherent, leading to isotropic efficiency proportional only to N. These features make it possible to separate coherent and incoherent processes occurring on a picoseconds time scale [66]. The picosecond pulse techniques can also be used to study inhomogeneous broadening of the vibrational transitions and its internal structure.

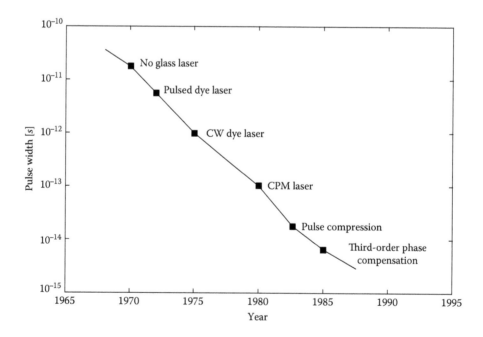

FIGURE 3.22   Historical development of the progress in generating ultrashort pulses. (With kind permission from **Springer Science+Business Media**: *Laser Spectroscopy: Basic Concepts and Instrumentation*, 1996, Demtröder W. With permission of Springer.)

Ultrashort optical pulses are related to ultrafast nonlinear optics: *ultrafast* is defined as referring to events that occur in a time of less than about 10 ps. The motivation for using ultrafast pulses can be either that they afford the time resolution necessary to resolve the process of interest, or they are required to obtain a high peak power at a relatively low pulse energy. Ultrafast pulses are not available at every wavelength. In fact, most of them have been obtained in the orange region of the spectrum where the most reliable sources exist. Some others have been obtained in the gallium arsenide semiconductor research region at 800 nm, and in the optical communications region at 1500 nm.

A major advance in the generation of ultrashort optical pulses has been the process of mode-locking. Another approach to generating short pulses is the process of pulse compression. Some other novel pulse-generation schemes have been recently developed. We will try to describe them briefly along with applications.

3.5.1 Mode-Locking

Pulses in the picosecond regime can be generated by mode-locking. The simplest way to visualize mode-locked pulses is as a group of photons clumped together and aligned in phase as they oscillate through the laser cavity. Each time they hit the partially transparent output mirror, part of the light escapes as an ultrashort pulse. The clump of photons then makes another round trip through the laser cavity before another pulse is emitted. Thus, the pulses are separated by the cavity round-trip time $2L/c$, where L is the cavity length and c is the speed of light.

Mode-locking requires a laser that oscillates in many longitudinal modes. Thus, it does not work for many gas lasers with narrow emission lines but can be used with argon or krypton ion, solid-state crystalline, semiconductor and dye lasers (which have exceptionally wide-gain bandwidth). The pulse length is inversely proportional to the laser oscillating bandwidth, so dye lasers can generate the shortest pulses because of their exceptionally broad-gain bandwidths.

The number of modes oscillating is limited by the bandwidth Δv over which the laser gain exceeds the loss of the resonator, as shown in Figure 3.23. Unless some mode-selecting element is placed in the laser resonator, the output consists of a sum of frequency components corresponding to the oscillating modes. The electric field may be written as

$$E(t) = \sum_n \alpha_n \exp i\left[(\omega_0 + n\delta\omega)t + f_n\right]$$

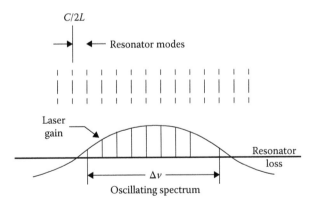

FIGURE 3.23 Resonator modes. The number of modes oscillating is limited by the bandwidth Δv and determined by the gain profile and the resonator loss.

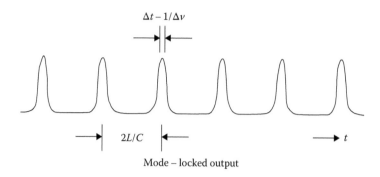

Mode – locked output

FIGURE 3.24   Train of pulses obtained at the output of the laser with all modes locked in the proper phase.

where α_n is the amplitude of the nth mode and $\delta\omega$ is the mode spacing. In general, the laser output varies in time, although the average power remains relatively constant. This is because the relative phases between the modes are randomly fluctuating. However, if the modes are forced to maintain a fixed phase and amplitude relationship, the output of the laser will be a well-defined periodic function of time. In this way, we say that the laser is "mode-locked."

Both continuous and pulsed lasers can be mode-locked. In either case, mode-locking produces a train of pulses separated by the cavity round-trip time, as shown in Figure 3.24.

In this case, the picture corresponds to a single pulse traveling back and forth between the laser resonator mirrors. It is also possible to produce mode-locking with N pulses in the cavity, spaced by a multiple of $c/2L$. The pulses have a width $\Delta\tau$, which is approximately equal to the reciprocal of the total mode-locked bandwidth Δv, and the temporal periodicity is given by $Tp = 2L/c$. The ratio of the pulse width to the period is approximately equal to the number of locked modes.

For pulsed lasers, the result is a series of pulses that fall within the envelope defined by the normal pulse length. Single mode-locked pulses can be selected by passing the pulse train through grating modulators that allow only a single pulse to pass.

The modulation can be active, by changing the transmission of a modulator, or passive, by saturation effects. In either case, interference among the modes produces a series of ultrashort pulses.

Active Mode-Locking

As to date, a shutter than can be inserted into the optical cavity, which opens and closes at the approximate frequency to obtain the desired pulse duration, has not yet been built; some approximations have been developed. For example, an intracavity phase or loss modulator has been inserted into the optical cavity driven at the frequency corresponding to the mode spacing [67]. The principle of active mode-locking by loss modulation is as follows: an optical pulse is likely to form in such a way as to minimize the loss from the modulator. The peak of the pulse adjusts in phase to be at the point of minimum loss from the modulator. However, the slow variation of the sinusoidal modulation provides only a weak mode-locking effect, making this technique unsuitable for generating ultrashort optical pulses. Similarly, phase modulation can also produce mode-locking effects [68].

Active mode-locking is particularly useful for mode-locking Nd: YAG and gas lasers, such as the argon laser [69,70]. Recently, a 10 fs pulse has been generated from a unidirectional Kerr-lens mode-locked Ti:sapphire ring laser [71].

Passive Mode-Locking

Passive mode-locking works by the insertion of a saturable absorbing element inside the optical cavity of a laser. The saturable absorber can be an organic dye, a gas, or a solid, but the first one is the more common. The first optical pulses in the picoseconds time domain [62], as well as the first optical pulses in the femtosecond time domain [72], were obtained by this method.

Passively mode-locked lasers can be divided in two groups: (a) giant pulse lasers and (b) continuous or quasi-continuous lasers. Passive mode-locking was first observed for the first group in ruby lasers [73] and in Nd:glass lasers [62]. Continuous passive mode-locking is observed primarily in dye lasers and was first theoretically described by New in 1972.

1. *Giant Pulse Lasers*

    The optical configuration for a mode-locked giant pulse laser is shown in Figure 3.25. The dye cell is optically contacted with the laser mirror in one end of the cavity in order to reduce the problem of satellite pulses [74–76]. In designing the cavity, it is important to eliminate subcavity resonances and spurious reflections, which may cause the formation of subsidiary pulse trains.

    For a giant pulse to occur, the upper laser level lifetime must be long (as in ruby or Nd:glass lasers), typically hundreds of microseconds. Pulse generation occurs in a highly transient manner in a time much shorter than the upper level population response. The "fluctuation model" proposed by Letokhov in 1969 describes the operation of these lasers. Briefly, the operation is as follows [77]: At the start of the flashlamp pumping pulse, spontaneous emission excites a broad spectrum of laser modes within the optical cavity. Since the modes are randomly phased, a fluctuation pattern is established in the cavity with a periodic structure corresponding to a cavity round-trip time $T = 2L/c$. When the gain is sufficient to overcome the linear and nonlinear losses in the cavity, the laser threshold is reached and the fields in the cavity initially undergo linear amplification. At some point, the field becomes intense enough to enter a phase where the random pulse structure is transformed by the nonlinear saturation of the absorber and by the laser gain saturation. As a result, one of the fluctuation spikes grows in intensity until it dominates and begins to shorten in time. As the short pulse gains intensity, it reaches a point where it begins to nonlinearly interact with the glass host and the pulse begins to deteriorate. At the beginning of the pulse train, as recorded on an oscilloscope, the pulses are a few picoseconds in duration and nearly bandwidth limited [79,80]. Later, pulses in the train undergo self-modulation of phase and self-focusing, which leads to temporal fragmentation of the optical pulse.

    The role of the saturable absorber in the fluctuation model is to select a noise burst that is amplified and ultimately becomes the mode-locked laser pulse. As a consequence, the relaxation time and the absorber, T_a, sets an approximate limit to the duration of mode-locked pulses [80,75]. Kodak dyes A9740 and A9860, with lifetimes of 7 and 11 picoseconds, respectively, are typically used. However, other dyes with shorter lifetimes have been investigated [81,82].

2. *Continuous Lasers*

    In 1981, Fork et al. [83] described the generation of sub-100 femtosecond pulses for the first time, and coined the term "colliding-pulse-mode-locked laser" or CPM laser. The initial CPM laser consisted of a seven-mirror ring cavity with Rhodamine 6G in the gain jet and the dye DODCI in the absorber jet. These same two dyes have been used since 1972 in passively mode-locked cw dye lasers, but only generated picosecond pulses [84].

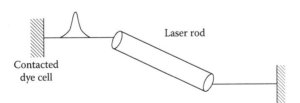

FIGURE 3.25 Optical configuration for a mode-locked giant pulse laser. (With kind permission from **Springer Science+Business Media**: *Ultrashort Laser Pulses, Generation and Applications*, Vol. 60, 1993, Kaiser W [ed.]. With permission of Springer.)

Passive mode-locked continuous lasers involve a very different physics of pulse formation from that of giant pulse lasers. Random noise fluctuations due to longitudinal mode beating occur in the laser until one of the noise spikes is large enough to saturate the absorber. This pulse sees an increased transmission through the absorber jet due to this saturation and then encounters the gain jet. Here, it saturates the gain slightly and reduces the gain for the noise spikes that follow. This selection process continues until only one pulse survives. The pulse then shortens further due to the same saturable effects. Saturable absorption selectively removes energy from the leading edge of the pulse, while saturable gain steepens the trailing edge. The pulse continues to shorten until a pulse-broadening effect, such as dispersion, can balance it [85].

The mode-locked pulse duration is typically much shorter than either the lifetime of the amplifying or gain medium, or the saturable absorber recovery time. In 1972, New [86] described for the first time, the conditions of pulse formation in continuous passively mode-locked lasers. Later, analytical and numerical techniques were applied to describe the transient formation of an ultrashort optical pulse [87–89], while Haus obtained a closed formed solution by assuming a cavity bandwidth and a hyperbolic secant pulse shape [90].

The shortest light pulses reported up to now are only 6 fs long [91]. This corresponds to about three oscillation periods of visible light at $\lambda = 600$ nm.

Table 3.8 shows a short summary of different mode-locking techniques and the typical pulse duration obtained.

Now, we will see some examples of mode-locked pulses generated by different kinds of lasers.

Neodymium lasers. Neodymium lasers can generate mode-locked pulses (generally with continuous excitation), usually of tens to hundreds of picoseconds. The shorter the pulse, the higher the peak power for a given laser rod and pump source; however, shrinking pulse length beyond a certain point can reduce the pulse energy. Pulse lengths are listed in Table 3.9 based on data from industry directories [92].

Because glass can be made in larger sizes and has lower gain than Nd-YAG, glass lasers can store more energy and produce more energetic pulses. However, its thermal problems limit repetition rate and thus lead to low average power. Nd-YLF also can store more energy than Nd-YAG because of its lower cross-section, but it cannot be made in pieces as large as glass. Mode-locked lasers emit series of short pulses, with spacing equal to the round-trip time of the laser cavity. Many of these pulses may be produced during a single flashlamp pulse, and Q-switches can select single mode-locked pulses. Glass lasers have lower repetition rates than Nd-YAG or Nd-YLF, and commercial models typically produce no more than a few pulses per second. Repetition rates may be as low as one pulse every few minutes, or single-shot for large experimental lasers.

TABLE 3.8 Short Summary of Different Mode-Locking Techniques

Technique	Mode Locker	Laser	Typical Pulse Duration	Typical Pulse Energy
Active mode-locking	Acousto-optic modulator	Argon cw	300 ps	10 nJ
		He–Ne cw	500 ps	0.1 nJ
	Pockels cell	Nd:YAG pulsed	100 ps	10 μJ
Passive mode-locking	Saturable absorber	Dye cw	1 ps	1 nJ
		Nd:YAG	1–10 ps	1 nJ
Synchronous pumping	Mode-locked pump laser and matching or resonator length	Dye cw	1 ps	10 nJ
		Color center	1 ps	10 nJ
CPM	Passive mode-locking and eventual synchronous pumping	Ring dye laser	< 100 fs	≈ 1 nJ

TABLE 3.9 Duration and Repetition Rates Available from Mode-Locked Pulsed Neodymium Lasers Operating at 1.06 μm

Type Length	Modulation	Excitation Source	Typical Repetition Rate	Typical Pulse
Nd-glass	Mode-locked	Flashlamp	Pulse trains[a]	5–20 ps
Nd-YAG	Mode-locked	Flashlamp	Pulse trains[a]	20–200 ps
Nd-YAG	Mode-locked	Arc lamp	50–500 MHz	20–200 ps
Nd-YAG	Mode-locked and Q-switched	Arc lamp	Varies[b]	20–200 ps[c]
Nd-YLF	Mode-locked	Arc lamp	Same as YAG	About half YAG
Nd-YLF[d]	Mode-locked	Diode	160 MHz	7 ps
Nd-YLF/glass	Mode-locked and chirped pulse amplification	Lamp	0–2 Hz	1–5 ps

[a] Series of 30–200 ps pulses, separated by 2–10 ns, lasting duration of flashlamp pulse, on the order of a millisecond.
[b] Depends on Q-switch rate; mode-locked pulses.
[c] In pulse trains lasting duration of Q-switched pulses (100–700 ns).
[d] Laboratory results (Juhasz et al. 1990).

Both Nd-glass and Nd-YLF can generate shorter mode-locked pulses, because they have broader spectral bandwidths than Nd-YAG. The shortest pulses are produced by a technique called *chirped pulse amplification,* which is similar to that used to generate femtosecond pulses from a dye laser. A mode-locked pulse, typically from an Nd-YLF laser, is passed through an optical fiber, where nonlinear effects spread its spectrum over a broader range than the laser generates. Then, that pulse is compressed in duration by special optics, and amplified in a broad-bandwidth Nd-glass laser. The resulting pulses are in the 1–3 picosecond range, and when amplified can reach the terawatt (10^{12} W) level in commercial lasers (although their brief duration means that pulse energy is on the order of only a joule). Pulses of 3×10^{12} W have been reported and work has commenced on a 10×10^{12} W laser [93].

Dye lasers. Synchronous mode-locking of a dye laser to a mode-locked pump laser (either rare gas ion or frequency-doubled neodymium) can generate pulses as short as a few hundred femtoseconds. Addition of a saturable absorber can further reduce pulse length. Alternatively, a saturable absorber can by itself passively mode-lock a dye laser pumped by a cw laser. In each case, the saturable absorber allows gain only briefly while it is switching between off and on states.

The commercial dye laser that generates the shortest pulses is a colliding-pulse mode-locked ring laser. The schematic diagram is shown in Figure 3.26. This type of ring laser does not include components to restrict laser oscillation to one direction, and produces pulses going in opposite directions around the ring. One-quarter of the way around the ring from the dye jet, the cavity includes a saturable absorber, which has lowest loss when the two opposite-direction pulses pass through it at the same time (causing deeper saturation and hence lower loss). Pulse lengths can be under 100 fs in commercial versions and tens of femtoseconds in laboratory systems.

Pulses from a dye laser can be further compressed by a two-stage optical system, which first spreads the pulse out in time (by passing it through a segment of optical fiber) and then compresses it spatially (by passing it between a pair of prisms or diffraction gratings). This requires extremely broad bandwidth pulses but can generate pulses as short as 6 fs [91].

The wavelength range available from passively mode-locked dye lasers has been extended with the use of different gain and absorber dyes, and subpicosecond pulses can be generated from below 500 nm to nearly 800 nm [94–96].

Ion lasers. Mode-locked ion lasers can produce trains of picosecond pulses from lasers emitting in multiple longitudinal modes. With the typical 5 GHz line-width of ion laser lines, the resulting pulses are about 90–200 picoseconds long, produced at repetition rates in the 75–150 MHz range.

Diode lasers. Diode lasers can be mode-locked in external cavities, and experimental diode lasers have been made with internal structures that generate mode-locked pulses [92]. For example, a passively

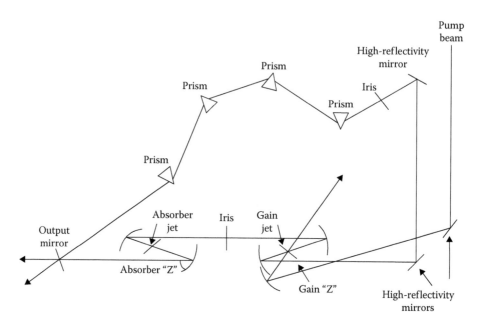

FIGURE 3.26   Schematic of a colliding-pulse mode-locked ring laser. (From Clark Instrumentation Inc.)

mode-locked two-section GaAlAs multiple-quantum-well laser has generated 2.4 ps pulses at 108 GHz. However, mode-locked diode lasers are not available commercially.

Colliding-pulse mode-locking techniques have been applied to form a monolithic quantum-well semiconductor laser [97].

Passive mode-locking with the help of bulk semiconductors and quantum wells was applied to generate femtosecond pulses in color-center lasers [98,99].

Most of the experiments on femtosecond pulses performed up to now have used dye lasers, Ti:sapphire-lasers, or color-center lasers. The spectral ranges were restricted to the regions of optimum gain of the active medium. New spectral ranges can be covered by optical mixing techniques. One example is the generation of 400 fs pulses in the mid-infrared around $\lambda = 5$ μm by mixing the output pulses from a colliding-pulse mode dye laser at 620 nm with pulses of 700 nm in a $LiIO_3$ crystal [100].

During recent years, new techniques have been developed in order to generate still shorter pulses. One of these is *pulse compression.*

3.5.2 Pulse Compression

More than 30 years ago, Gires and Tournois [101] proposed that optical pulses could be shortened by adapting microwave pulse compression techniques to the visible spectrum. The shortest pulse duration that can be produced by a laser oscillator is generally limited by the wavelength of the gain medium and the group velocity dispersion into the cavity. However, if we give enough initial peak power, the technique of pulse compression can produce pulses one order of magnitude shorter. The physics involved in pulse compression, called *self-phase modulation,* also plays an important role in the majority of the ultrashort laser oscillators.

The general principles of pulse compression were first applied to radar signals and later (in the 1960s) to optical signals [102] (see also Johnson and Shank [103], for a review of pulse compression). Pulse compression technique consists of two steps: in the first step, a frequency sweep or "chirp" is impressed on the pulse; in the second step, the pulse is compressed by using a dispersive delay line.

Single-mode optical fibers were first used to create a frequency chirp [104]. The chirp can be impressed on an intense optical pulse by passing the pulse through an optical Kerr medium [105]. When an intense optical pulse is assed through a nonlinear medium, the refractive index, n, is modified by the electric field, E, as follows [106]:

$$n = n_o + n_2 \langle E^2 \rangle + \dots$$

A phase change, δf, is impressed on the pulse:

$$\delta f \approx n_2 \langle E^2 \rangle (\omega z / c),$$

where ω is the frequency, z is the distance traveled in the Kerr medium, and c is the velocity of light.

As the intensity of the leading edge of the optical pulse rises rapidly, a time-varying phase or frequency sweep is impressed on the pulse carrier. Similarly, a frequency sweep in the opposite direction occurs as the intensity of the pulse falls on the trailing edge. This frequency sweep is given by

$$\delta \omega \approx (\omega z n_2 / c) d/dt \langle E^2(t) \rangle,$$

For a more rigorous approach to this problem, see Shank [106], pp. 26–29.

As optical fiber waveguides are a nearly ideal optical Kerr medium for pulse compression [107,108], one of the most common configurations of a pulse compressor uses self-phase modulation in an optical fiber to generate the chirped pulse and a pair of gratings aligned parallel to each other as the dispersive delay line.

Consider a Gaussian pulse

$$E(t)_{\text{input}} = A \exp\left[-(t/b)^2\right] \exp(i \omega_0 t)$$

The effect of the pulse compressor on the Gaussian pulse has been calculated in the limited of no group dispersion velocity by Kakfa and Baer in 1988. The self-phase modulation in the fiber transforms this pulse to

$$E(t)_{\text{fiber}} = A \exp\left[-(t/b)^2\right] \exp i \left[\omega_0 t + \Theta(I)\right],$$

where $\Theta(I)$ is an intensity-dependent shift in the phase of the carrier. We can Fourier transform the pulse $E(t)$ to the frequency domain $E(\omega)$ in order to apply the grating operator. This operator causes a time delay, which depends on the instantaneous frequency of the pulse:

$$E(\omega)_{\text{compressed}} = E(\omega)_{\text{fiber}} \exp[-if(\omega)].$$

The final pulse $E(t)_{\text{compressed}}$ is obtained by taking the Fourier transform of $E(\omega)_{\text{compressed}}$. The intensity can be calculated at any point by taking the square of this field.

The experimental set-up used for pulse compression in the femtosecond time domain is shown in Figure 3.27.

Mollenauer et al. [107] were the first who experimentally investigated pulse compression using optical fibers as a Kerr medium. They worked on the soliton compression of optical pulses from a color-center laser. As the wavelength of the optical pulscs at $\lambda = 1.3$ μm was in the anomalous or negative dispersion

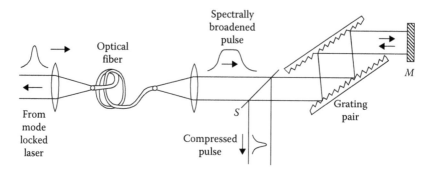

FIGURE 3.27   Diagram of the experimental arrangement for pulse compression in the femtosecond time domain.

region, a separate compressor was not needed, because the dispersive properties of the fiber material self-compressed the pulse. Using this compression technique, a 7 ps optical pulse was compressed to 0.26 ps with a 100 m length of single mode fiber [109].

Later in 1984, Mollenauer and Stolen extended the ideas of fiber soliton pulse compression to form a new type of mode-locked color-center laser: the soliton laser [110]. In the soliton laser, an optical fiber of the appropriate length is added to the feedback path of a synchronously pumped color-center laser. The soliton proeprties of the fiber feedback force the laser itself to produce pulses of a definite shape and width. With the soliton laser, pulses of less than 50 fs can be obtained.

The soliton mode-locking technique has been extended to the wavelength region where solitons cannot form. This new technique is called *additive pulse mode-locking* and has been described both theoretically [111] and experimentally [112–115].

Nakatsuka and Grischkowosky [104] worked in the positive group velocity dispersion regime ($\lambda \leq 1.3$ µm) using an optical fiber for the chirping process and anomalous dispersion from an atomic vapor as the compressor. Shank et al. [116] replaced the atomic vapor compressor with a grating pair compressor [117] and achieved pulse compression to 30 fs optical pulse width.

3.5.3 Applications of Ultrashort Optical Pulses

Measurement of Subpicosecond Dynamics

One application of optical pulses is to measure the relaxation of elementary excitations in matter, via the nonlinear optical response of the system under study. The time scale and form of the relaxation gives information concerning the microscopic physics of the dissipative forces acting on the optically active atom or molecule. The development of coherent light sources that produce pulses of a few tens of femtoseconds duration has greatly extended the possibility of realizing such measurements in liquid and condensed phases of matter, where relaxation time scales are of the order of pico- or femtoseconds [85].

There are several methods for the measurement of lifetimes of excited atomic or molecular levels, including phase-shift method [31,118,119] and the delayed-coincidence technique [120].

For measurements of very fast relaxation processes with a demanded time resolution below 10^{-10} s, the pump and probe technique is the best choice [121,31]. In this technique, the molecules under study are excited by a fast laser pulse on the transition from the base state to the first excited state, as shown in Figure 3.28. A probe pulse with a variable time delay τ against the pump pulse probes the time evolution of the population density $N_1(t)$. The time resolution is only limited by the pulse width ΔT of the two pulses but not by the time constants of the detectors.

Strong-Field Ultrafast Nonlinear Optics

The high peak powers obtained using short pulses have impacted significantly in the field of multiphoton atomic ionization and molecular dissociation. Three important applications are molecular

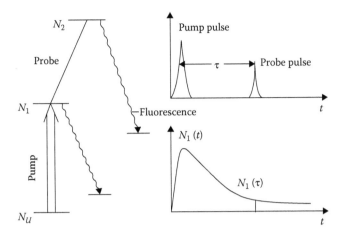

FIGURE 3.28 Pump and probe technique.

photodissociation [122,123], above-threshold ionization [124,125] and the generation of high-order harmonics [126–128].

Study of Biological Processes

Ultrafast laser techniques have been applied to the study of biological processes, such as heme protein dynamics, photosynthesis, and the operation of rhodopsin and bacteriorhodopsin [129].

For other examples of applications see Khoo et al. [37]. Recent applications include the determination of the magnitude and time response of the nonlinear refractive index of transparent materials using spectral analysis after nonlinear propagation [130]; the characterization of ultrafast interactions with materials through the direct measurement of the optical phase [131] and a proposal for the generation of subfemtosecond VUV pulses from high-order harmonics [132].

References

1. Wilson J, Hawkes JFB. *Optoelectronics: An Introduction*, 2nd edn, Prentice Hall International Series in Optoelectronics UK: Prentice Hall International, 1989.
2. Loudon R. *The Quantum Theory of Light*, 2nd edn, Oxford Science Publications, USA: Oxford University Press, 1982.
3. Walls DF. *Am. J. Phys.*, **45**, 952; 1977.
4. Walls DF, Milburn GJ. *Quantum Optics*, Berlin: Springer-Verlag, 1994.
5. Baldwin GC. *An Introduction to Nonlinear Optics*, New York, NY: Plenum Press, 1969.
6. Yariv A. *Quantum Electronics*, 3rd edn, Chichester: John Wiley & Sons, 1989.
7. Kleinman DA. *Phys. Rev.*, **126**, 1977; 1962.
8. Koechner W. *Solid State Laser Engineering*. New York, NY: Springer Science+Business Media, 2006, p. 588, 665.
9. Milonni PW, Eberly JH. *Laser Physics*, Hoboken, New Jersey: John Wiley & Sons, p. 479, 647, 652, 2010.
10. Zyss J, Ledoux I, Nicoud JF. Advances in molecular engineering for quadratic nonlinear optics, In *Molecular Nonlinear Optics: Materials, Physics and Devices*, Zyss, J. (ed), *Quantum Electronics, Principles and Applications*, New York, NY: Academic Press, 1994.
11. Bloembergen N. *Am. J. Phys.*, **35**, 989; 1967.
12. Penzkofer AA, Laubereau A, Kaiser W. *Quantum Electron.*, **6**, 55; 1979.
13. Mostowsky J, Raymer MG. Quantum statistics in nonlinear optics, In *Contemporary Nonlinear Optics*, Agrawal, G.P. and R.W. Boyd (eds), *Quantum Electronics, Principles and Applications*, New York, NY: Academic Press, 1992.

14. Fabelinskii L. *Molecular Scattering of Light*, New York, NY: Plenum Press; 1968.
15. Eckhardt G, Hellwarth RW, McClung FJ, Schwarz SE, Weiner D, Woodbury EJ. *Phys. Rev. Lett.*, **9**, 455; 1962.
16. Kaiser W, Maier M. In *Stimulated Rayleigh, Brillouin and Raman Spectroscopy* in *Laser Handbook*, Arecchi, F. T. and Schulz-Dubois EO (eds), Amsterdam: North Holland, 1972.
17. Raymer MG, Walmsley IA, *The Quantum Coherence Properties of Stimulated Raman Scattering* in *Progress in Optics*, Wolf E, (ed.) Vol. 28, Amsterdam: North-Holland, 1990.
18. Mitik-Dineva N, Stoddart PR, Crawford RJ, Ivanova E. Bacterial cell interactions with optical fiber surfaces, In *Fiber Lasers: Research, Technology and Applications*, M. Kimura (ed), 2009, p. 81, 82.
19. Haynes CL, Yonzon CR, Zhang X, Van Duyne RP. Surface-enhanced Raman sensors: Early history and the development of sensors for quantitative biowarfare agent and glucose detection, *J. Raman Spectrosc.*, **36**, 471; 2005.
20. Murphy T, Lucht S, Schmidt H, Kroenfeldt HD. Surface-enhanced Raman scattering (SERS) system for continuous measurements of chemicals in sea-water, *J. Raman Spectrosc.*, **31**(10), 943; 2000.
21. Stuart DA, Yuen JM, Shan N, et al. In vivo glucose measurement by surface-enhanced Raman spectroscopy, *Anal. Chem.*, **78**, 7211; 2006.
22. Pask HM, Myers S, Piper JA, Richards J, McKay T. *Opt. Lett.*, **28**, 435; 2003.
23. Keller MD, Kanter EM, Mahadevan-Jansen A. Raman spectroscopy for cancer diagnosis, *Spectroscopy*, **21**(11), 33–41; 2006.
24. Petry R, Schimitt M, Popp J. Raman spectroscopy—A prospective tool in the life sciences, *Chem. Phys. Chem.*, **4**(1), 14–30; 2003.
25. Carey PR. Raman spectroscopy, the sleeping giant in structural biology, awakes, *J. Biol. Chem.*, **274**(38), 26625–26628; 1999.
26. Schuster KC, Reese I, Urlaub E, Gapes JR, Lendl B. Multidimensional information on the chemical composition of single bacterial cells by confocal Raman microspectroscopy, *Anal. Chem.*, **72**(22), 5529–5534; 2000.
27. Gessner R, Rosch P, Kiefer W, Popp J. Raman spectroscopy investigation of biological materials by use of etched and silver coated glass fiber tips, *Biopolymers*, **67**(4–5), 327–330; 2002.
28. Rosch P, Harz M, Schmitt M, Peschke KD, Ronneberger O, Burkhardt H. Chemotaxonomic identification of single bacteria by micro-Raman spectroscopy: Applications to clean-room-relevant biological contaminations, *Appl. Environ. Microbiol.*, **71**(3), 1626–1637; 2005.
29. Xie C, Mace J, Dinno MA, Li YQ, Tang W, Newton RJ. Identification of single bacterial cells in aqueous solution using confocal laser tweezers Raman spectroscopy, *Anal. Chem.*, **77**(14), 4390–4397; 2005.
30. Boyd RW. *Nonlinear Optics*, Chapter 7, New York, NY: Academic Press, 1992.
31. Demtröder W. *Laser Spectroscopy: Basic Concepts and Instrumentation*, 2nd edn, Berlin: Springer-Verlag, 1996.
32. Ederer DL, McGuire JH. (ed.), *Raman Emission by X-Ray Scattering*, Singapore: World Scientific Publishing Company, 1996.
33. Barman I, Dingari NC, et al. Application of Raman Spectroscopy to Identify Microcalcifications and Underlying Breast Lesions at Stereotactic Core Needle Biopsy in *Cancer Res.* 73(11), p. 3206–3215, 2013.
34. Buckley K, Kerns JG, Gikas PD, et al. Measurement of abnormal bond composition *in* vivo using noninvasive Raman spectroscopy, IBMS BoneKEy 11, Article number: 602; 2014 | doi:10.1038/bonekey.2014.97 & 2014 International Bone & Mineral Society.
35. Chu B. *Laser Light Scattering: Basic Principles and Practice*, 2nd edn, New York, NY: Academic Press, 1991.
36. Bloembergen N. *Nonlinear Optics*, 4th edn, Singapore: World Scientific Publishing Company, 1996.
37. Khoo IC, Lam JF, Simoni F. (eds.), *Nonlinear Optics and Optical Physics, Series in Nonlinear Optics*, Vol. 2, Singapore: World Scientific Publishing Company, 1994.

38. Keller O. *Notions and Perspectives of Nonlinear Optics, Series in Nonlinear Optics*, Vol. 3, London: World Scientific Publishing Company, 1996.
39. Kaiser W, Garret CG. *Phys. Rev. Lett.*, **7**, 229; 1961.
40. Hopfield JJ, Worlock JM, Park K. *Phys. Rev. Lett.*, **11**, 414; 1963.
41. Bräunlich P. Multiphoton spectroscopy, In *Progress in Atomic Spectroscopy*, Hanle W, Kleinpoppen H (eds), New York, NY: Plenum, 1978.
42. Worlock JM. Two-photon spectroscopy, In *Laser Handbook*, Arecchi FT, Schulz-Dubois EO (eds), Amsterdam: North-Holland, 1972.
43. Shimoda K (ed.). *High Resolution Laser Spectroscopy*, Berlin: Springer-Verlag, 1976.
44. Letokhov VS, Chebotayev VP. *Nonlinear Laser Spectroscopy*, Berlin: Springer-Verlag, 1977.
45. Vasilenko LS, Chebotayev VP, Shishaev AV. *JETP Lett.*, **12**, 161; 1973.
46. Cagnac B, Grynberg G, Biraben F. *J. Phys.*, **34**, 845; 1973.
47. Grynberg G, Cagnac B, Biraben F. Multiphoton resonant processes in atoms, In *Coherent Nonlinear Optics: Recent Advances,* Feld MS, Letokhov VS (eds), *Serie Topics in Current Physcis*, Berlin: Springer-Verlag, 1980.
48. Biraben F, Cagnac B, Grynberg G. *Phys. Rev. Lett.*, **32**, 643; 1974.
49. Levenson MD, Bloembergen N. *Phys. Rev. Lett.*, **32**, 645; 1974.
50. Levenson MD. High-resolution imaging by wave-front conjugation, *Opt. Lett.*, **5**, 182; 1980.
51. Giacobino E, Biraben F, Grynberg G, Cagnac B. *J. Physique*, **83**, 623; 1977.
52. Morellec J, Normand D, Petite G. *Phys. Rev. A*, **14**, 300; 1976.
53. Letokhov VS (ed.). *Tunable Lasers and Applications*, Berlin: Springer-Verlag, 1976.
54. Yariv A. *Optical Electronics*, 4th edn, USA: International Edition, 1990.
55. Fisher RA. *Optical Phase Conjugation*, New York, NY: Academic Press, 1983.
56. Hellwarth RW. Third-order susceptibilities of liquids and gases, *Progr. Quant. Electron.*, **5**, 1; 1977.
57. Jain RK, Lind RC. Degenerate four-wave mixing in semiconductor-doped glass, *J. Opt. Soc. Am.*, special issue on phase conjugation, **73**, p. 647–653, 1983.
58. Bloom DM, Bjorklund GC. Conjugate wavefront generation and image reconstruction by four-wave mixing, *Appl. Phys. Lett.*, **31**, 592; 1977.
59. Hopf FA. Interferometry using conjugate-wave generation, *J. Opt. Soc. Am.*, **70**, 1320; 1980.
60. Hellwarth RW, Christensen P. Nonlinear optical microscope using SHG, *Appl. Opt.*, **14**, 247; 1974.
61. Pepper DM, AuYeung J, Fekete D, Yariv A. Spatial convolution and correlation of optical fields via degenerate four-wave mixing, *Opt. Lett.*, **3**, 7; 1978.
62. DeMaria AJ, Stetser DA, Heyman H. *Appl. Phys. Lett.*, **8**, 22; 1966.
63. Shapiro SL (ed.). *Ultrashort Laser Pulses*, Berlin: Springer-Verlag, 1977.
64. Von der Linde D, Laubereau A, Kaiser W. *Phys. Rev. Lett.*, **26**, 954; 1971.
65. Alfano RR, Shapiro SL. *Phys. Rev. Lett.*, **26**, 1247; 1977.
66. Feld MS, Letokhov VS (eds.). *Coherent Nonlinear Optics: Recent Advances*, Berlin: Springer-Verlag, 1980.
67. Hargrove LE, Fork RL, Pollack MA. *Appl. Phys. Lett.*, **5**, 4; 1964.
68. Harris SE, Targ R. *Appl. Phys. Lett.*, **5**, 205; 1964.
69. Smith PW. *Proc. IEEE*, **58**, 1342; 1970.
70. Harris SE. *Proc. IEEE*, **54**, 1401; 1966.
71. Kasper A, Witte KJ, 10-fs Pulse generation from a unidirectional Kerr lens mode-locked Ti:sapphire ringe laser" in *Generation, Amplification, and Measurement of Ultrashort Laser Pulses III*, White WE, Reitze DH (eds). Proc. SPIE **2701**, 2; USA: The International Society for Optical Engineering, 1996.
72. Fork RL, Greene BI, Shank CV. *Appl. Phys. Lett.*, **38**, 671; 1981.
73. Mocker H, Collins R. *Appl. Phys. Lett.*, **7**, 270; 1965.
74. Weber H. *J. Appl. Phys.*, **39**, 6041; 1968.
75. Bradley D, New GHC, Caughey S. *Phys. Lett.*, **30A**, 78; 1970.
76. Von der Linde D. *IEEE J. Quant. Electron.*, **QE-8**, 328; 1972.

77. Kaiser W. (ed.), *Ultrashort Laser Pulses, Generation and Applications, Topics in Applied Physics*, Vol. 60, 2nd edn, Berlin: Springer Verlag, 1993.
78. Von der Linde D, Bernecker O, Kaiser W. *Opt. Commun.*, **2**, 149; 1970.
79. Zinth W, Lauberau A, Kaiser W. *Opt. Commun.*, **22**, 161; 1977.
80. Garmire E, Yariv A. *IEEE J. Quant. Electron.*, **QE-3**, 222; 1967.
81. Kopinsky B, Kaiser W, Drexhage K. *Opt. Commun.*, **32**, 451; 1980.
82. Alfano RR, Schiller N, Reynolds G. *IEEE J. Quant. Electron.*, **QE-17**, 290; 1981.
83. Bloom DM, Liao PF, Economou NP. Observation of amplified reflection by degenerate four-wave mixing in atomic sodium vapor, *Opt. Lett.*, **2**, 58; 1978.
84. Ippen EP, Shank CV, Dienes A. *Appl. Phys. Lett.*, **21**, 348; 1972.
85. Walmsley IA, Kafka JD. Ultrafast nonlinear optics, In *Contemporary Nonlinear Optics*, Agrawal GP, Boyd RW (eds). New York, NY: Academic Press, 1992.
86. New GHC. *Opt. Commun.*, **6**, 188; 1972.
87. Garside B, Lim T. *J. Appl. Phys.*, **44**, 2335; 1973.
88. Garside B, Lim T. *Opt. Commun.*, **12**, 8; 1974.
89. New GHC, Rea DH. *J. Appl. Phys.*, **47**, 3107; 1976.
90. Haus H. *IEEE J. Quant. Electron.*, **QE-11**, 736; 1975.
91. Fork RL, BritoCruz CH, Becker PC, Shank CV. *Opt. Lett.*, **12**, 483; 1987.
92. Hecht J. *The Laser Guidebook*, 2nd edn, New York, NY: McGraw-Hill, 1992.
93. Perry MD. *Energy & Technology Review*, Livermore, California: Lawrence Livermore National Laboratory (review), 1988, p. 9.
94. French PMW, Taylor JR. *Ultrafast Phenomena V*, Berlin: Springer-Verlag, 1986, p. 11.
95. Smith K, Langford N, Sibbett W, Taylor JR. *Opt. Lett.*, **10**, 559; 1985.
96. French PMW, Taylor JR. *Opt. Lett.*, **13**, 470; 1988.
97. Wu MC, Chen YK, Tanbun-Ek T, Logan RA, Chin MA, Raybon G. *Appl. Phys. Lett.*, **57**, 759; 1990.
98. Islam MN, Sunderman ER, Soccoligh CE, et al. *IEEE J. Quant. Electron.*, **QE-25**, 2454; 1989b.
99. Islam MN, Sunderman ER, Bar-Joseph I, Sauer N, Chang TY. *Appl. Phys. Lett.*, **54**, 1203; 1989a).
100. Elsâsser TH, Nuss MC. *Opt. Lett.*, **16**, 411; 1991.
101. Gires F, Tournois P. *C. R. Acad. Sci. Paris*, **258**, 6112; 1964.
102. Giordmine JA, Duguay MA, Hansen JW. *IEEE J. Quant. Electron.*, **QE-4**, 252; 1968.
103. Johnson AM, Shank CV. *The Supercontinuum Laser Source*, Chapter 10, New York, NY: Springer-Verlag, 1989.
104. Nakatsuka H, Grischkowsky D. *Opt. Lett.*, **6**, 13; 1981.
105. Fisher RA, Kelly PL, Gustafson TK. *Appl. Phys. Lett.*, **37**, 267; 1969.
106. Shank CV. In *Ultrashort Laser Pulses. Generation and Applications*, Kaiser, W. (ed.), 2nd edn, *Topics in Applied Physics*, Vol. **60**, Berlin: Springer-Verlag, 1993.
107. Mollenauer LF, Stolen RH, Gordon JP. *Phys. Rev. Lett.*, **45**, 1095; 1980.
108. Nakatsuka H, Grischkowsky D, Balant AC. *Phys. Rev. Lett.*, **47**, 1910; 1981.
109. Mollenauer LF, Stolen RH, Gordon JP, Tomlinson WJ. *Opt. Lett.*, **8**, 289; 1983.
110. Mitschke FM, Mollenauer LF. *IEEE J. Quant. Electron.*, **QE-22**, 2242; 1986.
111. Ippen EP, Haus HA, Liu LY. *J. Opt. Soc. Am.*, **6**, 1736; 1989.
112. Goodberlet J, Wang J, Fujimoto JG, Schultz PA. *Opt. Lett.*, **14**, 1125; 1989.
113. French PMW, Williams JAR, Taylor JR. *Opt. Lett.*, **14**, 686; 1989.
114. Zhu X, Kean PN, Sibbett W. *Opt. Lett.*, **14**, 1192; 1989.
115. Zhu X, Sibbett W. *J. Opt. Soc. Am.*, **B7**, 2187; 1990.
116. Shank CV, Fork RL, Yen R, Stolen RH, Tomlinson WJ. *Appl. Phys. Lett.*, **40**, 761; 1982.
117. Treacy EB. *IEEE J. Quant. Electron.*, **QE-5**, 454; 1969.
118. Lakowvicz JR, Malivatt BP. *Biophys. Chem.*, **19**, 13; 1984a.
119. Lakowvicz JR, Malivatt BP. *Biophys. J.*, **46**, 397; 1984b.

120. O'Connor DV, Phillips D. *Time Correlated Single Photon Counting*, New York, NY: Academic Press, 1984.

121. Lauberau A, Kaiser W. Picosecond investigations of dynamic processes in polyatomic molecules and liquids, In *Chemical and Biochemical Applications of Lasers II*, Moore CB (ed.). New York, NY: Academic Press, 1977.

122. Scherer NF, Knee J, Smith D, Zewail AH. *J. Phys. Chem.*, **89**, 5141; 1985.

123. Dantus M, Rosker MJ, Zewail AH. *J. Chem. Phys.*, **37**, 2395; 1987.

124. Agostini P, Fabre F, Mainfray G, Petite G, Rahman NK. *Phys. Rev. Lett.*, **42**, 1127; 1979.

125. Gontier Y, Trahin M. *J. Atom. Mol. Phys. B*, **13**, 4381; 1980.

126. Rhodes CK. *Physica Scripta*, **717**, 193; 1987.

127. Li X, L'Huillier A, Ferray M, Lompre L, Mainfray G, Manus C. *J. Phys. B*, **21**, L31; 1989.

128. Kulander K, Shore B. *Phys. Rev. Lett.*, **62**, 524; 1989.

129. Hochstrasser RM, Johnson CK. Biological Processes Studied by Ultrafast Laser Techniques, in *Ultrashort Laser Pulses, Generation and Applications,* Topics in Applied Physics, Vol. **60**, 2nd edn, Berlin: Springer Verlag, 1993.

130. Nibbering ETJ, Lange HR, Ripoche JF, Le Blanc C, Chambaret JP, Franco M. In *Generation, Amplification and Measurement of Ultrashort Laser Pulses III,* White, W. E. and D. H. Reitze (eds), *Proc. SPIE,* **2701**, 229; USA: The International Society for Optical Engineering, 1996.

131. Clement TS, Rodriguez G, Wood WM, Taylor AJ, in *Generation Amplification and Measurement in Ultrashort Laser Pulses III,* White WE, Reitze DH (eds). *Proc. SPIE,* **2701**, 229; USA: The International Society for Optical Engineering, 1996.

132. Schafer KJ, Squier JA, Barty CPJ. Proposed generation of subfemtosecond VUV pulses from high-order harmonics" in *Generation, Amplification and Measurement of Ultrashort Laser Pulses III,* White, WE and Reitze DH (eds), *Proc. SPIE,* **2701**, 248; USA: The International Society for Optical Engineering, 1996.

Other References Mentioned but Not Cited

Bloch, D, Giacobino E, Ducloy M, Laser Spectroscopy V in *5th Int. Conf. Laser Spectrosc.*, McKellar AR, Oka T, Stoicheff BP (eds), Jasper, Alberta, Berlin, Heidelberg: Springer-Verlag, 1981.

Bodgan AR, Prior Y, Bloembergen N. Pressure-induced degenerate frequency resonance in four-wave light mixing, *Opt. Lett.*, **6**, 82; 1981.

Clements WRL, Stoicheff BP. *Appl. Phys. Lett.*, **12**, 246; 1968.

Cronin-Golomb M, Fischer B, White JO, Yariv A. Theory and applications of four wave mixing in photorefractive media, *IEEE J. Quant. Electron.*, **QE-20**(1), 12; 1984.

Dunning GJ, Lind RC. Demonstration of image transmission through fibers by optical phase conjugation, *Opt. Lett.*, 558; 1982.

Economou NP, Liao PF. Magnetic-field quantum beats in two-photon free-induction decay, *Opt. Lett.*, **3**, 172; 1978.

Eichler HJ. Laser-induced grating phenomena, *Opt. Acta.*, **24**, 631; 1977.

Fleischman M, Hendra PJ, McQuillan AJ. Raman spectra of pyridine adsorbed at a silver electrode, *Chem. Phys. Lett.*, **26**(2), 163–166; 1974.

Franken PA, Hill AE, Peters CW, Weinreich G. *Phys. Rev. Lett.*, **7**, 118; 1961.

Fu TY, Sargent III M. Effects of signal detuning on phase conjugation, *Opt. Lett.*, **4**, 366; 1979.

Fu TY, Sargent III M. Theory of two-photon phase conjugation, *Opt. Lett.*, **5**, 433; 1980.

Fujita M, Nakatsuka H, Nakanishi H, Matsuoka M. Backward echo in two-level systems, *Phys. Rev. Lett.*, **42**, 974; 1979.

Göppert-Mayer M. *Ann. Physik*, **9**, 273; 1931.

Hamilton DS, Heiman D, Feinberg J, Hellwarth RW. Spatial diffusion measurements in impurity-doped solids by degenerate four-wave mixing, *Opt. Lett.*, **4**, 124; 1979.

Hanbury-Brown R, Twiss RQ. *Nature*, **177,** 27; 1957. See also (1957) *Proc. R. Soc.*, **A242**, 300 (1957); *ibid.*, **A243**, 291 (1958).

Harter DJ, Boyd RW. Nearly degenerate four-wave mixing enhanced by the ac Stark effect, *IEEE J. Quant. Electron.*, **QE-16**, 1126; 1980.

Humphrey LM, Gordon JP, Liao PF. Angular dependence of line shape and strength of degenerate four-wave mixing in a Doppler-broadened system with optical pumping, *Opt. Lett.*, **5**, 56; 1980.

Juhasz T, Lai ST, Pessot MA. *Opt. Lett.*, **15**(24), 1458; 1990.

Kakfa JD, Baer T. *IEEE J. Quant. Electron.*, **QE-24**, 341; 1988.

Kato Y, Takuma H. *J. Chem. Phys.*, **54**, 5398; 1971.

Khoo IC. Degenerate four-wave mixing in the nematic phase of a liquid crystal, *Appl. Phys. Lett.*, **38**, 123; 1981.

Lam JF, Steel DG, McFarlane RA, Lind RC. Atomic coherence effects in resonant degenerate four-wave mixing, *Appl. Phys. Lett.*, **38**, 977; 1981.

Letokhov V. *Soviet Phys. JETP*, **28**, 562, 1026; 1969.

Levenson MD, Johnson KM, Hanchett VC, Chaing K. Projection photolithography by wavefront conjugation, *J. Opt. Soc. Am.*, **71**, 737; 1981.

Liao PF, Economou NP, Freeman RR. Two-photon coherent transient measurements of doppler-free linewidths with broadband excitation, *Phys. Rev. Lett.*, **39**, 1473; 1977.

Mollenauer LF, Stolen RH. *Opt. Lett.*, **9**, 13; 1984.

Moss SC, Lindle JR, Mackey HJ, Smirl AL. Measurement of the diffusion coefficient and recombination effects in Germanium by diffraction from optically-induced picosecond transient gratings, *Appl. Phys. Lett.*, **39**, 227; 1981.

Nilsen J, Yariv A. Nearly degenerate four-wave mixing applied to optical filters, *Appl. Opt.*, **18**, 143; 1979.

Nilsen J, Yariv A. Nondegenerate four-wave mixing in a Doppler-broadened resonant medium, *J. Opt. Soc. Am.*, **71**, 180; 1981.

Nilsen J, Gluck NS, Yariv A. Narrow-band optical filter through phase conjugation by nondegenerate four-wave mixing in sodium vapor, *Opt. Lett.*, **6**, 380; 1981.

Palmer AJ. Nonlinear optics in radiatively cooled vapors, *Opt. Commun.*, **30**, 104; 1979.

Raj RK, Bloch D, Snyder JJ, Carney G, Ducloy M. High-frequency optically heterodyned saturation spectroscopy via resonant degenerate four-wave mixing, *Phys. Rev Lett.*, **44**, 1251; 1980b.

Raj RK, Bloch D, Snyder JJ, Carney G, Ducloy M. High-sensitvity nonlinear spectroscopy using a frequency-offset pump, *Opt. Lett.*, **5**, 163, 326; 1980a.

Saikan S, Wataka H. Configuration dependence of optical filtering characteristics in backward nearly degenerate four-wave mixing, *Opt. Lett.*, **6**, 281; 1981.

Steel DG, Lind RC, Lam JF. Degenerate four-wave mixing in a resonant homogeneously broadened system, *Phys. Rev. A*, **23**, 2513; 1981b.

Steel DG, Lam JF, McFarlane RA. *5th International Confernce on Laser Spectroscopy*, Jasper, Alberta, 1981a.

Steel DG, Lam JF. Two-photon coherent-transient measurement of the nonradiative collisionless dephasing rate in SF6 via doppler-free degenerate four-wave mixing, *Phys. Rev. Lett.*, **43**, 1588; 1979.

Steel DG, Lind RC. Multiresonant behavior in nearly degenerate four-wave mixing: The ac Stark effect, *Opt. Lett.*, **6**, 587; 1981.

Wandzura SM. Effects of atomic motion on wavefront conjugation by resonantly enhanced degenerate four-wave mixing, *Opt. Lett.*, **4**, 208; 1979.

Woerdman JP, Schuurmans MFH. Effect of saturation on the spectrum of degenerate four-wave mixing in atomic sodium vapor, *Opt. Lett.*, **6**, 239; 1981.

Woodbury EJ, Ng WK. *Proc. IRE.*, **50**, 2367; 1962.

Yamada K, Fukuda Y, Haski T. Time development of the population grating in Zeeman sublevels in sodium vapor-detection of Zeeman quantum beats, *J. Phys. Soc. Jpn.*, **50**, 592; 1981.

4

Ray Tracing

Ricardo Flores-
Hernández
and Armando
Gómez-Vieyra

4.1 Introduction

The procedure of optical system layout is the initial determination of a set of optical components (lenses, mirrors, prisms, etc., as to type and power), and their arrangement or spacing to conform a desired optical system. This must be done in such a way that the system meets some basic operation requirements: the production of an image of desired size and orientation at the desired location, the complete system to fit into some available space, and so on. Bearing in mind from the beginning that the optical system will have the capability of providing the required resolution (spot size, MTF, etc.), once its final version is properly reached by means of some sort of optimization process.

This chapter is intended to provide the tools necessary to determine the ideal localization, size and orientation of the image formed by an optical system. These tools are the basic paraxial equations, which cover the relationships involved in the mentioned task. The word "paraxial" (almost axial) is synonymous with "first-order" and "Gaussian" used in other texts.

This chapter is a review of several paraxial ray tracing methods. First and although several books already exist that cover at least one conventional paraxial ray tracing method, in this chapter the span is somewhat wider: to present most if not all of the paraxial ray trace methods available in the literature. It is hoped that such wider knowledge will constitute a solid background and the skills necessary to understand and operate correctly modern optical design programs. Such a "launch pad" is highly required in regular professional optical design work. The main emphasis is on optical imaging ray tracing theory, provided the reader has some basic physics knowledge, typical for any engineering graduate.

4.1.1 Paraxial Rays

For the purpose of Geometrical Optics, the "Ray" concept becomes extremely useful, since many concepts and calculations are significantly simplified by means of it [1–4]. Along with it, we will frequently employ the expressions "pencils of rays" or "ray bundles" to denote sets of rays emitted from a point source or converging toward a point image. When either one of these points, or both, are located infinitely far away, these sets of rays will be parallel to each other. If additionally, associated object and/or image points are located on a common straight line and infinitely far away, bundle rays will be parallel to the given straight line. An optical image (a set of image points, usually on a plane surface) is the apparent reproduction of an object produced by a system of optical elements such as lenses, mirrors, prisms, fiber bundles, and so on.

Even though "ray" is an extremely simplified representation of light, there is no loss of generality since, as demonstrated by Lord Hamilton [1], Snell's law of refraction and reflection is fulfilled both by rays and wavefronts as long as the apertures traversed by them are much larger than the wavelength of light, which usually happens.

In general, both Paraxial Optics and Geometrical Optics are the fields of optics that employ the simplest representations of the interaction of light with matter. Both are powerful tools, which support all other branches of Optics, since by their application all initial calculations can be readily performed, whichever the experiment and whichever the optical system. In many cases, correct results are obtained through their simple equations and principles, exact enough to avoid going into more complicated studies of light propagation through optical media and instruments. In some cases (those strongly affected by coherence, polarization, or diffraction), after the appropriate geometrical optics approach, little additional analysis will be required. Some such cases are: laser cavity and laser systems design, coherent wavefront optical system design, extremely small aperture optical system design, polarized light optical system design, large aperture telescopes design (specially segmented and adaptive ones). In all these cases, the initial setup analysis is performed by means of simple paraxial methods. Subsequently, an intermediate analysis must be carried out by geometrical optics equations, including third-order aberration calculations, since they provide imaging behavior involving 80% (or more) of the luminous energy conveyed by the optical system onto each image point.

4.1.2 Snell's Paraxial Law

Most optical systems consist of a succession of reasonably homogeneous-isotropic media with constant refractive indices, separated by abrupt discontinuities (surfaces), for example, glass lenses surrounded by air. Refractive index is defined as

$$n = c / v, \tag{4.1}$$

where c = speed of light in vacuum, and v = speed of light in the medium.

When light passes through the interface between two different media, it undergoes a change in its direction of propagation, called Refraction, given mathematically by Snell's law. This means that given a boundary separating two dissimilar media, a ray will change its direction in going through the boundary unless it impinges normally to the surface. This change in direction is governed by Snell's law (Equation 4.2).

$$n \sin i = n' \sin i', \tag{4.2}$$

where the primed quantities are in the emergent space, i is the angle between the incident ray and the normal to the boundary at the ray intersection point, and i' is the angle between the emergent ray and the normal. Snell's law is the basic law in designing and analyzing optical systems. It is simple and powerful, and can be derived from Fermat's principle [3,4].

4.1.3 Paraxial Optics Validity Interval

Through optical systems, an infinite number of rays can propagate. Whatever their angle of incidence at the first surface, via Snell's law we can calculate exactly how each ray travels through any given optical system. However, even today, with the aid of ever faster computers, this is a costly operation, which led old time designers to envision a more efficient method called "Paraxial Optics" or "First-order optics," used even today as the appropriate tool for initial optical system design stage.

Paraxial optics takes only the first term of the series expansion of the sine function in Snell's law:

$$\sin u = u. \tag{4.3}$$

Therefore paraxial optics operates only with very small angles, such that the paraxial angular conditions $\sin u = u$ and $\tan u = u$ is fulfilled. Snell's law (Equation 4.2) then, for paraxial optics purposes only, can be written as:

$$nu = n'u', \tag{4.4}$$

where a universal convention, mentioned above, has been applied: "primes" denote the refracted ray and medium (designated here as "image space") magnitudes and "not primed" the incident ray and medium ("object space") magnitudes [3,5].

4.2 Perfect Optical System Properties

The span of Optical System Design goes from the conception (initial idea pictured by means of paraxial optics), through several optimization stages, (theoretical calculations improving geometrical optics system's performance), to its materialization (component fabrication and testing and optical device's assembly and evaluation).

All optical systems must be improved, that is, optimized, to reach its ideal "perfect optical system equivalent" or at least to come as close to it as possible. All human activities are limited by several factors, Optical design as well cannot escape from several limitation factors, such as

1. Budget (time and money)
2. Material's dispersion and Snell's law nonlinearities
3. Fabrication and/or assembly errors

Even if a "perfect" system could be built, it would end up being extremely expensive, unaffordable.

Optical system design practice begins by defining the desired system's specifications, such as limiting production cost, image quality, size and weight limitations, spectral range, and so on, the most frequent limitation for civil optical systems being their consumer price. A completely different panorama happens to military and space optical systems, where the complete project cost (mechanics plus electronics, etc.) makes almost irrelevant the particular cost of a single small fraction of it: the optical system itself. In such cases, the outstanding criteria are: highest reliability, low weight and optimum efficiency.

Independently of the restrictions constraining any real optical system, it is highly important to understand the properties of its equivalent Perfect Optical System [1,2,6], since they define the absolute optimum performances that the real system could be able to attain. These perfect optical system properties are the target that the optical designer tries to reach by means of computerized optimization processes and his experience, within the budgetary and technical specifications given [7–9].

4.2.1 Point Image

For every perfect optical system, every object point is in correspondence with a unique image point.

This statement means that every perfect optical system fulfills Fermat's principle: every ray emitted by an object point after traversing a perfect optical system reaches a single image point with an extreme optical path length, that is, constant for every ray bundle. Such situation is depicted schematically in Figure 4.1a and b. This first condition is extremely difficult to accomplish (as we'll see in forthcoming chapters) due to the non-linear character of the deviations all real rays undergo at every refraction (due to sine functions in Snell's law), schematically shown in Figure 4.1c.

To analyze and correct such undesired deviations, equations have been developed to represent them as "deviation contributions" at each refraction or reflection, called "ray aberrations." To facilitate even

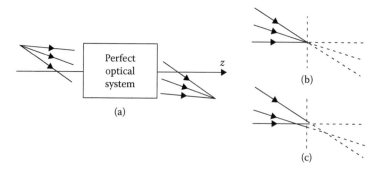

FIGURE 4.1 (a) Perfect point-to-point imaging through a perfect optical system. (b) Perfect image point formation. (c) A "point" image produced by a real optical system.

further their study and therefore their correction (minimization), ray aberrations are classified as Transverse and Longitudinal. Since rays are intimately connected to wavefronts, there are also equations representing Wavefront aberrations, that is, wavefront deviations from their desired perfectly spherical shape, the sole ones able to converge toward an image point. The aberrations most affecting the point imaging process (Figure 4.1) are called Spherical and Coma.

Every image where point objects correspond to actual point images is called "Stigmatic" and so is called the optical system able to do so. Such fortunate situations can happen in real optical systems, especially if we accept that produced images do not fulfill the second perfect imaging condition.

4.2.2 Plane Image

Given a plane object, a perfect optical system always produces a perfect plane image.

This condition constitutes an extension of the first one, a fulfillment of Fermat's principle for every object-image point pairs located on object-image planes. In real practice, given a plane object, real optical systems seldom produce plane images but somewhat toroidal image surfaces instead. Aberrations most affecting the plane imaging process are called Astigmatism and Petzval curvature.

Note in Figure 4.2 and especially in Figure 4.3 how coordinate axes are defined. The system's optical axis (z) is common both to Object and Image coordinate systems. Remaining (x) and (y) axes for object, system and image are parallel one to another and all three-axis coordinate systems constitute right-handed triads.

4.2.3 Volume, Three Dimensional Imaging

Every perfect optical system reproduces every object's symmetry (geometrical shape) into an image affected only by a scale factor called magnification, evenly in all directions.

This third condition is never accomplished by real optical systems, due to aberrations called Distortion and Longitudinal magnification (which is the square of the transverse magnification, as we will see soon). These two facts lead to severe deformations of the image across and along the z-axis, respectively (Figure 4.3b).

The paraxial optics version of every optical system represents its behavior as a Perfect Optical System. Such "perfect" systems constitute both the first step in the optical design process and the target that

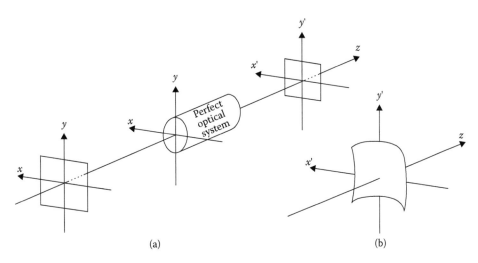

(a) (b)

FIGURE 4.2 (a) Perfect plane object to plane image formation by a perfect optical system. (b) Toroidal image surface produced by a real optical system.

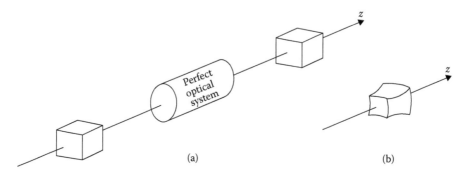

(a)   (b)

FIGURE 4.3   (a) 3D imaging by a perfect optical system. (b) 3D image formed by a real optical system.

the real optical system seeks to reach as the optimization process progresses. As mentioned before, there are economical and physical limitations that preclude such ideal and beautiful success.

Mathematically, the equivalent perfect optical system is described by the paraxial set of equations, specially by the paraxial Snell's law Equation 4.4, valid for very small angles (u, u'), and also the paraxial version of the sagitta of a spherical surface (Equation 4.5), measured from the plane tangent to its vertex, to the surface itself at the point (x,y), for very small heights $\left(x^2 + y^2\right)^{1/2}$ above the optical axis:

$$z = c\,\frac{x^2 + y^2}{2}. \tag{4.5}$$

So, we can define Paraxial optics as the branch of optics with deals with rays, which propagate infinitely close to the optical axis.

However, we take advantage of the simplicity inherent to linear paraxial optics equations, applying relatively large magnitudes such as $y = \text{height}\left[\text{mm or cm}\right]$ and $u = \text{angle}\left[\text{degrees}\right]$, but assuming that each one of them is multiplied by a factor like 10^{-10} or similar, to "transform" heights and angles into actual paraxial magnitudes.

The only real magnitudes handled 1:1 in paraxial optics are thicknesses $t\left[\text{mm or cm}\right]$ and curvatures

$$c_j = \frac{1}{r_j} = \left[\frac{1}{mm} \text{ or } \frac{1}{cm}\right], \tag{4.6}$$

where r_j is the surface's radius of curvature.

Note also that in paraxial optics, whichever the radius of curvature of a refracting or reflecting surface, at a small height above the system's optical axis (within a differential area around the surface-axis intersection, its vertex) we can assume the surface to be a plane surface perpendicular to the optical axis.

Very frequently the first stage in an optical system's design is performed defining it as an arrangement of "thin lenses." Thin lenses constitute helpful abstractions: they are assumed to be flat surfaces featuring some optical power (Equation 4.7).

$$\phi = \frac{1}{f} \tag{4.7}$$

surrounded by air $n = n' = 1$. Now we can add that, to paraxial optics every actual refracting or reflecting surface can also be represented as a thin lens of some power, where $n \neq n'$. Such power can be zero, if and only if the surface "is plane." So, with the aid of paraxial optics every initial conception of an optical system can be represented by means of a succession of thin lenses, and is convenient to begin this way especially while gaining experience as an optical designer.

4.3 Paraxial Ray Tracing

Typically, a real optical system consists of a number of reflecting and/or refracting surfaces. The overall nature of the image for such a system can be determined by using the image created by the first surface as the object of the second surface, and so on. Experience shows that equations for concatenated multi-element systems soon become very complicated and no gain (if any) is reached by deriving them. Optical engineers have developed powerful ray tracing methods to analyze complex optical systems in order to determine their basic first-order properties and to evaluate the aberrations present in the system.

4.3.1 Geometrical Optics Conventions

All of this is based in the field of geometrical optics, which is the study of the ray propagation of light, which can be described with a few simple geometrical relationships and principles:

- The law of rectilinear propagation.
    - The law of reflection that states that the outgoing ray lies in the plane of incidence and the angle of reflection is equal in magnitude to the angle of incidence (particular case of Snell's law for $n' = -n$).
- The law of refraction (Snell's Law).
- The law of reversibility.

In the field of paraxial optics, three simple computational methods have been developed, with which any rotationally symmetrical (centered) optical system can defined, analyzed, and even synthesized:

- Matrix method
- $y - nu$ method
- $y - \bar{y}$ method

4.3.2 Sign Conventions

Geometrical optics conventions (Figure 4.4) are universally valid to all its subdivisions: paraxial optics, third-order optics, fifth-order optics, wave optics, and so on.

Geometrical optics (paraxial rays or exact rays) deals mainly with ray propagation through optical systems, therefore all its sign conventions are defined around ray characteristics as they propagate from left to right (positive sense of ray propagation) from a given object plane, through the optical system until they reach the image plane. The POSITIVE definitions of the geometrical optics ray conventions, almost universally accepted, are shown in Figure 4.4.

1. Ray heights are positive when measured from the optical axis upward $\left(y_j > 0 \text{ and } y_{j+1} > 0 \right)$.
2. Ray propagation angles are positive when measured counterclockwise from the optical axis toward the ray ($u_j > 0$ and $u'_j > 0$).

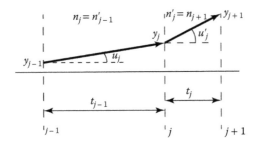

FIGURE 4.4 Geometrical optics sign conventions.

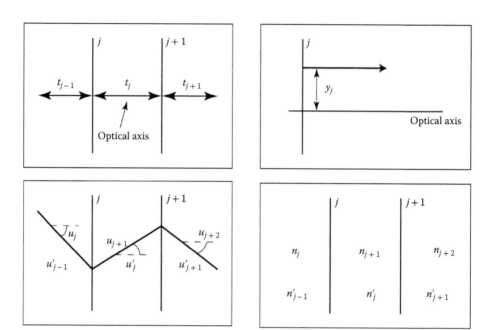

FIGURE 4.5   Notation conventions for Paraxial magnitudes.

3. Refractive indices are positive if rays propagate from left to right, i.e., in the positive sense $(n'_{j-1} = n_j > 0$ and $n'_j = n_{j+1} > 0)$.
4. Optical surface's radii of curvature are positive if their center of curvature is located on the right side of their vertex (surface-optical axis intersection).
5. Surface to surface distances are positive if the higher numbered surface locates rightwards from the low numbered one $(t_j = z_{j+1} - z_j > 0, t_{j-1} > 0$ and $t_j > 0)$.

All these positive conventions are shown in Figure 4.4, except convention "d."

Further, in Figure 4.5 all thicknesses t, heights y and refractive indices n are positive, angles u outside of surfaces $j, j+1$ are negative $(u_j < 0, u_{j+2} < 0)$, and the angle between them $u'_j = u_{j+1}$ is positive.

Note that refractive indices and angles of propagation possess two possible designations.

An important fact must be pointed out here. Objects and images can be either real or virtual (and these characteristics define their whole space's nature). Since object and image planes can happen everywhere along the system's optical axis, it is assumed that object and image spaces extend all the way from $-\infty$ to $+\infty$. Then, the real sequence of surfaces can be different from that shown in Figure 4.6 but the surface's order convention DOES NOT change. In Figure 4.6, the object distance $t_o = t_{0,1} > 0$ is positive and the image distance $t_I = t_{k,k+1} = t_{k,I} > 0$ is also positive since both are measured in the positive sense, from left to right. If object and image planes interchange places, their distances will be negative since in such configuration they will be measured in the negative sense, from right to left.

4.3.3 Axial and Meridional Rays

For every optical system layout, it will be mostly important, given an object's size h_o and position t_o, to know where the image plane is located (t_I) as well as the magnitude of the transverse magnification:

$$m = h_I / h_O. \tag{4.8}$$

To determine these parameters (t_I and h_I), two arbitrary rays must be traced.

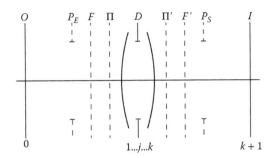

FIGURE 4.6 Surface order convention in optical systems.

An "axial ray" starts at the object plane-optic axis intersection with an arbitrary angle of propagation: $y_o = 0, n_o u_o$ = arbitrary (arbitrary = 0.1 for example). Such ray is traced through the optical system until it touches the optic axis again. The distance from the system's last surface (k) to the image space ray-axis intersection corresponds to the image plane position (t_I).

A "meridional ray" starts at the object's height h_o above or below the axis, with an arbitrary angle of propagation: $y_o = h_O, n_o u_o$ = arbitrary. Such ray is traced through the optical system until it comes out of the system's last surface's image side. The magnification then is computed as:

$$m = \frac{u_i}{u_o} = \frac{n'_k u'_k}{n_o u_o}\left(\frac{n_o}{n'_k}\right) = \frac{\omega'_k}{\omega_o}\left(\frac{n_o}{n'_k}\right). \tag{4.9}$$

Or more frequently it is traced to reach the previously found image plane position, to determine the image's height $y_I = h_I$, then magnification is given by (Equation 4.8).

4.3.4 Stops and Pupils

Assuming a particular surface within the optical system (preferably the diaphragm itself) has been chosen to be the system's Stop, as show in Figure 4.6, it is possible to define:

Entrance pupil (P_E) is the image of the Stop (D) as seen from the object space through the front refracting and/or reflecting optical components of the system (see Figure 4.6).

Exit pupil (P_S) is the image of the Stop (D) as seen from the image space through the rear refracting and/or reflecting optical elements of the system (see Figure 4.6).

Most frequently what happens is that the optical system's entrance pupil (P_E) is found after (to the right of) the first refracting surface, in such case $t_{(P_E,1)} < 0$ (in Figure 4.6, it is positive). Also, most frequently the exit pupil (P_S) is located inside of the optical system (to the left of the last surface), in such case $t_{(k,P_S)} < 0$ (in Figure 4.6, it is positive).

4.3.5 Surface Order Convention

When analysis, optimization or even initial definition of an optical system is being performed, it is highly important to have in mind a set of real and virtual surfaces associated to the optical system that follow a strict order. Such order must always be kept in mind in order to perform all ray propagation calculations correctly as well as to be able to communicate with other team designers. This universally accepted order is shown in Figure 4.6, where again, rays propagate in the positive sense, from left to right.

1. Object plane... O
2. Entrance pupil... P_E
3. Front focal plane... F
4. Front principal plane... Π

5. Some refracting surfaces...1,2,3... *j*-1
6. Stop = aperture diaphragm, with diameter... D_j
7. Some refracting surfaces...*j*+1, *j*+2...*k*
8. Back principal plane... Π'
9. Back focal plane... F'
10. Exit pupil... P_S
11. Image plane... *I*

4.3.5.1 Entrance and Exit Pupil Calculation

Assuming the diaphragm itself has been chosen to be the system's Stop, it is necessary to trace backwards a marginal ray from the object side of the Stop-axis intersection $y_D = 0, \omega_D =$ arbitrary (for example $\omega_D = 1$) to the first surface's object space side; the entrance pupil position is then

$$t_{(1,P_E)} = \frac{-y_1}{\omega_0}. \tag{4.10}$$

To calculate the entrance pupil size, a ray has to be traced from the edge of the diaphragm $\bar{y}_D = h_D$, $\bar{\omega}_D =$ arbitrary, (for example $\bar{\omega}_D = 0$) backwards, up to the newly found entrance pupil position; the entrance pupil height is

$$h_{P_E} = \bar{y}_1 + t_{(1,P_E)}\bar{\omega}_0. \tag{4.11}$$

The exit pupil location, as measured from the system's last surface (k) is found by tracing a marginal ray from the image side of the Stop-axis intersection $y_D = 0, \omega'_D =$ arbitrary (for example $\omega'_D = 1$), in the positive sense up to the last surface's image side, then, the exit pupil position is

$$t_{(k,P_S)} = \frac{-y_k}{\omega'_k}. \tag{4.12}$$

The exit pupil size is determined by a ray traced from the edge of the diaphragm $\bar{y}_D = h_D$, $\bar{\omega}'_D =$ arbitrary, (for example $\bar{\omega}'_D = 0$) in the positive sense, up to the newly found exit pupil position; the exit pupil height is

$$h_S = \bar{y}_k + t_{(k,P_S)}\bar{\omega}'_k. \tag{4.13}$$

4.3.6 The Field and Aperture of an Optical System

In the chapter on aberrations (Chapter 5), we will see that they increase their effect for point images away from the optic axis, and for larger system apertures. The further the image point from the axis, the wider the system's aperture, and the larger the aberration's effect.

Definition: **The semi-field of view.**

Image field cannot be infinite; every optical system is designed to form an image on a particular image plane format: a 35 mm film, a CCD, and so on. All this makes that both image size h_i and position t_i determine the system's semi-field of view through their relation with h_o and t_o as

$$1/2\text{FOV} = h_O/t_O. \tag{4.14}$$

Definition: **The system's aperture.**

For each detector or receptor type, a sensitivity or response capability is associated with the incident image-forming light falling onto it. Therefore, since initial design stages, the light throughput of the system has to be taken into account, which is straight forward associated to the optical system's effective aperture, usually and desirably determined by the Stop. In most systems, the stop is a circular arrangement of thin metallic pieces, articulated in such a way that by a lever displacement they create a variable, almost circular aperture.

A system's aperture most frequently is given in its image space, by a magnitude known as its F-number (F/#),

$$\text{F-number} = f' / D_{P_S}, \tag{4.15}$$

where f' is the image space effective focal length and D_{P_S} is the exit pupil's diameter.

Commercial photographic objectives display their F-number value and its effective focal length f as (FN/f or f#/f). However, for evaluation purposes at non-infinite conjugate, it is more convenient to take into account a new value, the so-called "Working F-Number": $WF\text{number} = t_{(P_S,I)} / D_{P_S}$, where $t_{(P_S,I)}$ is the exit pupil to image plane distance and $D_{P_S} = 2h_{P_S}$ is the exit pupil diameter. Such value is very approximately equal to the Airy disk's diameter at the system's image plane:

$$WF\text{number} = \frac{t_{(P_S,I)}}{D_{P_S}} \approx \frac{1.22\lambda f'}{D_{P_S}} = \text{Airy disk diameter} \tag{4.16}$$

4.3.7 Marginal and Principal Rays

To be able to compute the operating system aberrations, under its working object-image configuration, two particular axial and meridional rays must be traced: the marginal ray (y,ω) and the principal ray $(\bar{y},\bar{\omega})$, sometimes written also as (y_a,ω_a) and (y_b,ω_b), respectively.

Definition: **The marginal ray** (Equation 4.17) starts from the object plane-optic axis intersection, with an angle of propagation such that it reaches the entrance pupil's edge:

$$y_o = 0, \omega_o = \frac{y_{P_E}}{t_{(0,P_E)}} = \frac{h_{P_E}}{t_{(0,1)} - t_{(1,P_E)}}. \tag{4.17}$$

Definition: **The principal ray** (Equation 4.18) starts from the edge of the object plane and travels toward the entrance pupil-optic axis intersection:

$$\bar{y}_o = h_o, \bar{\omega}_o = \frac{-h_o}{t_{(0,P_E)}} = \frac{-h_o}{t_{(0,1)} - t_{(1,P_E)}}. \tag{4.18}$$

4.4 Paraxial Ray Tracing Equations

The simplest form of analysis and design of an optical system involves the use of the paraxial approximation, specifically the transfer and refraction paraxial equations [5, 10–14], which are presented and discussed in this section. Both equations are complementary and indispensable in paraxial ray tracing methods.

4.4.1 Paraxial Transfer (Translation) Equation

As rays propagate from one surface to another, they change their height as measured from the optical axis if they subtend a non-zero angle with respect to the same axis.

The paraxial transfer or translation equation is obtained from pure geometrical considerations, based on Figure 4.7.

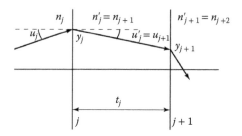

FIGURE 4.7   The paraxial transfer equation.

In Figure 4.7, the ray propagating from surface j to surface $j+1$ with an angle of propagation u'_j fulfills the following geometrical relationship:

$$u'_j = u_{j+1} = \frac{y_{j+1} - y_j}{t_j}.$$ (4.19)

We assume here that we know y_j, t_j and u'_j and that we need to know y_{j+1}, then

$$y_{j+1} = y_j + t_j u'_j$$ (4.20)

A more agile expression (as we will see ahead) is obtained by multiplying and dividing by n'_j as follows:

$$y_{j+1} = y_j + \frac{t_j}{n'_j} n'_j u'_j = y_j + \tau_j \omega'_j.$$ (4.21)

The magnitude $\tau = t/n$ is called "reduced thickness" and $\omega = nu$ "reduced angle."

4.4.2 Paraxial Refraction Equation

As already known, paraxial rays fulfill Snell's paraxial equation of refraction:

$$ni = n'i'.$$ (4.22)

From Figure 4.8 we have:

$$i_j = u_j + \theta_j,$$ (4.23)

where

$$\theta_j = \frac{y_j}{r_j} = y_j c_j,$$ (4.24)

and therefore Equation 4.23 becomes:

$$i_j = u_j + y_j c_j.$$ (4.25)

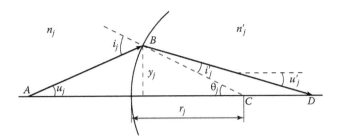

FIGURE 4.8 The paraxial equation of refraction.

Similarly, for the refracted ray angle we have

$$i'_j = u'_j + y_j c_j. \tag{4.26}$$

Note that according to the sign convention and the figure

$$u'_j = \tan\left(u'_j\right) < 0 \tag{4.27}$$

multiplying Equation 4.25 by n_j and Equation 4.26 by n'_j we obtain

$$\begin{aligned}
n_j i_j &= n_j u_j + n_j y_j c_j \\
n'_j i'_j &= n'_j u'_j + n'_j y_j c_j
\end{aligned} \tag{4.28}$$

These two equations constitute the left and right hand sides of Snell's law of refraction (Equation 4.22); therefore we can equate them as

$$n_j u_j + n_j y_j c_j = n'_j u'_j + n'_j y_j c_j, \tag{4.29}$$

where from we have for the refracted ray's angle of propagation

$$n'_j u'_j = n_j u_j + \left(n_j - n'_j\right) y_j c_j. \tag{4.30}$$

Spherical optical surfaces (refractive or reflective), lenses as well as complete optical system, have associated an "optical power." Above it was only mentioned that optical power equals the inverse of the corresponding effective focal length. Here we are dealing with a single optical surface that, as mentioned above can be paraxially represented by means of a thin lens located at its vertex.

Again, to be able to perform agile calculations, the power of a single refractive spherical surface is defined as

$$-\phi_j = c_j \left(n_j - n'_j\right). \tag{4.31}$$

Therefore, the more efficient version of the paraxial equation of refraction (Equation 4.30) is

$$n'_j u'_j = n_j u_j - y_j \phi_j \quad \text{or} \quad \omega'_j = \omega_j - y_j \phi_j. \tag{4.32}$$

4.4.3 Paraxial Invariants (Helmholtz, Lagrange)

In paraxial imaging theory, there are some system invariants that have the same value in front, within and behind the system, that is, they are constant as they pass through the system. The most important of these constants are the Helmholtz and the Lagrange invariants.

The Helmholtz invariant H (also called the linear light throughput equation) corresponds to the conservation of energy and information:

$$H = ny_a u_a = n'y'_a u'_a \tag{4.33}$$

$$H = nh_o u_b = n'h_i u'_b. \tag{4.34}$$

The marginal ray (Equation 4.33) is a paraxial quantity that gives a measure of how much light is transported through a system, since it is a measure of the cones of light diverging from the object plane and converging toward the image plane, coinciding their bases at the refracting surface with semi-diameters $y_a = y'_a$.

The principal ray (Equation 4.34) is a paraxial quantity that constitutes a measure of how much information is being transported by the same system, since all points on the object plane (h_o and below) are projected onto the image plane (h_i and below), affected only by the magnification factor m (Figure 4.9).

Inserting the invariant (Equation 4.34) into the definition of the lateral magnification (Equation 4.8), one obtains the magnification as a function of the aperture angles u_b and u'_b as was done in Equation 4.9:

$$m = \frac{h_i}{h_o} = \frac{n_o u_b}{n_i u'_b} = \frac{\bar{\omega}_0}{\bar{\omega}_I}. \tag{4.35}$$

Optical systems consist of some number of refracting and/or reflecting surfaces, therefore writing down the invariant H for a paraxial system consisting of two consecutive thin lenses (1 and 2), separated a distance d and taking into consideration the transfer of the marginal (a) and principal (b) rays. One obtains an invariant relationship involving both marginal and principal rays:

$$H = \frac{y_{a1}y_{b2} - y_{b1}y_{a2}}{d} \tag{4.36}$$

4.4.4 The Optical or Lagrange Invariant

Since paraxial optics are linear, the properties of a system are completely described by two different rays, usually and most conveniently the marginal and principal rays. From the knowledge of the parameters for these two rays, any third one can be calculated without having to calculate its propagation through each step. If two rays at any reference plane are of heights $y_a = y, y_b = \bar{y}$ with reduced angles

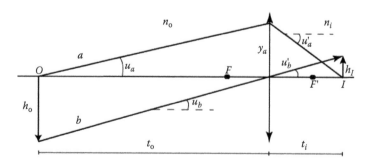

FIGURE 4.9 The Lagrange-Helmholtz invariants.

$nu_a = \omega$, $nu_b = \overline{\omega}$, then a new invariant can be defined (more useful for design purposes), called the "Lagrange" invariant, also called by some authors the "Optical Invariant":

$$\Lambda = yn\overline{u} - \overline{y}nu = y\overline{\omega} - \overline{y}\omega, \tag{4.37}$$

which maintains always its value not only on both sides of an optical surface of any kind, but across the whole object space, the optical system and the image space.

In the past, when logarithms, mechanical, electric or slide rule calculators were used, this invariant allowed a quick verification of correctness of the ray tracing of any pair of rays, particularly the marginal (y, ω) and principal $(\overline{y}, \overline{\omega})$ rays. Besides, the Lagrange invariant takes part as a variable in some expressions of third-order aberrations.

4.5 Matricial Paraxial Ray Trace

This section introduces the use of 2×2 matrices for paraxial ray tracing, to find the first-order properties of any optical system. Matrix methods also offer additional insight into the physical meaning of the marginal (full aperture) and chief or principal (full field) rays [10–13]. In addition, matrices will be used ahead in applications that would otherwise be cumbersome to solve (for example off-axis and/or tilted element representation).

The so-called system matrix contains all of the first-order properties of a system; however, it does not tell us directly such things as which surfaces are the aperture or the field stops or where the pupils and windows are located; all these characteristics are included but mixed in it. Matrix ray tracing method is a powerful tool in the design of laser cavities [14–16].

The biggest problem with matrix methods performed manually is that the process of writing down the various refraction and translation matrices and multiplying them together is lengthy and prone to error, however the development of computer programs minimizes these disadvantages.

The ray trace matricial method was introduced by Carl Frederick Gauss (Germany) around 1800. It is very specific, since each and every optical surface, space between surfaces, optical components, and all paraxial properties of complete optical systems can be represented by means of simple 2×2 matrices. This method constitutes the physical representation of geometrical optics paraxial ray propagation.

4.5.1 The Ray Vector

In matrix algebra, the elements onto which matrices operate are called vectors, denoted by column matrices 1×2. In paraxial optics, marginal and principal ray vectors arriving onto surface j object space side, respectively are defined as

$$\mathbf{R}_j = \begin{pmatrix} y_j \\ n_j u_j \end{pmatrix} = \begin{pmatrix} y_j \\ \omega_j \end{pmatrix}, \quad \overline{\mathbf{R}}_j = \begin{pmatrix} \overline{y}_j \\ n_j \overline{u}_j \end{pmatrix} = \begin{pmatrix} \overline{y}_j \\ \overline{\omega}_j \end{pmatrix} \tag{4.38}$$

and on the same surface's image space side, as they leave the surface after refraction (or reflection), are denoted respectively as

$$\mathbf{R}'_j = \begin{pmatrix} y_j \\ n'_j u'_j \end{pmatrix} = \begin{pmatrix} y_j \\ \omega'_j \end{pmatrix}, \quad \overline{\mathbf{R}}'_j = \begin{pmatrix} \overline{y}_j \\ n'_j \overline{u}'_j \end{pmatrix} = \begin{pmatrix} \overline{y}_j \\ \overline{\omega}'_j \end{pmatrix}. \tag{4.39}$$

Note that we are dealing with two rays referred at the same surface j, the first pair of these expressions represent these rays arriving onto the surface and the second pair represent the same rays departing from it, in both cases propagating in the positive sense, from left to right. Upon reflection or refraction, they possess the same height above or below the optic axis.

4.5.2 Transfer and Refraction Matrices

Rays propagate in rectilinear fashion through homogeneous isotropic media, changing their height from one reference plane to another (Equation 4.21) but keeping their propagation angle. From Equation 4.21 and $y_{j+1} = y'_j$, the matricial expression for ray propagation is performed as $\mathbf{R}_{j+1} = T_j \mathbf{R}'_j$, defining the translation matrix T_j as

$$T_j = \begin{pmatrix} 1 & t_j / n'_j \\ 0 & 1 \end{pmatrix} = \begin{pmatrix} 1 & \tau_j \\ 0 & 1 \end{pmatrix}. \tag{4.40}$$

Traversing an interface between two media (n_j and n'_j), rays change their propagation angle but since refraction is a phenomenon taking place at one particular point on the interface (Equation 4.32), they keep the same height $\left(y'_j = y_j \right)$. Then, we can perform it by matrix methods as $\mathbf{R}'_j = R_j \mathbf{R}_j$ defining the refraction (power) matrix as

$$R_j = \begin{pmatrix} 1 & 0 \\ -\phi_j & 1 \end{pmatrix} = \begin{pmatrix} 1 & 0 \\ c_j \left(n_j - n'_j \right) & 1 \end{pmatrix}. \tag{4.41}$$

Notation: ϕ= thin lens or optical surface power. Φ = optical system power

4.5.3 The Characteristic Matrix M_{k1}

Note that translation and refraction matrices are unitary and that both paraxial translation and refraction equations, and the paraxial Snell's law are first-order linear equations.

Then it is possible to consecutively multiply refraction and translation matrices (from right to left, as defined by Gauss), in the order that the interfaces and spaces (from left to right) are established in a proposed optical system, to obtain a single 2 × 2 matrix.

This final matrix result is called the System Characteristic Matrix, which will contain all the physical (however paraxial) properties of the system, defining the way every incoming ray will propagate through it in the positive sense, from left (object space) to right (image space). No translation t_o is included from the object to the system's first surface or from the system's last surface to image t_i. The system matrix from its first (1) to its last (k) surfaces is calculated then as

$$M_{k1} = R_k T_{k-1} R_{k-1} \ \dots \ R_2 T_1 R_1 = \begin{pmatrix} a_{k1} & b_{k1} \\ c_{k1} & d_{k1} \end{pmatrix} \tag{4.42}$$

Each of the elements of the system matrix have special meaning, as is shown in the following section.

4.6 Cardinal Points and Planes

Most, if not all, optic systems feature three pairs of axial points: principal (Π), nodal (N) and focal (F), called "cardinal points." Perpendicular to the optic axis at these points and with the same name and literals are defined three pairs of planes.

These axial points and planes associated are very important for understanding their optical system's behavior, as well as are very useful for a wide number of calculations. Cardinal points and "focal," "principal," and "nodal" (F, Π, N) planes feature the following characteristics:

1. Focal points and planes are images of object planes located at infinity. Image space focal point F′ is an image of an axial object point located at −∞; it therefore is located on the optic axis and through it passes the rear focal plane, perpendicular to the optics axis. Object space focal point F is an image of an axial object point located at +∞ formed by rays traveling in a negative sense and the front focal plane is perpendicular to the optic axis at F.
2. Principal planes Π and Π′ are Unitary Transverse Magnification Planes, meaning that whatever the angle at which a ray arrives at plane Π, it departs from plane Π′ at the same height but (perhaps) with a different propagation angle.
3. Nodal points N and N′ are unitary angular magnification points on the optic axis, meaning that whichever its propagation angle, any ray arriving at point Π will leave the system as if emitted from point Π′ under exactly the same propagation angle.

Principal and nodal points can coincide, as they most frequently do, as we will see ahead. The contrary (principal and nodal points at some distance from one another) happens only when object space and image space possess different refractive index values.

Assume an optical system in air, and an axial ray arriving from infinity onto the system's first surface with unit height ($y_1 = 1$). After propagating through the system it will intersect the optic axis at the system's rear focal point F′, whose position is obtained from the matrix expression

$$\mathbf{R}_{F'} = T_{dfp} M_{k1} \mathbf{R}_\infty. \tag{4.43}$$

Solving this matrix operation we arrive to the following useful results, involving the back focal distance (*bfd*), which corresponds to the image space distance from the system's last surface to the rear focal point (F′) and the effective rear focal length f'.

$$0 = a_{k1} + c_{k1} t_{(k,F')} \implies a_{k1} = \frac{t_{(k,F')}}{f'} = \frac{bfd}{f'} \tag{4.44}$$

$$u'_k = c_{k1} = -\Phi + 0 \implies c_{k1} = \frac{-1}{f'}, \tag{4.45}$$

where $\Phi = 1/f$ is the optical system power.

The characteristic matrix is unitary; therefore, its inverse is easily employed to trace paraxial rays in the reverse order (from right to left). In this way, it is possible find the front focal distance (*ffd*) with $y_F = 0$, assuming a ray comes from infinity in image space ($+\infty$), with unit height. The matrix operation applied to a ray arriving at unit height from infinity in image space is onto the system's rear surface k, is given by

$$\mathbf{R}_F = T_{ffd} M_{k1}^{-1} \mathbf{R}'_\infty, \tag{4.46}$$

which once solved leads to

$$0 = d_{k1} - c_{k1} t_{(1,F)} \implies d_{k1} = \frac{t_{(1,F)}}{f} = \frac{ffd}{f}, \tag{4.47}$$

and

$$-c_{k1} = \frac{1}{f}, \tag{4.48}$$

where the front focal distance is defined: $ffd = t_{1F}$ and f is the system's front effective focal length. Note that when $n_o = n_i$ both effective focal lengths are equal: $f = f'$.

4.6.1 Principal Planes Matrix $M_{\Pi'\Pi}$

Principal planes feature unitary magnification ($m = +1$); they are conjugates (are mutually image one of another). Therefore, a ray incident on the system's first surface, trying to reach the front principal plane at height y_Π, will leave the system's last surface as if it were coming from height $y_{\Pi'} = y_\Pi$ at the rear principal plane Π'.

Since principal planes are conjugates and feature $m_\Pi = +1$, the matrix operation linking object space principal plane (Π) and image space principal plane (Π') is

$$M_{\Pi'\Pi} = T_{\Pi'} \, M_{k1} T_\Pi, \tag{4.49}$$

which means

$$M_{\Pi'\Pi} = \begin{pmatrix} a_{k1} + c_{k1}t'_{(\Pi',k)} & b_{k1} + t_{(1,\Pi)}\left(a_{k1} + c_{k1}t'_{(\Pi',k)}\right) + d_{k1}t'_{(\Pi',k)} \\ c_{k1} & d_{k1} + c_{k1}t_{(1,\Pi)} \end{pmatrix} = \begin{pmatrix} 1 & 0 \\ -\Phi & 1 \end{pmatrix}, \tag{4.50}$$

wherefrom the position of the front principal plane as measured from the system's first surface is given by

$$1 = d_{k1} + c_{k1}t_{(1,\Pi)} \quad \Rightarrow \quad t_{(1,\Pi)} = \frac{1 - d_{k1}}{c_{k1}}, \tag{4.51}$$

and the position of the rear principal plane as measured from the system's last surface is given by

$$1 = a_{k1} + c_{k1}t'_{(\Pi',k)} \quad \Rightarrow \quad t'_{(\Pi',k)} = \frac{1 - a_{k1}}{c_{k1}}. \tag{4.52}$$

4.6.2 Nodal Points Matrix $M_{N'N}$

Nodal points feature unitary angular magnification ($m_\alpha = 1$) and are located on the optic axis, that is, $y_N = y_{N'} = 0$. This means that for optical systems with equal object and image spaces refractive indexes $n_o = n_i$, every ray $\mathbf{R}_N$ propagating in the positive sense, arriving onto the system's first surface, traveling toward the front nodal point (N) with geometrical propagation angle u_N, will leave the system's last surface as ray $\mathbf{R}_{N'}$ as if it were emitted from the rear nodal point (N') with the same geometrical angle of propagation $u_N = u_{N'}$.

Assume now an optical system with dissimilar object space and image space refractive indices $n_o \neq n_i$ then. Based on the paraxial Snell's law, the matrix operation

$$\mathbf{R}_i = T_{N'} M_{k1} T_N \mathbf{R}_o \tag{4.53}$$

can be interpreted as $\mathbf{R}_i = M_{N'N}\mathbf{R}_o$ behaving as

$$\mathbf{R}_i = \begin{pmatrix} m_\alpha & 0 \\ -\Phi & 1/m_\alpha \end{pmatrix}\mathbf{R}_o = \begin{pmatrix} m_\alpha & 0 \\ -\Phi & 1/m_\alpha \end{pmatrix}\begin{pmatrix} 0 \\ u_{No} \end{pmatrix} = \begin{pmatrix} 0 \\ \dfrac{n_o}{n_i}u_{No} \end{pmatrix}, \tag{4.54}$$

wherefrom the angular magnification is

$$m_\alpha = \frac{n_o}{n_i}. \tag{4.55}$$

In this case, the position of both nodal points, as measured from the system's first surface (for N) and from the system's last surface (for N') respectively, are found from the matrix operation

$$\mathbf{R}_{N'} = T_{N'k}T_{N'}M_{k1}T_NT_{N1}\mathbf{R}_N = T_{N'k}M_{N'N}T_{N1}\mathbf{R}_N, \tag{4.56}$$

and then

$$M_{N'N} = \begin{pmatrix} 1 & t_{(N',k)} \\ 0 & 1 \end{pmatrix}\begin{pmatrix} m_\alpha & 0 \\ -\Phi & 1/m_\alpha \end{pmatrix}\begin{pmatrix} 1 & t_{(N,1)} \\ 0 & 1 \end{pmatrix}$$

$$= \begin{pmatrix} m_\alpha - \Phi t_{(N',k)} & 1/m_\alpha\, t_{(N',k)} + t_{(N,1)}\left(m - \Phi\right)t_{(N',k)} \\ -\Phi & 1/m_\alpha - \Phi t_{(N,1)} \end{pmatrix}. \tag{4.57}$$

Then, to determine the position of the front nodal point N, consider a ray $\mathbf{R}_N$ propagating in the positive sense:

$$\begin{pmatrix} 0 \\ u'_{N'} \end{pmatrix} = \begin{pmatrix} m_\alpha - \Phi t_{(N',k)} & 1/m_\alpha\, t_{(N',k)} + t_{(N,1)}\left(m - \Phi\right)t_{(N',k)} \\ -\Phi & 1/m_\alpha - \Phi t_{(N,1)} \end{pmatrix}\begin{pmatrix} 0 \\ u_N \end{pmatrix}$$

$$= \begin{pmatrix} 0 \\ u_N\left(1/m_\alpha - \Phi t_{(N,1)}\right) \end{pmatrix}, \tag{4.58}$$

and then the front nodal point position is

$$\frac{u'_{N'}}{u_n} = \frac{1}{m_\alpha} - \Phi t_{N1} \quad \Rightarrow \quad t_{N1} = \left(\frac{1}{m_\alpha} - 1\right)\frac{1}{\Phi}. \tag{4.59}$$

Similarly, applying the inverse of the conjugate matrix to a right to left propagating ray, arriving onto the last surface traveling toward the rear nodal point; for the rear nodal point position we have

$$\frac{u_N}{u'_{N'}} = m - \Phi t_{(N',k)} \quad \Rightarrow \quad t_{(N'k)} = \left(m_\alpha - 1\right)\frac{1}{\Phi}. \tag{4.60}$$

Statements given for principal planes and nodal points are valid for both senses of propagation; then for the negative sense of propagation, they can be shortly written as: $\Pi' \to \Pi$ and $N' \to N$. Also by Equations 4.49 and 4.53, it must be pointed out that principal points and nodal points are conjugates: images of one another.

4.6.3 Focal Points Matrix $M_{F'F}$

First of all, it must be remembered that focal points are not conjugates among themselves, instead they are conjugates of planes located infinitely far away from the system.

Performing the matrix operation

$$M_{F'F} = T_{F'k} M_{k1} T_{1F},$$
(4.61)

we have a front focal plane to rear focal plane matrix, and from its upper-right element, using previous results for a_{k1}, b_{k1} and c_{k1}, we have

$$
M_{F'F} = \begin{pmatrix} 1 & t_{(F',k)} \\ 0 & 1 \end{pmatrix} \begin{pmatrix} a_{k1} & b_{k1} \\ c_{k1} & d_{k1} \end{pmatrix} \begin{pmatrix} 1 & t_{(1,F)} \\ 0 & 1 \end{pmatrix}
$$

$$
= \begin{pmatrix} a_{k1} + t_{(F',k)} c_{k1} & b_{k1} + t_{(F',k)} d_{k1} + t_{(1,F)} a_{k1} + t_{(1,F)} t_{(F',k)} c_{k1} \\ c_{k1} & d_{k1} + t_{(1,F)} c_{k1} \end{pmatrix}
$$
(4.62)

$$
= \begin{pmatrix} A_{F'F} & B_{F'F} \\ C_{F'F} & D_{F'F} \end{pmatrix}.
$$

For the upper right element $B_{F'F}$ we have

$$b_{k1} + \mathrm{ffd}\left(a_{k1} + c_{k1}\mathrm{bfd}\right) + d_{k1}\mathrm{bfd} \;\Rightarrow\; b_{k1} = f' + \frac{t_{(F',k)} t_{(1,F)}}{f'}.$$
(4.63)

4.7 The Conjugate Matrix M_{io}

A conjugate matrix is one that connects the object plane with the image plane as

$$\mathbf{R}_i = T_i M_{k1} T_o \mathbf{R}_o = M_{io} \mathbf{R}_o.$$
(4.64)

Therefore the conjugate matrix has the form:

$$
M_{io} = \begin{pmatrix} 1 & \tau_i \\ 0 & 1 \end{pmatrix} M_{k1} \begin{pmatrix} 1 & \tau_0 \\ 0 & 1 \end{pmatrix} = \begin{pmatrix} m & 0 \\ \Phi & 1/m \end{pmatrix},
$$
(4.65)

where $m = \dfrac{h_i}{h_o}$ is the magnification (Equation 4.8) and Φ the system power (Equation 4.7).

To find the image size and location, assuming the object distance t_0 is given, (the distance from object plane to first system surface), we need to trace a ray from the object to the last surface of the system as

$$\mathbf{R}'_k = M_{k1} T_o \mathbf{R}_o = M_{ko} \mathbf{R}_o,$$
(4.66)

where

$$
M_{ko} = \begin{pmatrix} a_{k1} & b_{k1} \\ c_{k1} & d_{k1} \end{pmatrix} \begin{pmatrix} 1 & \tau_0 \\ 0 & 1 \end{pmatrix} = \begin{pmatrix} a_{k1} & a_{k1}\tau_0 + b_{k1} \\ c_{k1} & c_{k1}\tau_0 + d_{k1} \end{pmatrix} = \begin{pmatrix} A_{ko} & B_{ko} \\ C_{ko} & D_{ko} \end{pmatrix}.
$$
(4.67)

And then, performing the last surface to image plane translation:

$$M_{io} = T_i M_{k0} = \begin{pmatrix} 1 & \tau_i \\ 0 & 1 \end{pmatrix} \begin{pmatrix} A_{k0} & B_{k0} \\ C_{k0} & D_{k0} \end{pmatrix} = \begin{pmatrix} A_{k0} + \tau_i C_{k0} & B_{k0} + \tau_i D_{k0} \\ C_{k0} & D_{k0} \end{pmatrix} = \begin{pmatrix} m & 0 \\ -\Phi & 1/m \end{pmatrix}. \quad (4.68)$$

Then the image plane location is given by

$$\tau_i = \frac{-B_{k0}}{D_{k0}} = \frac{-(b_{k1} + a_{k1}\tau_o)}{d_{k1} + c_{k1}\tau_o}, \quad (4.69)$$

and the magnification by

$$\frac{1}{m} = D_{k0} = d_{k1} + c_{k1}\tau_o, \quad (4.70)$$

or by

$$m = A_{k0} + \tau_i C_{k0} = a_{k1} + c_{k1}\tau_i\tau_o. \quad (4.71)$$

Newton measured object and image distances from their corresponding focal points, using the focal planes matrix found above (Equation 4.62). The conjugate object-image planes matrix relationship becomes

$$R_i = Z_i M_{F'F} Z_o R_o = M_{io} R_o, \quad (4.72)$$

which produces

$$M_{io} = \begin{pmatrix} 1 & z_i \\ 0 & 1 \end{pmatrix} \begin{pmatrix} A_{F'F} & B_{F'F} \\ C_{F'F} & D_{F'F} \end{pmatrix} \begin{pmatrix} 1 & z_0 \\ 0 & 1 \end{pmatrix} =$$

$$= \begin{pmatrix} A_{F'F} + z_i C_{F'F} & B_{F'F} + z_i D_{F'F} + z_o A_{F'F} + z_o z_i C_{F'F} \\ C_{F'F} & D_{F'F} + z_o C_{F'T} \end{pmatrix} = \begin{pmatrix} m & 0 \\ \Phi & 1/m \end{pmatrix}, \quad (4.73)$$

whose matrix M_{io} upper-right term is equal to zero, and from Equation 4.63 leads to Newton's famous image formation equation:

$$z_o z_i = ff', \quad (4.74)$$

since $t_{(F',k)} = z_i$ and $t_{(1,F)} = z_o$.

4.7.1 Entrance and Exit Pupil Calculation

Matrix methods can also be used to determine which surface actually acts as the aperture stop within a diaphragm-free optical system. To achieve it, in the optical system, calculate step by step the succession of matrix multiplications that have to be performed to compute the system's characteristic matrix:

$$M_{21}, M_{31}, M_{41}, \dots. M_{k1},$$

taking apart at each stage their upper-left term:

$$A_{21}, A_{31}, A_{41}, ... A_{k1}.$$

Note that this procedure is equivalent to collecting at each surface the height achieved by a marginal "p" ray ($y_1 = 1$, $n_0 u_0 = 0$) as it propagates along the optical system.

Then, compare the clear radius planned for each surface ρ_j with these A_{j1} elements:

$$\frac{A_{21}}{\rho_1}, \frac{A_{31}}{\rho_3}, \frac{A_{41}}{\rho_4}, ... \frac{A_{k1}}{\rho_k}. \tag{4.75}$$

The smallest among these ratio values determines which surface actually acts as the system's aperture stop.

Almost all optical systems somewhere feature a diaphragm as a variable aperture stop to control the amount of light reaching the chosen image plane detector. So, the above procedure can be used to determine the required semi-diameter ρ_i for each and every other element within the system as to assure that the diaphragm-stop alone performs the light flux control through the system. Once this situation is attained, the pupils tell the rest of the story. Because pupils are images of the aperture stop, we can find both entrance and exit pupil's sizes and locations using matrix calculations.

For the entrance pupil (P_E), assuming $R_D = I$ (identity matrix) is the zero power "refraction" matrix that represents the Stop-Diaphragm (D), find the matrix

$$M_{D1} = R_D T_{D-1} R_{D-1} .... R_1. \tag{4.76}$$

Actually, we need the inverse of this matrix to perform the negative propagation sense object (D) to image (P_E), wherefrom the entrance pupil location is given by

$$\tau_{P_E 1} = \frac{-B_{D1}}{A_{D1}}, \tag{4.77}$$

and the entrance pupil magnification is

$$m_{P_E} = \frac{1}{A_{D1}}. \tag{4.78}$$

To find the exit pupil (P_S) data, we need to determine the matrix that takes us from the stop (D) to last surface (k) image space side:

$$M_{kD} = R_k T_{k-1} R_{k-1} .... T_D. \tag{4.79}$$

From this matrix, the exit pupil position is found:

$$\tau_{P_S k} = \frac{-B_{kD}}{D_{kD}}. \tag{4.80}$$

And the exit pupil magnification is

$$m_{P_S} = \frac{1}{D_{kD}}. \tag{4.81}$$

4.7.2 Actual Marginal and Principal Paraxial Rays

Once a particular surface is found within the system or one of them is defined as the system's Diaphragm, that is, as the system's Stop, and both pupils have been calculated, we are in the position to define the precise marginal and principal rays.

For the marginal ray, which will allow us to find the image plane, it is necessary to trace backwards a marginal ray from the object side of the Stop-axis intersection to the first surface's object space side (by means of the above found matrix M_{1D}), and an unknown first surface to entrance pupil distance matrix T_{P_E1}; the operation to perform is:

$$M_{P_E1} = T_{P_E1}^{-1}M_{1D}^{-1} = \begin{pmatrix} 1 & -\tau_{P_E1} \\ 0 & 1 \end{pmatrix} \begin{pmatrix} d_{1D} & -b_{1D} \\ -c_{1D} & a_{1D} \end{pmatrix}$$

$$= \begin{pmatrix} d_{1D} + c_{1D}\tau_{P_E1} & -b_{1D} - a_{1D}\tau_{P_E1} \\ -c_{1D} & a_{1D} \end{pmatrix} = \begin{pmatrix} m_{P_E} & 0 \\ \phi_{1D} & 1/m_{P_E} \end{pmatrix}. \tag{4.82}$$

Then, the entrance pupil position is

$$\tau_{1E} = \frac{-b_{1D}}{a_{1D}}. \tag{4.83}$$

The entrance pupil size is found tracing a ray from the edge of the diaphragm

$$\mathbf{R}_{D1} = \begin{pmatrix} y_D \\ \omega_D \end{pmatrix} = \begin{pmatrix} h_D \\ \text{arbitrary} \end{pmatrix} (\text{for example } \omega_D = 0), \tag{4.84}$$

up to the newly found entrance pupil position

$$\mathbf{R}_{P_E} = T_{P_E1}^{-1}M_{1D}^{-1}\mathbf{R}_D = M_{P_ED}^{-1} = \begin{pmatrix} h_{P_E} \\ \omega_{P_E} \end{pmatrix}, \tag{4.85}$$

and then

$$\begin{pmatrix} y_{P_E} \\ \omega_{P_E} \end{pmatrix} = \begin{pmatrix} D_{P_ED} & -B_{P_ED} \\ -C_{P_ED} & A_{P_ED} \end{pmatrix} \begin{pmatrix} y_D \\ \omega_D \end{pmatrix} = \begin{pmatrix} y_D D_{P_ED} - B_{P_ED}\omega_D \\ A_{P_ED}\omega_D - y_D C_{P_ED} \end{pmatrix}. \tag{4.86}$$

If $\omega_D = 0$ as proposed in Equation 4.84, the entrance pupil height is

$$h_{P_E} = y_{P_E} = D_{P_ED}y_D. \tag{4.87}$$

The exit pupil location, as measured from the system's last surface (k), is found by the matrix operation:

$$\overline{\mathbf{R}}_{P_S} = T_{P_Sk}M_{kD}\overline{\mathbf{R}}_D' = M_{P_SD}\overline{\mathbf{R}}_D', \tag{4.88}$$

where it is convenient that

$$\overline{\mathbf{R}}_D' = \begin{pmatrix} \overline{y}_D \\ \overline{\omega}_D' \end{pmatrix} = \begin{pmatrix} 0 \\ 1 \end{pmatrix}. \tag{4.89}$$

Then from

$$T_{P_S k} M_{kD} \overline{\mathbf{R}}_D = \begin{pmatrix} A_{P_S D} + C_{P_S D} \tau_{Sk} & B_{P_S D} + D_{P_S D} \tau_{P_S k} \\ C_{P_S D} & D_{P_S D} \end{pmatrix} \begin{pmatrix} \overline{y}_D \\ \overline{\omega}'_D \end{pmatrix}, \tag{4.90}$$

the exit pupil position is

$$\tau_{P_S k} = \frac{-B_{P_S D}}{D_{P_S D}}. \tag{4.91}$$

The exit pupil size is determined by a ray similar to Equation 4.84

$$\overline{\mathbf{R}}'_D = \begin{pmatrix} \overline{y}_D \\ \omega_D \end{pmatrix} = \begin{pmatrix} \overline{y}_D \\ 0 \end{pmatrix}, \tag{4.92}$$

traced with

$$\overline{\mathbf{R}}_S = M_{P_S D} \overline{\mathbf{R}}'_D, \tag{4.93}$$

that is, from the edge of the diaphragm in the positive sense, up to the newly found exit pupil position. Then, the exit pupil height is

$$h_{P_S} = \left(A_{P_S D} + C_{P_S D} \tau_{P_S k} \right) \overline{y}_D. \tag{4.94}$$

4.7.3 Object Plane to Image Plane Calculations

Computation of the actual marginal and principal ray data is required to be able to compute Longitudinal (LCh) and Transverse (TCh) chromatic aberrations, as well as geometrical third-order aberrations (Sph, Cma, Ast, Ptz, Dist).

In object space, at the object plane itself, marginal $\mathbf{R}_o = \begin{pmatrix} y_o \\ \omega_o \end{pmatrix}$ and principal $\overline{\mathbf{R}}_o = \begin{pmatrix} \overline{y}_o \\ \overline{\omega}_o \end{pmatrix}$ rays are defined respectively as

$$y_o = 0, \omega_o = \frac{y_E}{t_o - t_{1E}} \tag{4.95}$$

and

$$\overline{y}_o = h_o, \overline{\omega}_o = \frac{-h_o}{t_o - t_{1E}}. \tag{4.96}$$

All that has been defined up to this point refers to centered, that is, z-axis rotationally symmetrical, spherical-surface optical systems. However, sometimes bi-plane symmetrical configurations are required, producing image planes located at different positions and different magnifications, one for their y-z symmetry and another for their x-z symmetry, called "anamorphic."

4.8 Centered Anamorphic System 4 × 4 Matrices

Since matrix operations constitute a tool representing the operation with a number of simultaneous equations, two (refraction and translation) in the meridional plane, for centered optical systems as described in previous sections. Several researchers have developed ray trace matrix methods for a wider set of optical systems: tilted and shifted surfaces, tilted plane parallel plates, plane and spherical diffraction gratings, and so on. Professor Herbert Gross in his book *Handbook of Optical Systems* [17] Chapter 2, provides a wide assortment of 3 × 3, 4 × 4 and 5 × 5 matrices applicable to many optical system configurations.

3 × 3 and 5 × 5 matrices are used to represent tilted and decentered optical components. Here only 4 × 4 matrices are considered corresponding to anamorphic optical systems featuring cylindrical and/or toroidal surfaces, with their axes of symmetry coincident with the Cartesian coordinate axes x,y, with azimuth angle $\varphi = 0$ as measured with respect to the x-axis. (z-axis as usual corresponds to the system's optical axis).

Rays must be defined now as

$$\vec{R} = \begin{pmatrix} x \\ y \\ u \\ v \end{pmatrix} = \begin{pmatrix} \vec{r} \\ \vec{u} \end{pmatrix} \tag{4.97}$$

and 4 × 4 matrices are defined as

$$M = \begin{pmatrix} A_{xx} & A_{xy} & B_{xx} & B_{xy} \\ A_{yx} & A_{yy} & B_{yx} & B_{yy} \\ C_{xx} & C_{xy} & D_{xx} & D_{xy} \\ C_{yx} & C_{yy} & D_{yx} & D_{yy} \end{pmatrix} = \begin{pmatrix} A & B \\ C & D \end{pmatrix}. \tag{4.98}$$

Matrices representing non-rotationally symmetrical optical components, with azimuth angle $\varphi = 0$, refracting an arbitrary ray $(x \neq 0, y \neq 0, u \neq 0, v \neq 0)$ are

1. Refraction by a cylindrical thin lens or cylindrical refractive surface (cylinder axis along y-axis):

$$M_{Cylind}\vec{R} = \begin{pmatrix} 1 & 0 & 0 & 0 \\ 0 & 1 & 0 & 0 \\ -\phi_x & 0 & 1 & 0 \\ 0 & 0 & 0 & 1 \end{pmatrix} \begin{pmatrix} x \\ y \\ u \\ v \end{pmatrix} = \begin{pmatrix} x \\ y \\ u - x\phi_x \\ v \end{pmatrix} = \vec{R}', \tag{4.99}$$

which means that the ray has been refracted, changing only its angle of propagation on the x, z plane.

2. Refraction by a toric thin lens or toric refractive surface with $-\phi_x$, $-\phi_y$ and $\varphi = 0$:

$$M_{Toric}\vec{R} = \begin{pmatrix} 1 & 0 & 0 & 0 \\ 0 & 1 & 0 & 0 \\ -\phi_x & 0 & 1 & 0 \\ 0 & -\phi_y & 0 & 1 \end{pmatrix} \begin{pmatrix} x \\ y \\ u \\ v \end{pmatrix} = \begin{pmatrix} x \\ y \\ u - x\phi_x \\ v - y\phi_y \end{pmatrix} = \vec{R}'. \tag{4.100}$$

3. Translation 4×4 matrix:

$$\vec{TR_j} = \begin{pmatrix} 1 & 0 & \tau & 0 \\ 0 & 1 & 0 & \tau \\ 0 & 0 & 1 & 0 \\ 0 & 0 & 0 & 1 \end{pmatrix} \begin{pmatrix} x \\ y \\ u \\ v \end{pmatrix} = \begin{pmatrix} x + u\tau \\ y + v\tau \\ u \\ v \end{pmatrix} = \vec{R}_{j+1}. \tag{4.101}$$

4.9 The Paraxial Tabular *y-nu* (*yω*) Ray Trace

In this section the *y-nu* tabular ray trace technique is introduced. It makes it relatively easy to analyze every rotationally symmetrical optical system, tracing rays to find image size and position for even the most complicated centered optical system [4]. The *y-nu* ray tracing tabular method was developed by A.E. Conrady (England) at the beginning of the twentieth century [17,18] and later it was somewhat modified at the University of Rochester N.Y. (USA) [19]. It is a very compact and agile method for the analysis of the paraxial behavior of centered and rotationally symmetrical optical systems of any kind. The name *y-nu* comes from the names of the ray variables, the ray height y and the reduced angle *nu*, which are calculated at each surface.

4.9.1 The *y-nu* Table

The complete "*y-nu*" table is divided into five sections (see Figure 4.10).

The uppermost contains the optical system's construction parameters, such as curvatures, thicknesses and its materials' physical properties of interest: refractive indices, Abbe numbers V and relative dispersions dn/n.

In the second section, the optical characteristics of the system are calculated, such as surface powers and reduced thickness.

In the third section, two (or more) paraxial rays are loaded and calculated, providing in the first column of boxes their data at the starting surface, namely their height and angles of propagation. Usually, such starting surface is the object plane.

In the fourth section, angles of incidence and their quotient q are calculated.

Finally in the fifth section, the aberration contributions at each surface are calculated, such as first-order chromatic aberrations (*Lch*, *Tch*) as well as the Seidel third-order aberrations (*Sph*, *Cma*, *Ast*, *Ptz*, *Dis*) [18,20].

4.9.2 Positive Sense Ray Trace

The paraxial or *y-nu* ray tracing method in the positive sense of ray propagation consists of two basic steps (see Figure 4.11):

- Positive sense translation between two surfaces:

$$y_{j+1} = y_j + \frac{t_j}{n_j} n_j u_j \quad \rightarrow \quad y_{j+1} = y_j + \tau_j \omega_j \tag{4.102}$$

- Positive sense refraction or reflection at surface j

$$n'_j u'_j = n_j u_j - \phi_j y_j \quad \rightarrow \quad \omega_{j+1} = \omega_j - \phi_j y_j, \tag{4.103}$$

Surface						
Space						
r						
c						
t						
n						
V						

$-\phi$						
t/n						
dn/n						

y						
nu						
$\bar{y}$						
$n\bar{u}$						
Λ						

i						
$\bar{i}$						
$q = \bar{i}/i$						

AchrL						
AchrT						

SphT						
ComaS						
AstLS						
Ptz						
Dist						

FIGURE 4.10   Complete *y-nu* table.

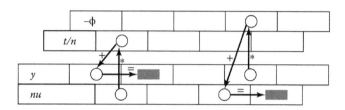

FIGURE 4.11   Positive sense ray trace.

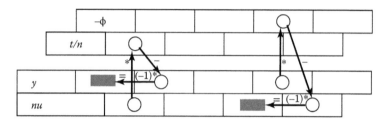

FIGURE 4.12   Negative sense ray trace.

where $-\phi_j = c_j\left(n_j - n_j'\right)$ is the surface optical power, and $\omega_j = n_j u_j = \omega_{j-1}'$ is called the reduced angle and $\tau_j = \dfrac{t_j}{n_j}$ the reduced thickness.

For a spherical mirror in air, the refraction equation we have is

$$-\phi_j = c_j\left(n_j - n_j'\right) = c_j\left(n_j - (-n_j)\right) = 2c_j n_j,\tag{4.104}$$

since in this case the "refracted" ray is actually reflected and therefore it propagates in the negative opposite direction, with $n' = -n$.

4.9.3 Negative Sense Ray Trace

In the *y-nu* table, it is also possible to perform ray propagation calculations in the opposite, negative sense (from right to left), Figure 4.12.

Negative sense translation between two surfaces:

$$y_{j-1} = -\left(-y_j + \frac{t_{j-1}}{n_{j-1}'}n_{j-1}'u_{j-1}'\right) = y_j - \tau_{j-1}\omega_{j-1}'\tag{4.105}$$

Negative sense refraction or reflection at surface *j*:

$$n_{j-1}'\,u_{j-1}' = -\left(-n_j u_j - \phi_j y_j\right) = \omega_j + \phi_j y_j = \omega_{j-1}'\tag{4.106}$$

4.10 Cardinal Points and Planes

4.10.1 Focal Points and Focal Distances

To find the image space focal point F', an axial ray from an axial object point at minus infinity has to be traced using the positive propagation sense equations up to the system's last (*k*) surface's image side, that is, up to the last refraction.

At the first surface such ray is denoted by $\left(y_1 = \text{arbitrary}, \omega_o = 0\right)$.
Then, the effective focal length equals to

$$f' = \frac{-y_1}{\omega'_k} \tag{4.107}$$

and the back focal length is given by

$$\text{bfl} = \frac{-y_k}{\omega'_k}. \tag{4.108}$$

The image space principal plane Π' location is given by

$$t_{(k,\Pi')} = \frac{y_k - y_1}{\omega'_k} = \text{bfl} - f' \tag{4.109}$$

and the image space nodal point location by

$$t_{(k,N')} = \text{bfl} - \frac{n_o}{n'_k} f'. \tag{4.110}$$

For the object space cardinal points, an axial ray starting form an axial image point at plus infinity has to be traced backwards, in the negative sense, through the whole system up to the object space of the first surface, that is, after refraction at surface 1.

At the last k surface, such ray is denoted as $y_k = \text{arbitrary}, \omega'_k = 0$.
The object space effective focal length is given by

$$f = \frac{-y_k}{\omega'_1} \tag{4.111}$$

and the front focal length

$$\text{ffl} = \frac{-y_k}{\omega'_1}. \tag{4.112}$$

The object space principal surface Π position is given by

$$t_{(1,\Pi)} = \frac{y_1 - y_k}{\omega'_1} = \text{ffl} - f \tag{4.113}$$

and the object space nodal point location by

$$t_{(1,N)} = \text{ffl} - \frac{n'_k}{n_o} f. \tag{4.114}$$

4.11 Entrance and Exit Pupil Calculation

To determine the properties of an optical system by means of the *y-nu* ray trace table, the most critical step is to choose the proper object space ray initial values. To find the location of an image, an arbitrary axial ray has to be traced, beginning at the point where the object plane crosses the optical axis, but for such axial ray to be the actual system's marginal ray, it has to propagate toward the entrance pupil's edge.

Assuming a particular surface within the system (the Diaphragm itself if it is given) has been chosen as the system's Stop, to determine the entrance pupil position it is necessary to trace backwards a principal-like ray. That is, from the object side of the Stop-axis intersection $(\overline{y}_D = 0, \overline{\omega}_D = \text{arbitrary})$, (for example $\omega'_D = 1$), through the frontal system elements and up to the first surface's object space side.

Once such calculation has been performed, the entrance pupil location then is given by

$$t_{(1,P_E)} = \frac{-\overline{y}_1}{\overline{\omega}_0}. \tag{4.115}$$

To calculate the entrance pupil size, another ray (this time marginal-like) has to be traced backwards from the diaphragm's edge $(y_D = h_D, \omega_D = \text{arbitrary})$, (e.g., $\omega_D = 0$), up to the found entrance pupil position:

$$h_{P_E} = y_1 + t_{(1,P_E)}\omega_o. \tag{4.116}$$

These results allow the actual marginal and principal rays to be calculated, as shown ahead.

To estimate the system image-forming characteristics up to third order, it is convenient to know its exit pupil position and size, since with them we'll be able to calculate the system's working F-number:

$$\text{WFN} = \frac{t_i - t_{P_S}}{2h_{P_S}} \approx \text{Airy disk diameter}, \tag{4.117}$$

where t_i is the last surface to image distance, t_{P_S} is the last surface to exit pupil distance and $2h_{P_S}$ is the exit pupil diameter.

The exit pupil location, as measured from the system's last surface (k) is found tracing a principal-like ray from the image side of the Stop-axis intersection $(\overline{y}_D = 0, \overline{\omega}'_D = \text{arbitrary})$ (e.g., $\overline{\omega}'_D = 1$), in the positive sense up to the last surface's image side, then

$$t_{(k,S)} = \frac{-\overline{y}_k}{\overline{\omega}'_k}. \tag{4.118}$$

The exit pupil size data is obtained by means of a marginal-like ray traced from the edge of the diaphragm $(y_D = h_D, \omega'_D = \text{arbitrary})$, (e.g., $\omega'_D = 0$), in the positive sense, up to the newly found exit pupil position and then

$$h_{P_S} = y_k + t_{(k,P_S)}\omega'_k. \tag{4.119}$$

4.11.1 Actual Meridional and Principal Paraxial Rays

Third-order aberration coefficients are calculated from paraxial actual marginal and principal ray data computed along and at every system surface.

From the entrance pupil data found above, the actual marginal and principal rays at the object plane are calculated respectively:

$$\mathbf{R}_0 = \begin{pmatrix} y_0 \\ \omega'_0 \end{pmatrix} = \begin{pmatrix} 0 \\ \dfrac{y_{P_E}}{t_o - t_{P_E}} \end{pmatrix}, \quad \overline{\mathbf{R}}_0 = \begin{pmatrix} \overline{y}_0 \\ \overline{\omega}'_0 \end{pmatrix} = \begin{pmatrix} h_0 \\ \dfrac{-h_0}{t_o - t_{P_E}} \end{pmatrix}, \tag{4.120}$$

where $t_o = t_{(0,1)}$ is the distance from the object plane to the first surface, h_o is the object height, y_{P_E} is the the pupil's height and $t_{P_E} = t_{(1,P_E)}$ is the distance from the first surface to the entrance pupil.

4.11.2 Ray Tracing, Object to Image Tabular Calculation

Tracing the marginal ray first, the image location is determined by tracing the marginal ray through the system up to the optical system's last surface's image side:

$$t_i = t_{(k,i)} = n'_k \frac{-y_k}{n'_k u'_k} = n'_k \frac{-y_k}{\omega'_k}. \tag{4.121}$$

For finding the image size, the principal ray is traced, from the edge of the object, through the optical system up to the image plane position just found, that is, by a translation equation applied to the image distance:

$$h_i = y_{k+1} = y_k + \tau_k n'_k \overline{u}'_k = y_k + \tau_k \overline{\omega}'_k. \tag{4.122}$$

4.12 The Delano Diagram (*y-y Bar* Diagram)

Both previous sections dealt with the "traditional" paraxial ray trace methods, developed by Gauss (matricial) and Conrady (tabular *y-nu*). Using these procedures, given the optical system construction specifications, its optical behavior is obtained (optical system ANALYSIS).

A technique by which system construction specifications are derived directly from the desired first-order specifications (optical system SYNTHESIS) is far preferable, and opens a broader space to optical designer's creativity. A further advantage of the method developed in the present section, is that the design internal parameters can be changed as required without upsetting its first-order general specifications.

The conventional procedure for laying out an optical system by matrix or *y-nu* methods, specially a complex one, involves considerable trial and error in adjusting the parameters of the system specifications until the desired first-order constraints are met. Moreover, altering any of the system variables (i.e., curvatures, and thicknesses) will in general upset all of the first-order properties, thereby introducing a serious complication in the process of modifying a design to improve its performance while maintaining its first order constraints.

On the contrary, and more beneficial, the technique by which system parameters are derived from the first-order constraints, and also able to alter the design without upsetting the first-order constraints, is the "*y-y bar* method." Such a technique was developed by Erwin Delano [21]. It is extremely elegant and powerful, and is revolutionizing the field of first-order optical design. Professor Roland V. Shack, at Optical Sciences Center (Tucson, Arizona, USA), gave a great impulse to it [22–24].

Delano's *y-y bar* method variables are the marginal (y) and chief ray $(\bar{y})$ heights at each surface, the refractive indices and the Lagrange invariant, which operate as a scale factor. The equations and procedures to be given here will deliver, from these simple variables, the optical system's parameters: curvatures, thicknesses, apertures, pupils data. This means that the *y-y bar* method constitutes a Synthesis method [25–28].

Delano's method is based on two fundamental equations:

Reduced Thickness

$$\tau_j = \frac{1}{\Lambda} \begin{vmatrix} y_j & \bar{y}_j \\ y_{j+1} & \bar{y}_{j+1} \end{vmatrix} \tag{4.123}$$

and power (surface or thin lens)

$$\phi_j = \frac{1}{\Lambda} \begin{vmatrix} \omega_j & \bar{\omega}_j \\ \omega'_j & \bar{\omega}'_j \end{vmatrix} \quad (4.124)$$

where, once Equation 4.123 values have been calculated, the rays reduced angles of propagation between two consecutive surfaces are calculated as

$$\omega_j = \frac{y_{j+1} - y_j}{\tau_j}, \bar{\omega}_j = \frac{\bar{y}_{j+1} - \bar{y}_j}{\tau_j}. \quad (4.125)$$

On Equations 4.123 and 4.124, it becomes evident that the Lagrange invariant plays a scale factor role.

The (y) and $(\bar{y})$ values correspond to the Cartesian coordinates of intersection points of consecutive straight-line segments of a simple polygonal drawing: the projection of a "paraxial skew ray."

Modifying any segment intersection coordinates on a given $y - \bar{y}$ diagram, it is possible to control all first-order characteristics of the optical system represented by the diagram, that is, an element or its total focal length, pupil's size and position, magnification, and so on, as needed.

4.12.1 Construction of the Paraxial Pseudo Skew Ray

To start the study of Delano's method, let's consider first the simplest possible optical system: a positive thin lens forming a real image of a real object, that is, operating at finite conjugates.

First, consider the Principal $y_b = \bar{y}$ and Marginal $y_a = y$ rays traced on the Meridional plane yz as usual (see Figure 4.13).

Both "a" and "b" rays convey information through the optical system and describe the way the thin lens transforms the object space into the image space. However, it is more convenient to visualize the optical system behavior by means of a single ray.

Delano's method is based on the tracing a single paraxial pseudo skew ray through the system, created as a linear combination of both marginal and principal rays. This ray is called "pseudo", because in paraxial optics only meridional (yz plane) rays can be traced, since paraxial conventions are valid for rays propagating on a plane containing the optic axis, the meridional y,z plane.

This two-ray linear combination, will "propagate" out of any marginal plane, it represents a skew ray, which contains unaltered both marginal and chief ray information. This is possible thanks to the linear character of the transfer and refraction paraxial equations. Since we are assuming only axial rotationally symmetric systems, where both marginal and chief rays are meridional rays, they can be represented on two mutually orthogonal meridional planes, to build the three-dimensional skew ray as their linear combination.

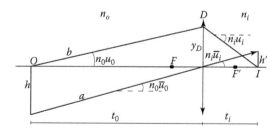

FIGURE 4.13 Conventional meridional ray plots.

The reverse is also true: the skew ray can be projected onto two mutually orthogonal planes as the proper marginal and chief rays. By convention, in a usual right-handed Cartesian three-dimensional coordinate system, Delano chose the vertical (y, z) plane for the marginal ray and the horizontal $(\bar{y}, z)$ plane for the chief ray; hence, the name "*y-y bar* diagram" was given by Delano to the skew ray projection. Now a single skew ray conveys all the information of how the optical system performs the object space to the image space transformation. Figure 4.14 shows how the paraxial skew ray is constructed.

In this first stage development of Delano's method y and $\bar{y}$ heights and the axial position of the plane perpendicular to the optical axis where they take place, constitute the three-dimensional coordinates of a point on the skew ray. Note that in Figures 4.13 and 4.14 the positive thin lens mount constitutes its diaphragm.

Delano then projected the so obtained paraxial skew ray onto an arbitrary plane perpendicular to the optical axis, which for this simple example is shown in Figure 4.15, which constitutes the Delano *y-y bar* diagram for this system. Note that such projection is invariant with respect to the location of the *y-y bar* projection plane along the z-axis.

To widen the understanding of this key procedure, let us consider a somewhat more enriched example. Figure 4.16 shows the way the *y-y bar* diagram was obtained for a thin lens telephoto. First, the principal ray meridional plane trace was rotated clockwise about the optic axis, to fall onto the horizontal $(\bar{y}, z)$ plane, the marginal ray trace was left undisturbed on the (y,z) plane. Then the skew ray was built by means of the linear combination of principal and marginal ray heights at each surface to create points $(y, \bar{y})$ in three dimensional space. Next these points were interconnected by means of straight lines to build the three-dimensional skew ray. Finally, the skew ray was projected onto the *y-y bar* plane shown at the extreme left end of Figure 4.16.

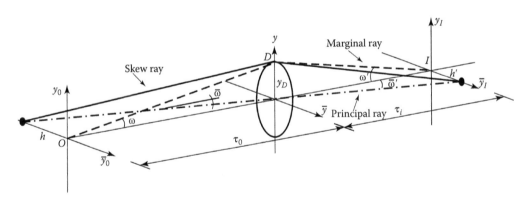

FIGURE 4.14 Skew ray construction: principal ray, marginal ray and skew ray.

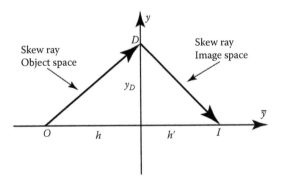

FIGURE 4.15 The projected skew ray: the *y-y bar* diagram.

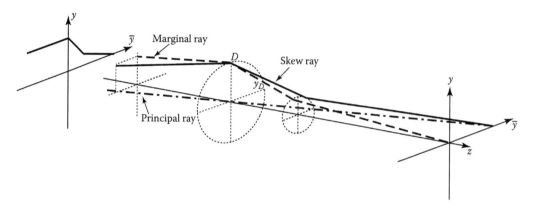

FIGURE 4.16 The *y-y bar* diagram construction for a thin lens telephoto.

Taking advantage of its representation in orthogonal two-dimensional coordinates, all of the *y-y bar* properties can be expressed by means of vector algebra and the optical properties of the system it represents, as will be given ahead.

4.12.2 Point Vector z and Reduced Distance τ

Every point on a Delano diagram line corresponds to a particular plane perpendicular to the optic axis of the system it represents. The end points of each straight-line segment therefore correspond to two such surfaces: the beginning and the end of a given distance along the optic axis; these surfaces are usually optically active, thin lenses or refractive surfaces. Each of these points then can be represented by means of a point vector $\mathbf{z}_j = \left(\bar{y}_j, y_j\right)$ starting at the coordinate origin end ending at the point itself.

The equation given above for the reduced distance τ between two planes along the optic axis (Equation 4.123) can be interpreted now as the vector product between the point vectors representing these planes, as the beginning and the end of a Delano diagram segment:

$$\tau_j = \mathbf{z}_j \times \mathbf{z}_{j+1} = - \begin{vmatrix} \bar{y}_j & y_j \\ \bar{y}_{j+1} & y_{j+1} \end{vmatrix} \hat{\mathbf{i}}_z. \tag{4.126}$$

But this vector needs to be scaled for this operation to represent the actual distance value between planes represented by the pair of point vectors; the Lagrange invariant performs such scaling:

$$\tau_j = \left| \mathbf{z}_j \times \mathbf{z}_{j+1} \right| = \frac{-1}{\Lambda} \begin{vmatrix} \bar{y}_j & y_j \\ \bar{y}_{j+1} & y_{j+1} \end{vmatrix}. \tag{4.127}$$

Note the column transposition and the minus sign in these two equations, performed to obey right-handed vector algebra conventions.

In his paper, E. Delano defined a pure scalar determinant operation, given above, at the introduction of this section (Equation 4.123). Positive physical distances τ correspond to vectors τ pointing in the positive z-axis direction and negative distances between surfaces to vectors pointing in the negative z-direction, as evident from Equation 4.126. Remember also that the magnitude of a vector product of two vectors is equal to the area of the associated rhomboid parallelogram spanned by them. To simplify the interpretation of a Delano diagram, the magnitude of the distance between two consecutive surfaces, represented by the line segment connecting their diagram points, is interpreted as one half the rhomboid area, that is, as the area of the triangle formed by both point vectors and the line segment defined by them.

4.12.3 Line Vector w and Power φ

Also, each straight line defined by the beginning and end points of a Delano diagram segment can be represented by a line vector $\mathbf{w}_j$, beginning at surface $j - 1$ and ending at surface j:

$$\mathbf{w}_j = \left(\bar{\omega}'_{j-1}, \omega'_{j-1}\right) = \left(\bar{\omega}_j, \omega_j\right). \tag{4.128}$$

This is the skew ray departing from surface $j - 1$ and incident on surface j. Consider the consecutive diagram line vector

$$\mathbf{w}'_j = \left(\bar{\omega}'_j, \omega'_j\right) = \left(\bar{\omega}_{j+1}, \omega_{j+1}\right), \tag{4.129}$$

which begins at their common point j and ends at the consecutive point $j+1$, which constitutes the skew ray refracted at surface j.

Assuming in the Delano diagram that point j represents a thin lens ϕ_j, these line vectors will feature dissimilar slopes, since the lens power changes the angles of propagation of both rays, marginal ω and principal $\bar{\omega}$.

Once more, we can apply a vector product to these two line vectors (Equations 4.128 and 4.129) to find the magnitude of the thin lens power, which provoked the change of slope between them:

$$\vec{\phi}_j = \mathbf{w}_j \times \mathbf{w}'_j = \begin{vmatrix} \bar{\omega}_j & \omega_j \\ \bar{\omega}'_j & \omega'_j \end{vmatrix} \hat{\mathbf{i}}_z. \tag{4.130}$$

Therefore, positive powers are also vectors pointing in the positive z-axis direction, and negative powers to vectors $\vec{\phi}$ pointing in the negative z-axis direction.

But again, this vector result needs scaling, to obtain the actual scalar power value associated to the thin lens (in this case) or to the refracting/reflecting surface represented by the diagram point j. Note that in Equation 4.131 determinant columns have been commuted, therefore there is a negative sign, with respect to the power Equation 4.124 given in Delano's paper.

$$\phi_j = \frac{-1}{\Lambda}\left|\mathbf{w}_j \times \mathbf{w}'_j\right| = \frac{-1}{\Lambda} \begin{vmatrix} \bar{\omega}_j & \omega_j \\ \bar{\omega}'_j & \omega'_j \end{vmatrix} \tag{4.131}$$

4.12.4 The Lagrange Invariant as a Vector

Even apart of this flexible representation, the *y-y bar* diagram is in itself ambiguous. Not until all the refractive indices and the Lagrange invariant have been specified does it become possible to synthesize a specific, physically real optical system prescription from its intended Delano diagram. By changing the values of the indices and the Lagrange invariant, one may obtain a variety of optical systems all of which have the same *y-y bar* diagram. Far from being a defect, this ambiguity is in fact highly advantageous in the application of the *y-y bar* diagram to specific design problems.

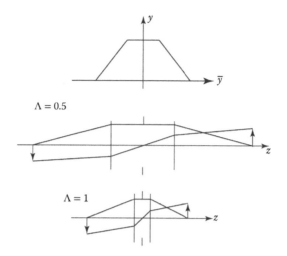

$\Lambda = 0.5$

$\Lambda = 1$

FIGURE 4.17   Scaling effect of the Lagrange invariant.

Changing the Lagrange invariant for a given Delano diagram obviously has no effect on the transverse ray heights along the system. From Equation 4.127, it is evident that the Lagrange invariant constitutes a scale factor relating the longitudinal dimensions (thickness, curvature radii, focal lengths) to the transverse ray properties $(\bar{y}, y)$.

Reduced angles $(\bar{\omega}, \omega)$ are therefore affected, as shown in Figure 4.17.

The Lagrange invariant in Delano's method also constitutes a vector pointing in the positive z-axis direction, calculated as the vector product of a point and a line vector associated to the common point of two consecutive diagram segments:

$$\vec{\Lambda} = -\left| \mathbf{z}_j \times \mathbf{w}_j \right| = - \begin{vmatrix} \bar{y}_j & y_j \\ \bar{\omega}'_j & \omega'_j \end{vmatrix} = y_j \bar{\omega}'_j - \bar{y}_j \omega'_j. \tag{4.132}$$

Note that due to its basic Equations 4.127, 4.131, and 4.132, a properly drawn Delano diagram constitutes a clock-wise progressing succession of line segments such that neither the segments themselves, nor their extensions touch the origin of coordinates or can be drawn beyond it.

4.12.5 Correct Definition of Delano Diagrams

For an optical system to be correctly represented by means of a Delano diagram, three conditions need to be accurately satisfied:

1. Refractive indices n_j must be defined for each and all line segments.
2. No diagram line or their extensions should pass through the origin.
3. The Lagrange invariant must be defined to specify system's axial dimensions.

Delano's representation in fact is extremely compact and versatile; one could say that any set of interconnected straight-line segments constitute the paraxial representation of a particular optical system, as long as those three precautions are taken into account.

Specially: No line passing through the origin can be a skew ray projection, because it would mean that a pupil and an image coincide at the same axial position, which is impossible.

Even worse, if a line segment (positive refractive index) or its extension can be drawn beyond the coordinate origin, it will imply that both marginal and principal rays propagate in the negative sense instead of the assumed positive one.

4.13 The Lagrange Invariant as a Bilinear Form and Ω–Ω Bar Diagram

Another field of science and technique comes in to enrich this new method of optical system synthesis; it is the so-called projective geometry. In projective geometry, a "bilinear form" is defined, where two three-dimensional orthogonal spaces (x, y, z) and (u, v, w) are interconnected by means of a linear equation linking both spaces, such as

$$1 = ux + vy + wz. \tag{4.133}$$

Under such projective geometry interconnection among these spaces, it can be demonstrated that a line in one of them becomes a point in the second one and a point in one of them becomes a line in the other, called its "dual." The line's slope in one space is equal to the point's position vector slope in the second one, and the converse is also true. Then, the Lagrange invariant (Equation 4.132) if normalized ($\Lambda = 1$) under the projective geometry scope, becomes a two-dimensional bilinear form:

$$1 = y_j \frac{\overline{\omega}'_j}{\Lambda} - \overline{y}_j \frac{\omega'_j}{\Lambda} = y_j \overline{\Omega} - \overline{y}_j \Omega. \tag{4.134}$$

This bilinear normalized Lagrange invariant then gives origin to a new space, dual to the original *y-y bar*: the Ω-Ω bar space, which contains the same information about the optical system, but as a new dual space.

In this new diagram, the line joining points $\mathbf{W}_j = \left(\overline{\Omega}_j, \Omega_j\right)$ and $\mathbf{W}'_j = \left(\overline{\Omega}'_j, \Omega'_j\right)$ correspond to the vector representation of the thin lens (ϕ_j), which provoked the change in magnitude of the angles of propagation of the marginal and principal rays when refracted by it:

$$\vec{\phi}_j = \mathbf{W}_j \times \mathbf{W}'_j. \tag{4.135}$$

Together these vectors define a triangle whose area constitutes a graphical representation of the thin lens (or refractive surface) power.

This new space is mostly unexplored, constituting a "virgin land" awaiting to give us its riches.

4.14 Conjugate Lines, Object-Image Conjugate Sets

A remarkable property of the Delano diagram is that it naturally contains all possible object-image conjugate pairs, each pair always located on a common straight line joining three points: object point, coordinate origin and image point.

In previous diagrams, only two such lines, called conjugate lines, have been used: the vertical y axis (where the Stop and its images, both pupils, are allocated) and the horizontal $\overline{y}$ axis, where the *y-y bar* diagram specifies the object and the image points as diagram intersections with the $\overline{y}$ axis.

Figure 4.18 shows a typical thin lens triplet (thin lenses A, B, C).

On it several additional conjugate lines are shown:

1. The back focal point F', image of object plane at $-\infty$, is marked on the image space line by a line parallel to the object space line and passing through the coordinate origin.
2. The front focal point F, image of object plane at $+\infty$, is marked on the object space line by a line parallel to the image space line and passing through the coordinate origin.

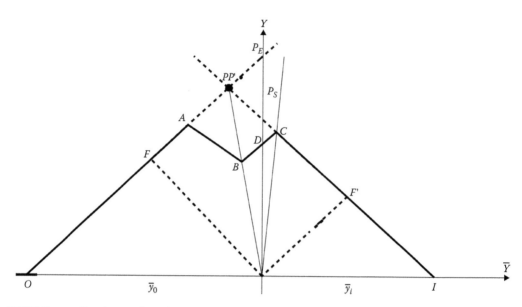

FIGURE 4.18 Thin lens triplet and conjugate lines.

3. The image of lens B as seen from object space through lens A is marked by the intersection of the extended object space line and the conjugate line passing through lens B and the origin.
4. The image of the stop, as seen from the object space (the entrance pupil E) through lenses A and B is marked by the intersection of object space line and the vertical *y* axis, i.e., the stop-pupils conjugate line.
5. The image of lens C as seen from object space, through lenses A, B, and the stop (D) is marked by the intersection of the object space line extension and the lens C conjugate line, passing through the coordinate origin and through point C.
6. The image of the stop, as seen from image space (the exit pupil S) through lens C is marked by the intersection of image space line and the vertical *y* axis, i.e., the stop-pupils conjugate line.
7. The "single thin lens equivalent" of the whole triplet is indicated as a thick dot at the intersection of object and image space lines, which also corresponds to the positions of both principal planes, therefore titled PP′ on the triplet diagram.

4.15 Vignetting and the Vignetting Diagram

By the word vignetting, it is meant any partial obstruction of the marginal or principal pencils of rays, due to an insufficient diameter of the lens mounts or auxiliar surfaces of the optical system affected by it.

4.15.1 Physical Meaning of Vignetting

Somewhat easier to explain is the contrary, a system free of vignetting. For an un-vignetted optical system, all pencils of rays are free to pass through all optical components and the stop to reach the image plane without any loss of energy and/or information (see the section on the Helmholtz invariant). The whole set of possible pencils of rays (actually cones of rays) emitted from the object plane are defined by two particular ones:

1. The marginal cone of rays starts at the axis-object plane intersection; its angular span with respect to the optical axis is defined by the marginal ray's angle of propagation.

2. The principal cone of rays has its vertex at the meridional object plane's edge point, the principal ray itself constitutes its core ray and its base covers the whole entrance pupil.

In a vignette-free system, as propagating through the system, both pencils span the whole aperture of the stop and reach the image plane unperturbed by the remaining apertures.

4.15.2 Vignetting Diagram Construction

Similar to Section 4.14, the development of the vignetting diagram is created by the succession of images of the internal lens mountings, seen from the image plane, projected now onto the whole image space line. But the image point $h_i = \bar{y}_i$ in a *y-y bar* diagram is nothing else than the edge of the imaged field, so onto that point the most critical pencil of rays will converge: the principal ray's pencil, most prone to suffer from vignetting. Conjugate lines for all system elements identify the location on the "image space line" of their image planes as seen from the edge of the system's image plane. Such image space is virtual; to convert its meaning into a scaled-down real space, images positions must be projected down onto the diagram's $\bar{y}$-axis, the one spanning from the object plane O to image plane I and equivalent, therefore, to the system's optic axis.

The existence or absence of vignetting depends upon the ratio of each lens mounting's image size to stop's image size, as given in Section 4.7.1, Equation 4.75. There will be vignetting if such ratio is less than unity. Such ratio can be constructed graphically by means of the Delano diagram performing the following steps (see Figure 4.19):

1. Determine on the image space line the images of all system elements by means of their conjugate lines.
2. Sufficiently below, draw a horizontal line parallel to the $\bar{y}$-axis.
3. Project downwards the whole set of image points found (1), onto line (2).
4. From the projection of the image point $h_i = \bar{y}_i$ onto line (2), draw upwards a vertical line, with length equal to $h_i = \bar{y}_i$.
5. From the edge of this vertical $h_i = \bar{y}_i$ line, trace lines to each one of the projected images' points (3) on line (2).

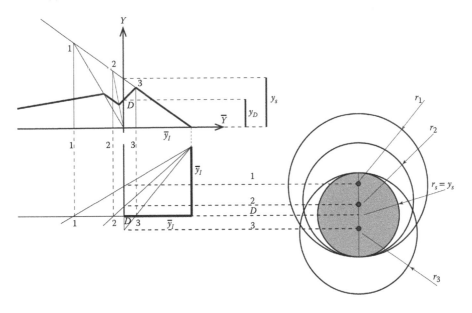

FIGURE 4.19 Construction of a vignetting diagram.

6. Some lines drawn in (5) intersect the extended y-axis, marking the positions of each and every axis-component intersection as if "photographed" onto a photographic plate located at the stop position, since the stop point $\mathbf{z}_S$ locates on the y-axis.
7. Draw a new vertical line away to the right of the $h_i = \bar{y}_i$ line drawn in (4).
8. Draw horizontal lines from intersections (6) to this new line (7).
9. With the center at the extension of the "photographed" position of the stop on vertical line (7), draw a radius $r_S = h_S = y_S$ circle.
10. With the center at each "photographed lens mount center" (8), draw a circle tangent to either the upper or the lower portion of "stop circle" (9). These circles radii $r_1, r_2, r_3, \dots$ feature identical scale as r_S (9).
11. The lens mount diameters, required for the triplet to be free of vignetting, are calculated by means of the following operations, equivalent to Equation 4.75:

$$D_1 = D_D \frac{r_1}{r_S}, D_2 = D_D \frac{r_2}{r_S}, D_3 = D_D \frac{r_3}{r_S}, \dots \quad \left(D_D = 2 y_D\right). \tag{4.136}$$

4.16 Y-Y Bar Optical System Synthesis Procedure

Given the way it is wanted for the marginal and chief ray to behave as they propagate through the desired optical system (i.e., the Delano y-y bar diagram), its design parameters can be easily calculated by applying the simple formulas given at the beginning (Equations 4.123 through 4.125). Let us assume that the desired optical system has been schematized as a thin-lens arrangement in air and the Lagrange invariant has been specified as well.

Then the distances between successive surfaces are given by Equation 4.123:

$$\tau_0 = \frac{-1}{\Lambda} \begin{vmatrix} \bar{y}_0 & y_0 \\ \bar{y}_1 & y_1 \end{vmatrix}; \quad \tau_1 = \frac{-1}{\Lambda} \begin{vmatrix} \bar{y}_1 & y_1 \\ \bar{y}_2 & y_2 \end{vmatrix}. \tag{4.137}$$

Then, marginal and principal ray reduced angles of propagation are calculated (Equation 4.125):

$$\omega_0 = \frac{y_1 - y_0}{\tau_0}, \bar{\omega}_0 = \frac{\bar{y}_1 - \bar{y}_0}{\tau_0}; \omega_1' = \frac{y_2 - y_1}{\tau_1}, \bar{\omega}_1' = \frac{\bar{y}_2 - \bar{y}_1}{\tau_1}. \tag{4.138}$$

The power of the first thin lens therefore is calculated (Equation 4.124):

$$\phi_1 = \frac{-1}{\Lambda} \begin{vmatrix} \bar{\omega}_0 & \omega_0 \\ \bar{\omega}_1' & \omega_1' \end{vmatrix} \tag{4.139}$$

and similarly for all remaining optical power elements.

The distances and thin lens powers are this way calculated to synthesize the whole optical system as a succession of thin lenses of powers $\phi_1, \phi_2, \phi_3, \dots \phi_k$ with reduced distances $\tau_0, \tau_1, \tau_2, \tau_3, \dots, \tau_k$.

The completed y-y bar is shown in Figure 4.20, where variables and aberrations are included as defined Flores-Hernandez and Stavroudis [25].

	Λ=																						
	$\bar{Y}$	Y	τ	$\bar{\omega}$	ω	ϕ	n	V	t	c	r	dn/n	i	$\bar{i}$	q	Sig	$\overline{\text{Sig}}$	Lch	Tch	Sph	Cma	Ast	Ptz	Dis
0																								
1																								
2																								
3																								
4																								
5																								
6																								

FIGURE 4.20 Complete *y-y bar* tabular table.

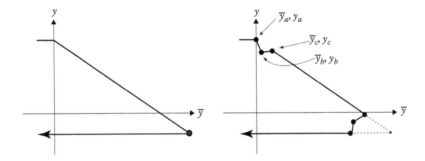

FIGURE 4.21 Thin-to-thick doublet Keplerian telescope transformation.

4.16.1 Optical System Synthesis, Thin to Thick Lens Transformation

It has been mentioned that the Delano power equation can be applied to both thin lenses and real lens elements as well. Transformation of a thin lens system, for example a Keplerian telescope, onto a thick-components system is a two-step procedure.

First step: each diagram point $\mathbf{z}_j$ has to be broken into two $\mathbf{z}_{ja}$, $\mathbf{z}_{jb}$ to represent a thick singlet or into three $\mathbf{z}_{ja}$, $\mathbf{z}_{jb}$, $\mathbf{z}_{jc}$ to represent a thick cemented doublet.

For example, the Keplerian's telescope (Figure 4.21) thin lens objective $\mathbf{z}_1$ is transformed onto a thick cemented doublet as

$$\mathbf{z}_{1a} = \mathbf{z}_1 + \Delta_a \mathbf{z} = \left(\bar{y}_1 + \Delta \bar{y}_a, y_1 + \Delta y_a\right) = \left(\bar{y}_{1a}, y_{1a}\right)$$

$$\mathbf{z}_{1b} = \mathbf{z}_1 + \Delta_b \mathbf{z} = \left(\bar{y}_1 + \Delta \bar{y}_b, y_1 + \Delta y_b\right) = \left(\bar{y}_{1b}, y_{1b}\right) \tag{4.140}$$

$$\mathbf{z}_{1b} = \mathbf{z}_1 + \Delta_c \mathbf{z} = \left(\bar{y}_1 + \Delta \bar{y}_c, y_1 + \Delta y_c\right) = \left(\bar{y}_{1c}, y_{1c}\right)$$

In order to maintain the original thin-lens characteristics of the system, the first point $\mathbf{z}_{1a}$ must always belong to the $\mathbf{z}_1$ thin lens object space line and the third point $\mathbf{z}_{1c}$ must always lie on the image space line of the same thin lens objective $\mathbf{z}_1$.

The intermediate point $\mathbf{z}_{1b}$ is placed somewhere in between a and b points but a little bit below the line connecting them, in order for the cemented surface it represents to behave as a negative power surface.

Second step : the refractive indices of the thick lens straight-line segments, corresponding to glass, of the new diagram, must be defined.

For a crown in front of the cemented doublet, thin-to-thick transformation as defined by Equation 1.140 means for example:

$$n_0 = 1, n_{1a} = n_{N-BK7} = 1.5168, n_{1b} = n_{SF6} = 1.80518, n_1' = 1$$

The axial physical thicknesses of the doublet objective are

$$t_{1a} = n_{1a}\tau_{1a} = n_{1a} \frac{-1}{\Lambda} \begin{vmatrix} \overline{y}_{1a} & y_{1a} \\ \overline{y}_{1b} & y_{1b} \end{vmatrix}, t_{1b} = n_{1b}\tau_{1b} = n_{1b} \frac{-1}{\Lambda} \begin{vmatrix} \overline{y}_{1b} & y_{1b} \\ \overline{y}_{1c} & y_{1c} \end{vmatrix}. \tag{4.141}$$

The reduced angles of propagation of the rays for the thick objective lenses are

$$\omega_{1a}' = \frac{y_{1b} - y_{1a}}{\tau_{1a}}, \overline{\omega}'_{1a} = \frac{\overline{y}_{1b} - \overline{y}_{1a}}{\tau_{1a}}, \omega_{1b}' = \frac{y_{1c} - y_{1b}}{\tau_{1b}}, \overline{\omega}'_{1b} = \frac{\overline{y}_{1c} - \overline{y}_{1b}}{\tau_{1b}}. \tag{4.142}$$

In turn, the surface powers are

$$\phi_{1a} = \frac{-1}{\Lambda} \begin{vmatrix} \overline{\omega}_0 & \omega_0 \\ \overline{\omega}'_{1a} & \omega'_{1a} \end{vmatrix}, \phi_{1b} = \frac{-1}{\Lambda} \begin{vmatrix} \overline{\omega}_{1a} & \omega_{1a} \\ \overline{\omega}'_{1b} & \omega'_{1b} \end{vmatrix}, \phi_{1c} = \frac{-1}{\Lambda} \begin{vmatrix} \overline{\omega}_{1b} & \omega_{1b} \\ \overline{\omega}'_1 & \omega'_1 \end{vmatrix} \tag{4.143}$$

and finally, the surface curvatures are calculated as

$$c_{1a} = \frac{\phi_{1a}}{n_{1a} - n_0} = \frac{\phi_{1a}}{n_{1a} - 1}, c_{1b} = \frac{\phi_{1b}}{n_{1b} - n_{1a}}, c_{1c} = \frac{\phi_{1c}}{n_1' - n_{1b}} = \frac{\phi_{1c}}{1 - n_b}. \tag{4.144}$$

A similar process has to be performed to synthesize the second (eyepiece) cemented or even air spaced doublet.

4.17 Exact Ray Tracing

Since ray tracing can be automated with computers, it is nowadays the most important tool for complete optical systems design [3,5,11]. For modern optical systems, it is even today with the help of modern computers not possible to replace ray tracing (sequential or non-sequential) by pure paraxial ray tracing or wave-optical methods. For modern design, ray tracing in combination with wave-optical evaluation methods like the calculation of the point spread function [3], assuming that only the exit pupil of the system introduces diffraction, is a sufficiently accurate method to analyzed imaging systems.

The first expressions for exact ray tracing were developed by the Hungarian scientist Josef Petzval in 1857; unfortunately, they were lost in a fire before publication. Through time, exact ray tracing methods have experienced several changes in form but not in content, searching always for better efficiency and according with the calculus methods and instruments available in each epoch: trigonometric and logarithmic tables, mechanical calculators, slide rules, and so on, until the advent of electronic computers.

The exact ray tracing methods can be divided into: the purely geometric method [29,30], the method based on differential geometry [4,31] and the vector method [32]. They have been adapted to commercial optical design programs. However, for an optical engineer, it is important to have the basic knowledge of each method, but in their common work life may not implement them. Furthermore, these methods are suitable according to the needs as in GRIN materials [33,34], diffraction gratings [35,36], and other devices.

A precondition for ray tracing is that the optical system is known very well. It is not sufficient to know some paraxial parameters but it is necessary to know the following data of the surfaces as well as the materials:

- Type of the surface, e.g., plane, spherical, parabolic, cylindrical, toric or other aspheric surface.
- Characteristic data of the surface itself, e.g., the radius of curvature in the case of a spherical surface or the aspheric coefficients in the case of an aspheric surface.
- Shape and size of the boundary of the surface, e.g., circular with a certain radius, rectangular with two side lengths or annular with an interior and an outer radius.
- Position and orientation of the surface in all three directions of space.
- Refractive indices of all materials and their dependence on wavelength.

The tracing of a given ray through an optical system has the following general structure:

1. Determine the point of intersection of the ray with the following optical surface. If there is no point of intersection or if the hit surface is absorbing, mark the ray as invalid and finish the tracing of this ray. Depending on the type of ray tracing, it may also be necessary in this case to leave the ray unchanged and to go to step 4. If there is a point of intersection, go to step 2.
2. Calculate the normal surface in the point of intersection.
3. Apply the law of refraction or reflection. Then, the new direction of the deflected ray is known and the point of intersection with the surface is the new starting point of the ray.
4. If there is another surface in the optical system, go back to step 1 or, if not, then finish the tracing of this ray.

References

1. Born M, Wolf E. *Principles of Optics*. Cambridge: Cambridge University Press, 1999.
2. Walther A. *The Ray and Wave Theory of Lenses*. Cambridge: Cambridge University Press, 1995.
3. Malacara-Hernández D, Malacara-Hernandez Z. *Handbook of Optical Design*. Boca Raton, FL: CRC Press, 2013.
4. Stavroudis ON. *The Optics of Rays, Wavefronts, and Caustics*. Cambridge, MA: Academic Press, 1972.
5. Smith WJ. *Modern Optical Engineering*. Bellingham, WA: SPIE Press, 2008.
6. Begunov BN., Zakaznov NP. *Teoría de Sistemas Ópticos*. Moscu: Mir, 1976.
7. Kasunic KJ. *Optical Systems Engineering*. New York, NY: McGraw-Hill, 2011.
8. Fischer RE., Tadic-Galeb B, Yoder PR. *Optical System Design*. New York, NY: McGraw-Hill, 2008.
9. Dereniak EL, Dereniak TD. *Geometrical and Trigonometric Optics*. Cambridge: Cambridge University Press, 2008.
10. Welford WT. *Aberrations of Optical Systems*. London: Adam Hilger, 1986.
11. Brouwer W. *Matrix Methods in Optical Instrument Design*. New York, NY: Benjamin, 1964.
12. Gerrard A, Burch JM. *Introduction to Matrix Methods in Optics*. London: Wiley, 1975.
13. Kloos G. *Matrix Methods for Optical Layout*. Bellingham, WA: SPIE, 2007.
14. Nussbaum A, Phillips R. *Contemporary Optics for Scientists and Engineers*. Englewood Cliffs, NJ: Prentice-Hall, 1976.
15. Kogelnik H, Li T. Laser beams and resonators, *Appl. Opt.*, **5**, 1550–1567, 1966.
16. Siegman AE. *Lasers*. Mill Valley, CA: University Science Books, 1986.

17. Gross H. *Handbook of Optical Systems*, Vol. 1. New York, NY: Wiley, 2005.
18. Conrady AE. *Applied Optics and Optical Design*. Dover Publications, 1957.
19. Hopkins R, Hanau R, Osenberg H et al. *Military Standardized Handbook 141*. US Government Printing Office , 1962.
20. Emsley HH. *Aberrations of Thin Lenses*. Great Britain: Constable and Company, 1956.
21. Delano E. First-order design and the $y - \bar{y}$ diagram, *Appl.Opt.*, **2**, 1251–1256, 1963.
22. Shack RV. Optical systems layout, New Methods in Optical Design and Engineering Optical Science Center, University of Arizona, 1972.
23. Shack RV. Analytic system design with pencil and ruler—The advantages of the $y - \bar{y}$ diagram, *Proc. SPIE*, **39**, 127–140, 1973.
24. Lopez-Lopez F. Analytical aspects of the $y - \bar{y}$ diagram, *Proc. SPIE*, **39**, 151–164, 1973.
25. Flores-Hernández R, Stavroudis ON. The Delano $y - \bar{y}$ diagram applied to optical system synthesis and optimization, $y - \bar{y}$ diagram point displacements, *Optik*, **102**, 171–180, 1996.
26. Flores-Hernández R, Stavroudis ON. Real time manipulations of the Delano diagram as a new linear tool for optical system optimization, *Optik*, **103**, 141–147, 1996.
27. Besenmatter W. Design zoom lenses aided by Delano diagram, *Proc. SPIE*, **273**, 242–255, 1980.
28. Zhang Y, Zhao J, Zhao X et al. High power excimer laser image relay system analysis using Delano diagram, *Proc. SPIE*, **8796**, 879622, 2013.
29. Hopkins RE, Hanau R. Fundamental methods in ray tracing, In *Military Standardization Handbook: Optical Design, MIL-HDBK 141*. Washington, DC: U.S. Defense Supply Agency, 1962.
30. Feder DP. Differentiation of ray tracing equations with respect to construction parameters of rotationally symmetric optics, *JOSA*, **58**, 1494–1505, 1968.
31. Spencer GH, Murty MVRK. General ray tracing procedure, *J. Opt. Soc. Am.*, **52**, 672–678, 1962.
32. Rodionov S. Automation of Optical Systems, Moscow: Mashinostroenie, 1982.
33. Sharma A, Kumar DV, Ghatak AK. Tracing rays through graded-index media: A new method, *Appl. Opt.*, **21**, 984–987, 1982.
34. Sharma A. Computing optical path length in gradient-index media: A fast and accurate method, *Appl. Opt.*, **24**, 4367–4370, 1985.
35. Smith RW. A note on practical formulae for finite ray tracing through holograms and diffractive optical elements, *Opt. Commun.*, **55**, 11–12, 1985.
36. Welford WT. A vector ray tracing equation for hologram lenses of arbitrary shape, *Opt. Commun.*, **14**, 322–323, 1975.

Appendix 4.1 Optics Terminology

Object: Anything from which electromagnetic radiation can be emitted or scattered.

Image: An electromagnetic reconstruction of any object due to the propagation through an optical system.

Optical component or optical element: Every lens, mirror, prism, beam splitter, and so on.

Lens: Optical component, usually made of glass, defined by two spherical surfaces assumed to be perfectly polished.

Optical axis: Imaginary straight line passing through both centers of curvature of a lens; this axis constitutes the axis of rotational symmetry of the lens.

Spherical mirror: Optical component, usually made of glass or metal, with just one spherical surface, assumed to be perfectly polished. A spherical surface does not have a unique axis of symmetry.

Front plane mirror: Optical component, usually made of glass, with a plane surface assumed perfectly polished and usually metallized or coated with dielectric multi-layers to produce a reflection close to 100% efficiency. A plane mirror has axis anywhere on its surface. In optical systems, the use of "rear surface mirrors" (regular ones) should be avoided.

Prism: Optical component, most frequently made of glass, and defined by a set of plane surfaces. Some prisms are used to deviate ray bundles, steering them into a new desired direction of propagation, usually to maintain the optical systems within reasonable overall dimensions or to invert and/or revert an image.

Optical system: An ordered set of optical components, usually designed to produce an image. Equivalent words that are also used would be "optical device" or "optical arrangement."

Centered optical system: An optical system where all its elements are aligned such that their individual optical axes constitute a single optical axis, the system's optical axis.

5

Optical Design and Aberrations

Armando Gómez-
Vieyra and Daniel
Malacara-Hernández

5.1 Introduction

Over the last 25 years, the field of optical design has been transformed by the increased availability and improved capabilities of optical design software. Even though optical systems, such as telescopes, microscopes, electronic projectors, and so on, can sometimes be constructed with simple means, nowadays optical design is unthinkable without the aid of a commercial software package. Furthermore, changing market requirements of a large part of the optical industry, in size, weight and cost, have marked an evolution in manufacturing methods, materials, and aspheric surfaces in some cases. Despite this, predominantly relying on traditional design techniques, they are still used today.

The optical design, as part of the optical engineering, is a process of selecting optical elements and putting them into a special order to the development of any simple or complex optical system that satisfies a customer's request. Much of the success of a good design depends on the skills, knowledge and experience of the designer. Optical design is the combination of art, science, a hard job and experience. It is a very serious mistake to think that design programs can create an optical system from its optimization subroutines. To perform an appropriate design, a solid base is needed in geometric optics and physical optics, to understand in depth the theory of aberrations and especially the experience acquired over time through the trials and errors. It is very common for beginner designers to omit the appearance that all design should be built, so it is always necessary to know the technical limitations of the real model construction, proper defining physical parameters, and tolerances required. It is understandable and highly recommended to request technical opinion on the optics manufacture before assuming that the job is finished.

5.2 History of Optical Design

Optical design is as old as the idea of gradually improving optical instruments, such as lenses [1], telescopes [2], microscopes [3], and so on. Until 1840, the major advances in this field corresponded to the development of the geometric theory of light [4]. In 1840, Joseph Petzval, a Hungarian mathematician, was the first to apply mathematics to the general problem of designing a lens. Seidel, in 1856, published the first complete mathematical treatment of geometrical aberration theory [5]. In the late nineteenth century, the impetus and vision of Carl Zeiss, with the help of Ernest Abbe and Otto Shott, would completely change the methods for the design and manufacture of optical instruments [4]. The German industry and academy widely developed the analytical theory of the optical systems design as well as the compensation of optical aberrations [6].

At the beginning of the twentieth century, Conrady was able to take the arcane and rather disorganized discipline of optical design and establish it on a systematic, didactic basis, and apply many of his newly devised procedures and theoretical insights [7,8]. The first 40 years of the twentieth century, optical design was done using a mixture of Seidel theory, a little ray tracing, and a great deal of experimental work. Still with the appearance of mechanical calculators (Marchand and Frieden) in the late 1930s, the procedure for the optical systems design remained the use of logarithmic tables in Europe. The reason appears to be that techniques for the use of log tables were developed to the point that an experienced "computer," usually a young lady, could actually trace rays faster than with calculators [9]. In the United States of America, the mechanical calculators supported the development of the young optical designer community. World War II provided impetus to the development of more efficient methodologies for ray tracing and optical aberrations calculus [10].

In 1944, Baker used the Mark I Calculator at Harvard to trace some skew rays [11]. Computers were first used in the UK for optical design in 1949, when C.G. Wynne had some ray tracing done at Manchester University [12]. At the Institute of Optics of the University of Rochester, the first use of a computer to trace rays was made in 1953 for the IBM 650, by Robert E. Hopkins. Another leader in this field, at the Eastman Kodak Co., was Donald P. Feder. This computer work saved a lot of time in the design process, but the programs were just tools to make the whole process faster. The use of computers for ray tracing did not really become practical until about late 1950s. Feder [13] presented one of the earliest papers that describes suitable ray tracing equations for use on large computers. The use of a computer for ray tracing did enable the design of lenses that would have been impractical otherwise, but this was an exceptional situation. However, rapid ray tracing, in itself, was not in most cases sufficient to enable the design of more advanced lenses.

The major impact of computers came when optimization techniques began to be used in lens design [14]. The dominant method has become that of damped least-squares (DLS), proposed by Wynne [12] in the UK and Girard [15] in France. Some designers have also used other methods with success, such as the adaptive method of Glatzel [16] and Rayces [17].

In the early 1980s, personal computers came to give a new impetus to the development of optical design, allowing for the development of accessible optical systems, through commercial software programs. In addition, the Internet currently allows unlimited access to scientific information and patent stocks further easing the development of optical design.

Optical design software has the capacity to evaluate the attributes of an optical system, but as long as humans are making decisions, the designers will have to decide how the many trade-offs will be made. The optical designer always emphasizes the need to understand why a particular optical component works, rather than to blindly hope that the optimization software will find a miraculous solution.

5.3 The Optical Design Process

In the field of optical engineering, it is common to find commercial optical systems; however, not ones that always satisfy customer needs or product. At this point, one has the need to design a new optical system. The process of designing a new optical system consists of establishing a set of optical components with optical properties, including materials (refraction indices, dispersions), surface curvatures, element thicknesses and spacing, surface shape, thin films, and so on. The design of an optical system is comprised of several distinct components, which include mainly

- The first-order optical layout
- The optimization of the optical system
- Performing a tolerance analysis of the optical system
- Performing stray light analysis
- Performing manufacturing analysis

All of the components are related to each other.

Before starting with the first-order optical layout, it is highly recommended to develop a pre-design step. In this step, the designer has to understand and figure out the issues and requirements of the customer before beginning the design; this is a critical step. All the necessary information and design specifications have been provided and any missing, incomplete, or inconsistent requirements must be discussed. Any doubt, omission, or improper assumption could turn into redesigns and costly delays. At this time, the designer can also make a reasonable determination as to whether or not the desired system is physically realizable.

Shannon [18] lists more than 100 parameters or areas that the designer needs to be concerned about, including items such as patentability and environmental hazards. Such considerations indeed affect the design quite often. In this way, the designer should be considering optical, mechanical, environmental, and optomechanics poverty. The type of instrument (polarimeter, spectrometer, non-linear imaging, etc.) imposes limitations and restrictions, such as polarization control, ghost control, special components (aspheric, diffraction gratings, etc.), special materials, and so on. In this way, the design must begin by considering the most critical component(s) first, understanding that they impose on the rest of the system.

The designer needs to attack those parts of the system first that are the most critical, even if they are less familiar. Postponing to the end, the subsystem with the critical components that require specialized knowledge could be a mistake with serious consequences.

5.3.1 The First-Order System Layout

In the first-order system layout step, the designer aims to find the first approximation of the real optical system. This first system approximation has the suitable first-order properties, specifically, magnification, resolution, placement of image, placement of pupils, desired radiometric and photometric properties, and any other special requirements. At this stage, the designer considers the fundamental properties of optical components (prism, lenses and mirrors), the paraxial ray tracing methods, the aberration analysis and the possibility of using any existing design as a starting point.

As is well known, any optical system may be constructed as a refractive, reflective or catadioptric system. The type of design form that is determined by the optical requirements actually raises the design shape. The aperture and the field size determine the difficulty/complexity of a design and are used to identify an appropriate starting design form. The selection of the initial design form is a key role for the optical design. Modern designs perform these basic forms, often incorporating aspheres to improve performance, reduce package size, and other physics variables.

The starting design, as described by Hilbert et al. [19], may be set up in several different ways:

1. Similar existing designs in the technical literature. The books by Cox [20], Laikin [21], Smith and Genesee Optics Software [22], and many others are good sources, as commercial patents.

2. Scaling of an existing design, with the same f-number and similar main characteristics.
3. Substantially modifying a design with characteristics at least close to the ones desired.
4. First-order designing of the system from scratch and then using third-order aberration theory to obtain an approximate design.

Considering the last point, we could start the first-order approach system with thin lenses. Continuing the development until we find an appropriate thick lenses system. However, in the case of mirror systems, the thin-lens approach is insufficient; the designer normally starts with a known mirror system that satisfies the first-order requirements.

When performing the refinement of the first optical system, it is necessary and essential to control the aberrations. Traditionally, the way to obtain the Seidel and chromatic aberrations are while the system is symmetrical and centered. However, when the system has an oblique incidence configuration, the matter is complicated. It is always possible to work in algebraic or tabular form, although recently the use of software has gained ground.

5.3.2 The Optimization of the Optical System

The designer uses the lens design program to determine the optical parameters necessary to meet the required performance specification. To achieve this aim, design programs have implemented different optimization methods. The key part of the optimization process is the merit function, as this restricts the operands (the terms of the merit function) over which the optimization algorithm works. Then, the optical design program takes the information provided by the designer (an optical design starting point) to find the combination of parameter values that best meet the performance goals.

The merit or error function in automatic optimization is a function of the parameters that describe the quality of the system (how close it is to the ideal solution). The greater the merit function, the worse the system is. The goal of all optimization methods is to reduce this merit function as much as possible. In this sense, the experience of the designer is essential, because the success depends on

- The starting design (first-order system layout)
- The choice of variables
- The operands in the merit function
- The number of iterations
- The optimization method

The starting design is a fundamental part of the optimization process. A good starting design election, that is, the better approach to the required characteristics, finds that the optimal solution can be faster and easier. Then the phrase "A good beginning is a half of ending" applies here.

As a next step, the designer needs to identify the optical variables (radii, thicknesses, airspaces, and material parameters) that are allowed to vary during optimization; all other lens parameters are kept frozen. The variables election is very much the responsibility of the designer's experience, as well as the designer's ability (in a few words, it is an art).

The definition of error function is linked to the number of selected variables. Each selected variable provides a greater freedom in the merit function, however, many variables have no guarantees to get an optimal solution. This means, as the optimization subroutines are mathematical functions, which do not distinguish the physical and engineering parameters, the optimization process can find an infeasible solution.

The merit function may be defined in many different manners, for example, by

1. The geometrical spot size or mean square size of the image
2. The root-mean-square wavefront deviation
3. The modulation transfer function (MTF), optimized at some desired spatial frequencies range, as explained by Rimmer et al [23]
4. An appropriate linear combination of primary and higher-order aberration coefficients

The decision about the type of merit function to be used depends on the application of the optical system being designed, as well as on the personal preferences and experience of the optical designer. It should be noticed that in a perfect (diffraction-limited) system, all definitions of the merit function are simultaneously minimized to a zero value. In a real imperfect system, the choice of the merit function affects the final performance of the system. The designer must keep in mind that the functions forming the merit function are not linear.

A useful tool is the choice of the number of iterations; choosing a small number of iterations allows us to observe the process of optimization, which will alter the shape of the design. Then, the designer could evaluate whether the variables and the merit function were properly chosen. In case you do not obtain the desired results, you can return to the starting design and re-define all the parameters for the optimization. Releasing the number of iterations and choosing a large number of variables are not the best options in the optimization process, because the designer loses control of the performance in the final system.

Commercial software programs have some optimization algorithms to find minima in the merit function. These algorithms are classifying in local and global optimization algorithms. Local optimizers use gradient methods to find the local minimum. The starting variable values and the merit function determines which local minimum is found. Global optimizers are some form of search in the entire parameter space that can potentially locate a global minimum of the merit function. Thus, it raises the possibility of a lens designed almost entirely by the computer, with minimum intervention by the designer.

Many different schemes have been developed for optical design optimization, with different advantages and disadvantages. The most common local optimization methods are

- The optimum gradient method, the steepest descent method, or the zigzag method
- The conjugate gradient method [24]
- Grey's optimization method [25]
- The least-squares method [26]
- The Glatzel adaptive method [16,27]
- The damped least-squares method [28,29]

The last two methods are the most common methods used for lens design.

The Glatzel adaptive method, described in detail by Glatzel and Wilson [16], has been independently proposed by Glatzel [27]. This method resembles the path followed by a designer before the lens optimization programs were used. A merit function is not used. Instead, the individual aberrations are corrected, a few at a time, until all are corrected. The designer selects a small number of aberrations, not exceeding the number of independent variables and assigns a target value to them. For the other aberrations, only a maximum limit is assigned.

The damped least-squares method is the most common local method. This method is a modification of the least-squares method, in which the derivatives of the merit function with respect to the variables are calculated, indicating in which direction the variable must move to minimize the merit function. This method was originally proposed by Levenverg in 1944. Since there is not normally a unique solution to the mathematical problem, the process is iterative and performed one small step at a time. This method eventually converges to a minimum of the merit function, but normally, only a local minimum. This is the most common method used in commercial optical design programs.

The global optimization [30] seeks to perform as best as possible for the optical system, minimizing the global merit function considering all possible variables and operands. These methods have the advantage that the solution does not depend on the starting point. There are many procedures to perform global optimization, which require extensive computation time, but only quite recently is there computer power available for this task. The most common global optimization methods are as follows:

- The grid search, in which the merit function is evaluated at equidistant points on a regular grid on the variable space. Once the point with the minimum merit function is located, the minimum on this region is found.

- The simulated annealing algorithm, first used by Bohachevsky et al. [31]. In this algorithm, the variable space is sampled with a controlled random search.
- Genetic Algorithm (GA) is based upon this process of natural selection. In this algorithm, a numerical sequence selectively chooses genes from the best-performing parents to create a new system.

While practically all optical design codes utilize damped least-squares for local optimization, they employ different global methods of optimization, so the designer needs to consult the relevant software manuals for details and references in each case.

5.3.3　Tolerance Analysis of the Optical System

Tolerancing is one of the most complex aspects of optical design and engineering. Tolerance analysis is a statistical process during which changes or perturbations are introduced into the optical design to determine the performance level of an actual design manufactured and assembled to a set of manufacturing tolerances. The basic concepts involved in establishing a tolerance budget have been summarized by Smith [5,32].

In this step, the optical designer must consider that

- No optical surface will be polished to perfect curvature and figure.
- No optical glass will be perfect (index, dispersion, and a minimum inhomogeneity).
- No mechanical mount will be perfectly machined.
- No mechanical and optical component will be perfectly athermalized.
- No component will be perfectly positioned or aligned.
- Many other error sources will serve to degrade the performance of the assembled system.

Different considerations apply to reflective, refractive, or catadioptric systems [18]. Understanding the capabilities of the optical fabrication shop as well as the mechanical fabrication shop is a good starting point for the parameter ranges.

The tolerance figure of merit describes the impact of each perturbation on the system performance. During the tolerance process, each parameter is perturbed individually. This allows the optical engineer to determine which parameters are most sensitive. Parameters that perturb the design outside a portion of the error budget available for fabrication errors need to be more tightly constrained. Tightening constraints adds significantly to the manufacturing costs.

An optical design is not complete until a tolerance analysis has been carried out. The optical design needs to be manufacturable with some realistic range of parameter specifications to be useful to the customer. The best design is not necessarily the design that best matches the design specification.

5.3.4　Stray Light Analysis

Before an optical design is finalized, it may be necessary to perform a stray light to simulate the real working environment more completely and insure that the limitations imposed by sequential ray tracing do not have hidden significant design flaws.

The analysis of stray light suppression is the study of all unwanted sources that reduce contrast or image quality. Primary sources of stray light include reflections from uncoated or poorly coated surfaces, surface or bulk scattering, thermal emission, bright out-of-field objects, glints from mechanical structures, diffracted light from edges, and unwanted diffraction orders from gratings.

A stray light analysis determines both the amount of stray light at the detector and the different paths that it took to get there. If the stray light is large enough to cause a problem, the analysis should also identify any optical and/or optomechanical design changes needed to reduce the stray light. During stray light evaluation, all the surfaces that light might encounter must be modeled. This includes the lens mountings, the inside of the tubes or other instrument enclosures, and even structures outside the instrument aperture that can reflect into it.

Stray light can be divided into

- Ghost images
- Veiling glare

Veiling glare occurs when object field point rays and rays from objects outside the geometrical field of view of the optical system strike a lens surface and are thereby scattered into the optical system. Veiling glare produces a diffuse background haze on the image plane, resulting primarily in a loss of overall image contrast.

Ghost analysis is relatively simple in the sense that it involves the same imaging and ray tracing algorithms as before. Ghosts can arise from reflections at air/glass interfaces that are not sufficiently suppressed through antireflection coatings. A very important source of ghosts can also be the detector surface, which often reflects more light back into the system than it absorbs. Ghost images have structure (often with sharp edges) and create local image-quality disruptions.

Stray light calculations can take many hours or days of computing time.

5.3.5 Manufacturing Analysis

The designer has the obligation to make a final analysis before concluding his work. This is to be sure that his design can be manufactured. The design can be very good on paper, but may be impossible to construct. Sometimes, there are technological, physical, and budget limitations. By delivering the plane of construction, one must be take into account the manufacturing technology to be used (CNC, magneto rheological, traditional work shop, etc), finish specifications, type of materials requested in each component, and so on. This is the performing manufacturing analysis [33,34].

5.4 Monochromatic Optical Aberrations

In the process of designing and building optical instruments, one of the fundamental steps is the analysis and compensation of aberrations. The description of the physical concept of an optical aberration is "The optical path difference between a real wavefront (or real-time image position) and ideal wavefront (or the position of a real-time image)," as shown in Figure 5.1.

Defining the concept of aberration from a mathematical expression, which results in graphic interpretations, is a key element to analyzing and compensating an optical system. That is, the correct interpretation of the mathematical approach to describe an optical aberration allows us to understand how

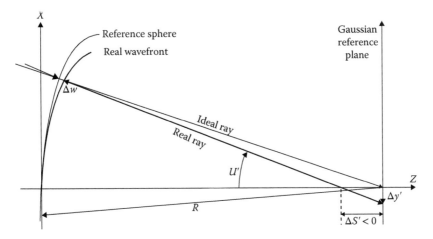

FIGURE 5.1 Optical aberrations (ΔW = wavefront aberration, ΔS = longitudinal ray aberration and Δy = transverse ray aberration).

imperfect the developing of the optical system is. The different approaches of defining the aberration function depend on

- The reference taken (the pupil or the image plane)
- The type of system (center, oblique incidence, anamorphic, type of pupil, etc.)
- The tool for evaluation (single pencil calculation, accurate approach of ray tracing, etc.)
- The level of demand of the calculation (third order, fifth order, etc.)

Throughout history, several notations have been defined to represent the same phenomenon, in the specific case of aberrations we have:

- An elegant and formal style from a mathematical point of view is to employ the Hamiltonian theory [35].
- A very practical and convenient method for a rapid calculation was proposed by Seidel and perfected by Conrady [7].
- There are more analytical methods (based on analytic geometry and differential geometry) mainly developed by HH Hopkins, Welford, Stravrodis and Sasian, which have been particularly interested in applications in imaging theory (physical optics).
- In the present, the digital age demands have given rise to the popularization of two representations, the one of Buchdalh, used in the optical software design and the Zernike polynomials, although it has great limitations, it is very typical for most real cases.

In these four categories, all mathematics developed for the aberration functions representation are included. It is important to keep in mind that all of these ways of representing the aberrations are virtually equivalent. Most design programs allow you to choose any combination of these, as well as the reference system with which you want to work (left or right).

The aberration of an optical system is defined as the difference between something ideal and something real, whereby to define the quantity to be measured and reference, we have:

- Wavefront aberrations measured in the system pupil (Figure 5.2)
- Ray aberrations
    - Transversal (Figure 5.3)
    - Longitudinal (Figure 5.4)
- Angular aberrations (Figure 5.5)

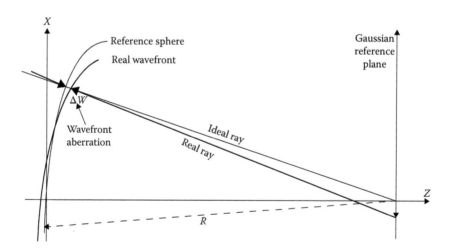

FIGURE 5.2 The wavefront aberration is defined as the optical deviation of the real wavefront from a reference wavefront measure along a ray (ΔW commonly used as W).

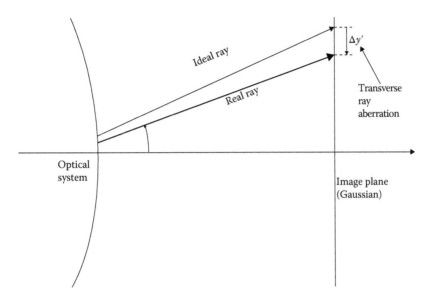

FIGURE 5.3 The transverse aberrations are the displacement component between the ideal image and the real position, defined in the Gaussian plane.

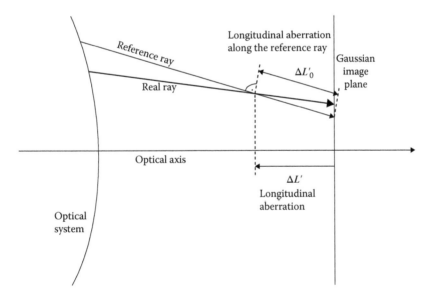

FIGURE 5.4 Longitudinal aberration is the distance from the intersection point with a reference ray to a reference plane.

The most popular are the transverse aberration and the wavefront aberration, which are related in general by approximations:

$$TA_y = -\frac{R}{n_i}\frac{\partial W}{\partial y}$$

$$TA_x = -\frac{R}{n_i}\frac{\partial W}{\partial x},$$

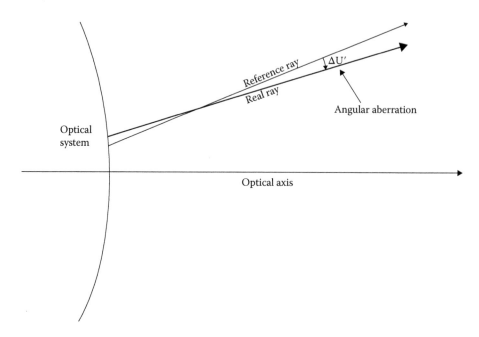

FIGURE 5.5 The angular aberration of a real ray is equal to the relative deviation of its direction of propagation from the reference ray direction.

where n_i is the refractive index between the pupil and the image plane, and R is the ideal radius of curvature from the image point to the pupil. These expressions are derived rigorously by Welford [36] and Born and Wolf [37]. A couple of exact equations from its definition are those derived by Rayces [38].

The aberration function, either the ray, wavefront, or angular, can be expanded into a power series in terms of aperture and field [39,40]. Each term of function is multiplied by a calculated coefficient, which is proportional to the amount of aberration present.

In the literature there are some families of sets defined in order to represent the wavefront aberration or the ray aberration. The first family was defined by Seidel, which considered only the third-order aberrations (to be defined and measured in the image plane and considered only the second term in the Sinus power series). With the development of more complex mathematical methods this notation evolved, mainly driven by Conrady [7], Kingslake [41] and others. With the advent of the computer age, the work of Buchdahl arrived [42], which introduced the concept of higher-order aberrations in the design process. Buchdahl developed an iterative method to obtain the higher-order aberrations, but the complexity limits the audience able to understand the complicated mathematics.

Work on aberration theories that are employed more in the community were developed by H. H. Hopkins. Wave Theory of Aberrations is central to all modern optical design and provides the mathematical analysis, which enables the use of computers to create the wealth of high quality optical systems available today. This has been newly updated and extended by Sassian.

Recently, the Zernike polynomials have been taken as the family of polynomials to represent the wavefront aberrations. These are defined for a continuous circular pupil and have the advantage that each term is orthogonal to the others, but do not consider the field. Also they have other disadvantages, which will be discussed later.

It must be considered that these mathematical tools are not visual or performing for the designer, who commonly uses a computer, a method of exact ray tracing (in which he does not perform any approximation) and his experience (the optical design is an art and science). The results delivered by the computer can be presented graphically as

- Spot diagram (Figure 5.6)
- Transverse ray aberration plot (Figure 5.7)
- Longitudinal ray aberration plot (Figure 5.8)
- Optical path difference (Figure 5.9)
- Field aberration plot (Figure 5.10)
- 3D wavefront plot
- Modulation Transfer Function (MTF, PSF, and OTF)

They are the real designer tools. We will now discuss in more detail the aberration wavefront representations, which are basic in optical design.

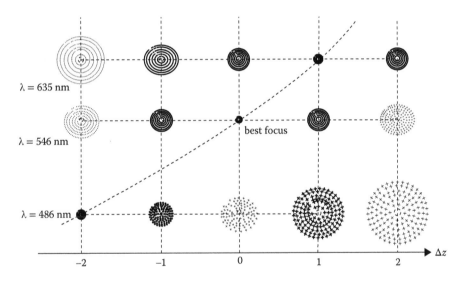

FIGURE 5.6 The spot diagram is formed by intersection points with the image plane when all rays are traced, as function of wavelength and defocus.

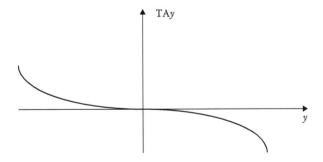

FIGURE 5.7 Transverse ray aberration plot describes the displacement of the (tangential or/and sagittal) rays' intersection points in image plane as a function of the aperture or the pupil coordinates.

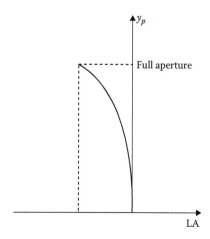

FIGURE 5.8 The longitudinal aberration plot represents the distance from the intersection point with a reference ray to a reference plane (Figure 5.6) dependent on the aperture and pupil.

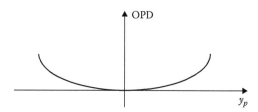

FIGURE 5.9 The optical path difference plot shows slices of the full wavefront error of an optical system and is typically a measurement in units of wavelength.

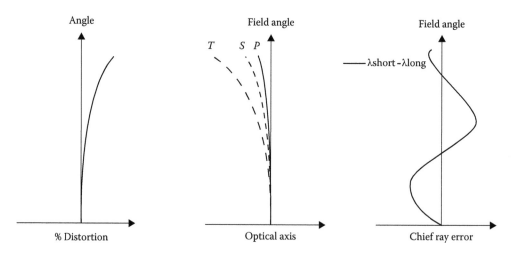

FIGURE 5.10 The field plots present information on certain aberrations across the entire field. The three field plots most often used are distortion, field curvature, and lateral color.

5.5 Wavefront Aberration Polynomial

Historically, lens designers used left-handed Cartesian coordinates with positive slopes of rays propagating downward [43]. This was done for computational convenience and error mitigation when doing manual computations. Most current optical design and analysis software packages use right-handed coordinates and have been adopted as standard by the American National Standard. That notation is used in this chapter.

The wavefront aberration of a general optical system with a point object can be represented in Cartesian coordinates as

$$W(x,y) = \sum_{i=0}^{k} \sum_{j=0}^{i} c_{ij} x^{i-j} y^{j} \tag{5.1}$$

including any symmetries and higher-order aberration terms, where k is the degree of the polynomial. This is a Taylor wavefront representation [44]. Defining the polar coordinates

$$x = \rho \cos\theta \tag{5.2}$$

and

$$y = \rho \sin\theta, \tag{5.3}$$

where the angle θ is measured with respect to the x-axis. Then, the wavefront shape may be written as

$$W(\rho,\theta) = \sum_{n=0}^{k} \sum_{l=0}^{n} \rho^{n} \left(a_{nl} \cos^{l}\theta + b_{nl} \sin^{l}\theta \right), \tag{5.4}$$

where the $\cos\theta$ and $\sin\theta$ terms describe the symmetrical and antisymmetrical components of the wavefront, respectively [43,44]. Nonetheless, not all possible values of n and l are permitted, hence n and l must be both odd or both even, for this system. Moreover, the aberration function can be written in terms of Hamilton, a three-rotational invariant for an optical system with rotational symmetry [37,42], which has disadvantages from the engineering point of view.

5.5.1 Wavefront in a Rotationally Symmetric Optical System

Let us assume that the optical system is a centered one; that is, it is a system where all reflecting or refracting spheres have their centers of curvature lying along a straight line called the optical axis. The optical surfaces are assumed to be rotationally symmetric; that is, they are spheres or conicoids, such as ellipsoids, hyperboloids, or paraboloids.

In this centered or axially symmetric system, if the point object is displaced from the optical axis in the y direction, the aberrated wavefront function W(x, y) has a lateral symmetry about their axis that can be expressed as

$$W(x,y) = W(-x,y). \tag{5.5}$$

By definition, the symmetric optical system is invariant under an arbitrary rotation about its optical axis and under reflection about any plane containing the optical axis. Then, restricting to the case of a wavefront produced by an axially symmetric optical system, with a point object displaced along the y-axis, the wavefront is symmetric about the tangential or meridional plane, obtaining

$$W(\rho,\theta) = \sum_{n=0}^{k} \sum_{l=0}^{n} a_{nl} \rho^n \cos^l \theta, \tag{5.6}$$

where the object height is different from zero and oriented along the y-axis. This is implicit in the value of the coefficients a_{nl}. The rotational symmetry of the system about the optical axis and the wavefront lateral symmetry about the meridional plane makes it that only even values of $l+n$ are permitted. Hence, we can write

$$W(\rho,\theta) = a_{00}$$
$$a_{11}\rho\cos\theta$$
$$a_{20}\rho^2 + a_{22}\rho^2\cos^2\theta \tag{5.7}$$
$$a_{31}\rho^3\cos\theta + a_{33}\rho^3\cos^3\theta$$
$$a_{40}\rho^4 + a_{42}\rho^4\cos^2\theta + a_{44}\rho^4\cos^4\theta + \dots$$

The first fifteen terms of the series Equation 5.6, known as primary monochromatic aberrations, are relevant, described as in Table 5.1.

At this point, the wavefront aberration has the image height value implicitly included in the value of some coefficients a_{nl}. If desired, the explicit value of this Gaussian image height h (field) can be included in the wavefront representation; we obtain an expression first proposed by H.H. Hopkins [39] as follows:

$$W(h,\rho,\theta) = \sum_{k=0}^{\infty} \sum_{n=0}^{\infty} \sum_{l=0}^{n} \alpha_{knl} h^k \rho^n \cos^l \theta \tag{5.8}$$

with $i=k+n$ defining the aberration order. Equation 5.8 defines the symmetrical aberrations present in centered optical systems. The description for the $i=4$ order corresponds to representation of the wavefront aberration in primary coefficients and i=6 corresponds to secondary wavefront aberration coefficients.

The primary aberrations are those of a centered system with axial rotational symmetry, up to the fourth power, which can be written in ANSI-OSA (Mahajan and Maroulis, 2003) convention as

$$W(h,\rho,\theta) = a_{040}\rho^4 + a_{131}h\rho^3\cos\theta + a_{222}h^2\rho^2\cos^2\theta + a_{220}h^2\rho^2 + a_{311}h^3\rho\cos\theta \tag{5.9}$$

These aberrations are commonly called spherical, coma, astigmatism, field curvature, and distortion, respectively, and have different notations depending upon the author.

TABLE 5.1 Description of Taylor Coefficients

a_{00}	Piston coefficient
a_{10} and a_{11}	Tilt coefficient
a_{20}, a_{21}, and a_{22}	Defocus and astigmatism coefficient
a_{30}, a_{31}, a_{32}, and a_{33}	"Coma-like" coefficient
a_{40}, a_{41}, and a_{42}	"Spherical-like" coefficient

When using any one of them to compute primary aberrations, care should be taken to understand whether the values obtained are the coefficients for the transverse aberrations, longitudinal aberrations, or wavefront aberrations. Additionally, Equation 5.9 can be expressed in Cartesian coordinates as

$$W(h,x,y)=a_{040}\left(x^2+y^2\right)^2+a_{131}h\,x\left(x^2+y^2\right)+a_{222}h^2x^2+a_{220}h^2\left(x^2+y^2\right)+a_{311}h^3x, \qquad (5.10)$$

which is a primary wavefront representation of a centered optical system (Mahajan and Maroulis, 2003). Since in Equation 5.10 both astigmatism and distortion terms are x dependent, the following can be pointed out:

- This is a representation of laterally symmetrical wavefronts.
- The center of curvature of its reference sphere is located on the sagittal image surface.

This representation is related to Seidel transverse aberrations, but it does not directly represent the Seidel wavefront aberration as shown in Section 5.5.2.

5.5.2 Seidel Wavefront Aberrations

The primary aberration terms of Seidel (fourth-order wave aberration terms or third-order ray aberration terms) are important in optical design. Seidel published his work on a systematic method for computing third-order longitudinal aberrations (primary ray aberrations) and provided explicit formulas [45]. These aberrations are commonly referred to as the Seidel aberrations and are denoted in order: spherical, coma, astigmatism, Petzval (field curvature), and distortion, which have different notations depending on the author. When using any computation scheme to determine the Seidel aberrations, one should care to understand whether the values are coefficients for transverse, longitudinal, or wavefront aberrations. Then, the Seidel wavefront aberration is represented by

$$W(h,x,y)=\frac{1}{8}S_I\left(x^2+y^2\right)^2+\frac{1}{2}S_{II}h\,x\left(x^2+y^2\right)+\frac{1}{4}S_{III}h^2\left(3x^2+y^2\right)$$
$$+\frac{1}{4}S_{IV}h^2\left(x^2+y^2\right)+\frac{1}{2}S_V h^3x \qquad (5.11)$$

whose spherical reference wavefront has its center of curvature at the Gaussian image given by $3x^2+y^2$. Thus S_I, S_{II}, S_{III}, S_{IV}, S_V are the Seidel sums (spherical, coma, astigmatism, Petzval curvature, and distortion) The Seidel sums S_i's are represented by G_i by some other authors. Equation 5.10 can be rewritten as Equation 5.11, to change its reference sphere to the sagittal image surface. We obtain the primary wavefront aberrations described by Hopkins, commonly written as

$$W(h,x,y)=\frac{1}{8}S_I\left(x^2+y^2\right)^2+\frac{1}{2}S_{II}h\,x\left(x^2+y^2\right)+\frac{1}{2}S_{III}h^2x^2$$
$$+\frac{1}{4}(S_{III}+S_{IV})h^2\left(x^2+y^2\right)+\frac{1}{2}S_V h^3x \qquad (5.12)$$

The above mentioned change of representative sphere is evidenced by the change of

$$\frac{1}{4}S_{III}h^2\left(3x^2+y^2\right) \text{ to } \frac{1}{2}S_{III}h^2x^2$$

and

$$\frac{1}{4}S_{IV}h^2\left(x^2+y^2\right) \text{ to } \frac{1}{4}(S_{III}+S_{IV})h^2\left(x^2+y^2\right).$$

TABLE 5.2   Different Representations of Primary Wavefront Aberrations

W_{Seidel}		W_{Hopkins}		
$\frac{1}{8}S_I$	$\left(x^2+y^2\right)^2$	$\frac{1}{8}S_I$	$\left(x^2+y^2\right)^2$	Spherical
$\frac{1}{2}S_{II}h$	$x\left(x^2+y^2\right)$	$\frac{1}{2}S_{II}h$	$x\left(x^2+y^2\right)$	Coma
$\frac{1}{4}S_{III}h^2$	$\left(3x^2+y^2\right)$	$\frac{1}{4}S_{III}h^2$	x^2	Astigmatism
$\frac{1}{4}S_{IV}h^2$	$\left(x^2+y^2\right)$	$\frac{1}{4}(S_{III}+S_{IV})h^2$	$\left(x^2+y^2\right)$	Field of curvature
$\frac{1}{2}S_V h^3$	x	$\frac{1}{2}S_V h^3$	x	Distortion

At this point, it is important to emphasize that both representations are primary wavefront aberration representations for centered optical systems with respect to different reference spheres (Table 5.2).

Whereas the first three terms of Ec.11 (spherical, coma, and astigmatism) are point aberrations (could be represented in Zernike polynomials) that generate a blurred image point, the last two terms just cause a shift of the image point relative (field) to the ideal paraxial image point but the image point itself would be sharp.

5.5.2.1 Spherical Aberration

Spherical aberration occurs for object points on the optical axis of a rotationally symmetric optical system. Spherical aberration causes rays with a large height *on* exit pupil of the optical system to be refracted/reflected more strongly, so that they intersect the optical axis in front of the paraxial image plane (Figure 5.11).

Spherical aberration is the only one of the third-order aberrations that appears on axis ($h=0$); it possesses a rotational symmetry and it has the same magnitude in all the fields ($H\neq0$). A typical property of spherical aberration is that it increases with the fourth power of the aperture:

$$W\left(h,\rho,\theta\right)=a_{040}\rho^4 .$$

An aplanatic system is a system that does not possess spherical aberration as a whole and satisfies the sine condition, consequently coma free. An aplanatic surface is one of the special cases described by Abbe, which is simultaneously free of spherical and coma aberrations, discussed by many authors, such as Malacara and Malacara [44].

5.5.2.2 Coma

Coma is an aberration that occurs only for off-axis points of a rotationally symmetric optical system. Coma is due to the variation of image location and size with the zonal radius in the pupil for an off-axis object point (Figure 5.12). It can also be regarded as the variation of magnification with pupil position since the size of the ring increases as the radial zone moves out to the edge of the pupil. This deformation of the image point gives it the name coma because in the paraxial image, it looks like a comet (Figure 5.13).

Tangential coma is defined as the distance from the principal ray intersection point to the tangential ray intersection point, and sagittal coma is defined as the distance from the principal ray intersection point to the sagittal ray intersection point. The tangential coma is always three times the sagittal coma.

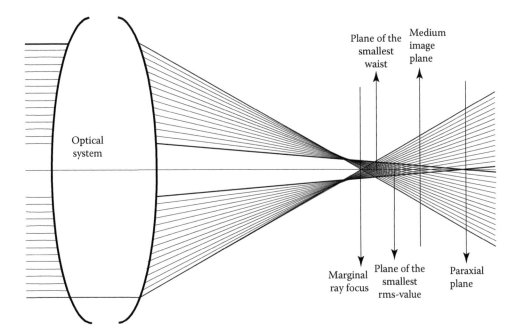

FIGURE 5.11 Spherical aberration is a variation in focus position between the paraxial rays and those through the edge of the optical system aperture.

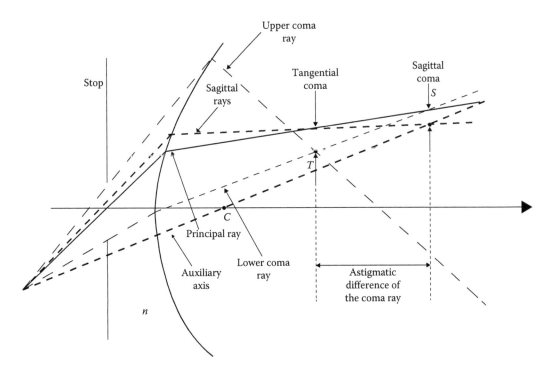

FIGURE 5.12 The extreme upper and lower rays of the marginal zone come to focus at T, while the extreme front and rear rays come to focus at S. S and T lie below the principal ray on our diagram, which indicates the presence of a negative coma.

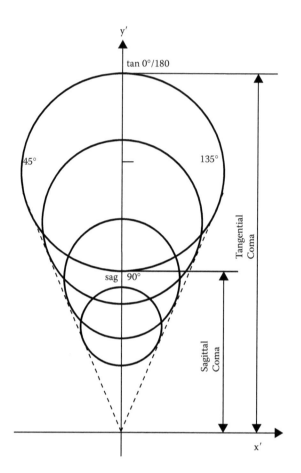

FIGURE 5.13   Schematic figure of a primary coma at a Gaussian image plane.

In a system with coma, most of the light energy is concentrated in the vicinity of the principal ray. The coma depends on the third power of the aperture and linearly on the image height as in the following expressions:

$$W(h,\rho,\theta) = a_{131}h\rho^3\cos\theta$$

This is the reason why coma especially occurs for large numerical apertures whereas astigmatism dominates for small numerical apertures and large object heights.

5.5.2.2.1 The Sine Condition

The sine condition, described by several authors, such as Conrady [7] and Welford [36], is an equation that connects the paraxial and real aperture angles in the object and image space. The sine condition describes the case when a constant magnification for all the aperture zones, that means the marginal amplification and the paraxial magnification, is the same, and as a consequence, the coma is zero.

$$\frac{\sin U}{u} = \frac{\sin U'}{u'},$$

where U and U are the fine aperture angles, and u and u are the paraxial aperture angles. In the practice, a deviation from this relationship is defined as an offense against the sine condition (OSC) widely described by Malacara and Malacara [44] and Kingslake [41].

5.5.2.3 Astigmatism

In the presence of astigmatism, rays in the meridional and sagittal planes are not focused at the same distance from the system pupil (Figure 5.14). Astigmatism can be treated as the variation of focus with pupil orientation. So, the geometrical shape of the image point is in general an ellipse. In two special planes, called the tangential and sagittal focal planes, the ellipses degenerate into two focal lines. The focal lines are perpendicular to each other. Between the tangential and the sagittal focal plane, there is another plane where the shape of the image point is a circle, but of course this circle is extended whereas an ideal image point is in geometrical optics.

There is no astigmatism on axis (h=0), and its magnitude increase with the field ($h\neq0$). The astigmatism of an optical system is proportional to the square of the aperture and the square of the image height:

$$W(h,\rho,\theta) = a_{222}h^2\rho^2\cos^2\theta.$$

Design programs plot the longitudinal fields; in the presence of astigmatism, the tangential image surface lies three times as far from the Petzval surface as from the sagittal surface, as shown in Figure 5.15. When there is no astigmatism, the sagittal and tangential and Petzval are the same.

Any marginal image can be found using the general Coddington Equation 5.7; the two particular cases are the tangential and sagittal images.

5.5.2.4 Field Curvature

The field curvature, known as Petzval curvature, is not a point aberration but a field aberration. The field curvature describes the departure of the image surface from the flat surface, as shown in Figure 5.15. An image point at any point of the field can be sharp but the position of the image point is shifted relative to the ideal Gaussian plane. In the presence of astigmatism, there are even two different spheres for rays in the tangential plane and in the sagittal plane. There is no field of curvature on the axis (h=0). Considering no astigmatism, the field curvature shows a quadratic dependence on the aperture and field:

$$W(h,\rho,\theta) = a_{220}h^2\rho^2.$$

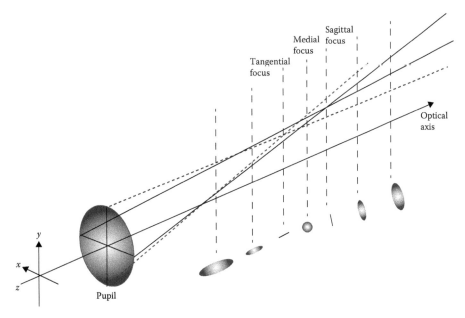

FIGURE 5.14 Astigmatic ray bundle.

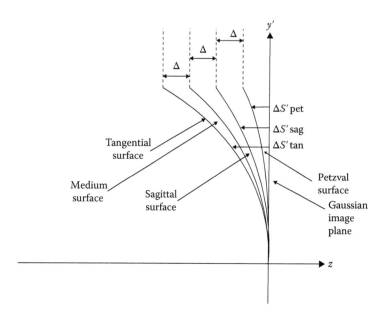

FIGURE 5.15   Astigmatic image surface locations.

FIGURE 5.16   The effects of distortion.

The Petzval surface is not a good image surface; it is the fundamental field curvature of the system. The Petzval curvature is inversely proportional to the product of the refractive index of the lens and its focal length.

5.5.2.5 Distortion

The last Seidel aberration is distortion, which is also a field aberration and not a point aberration. Distortion means that the lateral magnification for imaging is not a constant for all off-axis points but depends to some extent on the image height. Distortion is the variation of tilt with the field; it does not blur the image. Points are still imaged into points; however, straight lines are not imaged into straight lines. Distortion varies as the third power of the FOV as in the following expression:

$$W(h,\rho,\theta) = a_{311}h^3\rho\cos\theta.$$

Considering a regular grid as an object, when each image point moves radially toward the center of the image, this is called barrel distortion (negative distortion), as shown in Figure 5.16. When the distortion is positive, and each image point is displaced radially outward from the center, this distortion is called pincushion distortion (Figure 5.16).

5.6 Wavefront Aberrations Representation for Any Non-Centered and Non-Symmetrical Optical Systems

In the general case, when the optical system is not symmetric and/or it is not centered, the wavefront aberration representations given in the previous section are insufficient to describe this wavefront, because non-symmetrical aberrations and non-circular pupils could be present such in the systems. This leads to a more complete family of polynomials, for example the wavefront representation of Equation 5.1, which includes all wavefront conditions present in the system.

5.6.1 Zernike Polynomials

The preceding general representation (Equation 5.1 or 5.4) is not the most adequate for the analysis of an optical system. A grouping of two or more monomial components is normally used to represent certain common wavefront deformations called optical aberrations. For example, a spherical wavefront deformation is called defocus, a toroidal or cylindrical deformation is called astigmatism and a rotationally symmetric aspheric deformation with a fourth power radial dependence is called primary spherical aberration, and so on.

There are several practical reasons why orthogonal polynomials are useful and desirable, for example, an easier numerical manipulation and an easier interpretation in interferometrical applications. If the wavefront aberration polynomials are designed to form an orthogonal (in a circular pupil with unit semi-diameter) set of polynomials, they are called the orthogonal Zernike polynomials [46], and recently, the orthonormal Zernike wavefront representation has been adopted as standard [47].

Any continuous wavefront shape may be represented as

$$W(\rho,\theta) = \sum_{n=0}^{k} \sum_{m=0}^{n} A_{nm} U_{nm}(\rho,\theta),$$

where $U_{nm}(\rho,\theta)$ is each Zernike polynomial term and A_{nm} is the aberration amount coefficient. The description of Zernike polynomials has historically been defined orthogonally [46] and has recently adopted orthonormality as standard [47].

Both Zernike wavefront representations (Tables 5.3 and 5.4) have the following properties:

1. The reference sphere has a radius of curvature determined only by the defocus term. This sphere represents the best least-squares fit of the aberrated wavefront to a sphere. A consequence of this is that if the defocus term is simply removed, a plane surface would be the closest sphere to this aberration wavefront without the defocusing term, in the least-squares sense.

2. The closest reference sphere for each of the Zernike aberration terms is a flat surface with no tilt. This is achieved during the orthogonalization procedure by adding the proper linear combination of Zernike terms of lower order to each Zernike term.

In other words, the center of curvature of the reference sphere is not at the Gaussian image, but at a point close to the optimum image, located between the sagittal and tangential images (see Figure 5.15). We can say that this center of curvature is close to the medium focus (waist of the caustic only if the spherical aberration is present), but not exactly there, for two reasons:

1. There is a small axial displacement for the medium focus image, because it is defined for the best image considering paraxial transverse aberrations (sagittal and tangenital), while in the Zernike polynomials it is the best image considering the minimum wavefront deformations for the whole aperture (this includes paraxial as well as marginal rays).

TABLE 5.3   Wavefront Aberrations in Cartesian Coordinates

#	Zernike Orthogonal	Zernike Orthonormal	Name
1	1	1	**Piston**
2	η	2η	**y tilt**
3	ξ	2ξ	**x tilt**
4	$2\eta\xi$	$2\sqrt{6}\,\eta\xi$	**Primary astigmatism at 45°**
5	$2(\xi^2+\eta^2)-1$	$\sqrt{3}\left[2(\xi^2+\eta^2)-1\right]$	**Defocus**
6	$\xi^2-\eta^2$	$\sqrt{6}(\xi^2-\eta^2)$	**Linear astigmatism**
7	$\eta(3\xi^2-\eta^2)$	$2\sqrt{2}\left[\eta(3\xi^2-\eta^2)\right]$	**Trefoil y**
8	$3\eta(\xi^2+\eta^2)-2\eta$	$2\sqrt{2}\left[3\eta(\xi^2+\eta^2)-2\eta\right]$	**Coma y**
9	$3\xi(\xi^2+\eta^2)-2\xi$	$2\sqrt{2}\left[3\xi(\xi^2+\eta^2)-2\xi\right]$	**Coma x**
10	$\xi(\xi^2-3\eta^2)$	$2\sqrt{2}\left[\xi(\xi^2-3\eta^2)\right]$	**Trefoil x**
11	$4\eta\xi(\xi^2-\eta^2)$	$\sqrt{10}\left[4\eta\xi(\xi^2-\eta^2)\right]$	**Quatrefoil**
12	$8\eta\xi(\xi^2+\eta^2)-6\eta\xi$	$\sqrt{10}\left[8\eta\xi(\xi^2+\eta^2)-6\eta\xi\right]$	
13	$6(\xi^2+\eta^2)^2-6(\xi^2+\eta^2)+1$	$\sqrt{5}\left[6(\xi^2+\eta^2)^2-6(\xi^2+\eta^2)+1\right]$	**Spherical**
14	$4(\xi^4-\eta^4)-3(\xi^2-\eta^2)$	$\sqrt{10}\left[4(\xi^4-\eta^4)-3(\xi^2-\eta^2)\right]$	
15	$(\xi^2+\eta^2)^2-8\xi^2\eta^2$	$\sqrt{10}\left[(\xi^2+\eta^2)^2-8\xi^2\eta^2\right]$	**Quatrefoil**

TABLE 5.4   Wavefront Aberrations in Polar Coordinates

#	Zernike Orthogonal	Zernike Orthonormal	Name
1	1	1	**Piston**
2	$\rho\sin(\theta)$	$2\rho\sin(\theta)$	**y tilt**
3	$\rho\cos(\theta)$	$2\rho\cos(\theta)$	**x tilt**
4	$\rho^2\sin(2\theta)$	$2\sqrt{6}\,\rho^2\sin(2\theta)$	**Primary astigmatism at 45°**
5	$2\rho^2-1$	$\sqrt{3}(2\rho^2-1)$	**Defocus**
6	$\rho^2\cos(2\theta)$	$\sqrt{6}\left[\rho^2\cos(2\theta)\right]$	**Linear astigmatism**
7	$\rho^3\sin(3\theta)$	$2\sqrt{2}\left[\rho^3\sin(3\theta)\right]$	**Trefoil y**
8	$(3\rho^3-2\rho)\sin(\theta)$	$2\sqrt{2}\left[(3\rho^3-2\rho)\sin(\theta)\right]$	**Coma y**
9	$(3\rho^3-2\rho)\cos(\theta)$	$2\sqrt{2}\left[(3\rho^3-2\rho)\cos(\theta)\right]$	**Coma x**
10	$\rho^3\cos(3\theta)$	$2\sqrt{2}\left[\rho^3\cos(3\theta)\right]$	**Trefoil x**
11	$\rho^4\sin(4\theta)$	$\sqrt{10}\left[\rho^4\sin(4\theta)\right]$	**Quatrefoil**
12	$(4\rho^4-3\rho^2)\sin(2\theta)$	$\sqrt{10}\left[(4\rho^4-3\rho^2)\sin(2\theta)\right]$	
13	$1-6\rho^2+6\rho^4$	$\sqrt{5}(1-6\rho^2+6\rho^4)$	**Spherical**
14	$(4\rho^4-3\rho^2)\cos(2\theta)$	$\sqrt{10}\left[(4\rho^4-3\rho^2)\cos(2\theta)\right]$	
15	$\rho^4\cos(4\theta)$	$\sqrt{10}\left[\rho^4\cos(4\theta)\right]$	**Quatrefoil**

2. There is a small lateral displacement because the medium focus image is defined for paraxial rays, while the best focus for Zernike polynomials is for the whole aperture. The lateral displacement of the image is due to the presence of the odd aberrations, like coma.

On the other hand, the difference between the two Zernike representations is that the norm of the orthonormal representation is normalized. Zernike polynomials had become so popular that several types of orthogonal polynomials for non-circular pupils (ellipsoidal, hexagonal, square) based on Zernike family were proposed by Mahajan [47].

It has to be emphasized that there are no terms corresponding to curvature of field and distortion since the Zernike polynomials can only represent point aberrations and not field aberrations.

5.7 Chromatic Aberrations

The refractive index of any transparent material is a function of the wavelength (color) of the light. In other words, the focal length of a lens is a function of wavelength because the focal length is dependent on the dispersion of a material and the refractive index. This phenomena is called chromatic aberration of the optical system. The dispersion changes the paraxial parameters, such as the focal length of a lens, as shown in Figure 5.17. The Abbe number, defined as

$$V_d = \frac{n_d - 1}{n_F - n_C},$$

is used to characterize the dispersion of material; the subscript corresponds to the Fraunhofer lines. The designer must consider that the Abbe number is commonly defined in the visible, as was presented here, but can be defined for any range of the optical spectrum.

Chromatic aberrations are first-order aberrations and are present in any refractive optical system. The control and compensation can be proposed by using different types of optical materials (flint and crown) in all optical design. In literature, this has been described by many authors, including Cruickshank [48], Herzberger and Salzberg [49], and Herzberger and Jenkins [50]. This aberration may be obtained with strictly paraxial rays, that is, with only first-order (Gaussian) theory. The chromatic aberrations are as follows:

- Axial chromatic aberration causes light of different wavelengths to focus at different planes, resulting in a blur of the image in the paraxial image plane.
- Lateral chromatic aberration causes light of different wavelengths to focus at different heights in the paraxial image plane.

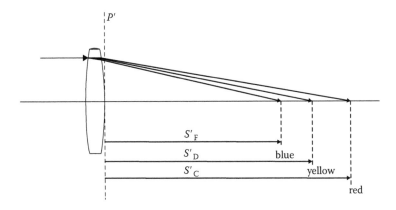

FIGURE 5.17 Chromatic aberration.

5.8 Final Summary

The optical design, as has been stated here, is more an art than a science. Therefore, it is impossible to give a full treatment of the subject. But, we recommend the following additional resources in order to dig deeper into the topic.

Theory Books

Gross H. *Handbook of Optical Systems, Fundamentals of Technical Optics* (Vol. 1). Hoboken, NJ, Wiley-VCH, 2005.

Gross H, Zügge H, Peschka M, Blechinger F. *Handbook of Optical Systems, Aberration Theory and Correction of Optical Systems* (Vol. 3). Hoboken, NJ, Wiley-VCH, 2007.

James JF. *Spectrograph Design Fundamentals*, Cambridge: Cambridge University Press, 2007.

Kidger MJ. *Fundamental Optical Design*. Bellingham, WA: SPIE Press, 2002.

Kidger MJ. *Intermediate Optical Design*. Bellingham, WA: SPIE Press, 2004.

Kingslake R, Johnson RB. *Lens Design Fundamentals* (2nd ed.). Academic Press/SPIE Press, 2010.

Korsch D. *Reflective Optics*. Cambridge, MA: Academic Press, 1991.

Schroeder DJ. *Astronomical Optics* (2nd ed.). Cambridge, MA: Academic Press, 2000.

Slyusarev GG. *Aberration and Optical Design Theory*. London: Adam Hilger, 1984.

Smith WJ. *Modern Optical Engineering* (4th ed.). New York, NY: McGraw-Hill, 2008.

Recourse Designs

Gross H, Blechinger F, Achtnaer. *Handbook of Optical Systems, Survey of Optical Instruments* (Vol. 4). Hoboken, NJ, Wiley-VCH, 2008.

Kidger MJ. *Intermediate Optical Design*. Bellingham, WA: SPIE Press, 2004.

Laikin M. *Lens Design* (4th ed.). Boca Raton, FL: CRC Press, 2007.

Smith WJ. *Modern Lens Design* (2nd ed.). New York, NY: McGraw-Hill, 2005.

Specialized Software

Code V (optics.synopsys.com)

OSLO (www.sinopt.com)

SYNOPSYS (www.osdoptics.com)

ZEMAX (www.zemax.com)

ASAP (www.breault.com)

OpTaliX (www.optenso.com)

OptiCAD (www.opticad.com)

References

1. Kingslake R. *A History of the Photographic Lens*. Boston, MA: Academic Press, 1989.
2. King HC. *The History of the Telescope*. New York, NY: Dover, 1979.
3. Croft WJ. *Under the Microscope a Brief History of Microscopy*. Singapore: World Scientific Publishing, 2006.
4. Mait JN. A history of imaging: Revisiting the past to chart the future, *Optic. Photon. News*, **17**(2), 22–27; 2006.
5. Smith GH. *Practical Computer-Aided Lens Design*. Richmond, VA: Willmann-Bell.

6. M. V. Rohr; 1998. *Geometrical Investigation of the Formation of Images in Optical Instruments*. London: Department of Scientific and Industrial Research, 1920.

7. Conrady AE. *Applied Optics and Optical Design*. New York, NY: Dover, 1957 [Part I] and 1960 [Part II].

8. Emsley HH. *Aberrations of Thin Lenses, an Elementary Treatment for Technicians and Students*. Great Britain: Constable and Company LDT, 1956.

9. Kinger MJ. *Intermediate Optical Design*. Bellingham, WA: SPIE Press, 2004.

10. Hopkins RE. Optical design 1937 to 1988…Where to from here? *Opt. Eng.*, 27(12), 1019–1026; 1988.

11. Baker JG. *Design and Development of an Automatically Focusing 40-inch f5.0 Distortionless Telephoto and Related Lenses for High-altitude Aerial Reconnaissance*, N.D.R.C., Section 16.1, Optical Instruments, 1944.

12. Wynne CG. Lens designing by electronic digital computer, I, *Proc. Phys. Soc.*, 72, 777; 1959.

13. Feder DP. Optical calculations with automatic computing machinery, *J. Opt. Soc. Am.*, 41(9), 630–635; 1951.

14. Fletcher R. *Practical Methods of Optimization*. Hoboken, NJ: John Wiley & Sons, 1987.

15. Girard A. Calcul automatique en optique géométrique, *Revue d'Optique*, 37, 225–241 and 397–424; 1957.

16. Glatzel E, Wilson R. Adaptive automatic correction in optical design, *Appl. Opt.*, 7, 265–276; 1968.

17. Rayces JL. Ten years of lens design with Glatzel's adaptive method, *SPIE Proc.*, 237, 75–81; 1980.

18. Shannon RR. *The Art and Science of Optical Design*. Cambridge, UK: Cambridge University Press, 1997.

19. Hilbert RS, Ford EH, Hayford MJ. *A Tutorial on Selection and Creation of Starting Points for Optical Design*, OSA Annual Meeting, Boston; 1990.

20. Cox A. *A System of Optical Design*. New York, NY: Focal Press, 1964.

21. Laikin M. *Lens Design*. New York, NY: Marcel Dekker, 1990.

22. Smith WJ, Genesee Optics Software. *Modern Lens Design. A Resource Manual*. New York, NY: McGraw-Hill, 1992.

23. Rimmer MP, Brugge TJ, Kuper TG. MTF optimization in lens design, *SPIE Proc.*, 1354, 83–91; 1990.

24. Feder DP. Automatic optical design, *Appl. Opt.*, 2, 1209–1226; 1963.

25. Cornwell LW, Pegis RJ, Rigler AK, Vogl TP. Grey's method for nonlinear optimization, *J. Opt. Soc. Am.*, 63, 576–581; 1973.

26. Rosen S, Eldert C. Least-squares method of optical correction, *J. Opt. Soc. Am.*, 44, 250–252; 1954.

27. Glatzel E. Ein neues Verfahren zur automatischen Korrektion optischer Systeme mit electronischen Rechenmaschinen, *Optik*, 18, 577–580; 1961.

28. Meiron J. Automatic lens design by the least-squares method, *J. Opt. Soc. Am.*, 19, 293–298; 1959.

29. Meiron J. Damped least-squares method for automatic lens design, *J. Opt. Soc. Am.*, 55, 1105–1109; 1965.

30. Kuper TG, Harris TI. A New Look at Global Optimization for Optical Design, *Photonics Spectra*, January 1992.

31. Bohachevsky IO, Viswanathan VK, Woodfin G. An intelligent optical design program, *Proc. SPIE*, 485, 104–112; 1984.

32. Smith W. *Modern Optical Engineering*. New York, NY: McGraw-Hill, 2008.

33. Bliedtner J, Gräfe G, Hector R. *Optical Technology*. China: McGraw-Hill, 2011.

34. Karrow HH. *Fabrication Methods for Precision Optics*. Hoboken, NJ: Wiley, 2004.

35. Synge JL. *Geometrical Optics*. New York, NY: Cambridge University Press, 2008.

36. Welford WT. *Aberrations of Optical Systems*. New York, NY: Adam Hilger, 1986.

37. Born M, Wolf E. *Principles of Optics*. Cambridge: Cambridge University Press, 1999.

38. Rayces JL. Exact relations between wave aberration and ray aberration, *Opt. Acta*, 11, 85–88; 1964.

39. Hopkins HH. *Wave Theory of Aberrations*. New York, NY: Oxford University Press, 1950.

40. Sasián J. *Introduction to Aberrations in Optical Imaging Systems*. Cambridge: Cambridge University Press, 2013.
41. Kingslake R. *Lens Design Fundamentals*. San Diego, CA: Academic Press, 1978.
42. Buchdahl HA. *Optical Aberration Coefficients*. New York, NY: Dover, 1968.
43. Johnson RB. Historical perspective on understanding optical aberrations. In W. J. Smith(Ed.), *Lens Design SPIE Critical Reviews of Optical Science and Technology* (Vol. CR24). Bellingham, WA: SPI Optical Engineering Press, 18–29, 1992.
44. Malacara D, Malacara Z. *Handbook of Optical Design*. Boca Raton, FL: CRC Press, 2013.
45. Mahajan V, Maroulis Aberrations. In R. G. Driggers(Ed.), *Encyclopedia of Optical Engineering* (Vol. 1). New York, NY: Marcel Dekker, 2003.
46. Malacara D, DeVore SL. Interferogram evaluation and wavefront fitting. In D. Malacara(Ed.), *Optical Shop Testing*. New York, NY: John Wiley, 1992.
47. Mahajan VN. Zernike polynomials and wavefront fitting in optical shop testing. In D. Malacara(Ed.), *Optical Shop Testing*. New York, NY: John Wiley, 2007.
48. Cruickshank FD. On the primary chromatic coefficients of a lens system, *J. Opt. Soc. Am.*, **36**, 103–107; 1946.
49. Herzberger M, Salzberg CD. Refractive indices of infrared optical materials and color correction of infrared lenses, *J. Opt. Soc. Am.*, **52**, 420–427; 1962.
50. Herzberger M, Jenkins FA. Color correction in optical systems and types of glass, *J. Opt. Soc. Am.*, **39**, 984–989; 1949.

6

Prisms and Refractive Optical Components

Daniel Malacara-
Hernández and
Duncan T. Moore

6.1 Prisms

In this chapter, we will describe only prisms made out of isotropic materials, such as glass or plastic [1,2]. There are two kinds of prisms: prisms that produce chromatic dispersion of the light beam and prisms that only deflect the light beam, changing its traveling direction and image orientation. Deflecting prisms usually make use of total internal reflection, which occurs only if the internal angle of incidence is larger than the critical angle (about 41°) for a material whose index of refraction is 1.5. If an internal reflection is required with internal angles of incidence smaller than the critical angle, the surface has to be coated with silver or aluminum.

A deflecting prism or system of plane mirrors does not only deflect the light beam but also changes the image orientation [3–6]. We have four basic image transformations: a reflection about any axis at an angle θ; a rotation by an angle θ; an inversion, which is a reflection about a horizontal axis; and a reversion, which is a reflection about a vertical axis. Any mirror or prism reflection produces a reflection transformation. The axis for this operation is perpendicular to the incident and the reflected beams. Two consecutive reflections may be easily shown to be equivalent to a rotation, with the following rule:

$$\text{reflection at } \alpha_1 + \text{reflection at } \alpha_2 = \text{rotation by } 2(\alpha_2 - \alpha_1).$$

These transformations are illustrated in Figure 6.1.

An image is said to be readable if it can be returned to its original orientation with just a rotation. It can be proved that an even number of inversions, reversions, and reflections are equivalent to a rotation, thus producing a readable image. On the other hand, an odd number of reflections always give a nonreadable image. Any single reflection of the observed light beam produces a reflection, reversion, or

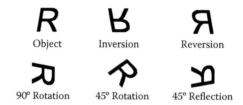

FIGURE 6.1   Image transformations.

inversion of the image. An important conclusion is that an image observed through a deflecting prism is readable if it is reflected an even number of times. Two consecutive transformations may be combined to produce another transformation, as in the following examples:

$$\text{Inversion} + \text{Reversion} = \text{Rotation by } 180°$$

$$\text{Inversion} + \text{Rotation by } 180° = \text{Reversion}$$

$$\text{Reversion} + \text{Rotation by } 90° = \text{Reflection at } 45°.$$

We may also show that if the axis of a reflection transformation rotates, the resulting image also rotates, in the same direction and with twice the angular speed. Thus, a practical consequence is that all inverting systems may be converted to reversing systems by a rotation of an angle of 90°.

Prisms and mirror systems with arbitrary orientations have many effects that must be taken into account when designing an optical system [1,2]. Among these effects we can mention the following:

1. A change in the direction of propagation of the light beam.
2. A transformation on the image orientation.
3. The image is displaced along the optical axis.
4. The finite size of their faces act as stops.
5. Some aberrations, mainly spherical and axial chromatic aberrations, are introduced.

The image displacement ΔL of the position on the image along the optical axis is given by

$$\Delta L = \left(n - \frac{\cos U}{\cos U'} \right) \frac{t}{n} \cong \frac{(n-1)}{n} t, \tag{6.1}$$

where U is the external angle of the meridional light ray with respect to the optical axis, U' is the internal angle of the meridional light ray with respect to the optical axis, t is the effective prism thickness, and n is its refractive index. The approximation is valid for paraxial rays.

Using this expression it is easy to conclude that the longitudinal spherical aberration (SphL) introduced to the system by the presence of the prism is given by

$$\text{SphL} = \frac{t}{n} \left(1 - \frac{\cos U}{\cos U'} \right) \cong \frac{t u^2}{2n}, \tag{6.2}$$

where u is the paraxial external angle for the meridional ray with the optical axis.

From Equation 6.1 we can see that the axial longitudinal chromatic aberration introduced by the prism is

$$l_C - l_F = \frac{\Delta n}{n^2} t. \tag{6.3}$$

Thus, any optical design containing prisms must take into account the glass added by the presence of the prism while designing and evaluating its aberration by ray tracing. All studied effects may easily be taken into account by unfolding the prism at every reflection to find the equivalent plane-parallel glass block. By doing this, we obtain what is called a tunnel diagram for the prism.

The general problem of the light beam deflection by a system of reflecting surfaces has been treated by many authors, including Pegis and Rao [4] and Walles and Hopkins [5]. The mirror system is described using an orthogonal system of coordinates x_o, y_o, z_o in the object space, with z_o being along the optical axis and pointing in the traveling direction of the light. For a single mirror we have the following linear transformation with a symmetrical matrix R:

$$
\begin{bmatrix} l' \\ m' \\ n' \end{bmatrix} = \begin{bmatrix} (1-2L^2) & (-2LM) & (-2LN) \\ (-2LM) & (1-2M^2) & (-2MN) \\ (-2LN) & (-2MN) & (1-2N^2) \end{bmatrix} \cdot \begin{bmatrix} l \\ m \\ n \end{bmatrix}, \tag{6.4}
$$

where (l, m, n) and (l', m', n') are the direction cosines of the reflected and incident rays, respectively. The quantities (L, M, N) are the direction cosines of the normals to the mirror. This expression may also be written as $\overline{l'} = A\overline{l}$, where $\overline{l'}$ is the reflected unit vector and $\overline{l}$ is the incident unit vector. To find the final direction of the beam, the reflection matrices for each mirror are multiplied in the order opposite to that in which the light rays are reflected on the mirrors, as follows:

$$
\overline{l_n'} = [R_N R_{N-1}, \ldots, R_2 R_1] \overline{l} \tag{6.5}
$$

On the other hand, as shown by Walles and Hopkins [5], to find the image orientation, the matrices are multiplied in the same order that the light strikes the mirrors.

Now, let us consider the general case of the deflection of a light beam in two dimensions by a system of two reflecting faces with one of these faces rotated at an angle θ relative to the other, as shown in Figure 6.2. The direction of propagation of the light beam is changed by an angle 2θ, independently of the direction of incidence with respect to the system, as long as the incident ray is in a common plane with the normals to the two reflecting surfaces.

In the triangle **ABC** we have

$$
\phi = 2\alpha + 2\beta \tag{6.6}
$$

and, since in the triangle **ABD**,

$$
\theta = \alpha + \beta, \tag{6.7}
$$

then we can find

$$
\phi = 2\theta. \tag{6.8}
$$

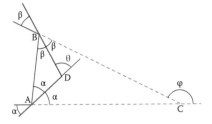

FIGURE 6.2　Reflection of a ray in a system of two reflecting surfaces.

Thus, if the angle between the two mirrors is θ, the light ray will deviate its trajectory by an angle φ, independently of the direction of incidence of the light ray.

Generalizing this result, we can prove that by means of three reflections in three mutually perpendicular surfaces, a beam of light may also be deflected by an angle of 180°, reflecting it back along a trajectory in a parallel direction to the incident light beam. This is called a retroreflecting system.

The prism with three mutually perpendicular reflecting surfaces is called a cube corner prism, which has been studied by Yoder [7], Chandler [8], and Eckhardt [9].

The general problem of prisms of systems of mirrors with a constant deviation independent of the prism orientation, also called stable systems, has been studied by Friedman and Schweltzer [10] and by Schweltzer et al. [11]. They found that in three-dimensional space this is possible only if the deflection angle is either 180° or 0°.

6.1.1 Deflecting Prisms

Besides transforming the image orientation, these prisms bend the optical axis, changing the direction of propagation of the light. There are many prisms of this kind as will now be described.

The *right angle* prism is the simplest of all prisms and in most cases it can be replaced by a flat mirror. The image produced by this prism is not readable, since there is only one reflection, as shown by Figure 6.3a. This prism can be modified to produce a readable image. This is accomplished by substituting the hypotenuse side by a couple of mutually perpendicular faces, forming a roof, to obtain an *Amici prism* (Figure 6.3b). Rectangular as well as Amici prisms can be modified to deflect a beam of light 45° instead of 90°, as in the prisms shown in Figure 6.4.

In the prisms previously described, the deflecting angle depends on the angle of incidence. It is possible to design a prism in which the deflecting angle is independent of the incidence angle. This is accomplished with two reflecting surfaces instead of just one, by using the property described above. The deflection angle is twice the angle between the two mirrors or reflecting surfaces.

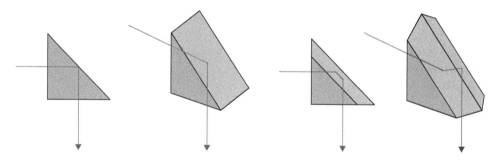

FIGURE 6.3   (a) Right-angle and (b) Amici prisms.

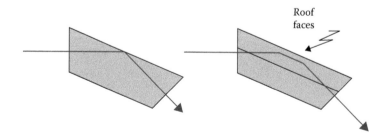

FIGURE 6.4   45° deflecting prism.

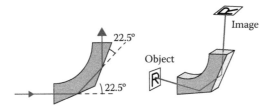

FIGURE 6.5 Wollaston prism.

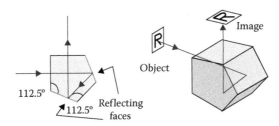

FIGURE 6.6 Pentaprism.

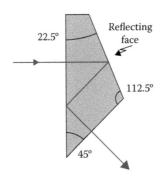

FIGURE 6.7 Constant 45° deflecting prism.

This property is used in the *Wollaston* prism (Figure 6.5) and in the *pentaprism* (Figure 6.6). In the Wollaston prism, both reflecting surfaces form a 45° angle and the deflecting angle is 90°. In the pentaprism, both surfaces form an angle of 135° and thus the deflection angle is 270° in these two prisms; the image is readable, since there are two reflections. The pentaprism is more compact and simpler to build. Although both prisms can be modified to obtain a 45° deflection, it results in an impractical and complicated shape. To obtain a 45° deflection independent of the incidence angle, the prism in Figure 6.7 is preferred. These prisms are used in microscopes, to produce a comfortable observing position.

6.1.2 Retroreflecting Prisms

A *retroreflecting prism* is a particular case of a constant deviation prism, in which the deflecting angle is 180°. A right-angle prism can be used as a retroreflecting prism with the orientation shown in Figure 6.8. In such a case, it is called a *Porro prism*. The Porro prism is a perfect retroreflector; the incident ray is coplanar with the normals to the surfaces.

Another perfect retroreflecting prism, made with three mutually perpendicular reflecting surfaces, is called a *cube corner prism*. This prism is shown in Figure 6.9.

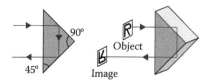

FIGURE 6.8   Rectangular prism in a two-dimensional retroreflecting configuration.

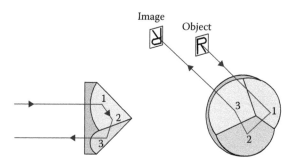

FIGURE 6.9   Cube corner prism.

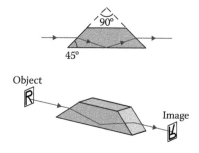

FIGURE 6.10   Dove prism.

Cube corner prisms are very useful in optical experiments where retroreflection is needed. Uses for the cube corner retroreflector are found in applications where the prism can wobble or jitter or is difficult to align because it is far from the light source. Applications for this prism range from the common ones like reflectors in a car's red back light to the highly specialized ones like the reflectors placed on the surface of the moon in the year 1969.

6.1.3 Inverting and Reverting Prisms

These prisms preserve the traveling direction of the light beam, changing only the image orientation. In order to produce an image inversion or reversion, these prisms must have an odd number of reflections. We will consider only prisms that do not deflect the light beam. The simplest of these prisms has a single reflection, as shown in Figure 6.10. This is a single rectangular prism, used in a configuration called a *dove prism*.

Although we have two refractions, there is no chromatic aberration since entrance and exit faces act as in a plane-parallel plate. These prisms cannot be used in strongly convergent or divergent beams of light because of the spherical aberration introduced, unless this aberration is compensated elsewhere in the system.

An *equilateral triangle prism* can be used as an inverting or reverting prism if used as in Figure 6.11. On this configuration, we have two refractions and three reflections. Like the dove prism, this prism cannot be used in strongly convergent or divergent beams of light.

Figures 6.12 through 6.14 show three reverting prisms with three internal reflections. The first one does not shift the optical axis laterally, while in the last two, the optical axis is displaced. These prisms can be used in converging or diverging beams of light. The first two prisms can be made either with two glass pieces or a single piece.

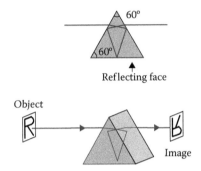

FIGURE 6.11 Inverting-reversing equilateral triangle prism.

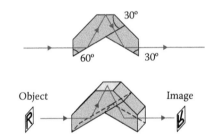

FIGURE 6.12 A reverting-inversing prism.

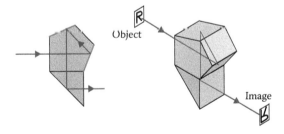

FIGURE 6.13 A reverting-inversing prism.

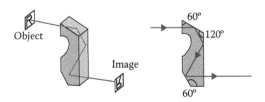

FIGURE 6.14 A reverting-inversing prism.

The *Pechan prism*, shown in Figure 6.15, can be used in converging or diverging pencils of light, besides being a more compact prism than the previous ones.

6.1.4 Rotating Prisms

A *half-turn rotating prism* is a prism that produces a readable image, rotated 180° as the real image produced by a convergent lens. A rotating prism can bring back the rotated image of a lens system to the original orientation of the object. These prisms are useful for monocular terrestrial telescopes and binoculars.

All reversing prisms can be converted to rotating prisms by substituting one of the reflecting surfaces by a couple of surfaces with the shape of a roof. With this substitution, the *Abbe prism*, the *Leman prism*, and the *Schmidt–Pechan prism*, shown in Figure 6.16, are obtained. This last prism is used in small hand telescopes. An advantage for this prism is that the optical axis is not laterally displaced.

A double prism commonly used in binoculars is the *Porro prism*, shown in Figure 6.17.

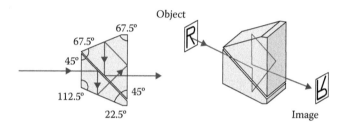

FIGURE 6.15 Pechan prism.

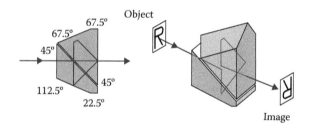

FIGURE 6.16 Schmidt–Pechan prism.

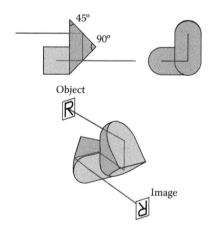

FIGURE 6.17 Porro prism.

6.1.5 Beamsplitting Prisms

These prisms divide the beam of light into two beams, with the same diameter as the original one, but the intensity is reduced for both beams that now travel in different directions. Beamsplitting prisms are used in amplitude division interferometers, binocular microscopes, and telescopes, where a single image must be observed simultaneously with both eyes. Basically, this prism is formed by a pair of rectangular prisms glued together to form a cube. One of the prisms has its hypotenuse face deposited with a thin reflecting film, chosen in such a way that, after cementing both prisms together, both the reflected and transmitted beam have the same intensity. Both prisms are cemented in order to avoid a total internal reflection. This prism and a variant are shown in Figure 6.18.

6.1.6 Chromatic Dispersing Prisms

The refractive index is a function of the light wavelength and, hence, of the light color. This is the reason why chromatic dispersing prisms decompose the light into its elementary chromatic components, obtaining a rainbow, or spectrum.

Equilateral Prism. The simplest chromatic dispersing prism is the equilateral triangle prism illustrated in Figure 6.19. This prism is usually made with flint glass, because of its large refractive index variation with the wavelength of the light.

As shown in Figure 6.19, ϕ is the deviation angle for a light ray and θ is the prism angle. We can see from this figure that

$$\phi = (\alpha - \alpha') + (\beta - \beta'); \tag{6.9}$$

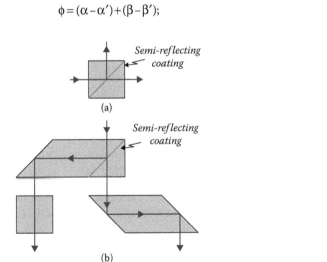

FIGURE 6.18 Binocular beamsplitting system. (a) Single cubic prism and (b) Complete binocular system.

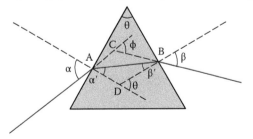

FIGURE 6.19 Triangular dispersing prism.

Also,

$$\theta = \alpha' + \beta', \tag{6.10}$$

From this we obtain

$$\phi = \alpha + \beta - \theta. \tag{6.11}$$

From Snell's law, we also know that

$$\frac{\sin \alpha}{\sin \alpha'} = n \tag{6.12}$$

and

$$\frac{\sin \beta}{\sin \beta'} = n. \tag{6.13}$$

From this we conclude that the deviation angle is a function of the incidence angle α, the apex angle θ, and the refractive index n. The angle ϕ as a function of the angle α for a prism with an angle $\theta = 60°$ and $n = 1:615$ is shown in Figure 6.20.

The deviation angle ϕ has a minimum magnitude for some value of α equal to α_m. Assuming, as we easily can, that there exists a single minimum value for ϕ, we can use the reversibility principle to see that this minimum occurs when $\alpha = \beta = \alpha_m$. It may be shown that

$$\sin \alpha_m = n \sin \theta/2. \tag{6.14}$$

Assuming that for yellow light $\alpha = \alpha_m$ in a prism with $\theta = 60°$ made from flint glass, the angle ϕ changes with the wavelength λ, as shown in Figure 6.21.

Let us now assume that the angle θ is small. It can be shown that the angle ϕ is independent from α and is given by

$$\phi = (n-1)\theta. \tag{6.15}$$

Pellin-Broca or constant deviation prism. Taking as an example the prism shown in Figure 6.22, we can see that the beam width for every color will be different and with an elliptical transverse section.

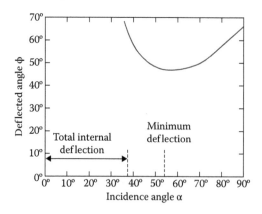

FIGURE 6.20   Angle of deflection vs. the angle of incidence in a dispersing prism.

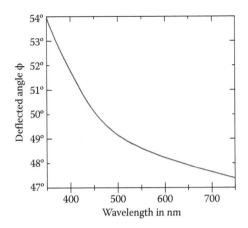

FIGURE 6.21   Deflection angle vs. the wavelength in a dispersing prism.

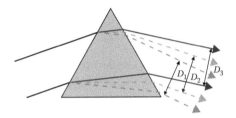

FIGURE 6.22   Variation in the beam deflection for different wavelengths in a triangular prism.

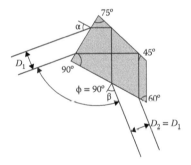

FIGURE 6.23   Pellin-Broca or constant deviation prism.

The minor semi-axis for the ellipse for the refracted beam will be equal to the incident beam only when the angle α is equal to the angle β.

For precise photometric spectra measurements, it is necessary that the refracted beam width be equal to the incident beam for every wavelength. This condition is only met when the prism is rotated so that $\alpha = \beta$ (minimum deviation). Usually, these measurements are uncomfortable, since both the prism and the observer have to be rotated.

A dispersing prism, which meets the previous condition with a single rotation of the prism for every measurement and does not require the observer to move, is the *Pellin-Broca* [12] or *constant deviation* prism, shown in Figure 6.23. This prism is built in a single piece of glass, but we can imagine it as the superposition of three rectangular prisms, glued together as shown in the figure. The deflecting angle ϕ is constant, equal to 90°. The prism is rotated to detect each wavelength. The reflecting angle must be 45° and, hence, angles α and β must be equal.

The Pellin-Broca prism has many interesting applications besides its common use as a chromatic dispersive element [13].

6.2 Lenses

In this section, we will study isotropic lenses of all kinds—thin and thick, as well as Fresnel and gradient index lenses.

6.2.1 Single Thin Lenses

A lens has two spherical concave or convex surfaces. The optical axis is defined as the line that passes through the two centers of curvature. If the lens thickness is small compared with the diameter, it is considered a thin lens. The focal length f is the distance from the lens to the image when the object is a point located at an infinite distance from the lens, as shown in Figure 6.24. The object and the image have positions that are said to be conjugate to each other and related by

$$\frac{1}{f} = \frac{1}{l'} - \frac{1}{l},$$

(6.16)

where l is the distance from the lens to the object and l' is the distance from the lens to the image, as illustrated in Figure 6.25. Some definitions and properties of the object and image are given in Table 6.1.

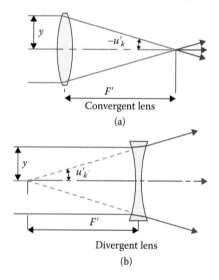

FIGURE 6.24 Focal length in a thin lens. (a) Convergent lens and (b) Divergent lens.

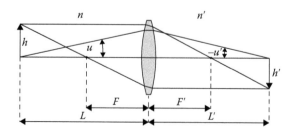

FIGURE 6.25 Image formation in a thin lens.

TABLE 6.1   Some Properties of the Object and Image Formed by a Lens

		Real		Virtual
Object	$l < 0$	Left of lens	$l > 0$	Right of lens
Image	$l' > 0$	Right of lens	$l' < 0$	Left of lens

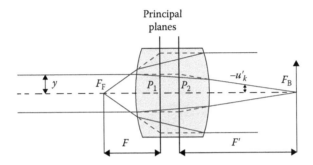

FIGURE 6.26   Effective and back focal lengths in a thick lens or system of lenses.

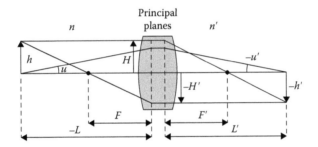

FIGURE 6.27   Image formation in a thick lens or system.

The focal length is related to the radii of the curvature r_1 and r_2 and the refractive index n by

$$\frac{1}{f} = (n-1)\left(\frac{1}{r_1} - \frac{1}{r_2}\right), \tag{6.17}$$

where r_1 or r_2 are positive if its center of curvature is to the right of the surface, and negative otherwise.

6.2.2 Thick Lenses and Systems of Lenses

A thick lens has a thickness that cannot be neglected in relation to its diameter. In a thick lens we have two focal lengths, that is, the effective focal length F and the back focal length F_B, as shown in Figure 6.26. The effective focal length is measured from the principal plane, which is defined as the plane where the rays would be refracted in a thin lens whose focal length is equal to the effective focal length of the thick lens. There are two principal planes: one when the image is at the right-hand side of the lens and another when the image is at the left-hand side of the lens (Figure 6.27).

In an optical system where the object and image media are air, the intersections of the principal planes with the optical axis define the principal or nodal prints. A lens can rotate about an axis, perpendicular to the optical axis and passing through the second modal point N_2 and the image from a distant object remains stationary. This is illustrated in Figure 6.28.

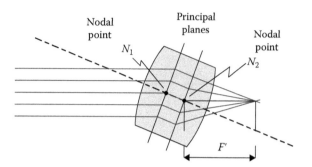

FIGURE 6.28 Rotating a thick lens system about the nodal point.

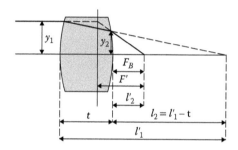

FIGURE 6.29 Light refraction in a thick lens.

In a thick lens or system of lenses, the object and image positions are given by

$$\frac{1}{f} = \frac{1}{L'} - \frac{1}{L}. \tag{6.18}$$

6.2.3 Thick Lenses

The effective focal length f of a thick lens is the distance from the principal plane $\mathbf{P}_2$ to the focus. It is a function of the lens thickness, and it is given by (see Figure 6.29)

$$\frac{1}{f} = (n-1)\left(\frac{l'}{r_1} - \frac{l}{r_2}\right) + \frac{(n-1)t}{nr_1r_2}. \tag{6.19}$$

The effective focal length is the same for both possible orientations of the lens. The back focal length is the distance from the vertex of the last optical surface of the system to the focus, given by

$$\frac{1}{F_B} = (n-1)\left|\frac{1}{r_1 - t\frac{(n-1)}{n}} - \frac{1}{r_2}\right| \tag{6.20}$$

This focal length is different for both lens orientations. The separation T between the two principal planes can be called the effective thickness and it is given by

$$T = \left|1 - \frac{F(P_1 + P_2)}{n}\right|t \approx (n-1)\frac{t}{n}, \tag{6.21}$$

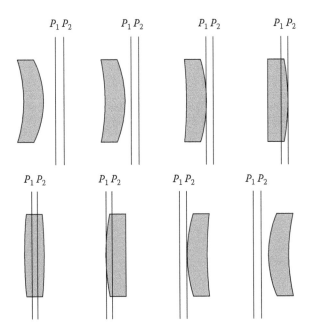

FIGURE 6.30 Principal planes' location in a thick lens.

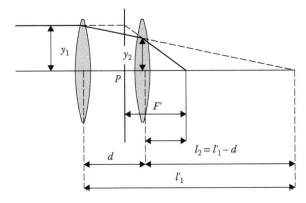

FIGURE 6.31 Light refraction in a system of two separated thin lenses.

where P_1 and P_2 are the powers of both surfaces, defined by

$$P_1 = \frac{(n-1)}{r_1}; P_2 = \frac{(n-1)}{r_2}. \tag{6.22}$$

We see that the separations between the principal planes are nearly constant for any radii of curvature, and gives the effective focal length, as in Figure 6.30.

6.2.4 System of Two Separated Thin Lenses

The effective focal length of a system of two separated thin lenses, as in Figure 6.31, is given by

$$\frac{1}{f} = \frac{1}{f_1} + \frac{1}{f_2} - \frac{d}{f_1 f_2}, \tag{6.23}$$

where d is the distance between the lenses. An alternative common expression is

$$F = \frac{f_1 f_2}{f_1 + f_2 - d}. \tag{6.24}$$

As in thick lenses, the effective focal length is independent of the system orientation. The separation T between the two principal planes is

$$T = [1 - F(P_1 + P_2)]d, \tag{6.25}$$

where the lens powers P_1 and P_2 are defined by

$$P_1 = \frac{1}{f_1}; P_2 = \frac{1}{f_2}. \tag{6.26}$$

6.2.5 Aspheric Lenses

Optical aberrations, such as spherical aberration, coma, and astigmatism, can seriously affect the image quality of an optical system. To eliminate the aberrations, several optical components, lenses, or mirrors have to be used so that the aberration of one element is compensated by the opposite sign of the others. If we do not restrict ourselves to the use of spherical surfaces, but use some aspherical surfaces, a better aberration correction can be achieved with fewer lenses or mirrors.

Hence, in theory, the use of aspherical surfaces is ideal. They are, in general, more difficult to manufacture than spherical surfaces. The result is that they are avoided if possible, but sometimes there is no other option but to use them. Let us now consider a few examples for the use of aspherical surfaces:

1. The large mirrors in astronomical telescopes are either paraboloids (Cassegrainian and Newtonian telescopes) or hyperboloids of revolution (Ritchey–Chretien , such as the Hubble telescope).
2. Schmidt astronomical cameras have at the front a strong aspheric glass plate to produce a system free of spherical aberration. Coma and astigmatism are also zero, because of the symmetry of the system around a common center of curvature.
3. A high-power single lens free of spherical aberration is frequently needed: for example, in light condensers and indirect ophthalmoscopes. Such a lens is only possible with one aspheric surface.

A rotationally symmetric aspheric optical surface is described by

$$z = \frac{cS^2}{1 + \sqrt{1 - (K+1)c^2 S^2}} + A_4 S_4 + A_6 S^6 + A_8 S^8 + A_{10} S^{10}, \tag{6.27}$$

where K is the conic contrast, related to the eccentricity e by $K = -e^2$. The constants A_4, A_6, A_8, and A_{10} are called apsheric deformation constants.

The conic constant defines the type of conic, according to the Table 6.2. It is easy to see that the conic constant is not defined for a flat surface.

6.2.6 Fresnel Lenses

A Fresnel lens was invented by Agoustine Fresnel and may be thought off as a thick plano-convex lens whose thickness has been reduced by breaking down the spherical face in annular concentric rings. The first Fresnel lens was employed in a lighthouse in France in 1822. The final thickness is approximately constant; therefore, the rings had different average slopes and also different widths. The widths decrease as the square of the semidiameter of the ring, as shown in Figure 6.32.

TABLE 6.2 Values of Conic Constants for Conic Surfaces

Type of Conic	Conic Constant Value
Hyperboloid	$K < -1$
Paraboloid	$K = -1$
Ellipse rotated about its major axis (prolate spheroid or ellipsoid)	$-1 < K < 0$
Sphere	$K = 0$
Ellipse rotated about its minor axis (oblate spheroid)	$K > 0$

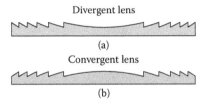

Divergent lens

(a)

Convergent lens

(b)

FIGURE 6.32 A Fresnel lens (a) Divergent lens and (b) Convergent lens.

In order to reduce the spherical aberration of a Fresnel lens, the slope of each ring is controlled in order to produce an effective aspherical surface. In a properly designed and constructed lens, the on-axis transverse aberration of all rings is zero; however, its phase difference is not necessarily zero, but random. Thus, they are not diffraction limited by the whole aperture but only by the central region.

Fresnel lenses are made by hot pressing an acrylic sheet [14,15]. Thus, a nickel master must first be produced.

6.3 Gradient Index Optics

A gradient index (GRIN) optical component is one where the refractive index is not constant but varies within the transparent material [16]. In nature, the gradient index appears on the hot air above roads, creating the familiar mirage. In optical instruments, gradient index lenses are very useful as will be shown here.

The variation in the index of refraction in a lens can be

1. In the direction of the optical axis, which is an axial gradient
2. Perpendicular to the optical axis, which is a radial gradient
3. Symmetric about a point, which is a spherical gradient

The spherical about the gradient is rarely used in optical components mainly because they are difficult to fabricate and because they are equivalent to axial gradients.

Gradient index components are most often fabricated by an ion exchange process. They are made out of glass, polymers, zinc selenide/zinc sulfide and germanium [17–19]. Optical fibers with gradient indices have been made by a chemical vapor deposition process.

6.3.1 Axial Gradient Index Lenses

Figure 6.33 shows a plano-convex lens with an axial gradient. If the lens is made with a homogeneous glass, it is well known that a large amount of spherical aberration occurs, making the marginal rays converge at a point on the optical axis closer to the lens than the paraxial focus. A solution to correct this aberration is to decrease the refractive index near the edge of the lens. If a gradient is introduced, decreasing the refractive index in the direction of the optical axis, the average refractive index

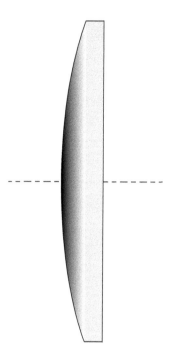

FIGURE 6.33   An axial gradient plano-convex lens.

is lower at the edge of the lens than at its center. The refractive index along the optical axis can be represented by

$$n(z) = N_{00} + N_{01}z + N_{02}z^2 + \dots, \tag{6.28}$$

where N_{00} is the refractive index at the vertex of the convex (first) lens surface. It can be shown [18] that for a single plane convex lens, a good correction is obtained with linear approximation. The gradient should have a depth equal to the sagitta of the convex surface. With these conditions it can be proven that the refractive index change Δn along the optical axis is given by

$$\Delta n = \frac{0.15}{\left(f / N_{00}\right)^2}, \tag{6.29}$$

where $(f/\#)$ is the f-number of the lens. Thus, the necessary Δn for sin $f/4$ lens is only 0.0094, while for an $f/1$ lens, it is 0.15.

6.3.2 Radial Gradient Index Lenses

Radial index gradients are symmetric about the optical axis and can be represented by

$$n(r) = N_{00} + N_{10}r^2 + N_{20}r^4 + \dots, \tag{6.30}$$

where r is the radial distance. A very thick plano-plano lens or rod, as shown in Figure 6.34, refractor in a curved sinusoidal path all rays enter the lens. The wavelength L of this wavy path can be shown by

$$L = 2\pi \left(-\frac{N_{00}}{2N_{20}} \right)^{1/2}. \tag{6.31}$$

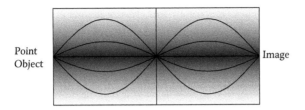

FIGURE 6.34   A radial gradient rod lens.

Thus, if the rod has a length L, an object at the front surface can be imaged at the rear surface with unit magnification without any spherical aberration. This property is used in relay systems, such as those in endoscopes or borescopes.

An interesting and useful application of gradient index optics rods is as an endoscopic relay, as described by Tomkinson et al. [20]. The great disadvantage is that endoscopes using these rods are rigid. In compensation, the great advantage is that they have a superior imaging performance in sharpness as well as in contrast.

6.4 Summary

Optical elements made out of isotropic materials, homogeneous as well as in homogeneous (gradient index), are the basic components of most optical systems.

References

1. Hopkins RE. Mirror and prism systems. In *Military Standardization Handbook: Optical Design, MIL-HDBK*, Vol. 141. Washington, DC: U.S. Defense Supply Agency, 1962.
2. Hopkins RE. Mirror and prism systems. In *Applied Optics and Optical Engineering*, Kingslake, R. (ed.), Vol. 3. San Diego, CA: Academic Press, 1965.
3. Berkowitz DA. Design of plane mirror systems, *J. Opt. Soc. Am.*, **55**, 1464–1467; 1965.
4. Pegis RJ, Rao MM. Analysis and design of plane mirror systems, *Appl. Opt.*, **2**, 1271–1274; 1963.
5. Walles S, Hopkins RE. The orientation of the image formed by a series of plane mirrors, *Appl. Opt.*, **3**, 1447–1452; 1964.
6. Walther A. Comment on the paper: Analysis and design of plane mirror systems by Pegis and Rao, *Appl. Opt.*, **3**, 543–543; 1964.
7. Yoder PR Jr. Study of light deviation errors in triple mirror and tetrahedral prism, *J. Opt. Soc. Am.*, **48**, 496–499; 1958.
8. Chandler KN. On the effects of small errors in angles of corner-cube reflectors, *J. Opt. Soc. Am.*, **50**, 203–206; 1960.
9. Eckhardt HD. Simple model of corner reflector phenomena, *Appl. Opt.*, **10**, 1559–1566; 1971.
10. Friedman Y, Schweltzer N. Classification of stable configurations of plane mirrors, *Appl. Opt.*, **37**, 7229–7234; 1998.
11. Schweltzer N, Friedman Y, Skop M. Stability of systems of plane reflecting surfaces, *Appl. Opt.*, **37**, 5190–5192; 1998.
12. Pellin P, Broca A. Spectroscope a déviation fix, *J. Phys.*, **8**, 314–319; 1899.
13. Moosmüller H. Brewster's angle Porro prism: A different use for a Pellin-Broca prism, *Opt. Photonics News*, **9**, 8140–8142; 1998.
14. Miller OE, McLeod JH, Sherwood WT. Thin sheet plastic Fresnel lenses of high aperture, *J. Opt. Soc. Am.*, **41**, 807–815; 1951.
15. Boettner EA, Barnett NE. Design and construction of Fresnel optics for photoelectric receivers, *J. Opt. Soc. Am.*, **41**, 849–857; 1951.

16. Marchand EW. *Gradient Index Optics*. New York, NY: Academic Press, 1978.

17. Moore DT. Design of a single element gradient-index collimator, *J. Opt. Soc. Am.*, **67**, 1137–1137; 1977.

18. Moore DT. Gradient-index materials. In *CRC Handbook of Laser Science and Technology, Supplement 1: Lasers*, Weber, M. J. (ed.). New York, NY: CRC Press, pp. 499–505; 1995.

19. Moore DT. Gradient index optics. In *Handbook of Optics*, Bass M, and Mahajan VN (eds.), Vol. 1, Chap. 24. New York, NY: McGraw Hill, 2010.

20. Tomkinson T H, Bentley JL, Crawford MK, Harkrider CJ, Moore DT, Ronke JL. Rigid endoscopic relay systems: A comparative study, *Appl. Opt.*, **35**, 6674–6683; 1996.

7

Reflective Optical Components

Daniel Malacara-
Hernández and
Armando
Gómez-Vieyra

7.1 Introduction

Technological developments in fields, such as astronomical instrumentation, lithography, spectroscopy, biomedical, and so on, have impacted the development and implementation of high-reflectance optical devices. Within these components are traditional mirrors, prisms, and beamsplitters, but deformable mirrors and digital micro-mirror devices (DMD) have also been introduced. In this chapter, the traditional and modern aspects of the reflective components will be discussed.

7.2 Mirrors

The mirrors as well as lenses can be used in optical systems to change beam directions and to form images. High reflection is achieved by special metallic or dielectric coatings of thin films. Like a lens, a mirror can introduce all five Seidel monochromatic aberrations. The main advantage of mirrors compared to lenses and prisms is the absence of two chromatic aberrations. The possible variables are the mirror shape, aperture, pupil position, and image and object positions.

7.2.1 Plane Mirrors

Plane mirrors are used for ray deflection. They are introduced to the folding of optical systems, beam delivery for optical material-processing systems through articulated arms, and many more applications. For example, special mirror forms are used in the polygons for scanning systems and segmented plane mirrors are used in DMD mirrors. The required accuracy of plane mirror surfaces is very high; this means surface errors of $\frac{\lambda}{10}$ usually.

7.2.2 Spherical Mirrors

The advantage of spherical mirrors compared to lenses is mainly that the deflection does not depend on the wavelength, which means complex imaging systems with long focal lengths and free of chromatic aberrations. However, the aperture and the image field are limited by spherical aberration. A spherical mirror is free of spherical aberration only when the object and the image are both at the center of curvature or at the vertex of the mirror, fulfilling the conditions of Abbe aplanatic [1].

Let us begin with the study of spherical mirror aberrations. The first-order parameters in spherical mirrors (concave or convex mirrors) with the object at a finite distance l and the stop at a finite distance in front of the mirror are presented in Figure 7.1.

Assuming a first-order theory in an optical system is centered, the focal length is

$$f = \frac{2}{r} \tag{7.1}$$

and the optical power is

$$\phi = \frac{r}{2}. \tag{7.2}$$

The expressions for the primary off-axis aberrations for a concave or convex spherical mirror are:

Primary longitudinal spherical aberration

$$SphL = -\left(\frac{r-1}{r-2l}\right)\frac{y^2}{r} \tag{7.3}$$

Sagittal coma

$$\text{Coma}_s = -\frac{\left(\frac{1}{r}-\frac{1}{l}\right)\left(1-\frac{\bar{l}}{r}\right)}{\left(1-\frac{\bar{l}}{l}\right)}\frac{y^2 h'}{r} \tag{7.4}$$

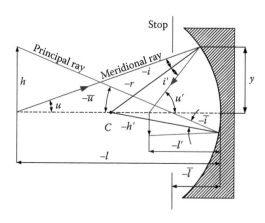

FIGURE 7.1 First-order parameters in a mirror with the object at finite distance from the mirror.

Longitudinal sagittal astigmatism

$$AstL_s = -\frac{\left(1-\dfrac{\bar{l}}{r}\right)^2}{\left(1-\dfrac{\bar{l}}{l}\right)^2}\frac{h^2}{r} \tag{7.5}$$

Petzval curvature

$$Ptz = \frac{h'^2}{r} = \frac{h'^2}{2f} \tag{7.6}$$

If the object is at infinity, as shown in Figure 7.2, $\left(l'=-f=\dfrac{r}{2}\right)$, the image is at the focus and the longitudinal spherical aberration has the value

$$SphL = -\frac{y^2}{4r} = \frac{D^2}{32f} = \frac{z}{2} \tag{7.7}$$

and, the transverse aberration is

$$SphT = \frac{y^3}{2r^2} = \frac{D^3}{64f^2}, \tag{7.8}$$

where D is the diameter of the mirror, z is the sagitta of the surface, and r is the radius of curvature. By integration of this expression, the wavefront aberration is given by

$$W(y) = \frac{y^4}{4r^3}. \tag{7.9}$$

This wavefront distortion is twice the sagitta difference between a sphere and a paraboloid. This is to be expected, since the paraboloid is free of spherical aberration when the object is at infinity.

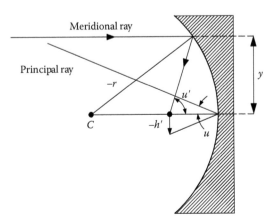

FIGURE 7.2 First-order parameters in a mirror with the object at infinite distance from the mirror.

The value of the Petzval curvature, that is, the field curvature in the absence of astigmatism, is Equation 7.6. Thus, the Petzval surface is concentric with the mirror. In a spherical mirror, when the object and the image are at the center of curvature, the spherical aberration is zero. If the stop is at the center of curvature, the coma and astigmatism aberrations are also zero. Then, only the Petzval curvature exists and coincides with the field curvature.

The value of the sagittal coma aberration is a function of the stop position, given by

$$\text{Coma}_s = \frac{y^2 h' \left(\bar{l} - r \right)}{r^3}. \tag{7.10}$$

If the object is located at infinity, the spherical mirror would be free of coma with the stop at the mirror, only if the principal surface (the mirror surface itself) is a sphere with a center of curvature at the focus, but obviously this is not the case. Then, when the stop is at the mirror $\left(\bar{l} = 0 \right)$, the value of the sagittal coma is

$$\text{Coma}_s = \frac{D^2 h'}{16 f^2}. \tag{7.11}$$

When the stop is at the center of curvature $\left(\bar{l} = r \right)$, the value of the sagittal coma becomes zero, as mentioned before.

The longitudinal sagittal astigmatism is given by

$$AstL_s = -\frac{h'^2}{r} \left(\frac{\bar{l} - r}{r} \right)^2, \tag{7.12}$$

which may also be written as

$$AstL_s = -\left(\frac{\bar{l} - r}{r} \right)^2 Ptz. \tag{7.13}$$

As pointed out before, we can see that $AstL_s = Ptz$ when $\bar{l} = 0$ and the field is curved with the Petzval curvature, as shown in Figure 7.3a. In general, it can be shown that the sagitta represented here by Best, the surface of the best definition located between the sagittal and the tangential surfaces, is given by

$$\text{Best} = \left[1 - 2 \left(\frac{\bar{l} - r}{r} \right) \right] Ptz. \tag{7.14}$$

If the stop is located at

$$\frac{\bar{l}}{r} = \pm \frac{1}{\sqrt{3}} + 1 = 0.42; 1.58, \tag{7.15}$$

the tangential surface is flat, as shown in Figure 7.3b. When the stop is placed at

$$\frac{\bar{l}}{r} = \pm \frac{1}{\sqrt{2}} + 1 = 0.29, 1.707, \tag{7.16}$$

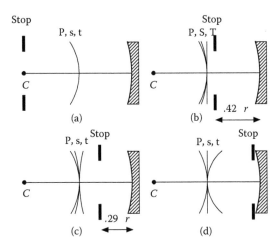

FIGURE 7.3 Anastigmatic surface for a spherical mirror with the stop at four different positions.

the surface of the best definition is a plane, as in Figure 7.3c. When the stop is at the mirror, as in Figure 7.3d, the sagittal surface is flat.

In the current retinal imaging systems, laser cavities or astronomical spectroscopes, there is the need to employ off-axis reflective systems. These systems often use different configurations where the object or image could be at finite distances or at infinity with respect to the spherical mirrors. Using the principle of the optical path difference principle between the sphere and reference surface (ellipsoid, paraboloid or hyperboloid), Gomez et al. [2–4] calculated the mathematical expressions of the third-order wavefront aberrations in a single off-axis spherical mirror. Considering the spherical mirror, with its pupil at the same plane as the mirror and with an off-axis point object. The reflected wavefront is aberrated. The aberrations produce an image with tangential, saggital, Petzval and least confusion (medium) foci, assuming a relatively small coma. Then, the spherical mirror is replaced by a reflecting reference surface, according to the case. In the case of real image and real object points, we use the ellipsoid as a reference surface, which then has one of its foci at the point object and the second focus near the images produced by the spherical mirror. Moreover, a spherical mirror with oblique incidence and an object or image at the infinite, needs a paraboloid as reference surface. Finally, when the mirror with oblique incidence works with an imaginary object or image, we must use a hyperboloid as reference surface. The wavefront produced by the reflecting reference surface is spherical (no aberration), hence, we can take this wavefront as a reference to measure the aberration for the wavefront produced by the spherical mirror.

The main constraint of classical off-axis reflecting imaging systems is the primary astigmatism. This, in off-axis spherical reflective imaging systems, can be eliminated by using the proper configuration based in the general Coddington marginal ray fan equation [5], as shown in Gomez et al. [6]. Furthermore, in off-axis reflecting spectroscopy systems, compensate coma is most important, as shown in Rosendahl [7].

7.2.3 Paraboloidal Mirrors

A paraboloidal mirror is an aspherical surface in which the conic constant is $K = -1$, as described in Chapter 6. The spherical aberration is absent if the object is located at infinity. However, if the object is at the center of curvature, the spherical aberration appears. Now, let us examine each of the primary aberrations in a paraboloidal mirror. The exact expression for the longitudinal aberration of the normals to the mirror is given by

$$SphL_{\text{normls}} = f \tan^2 \varphi \tag{7.17}$$

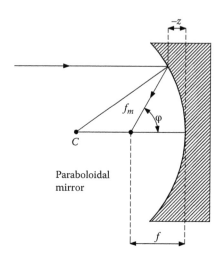

FIGURE 7.4   Paraboloidal mirror.

as illustrated in Figure 7.4. When the object is at the center of curvature, the spherical aberration of the paraboloid is approximately twice the aberration of the normals. Thus, we may write

$$SphL = 2f\frac{y^2}{r^2} = \frac{y^2}{r} = -\frac{D^2}{8f}. \tag{7.18}$$

For the spherical aberration of a spherical mirror with the object at infinity, the absolute values are different by a factor of four and are opposite in sign. In other words, the wavefront aberrations must have opposite signs and the same absolute values.

As for the sphere, if the object is located at infinity, the paraboloid would be free of coma with the stop at the mirror only if the principal surface is spherical with the center of curvature at the focus; but again, this is not the case. The principal surface is the parabolic surface itself. Thus, the value of *OSC* [1] is given by

$$OSC = \frac{f_M}{f} - 1, \tag{7.19}$$

where f_M and f are the marginal and paraxial focal lengths, measured along the reflected rays, as shown in Figure 7.4. For a paraboloid, we may show that

$$f_M = f \mp z \tag{7.20}$$

with the sagitta z given by

$$z = -\frac{y^2}{4f} = -\frac{D^2}{16f}. \tag{7.21}$$

Thus, the value of the sagittal coma can be shown to be

$$Coma_s = OSC \cdot h = -\frac{zh'}{f} = \frac{D^2 h'}{16f^2}, \tag{7.22}$$

The coma aberrations are the same for spherical and parabolic mirrors when they stop at the mirror.

In a paraboloid, the spherical aberration with the object at infinity is zero. Then, the contributions are:

$$SphL_{sphere} + SphL_{asphere} = 0. \tag{7.23}$$

Similarly, for longitudinal sagittal astigmatism:

$$AstL_{S\,total} = AstL_{S\,sphere} + AstL_{S\,asphere}. \tag{7.24}$$

As can be demonstrated, the astigmatism for a paraboloid and a sphere are related by

$$AstL_{S\,total} = SphL_{sphere}\left(\frac{\bar{i}}{i}\frac{\bar{y}}{y}\right)^2 \tag{7.25}$$

or

$$AstL_{S\,total} = AstL_{L\,sphere}\left[1 - \left(\frac{\bar{i}}{i}\frac{\bar{y}}{y}\right)^2\right]. \tag{7.26}$$

The astigmatism when the stop is at the mirror is equal to the astigmatism of a spherical mirror, which can also be written as

$$AstL_{S\,total} = \left[\frac{\left(\bar{l}-r\right)^2 - \bar{l}^2}{r^2}\right]Ptz. \tag{7.27}$$

We see that the surface of the best definition is flat when $\dfrac{\bar{l}}{r} = 0.25$ and not 0.29, as in the case of the spherical mirror.

7.2.4 Ellipsoidal Mirrors

Ellipsoidal mirrors, whose surface is generated by rotating an ellipse about its major axis, produce an image free of spherical aberration when the image and object are located at each of their foci [8]. As described in Chapter 6, the conic constant for these ellipsoidal surfaces is in the range $-1 < K < 0$. If the major and minor semi-axes are a and b, respectively, as in Figure 7.5, the separation between the two foci of the ellipsoid is $a^2 - b^2$ and $K = -\left(1 - \dfrac{b^2}{a^2}\right)$. Thus, these two foci are at distances d_1 and d_2 from the vertex given by

$$d_1, d_2 = a \pm \left(a^2 - b^2\right)^{1/2}, \tag{7.28}$$

which, in terms of the conic constant K, can be shown to be

$$d_1, d_2 = \frac{r}{(K+1)}\left(1 \pm \sqrt{-K}\right), \tag{7.29}$$

where r is the radius of curvature at the vertex of the ellipsoidal surface.

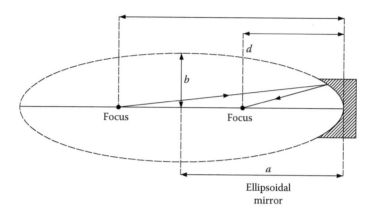

Ellipsoidal
mirror

FIGURE 7.5   An ellipsoidal mirror.

7.3 Mirror Systems

There are some optical systems that use only mirrors [9,10], such as the astronomical mirrors studied in Chapter 13 in this book. In this section, some other interesting systems are described.

7.3.1 Manguin Mirror

This mirror was invented in 1876 in France by Manguin as an alternative for the parabolic mirror used in search lights. It is made with a meniscus negative lens coated with a reflective film on the convex surface, as shown in Figure 7.6. The radius of curvature of the concave refracting surface and the thickness are the variables used to correct the spherical aberration. A bonus advantage is that the coma aberration is less than half that of a parabolic mirror. This system has two more advantages: first, the surfaces are spherical not parabolic, making construction a lot easier; second, the reflecting coating is on the back surface, avoiding air exposure and oxidation of the metal. A Manguin mirror made with crown glass (BK7) can be obtained with the following formulas:

$$r_1 = 0.1540T + 1.0079F \tag{7.30}$$

and

$$r_2 = 0.8690T + 1.4977F, \tag{7.31}$$

where T is the thickness, F is the effective focal length of the mirror, r_1 is the radius of curvature of this reflecting surface, and r_2 is the radius of curvature of the front surface.

7.3.2 Dyson System

Unit magnification systems are very useful for copying small structures or drawings; a typical application is in photolithography in the electronics industry. In general, these systems are symmetric, thus automatically eliminating coma, distortion, and magnification chromatic aberration. An example of these systems, illustrated in Figure 7.7, was designed by Dyson [11]. The system is concentric. A marginal meridional ray on the axis, leaving from the center of curvature, would not be refracted. Thus, spherical aberration and axial chromatic aberration are absent. The radius of curvature r_L of the lens is

$$r_L - \left(\frac{n-1}{n} \right) r_{M}, \tag{7.32}$$

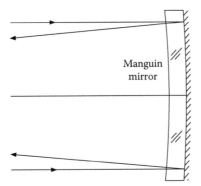

FIGURE 7.6 The Manguin mirror.

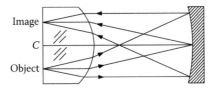

FIGURE 7.7 The Dyson system.

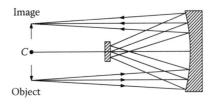

FIGURE 7.8 The Offner system.

where r_M is the radius of curvature of the mirror, in order to make the Petzval sum zero. The primary astigmatism is also zero, since the spherical aberration contribution of both surfaces is zero. However, the higher-order astigmatism appears not very far from the optical axis. Thus, all primary aberrations are corrected in this system.

It may be noted that since the principal ray is parallel to the optical axis in the object as well as in the image media, the system is both frontal and back telecentric.

7.3.3 Offner System

The Offner [12] system is another 1:1 magnification system, formed only by mirrors, as shown in Figure 7.8. The system is concentric and with zero Petzval sum, as in the case of the Dyson system. This system may be also corrected for all primary aberrations, but since higher-order astigmatism is large in this configuration, actual Offner systems depart from this configuration. Primary and higher-order astigmatism are balanced at a field zone to form a well-corrected ring where the sagittal and the tangential surfaces intersect.

7.4 Reflective Coatings

Reflecting coatings are an essential part of mirrors, beamsplitters, and prisms. These can be of many different types [13]. Some of them allow increased aluminum reflectance in visible spectrums using four alternating layers of silicon dioxide and titanium dioxide [14]. The best option also depends on the application and a large number of factors, such as the type of application, wavelength range, the environment around the optical system, cost limitations, and so on. Thin films are described in detail in Chapter 32. However, here we describe some of the many thin film coatings used in reflective optical systems. The experimental procedures used to produce them are not described here.

7.4.1 Silver

An advantage of silver films is that they can be deposited by an inexpensive chemical procedure, without any vacuum deposition chamber [15]. Silver films have a good reflectance in the whole visible spectrum and in the near-ultraviolet, as illustrated in Figure 7.9 [16].

A serious disadvantage is that silver oxidizes and becomes black because of atmospheric humidity. Actually, pure silver is nearly useless for surface reflections. In optical components with internal reflections (e.g. prisms), the back of the silver is protected with a plated layer of less active metal (e.g., copper, Inconel, or nickel) and a coat of paint. When it is necessary as the first surface, the silver is protected with a dielectric overcoat, which is deposited in a vacuum chamber. In general, the silver is the best metal coating in the visible and near-infrared, this means that it has the most viable option in research and instrumentation (biomedical, biological and chemical).

7.4.2 Aluminum

Aluminum coating has to be deposited in a vacuum deposition chamber by evaporation. The reflectivity of aluminum is worse than that of silver in the ultraviolet region, but better in the infrared region, as shown in Figure 7.9.

The most important characteristic of aluminum is that although it oxidizes quite rapidly, this oxide is transparent: it can be stained and easily scratched with dust. So, it is generally protected with a hard

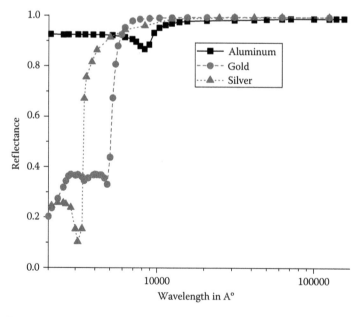

FIGURE 7.9 Reflectivity for aluminum, silver, and gold at different wavelengths. (From Lynch, D. W. and W. R. Hunter In E. D. Palik [ed.], *Handbook of Optical Constants of Solids*, Orlando, Academic Press, 1985.)

coating. Most optical mirrors are covered and protected with a layer of half a wave thick magnesium fluoride (MgF) or silicon monoxide (SiO). The great advantage is that a protected mirror can be easily be cleaned with isopropyl alcohol or acetone without damaging the mirror. Multilayer protective coatings are also used to enhance the reflectivity in the visible or in the ultraviolet regions. Its greatest application is currently in teaching laboratories.

7.4.3 Gold

Gold has to be deposited by vacuum evaporation. As expected, it is more expensive than aluminum and silver, but it has the great advantage of a good reflectivity in the infrared region, as shown in Figure 7.9. Optical elements for use in the infrared frequently use gold. Gold mirrors can also be protected with a silicon monoxide overcoating.

7.4.4 Rhodium

Rhodium is a noble metal that has a high performance in the ultraviolet, a wide range of angles of incidence and longevity. It is the most efficient and durable coating available for ellipsoidal reflectors and it is used for first mirrors in ITER [17,18].

7.4.5 Dielectric Films

Dielectric multilayer films are deposited by vacuum evaporation with a procedure that permits deposition of different kinds of materials with a precisely controlled thickness (see Chapter 32). Mirrors made in this manner are expensive. However, the great advantage is that the reflectivity at the desired wavelength range, the polarization state and the range of angles of incidence can be obtained as required. Unlike metal mirrors, they can reflect or transmit all of the incident light without absorbing anything; also, they can reflect some colors and transmit others. The dielectric mirror behavior is based on constructive interference between lights reflected from the individual thin film.

The dielectric coatings are remarkably hard, durable, and abrasion-resistant; that is, the coatings are more robust than metal mirrors. Mirrors must therefore be designed and fabricated to the specific purpose intended.

An interesting application of these films is in the concave mirrors used in projector lamps, where the reflectivity has to be as high as possible in the visible region but as low as possible in the infrared region to avoid heating the film.

7.5 Beamsplitters

Beamsplitters are partially reflecting and transmitting mirrors. They can be made with an extremely thin film of metal or with a stack of thin films.

The reflectance and transmittance of thin metal films, as shown in Figure 7.9, is almost flat over the visible wavelength range; therefore, the color of the reflected and transmitted light beam preserves their color. On the other hand, in a dielectric thin films beamsplitter, the reflected and transmitted beams have complementary colors. They preserve their original colors only if the reflectivity is constant for the entire visible spectrum. Another important difference is that metal films absorb a small part of the luminous energy.

Polarizing beamsplitters reflect a large percentage of the S polarized beam while transmitting the P polarized beam. Thus, if the incident beam is unpolarized, the reflected and the transmitted beams are linearly polarized, in orthogonal directions.

7.6 Digital Micro-Mirror Device

A digital micro-mirror device (DMD), introduced by Texas Instruments, is both a microelectronic mechanic system (MEMS) and a spatial light modulator (SML). It is a semiconductor based light switch array of thousands of individual addressable, tiltable, mirror-pixels, which are controlled by underlying CMOS electronics. One of the advantages of this device is its broadband capability and consequently, it has managed to be inserted in fields, such as lithography, microscopy, 3D projection, and so on.

7.7 Deformable Mirrors

In the last 20 years, it has increased the use of deformable mirrors (DM), whose origin is in astronomical instrumentation, occupying a very important place in the biomedical systems. Similarly, its technological evolution has been continued progress. Gone are the mechanical pistons and piezoelectric systems. Currently, most of the adaptive optics biomedical applications are based on MEMS and magnetic actuators. Reflecting this has been the progress in the electronic stabilization of these devices. The incorporation of various reflective films, the diversity of actuator sizes and the actuators number per array. Most of DMs consist of a thin, continuous face sheet attached to actuators, which translate an electrical signal into a mechanical deformation of the reflective surface. The application of the DM determines the number of actuators, the actuators pitch, the actuator stroke and the speed of the DM. It is important to consider in some cases, the hysteresis and the percentage of non-functional actuators.

References

1. Malacara D, Malacara Z. *Handbook of Lens Design*. New York, NY: Marcel Dekker, 1998.
2. Gómez-Vieyra A, Malacara-Hernández D. Geometric theory of wavefront aberrations in an off-axis spherical mirror, *Appl. Opt.*, 50, 66–73; 2011.
3. Gómez-Vieyra A, Vargas CA, Malacara-Hernandez D. Calculation of wavefront aberrations in off-axis spherical mirror with object or image at the infinite, *Opt. Eng.*, 51, 083002; 2012a.
4. Gómez-Vieyra A, Malacara-Hernandez D, Hidalgo-Gonzalez JC, Vargas CA. The wavefront aberrations in off-axis spherical mirror with object point or image point, *Proc. SPIE 8550, Opt. Syst. Des.*, 85501L; 2012b.
5. Malacara-Hernández Z, Malacara-Hernández D, Gómez-Vieyra A. A new derivation of a general Coddington equation from the sagittal and tangential Coddington equations, *Optik*, 124, 1627–4628; 2013.
6. Gómez-Vieyra A, Malacara-Hernández D. Methodology for the third-order astigmatism compensation in off-axis spherical reflective systems, *Appl. Opt.*, 50, 1057–1064; 2011.
7. Rosendahl GR. Contributions to the optics of mirror systems and gratings with oblique incidence. III. Some applications, *J. Opt. Soc. Am.*, 52(4), 412–415; 1962.
8. Nielsen JR. Aberrations in ellipsoidal mirrors used in infrared, *J. Opt. Soc. Am.*, 39, 59–63; 1949.
9. Churilovskii VN, Goldis KI. An apochromatic catadioptric system equivalent to a parabolic mirror, *Appl. Opt.*, 3, 843–846; 1964.
10. Erdos P. Mirror anastigmat with two concentric spherical surfaces, *J. Opt. Soc. Am.*, 49, 877–886; 1956.
11. Dyson J. Unit magnification system without Seidel aberrations, *J. Opt. Soc. Am.*, 49, 713; 1959.
12. Offner A. New concepts in projection mask aligners, *Opt. Eng.*, 14, 131; 1975.
13. Hass G. Mirror coatings. In *Applied Optics and Optical Engineering*. New York, NY: Academic Press, 1965.
14. Carniglia CK. Mirrors: Coating choices make a difference. In *The Photonics Design and Applications Handbook*, 46th edn., Vol. 3, pp. 307–310. Pittsfield, MA: Laurin Publishing, 2000.
15. Strong J. *Procedures in Experimental Physics*. Englewood Cliffs, NJ: Prentice-Hall, 1938.

16. Lynch DW, Hunter WR. In E. D. Palik (ed.), *Handbook of Optical Constants of Solids*, pp. 275–367. Orlando, FL: Academic Press, 1985.
17. Eren B, Marot L, Litnovsky A, et al. Reflective metallic coatings for first mirrors on ITER, *Fusion Eng. Des.*, 86, 2593–2596; 2011.
18. Marot L, Arnoux G, Huber A, et al. Optical coatings as mirrors for optical diagnostics, *J. Coat. Sci. Technol.*, 2, 72–78; 2015.

8

Diffractive Elements

Cristina Solano

8.1 Introduction

Considering an optical system as a device that transforms input wavefronts into output wavefronts, the amount of transformations is quite limited in the case of a refractive–reflective optical system. A considerable increase of possibilities can be obtained by using diffracted elements.

In general, the term "diffractive element" (DE) covers the range of elements that use the wave nature of light to control light by means of diffraction, [1]. Although the following specific terms can be used:

- **Diffractive lenses**: elements that perform functions similar to conventional refractive lenses, i.e., they form images.
- **Kinoforms**: diffractive elements whose phase modulation is introduced by a surface relief structure.
- **Binary optics**: kinoforms produced by means of lithography using a number of phase controlling surfaces.
- **Holographic optical elements**: recorded by the interference of two waves or by calculating it using a computer.

Regardless of the name or method of fabrication, diffractive optics can be understood with just a few basic tools. The properties of diffractive optics that are shared with conventional elements (focal length, chromatic dispersion, aberration contributions, etc.) do not depend on the specific type of diffractive element. Given a phase function or, equivalently, a grating spatial frequency distribution, the influence of the diffractive element or an incident ray for a specified diffraction order is found via the grating equation,

$$\lambda = d(\sin\theta_1 + \sin\theta_2), \tag{8.1}$$

where θ_1 and θ_2 are the angles of incidence and diffraction, respectively, and d is the period of the diffraction grating and λ is the used wavelength.

The specific type involved (kinoform, binary lens, HOE, etc.) only influences the diffraction efficiency. For this reason, throughout this chapter the general term of diffractive elements (DEs) will be used.

One factor that has encouraged much of the recent interest in diffractive optics has been the increased optical performance of such optical elements. This allows the fabrication of optical elements that are smaller, lighter, and cheaper to fabricate, and are more rugged and have superior performance than the conventional optical components they often replace. In addition, the currently available design capabilities for binary optics make the design and manufacture of components having optical properties never before produced possible.

8.2 Design Issues

Diffractive optical elements introduce a controlled phase delay into an optical wavefront by changing the optical path length, either by varying the surface height, or by modulating the refractive index as a function of position. Because of their unique characteristics, diffractive elements (DEs) present a new and very useful tool in optical design.

Innovative diffractive components have been applied in a number of new systems, such as laser-based systems, in which unusual features of these elements are utilized. Due to the great number of parameters that can be used to define a diffractive component, the efficient handling of this degree of freedom presents a technical challenge.

One of the best ways to describe the design procedure was presented by Joseph Mait [2] who divided it into three basic stages: analysis, synthesis, and implementation.

1. *Analysis.* There are two important aspects. First, it is necessary to understand the physics of the image formation required by the proposed DE, which will determine the method to be used (Figure 8.1). Among the methods available are scalar or vector diffraction theory, Fresnel and Fourier transform, convolution, correlation, and rigorous theory. The choice of method depends on the required diffraction properties of the DE and will affect the complexity of the design algorithm and the definition and value of the measured performance.

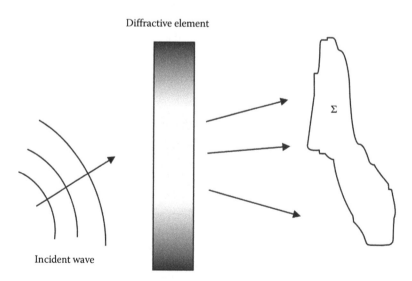

FIGURE 8.1 Basic geometry of diffractive optics (transmission mode). The incident wave is diffracted by the diffractive element (DE). The resultant amplitude distribution Σ must satisfy the requirements of the specific design.

Secondly, it is important to take into account the fabrication model (i.e., the degree of linearity of the material and the possible errors in its fabrication) in which the data generated by the computer will be recorded.

2. *Synthesis.* Identifying the appropriate scheme that mathematically represents the underlying physical problem, the appropriate optimization techniques, design metrics (i.e., diffraction efficiency, quantization or reconstruction error, modulation transfer function, aberrations, undesired light, glare, etc.), and their degree of freedom.

Among the proposed procedures for optimization of the problem are the quantization, steepest descent, simulated annealing, and iterative Fourier algorithm, which in general can be classified as direct and indirect approaches.

In the direct method, the performance of the primary metric is optimized. This method, although simple, can be time consuming. For indirect approaches, the optimization of an alternate metric for solving the design problem is necessary.

3. *Implementation.* This step considers the execution of the design, and the fabrication and testing of the resulting diffracting element. It is an iterative procedure. The DE performance can be improved by using the data collected during the testing, or by introducing more data on the material response into the design.

8.2.1 Modeling Theories

The theoretical basis for modeling diffractive optics can be divided into three regimes: geometrical optics, scalar, and vector diffraction. The main features of each regime are described below.

8.2.1.1 Geometrical Optics

In this case, rays are used to describe the propagation of the diffracted wavefront but neglect its amplitude variations; that is, geometrical optics can predict the direction of diffraction orders but not their relative intensities. Despite these limitations, if a diffractive element is used in an application, which is normally designed by tracing rays, then ray tracing coupled with simple efficiency estimates will usually be sufficient. The majority of these applications are conventional systems (imaging systems, collimating or focusing optics, laser relay systems, etc.). In such systems, the diffractive element corrects residual aberrations (chromatic, monochromatic, or thermal) or replaces a conventional optic (e.g., a Fresnel lens replacing a refractive lens). In most of these cases, the diffractive optic is blazed for a single order and can be thought of as a generalized grating—one in which the period varies across the element.

Two methods are used: the grating equation and the Sweat model. In the grating equation [3] the diffraction of an incident wave is calculated on the grating point by point, and the propagation of the diffracted wave through the rest of the optical system. This diffracted beam is calculated with the grating equation, which is the diffractive counterpart to Snell's law or the reflection law. Deviation of the observed wavefront from a spherical shape constitutes the system's aberrations. This approach is very useful for the analysis of these aberrations. Usually a thin hologram has been considered, with an amplitude transmission during reconstruction that is proportional to the intensity during the recording process. Other assumptions lead to additional diffraction orders and different amplitude distributions. For a given diffraction order, the imaging characteristics are the same, regardless of the diffracting structure assumed.

Sweat [4,5] showed that a diffracting element is mathematically equivalent to a thin refracting lens in which the refractive index approaches infinity and the lens' curvatures converge to the substrate of the diffracting lens. Snell's law is then used to trace rays through this refractive equivalent of the diffractive element. As the index approaches infinity, the Sweat model approaches the grating equation. Almost all commercially available ray tracing software can handle either of these models.

8.2.1.2 Scalar Diffraction Theory

This approach must be used when the amplitude variations are not negligible if the value of the system's diffraction efficiency cannot be separated from the rest of the design, or if the diffraction element cannot be approximated by a generalized grating. There are two fundamental assumptions usually involved in the application of scalar diffraction theory to the design and analysis of DEs. The first is that the optical field just past the DE can be described by a simple transmission function. In this case, the thin element approximation and, often, the paraxial approximation are used, along with the treatment of the electromagnetic wave as a scalar phenomenon. This ensures that the design problems are comparatively simple to solve. The second assumption is the choice of the propagation method to transform this field to the plane of interest. On this point, it is possible to use the mathematical formulation of Fourier optics when the image is in the far field elements, or Fresnel optics when it is in the near field.

Scalar diffraction theory is a simple tool for the analysis of diffractive optical elements. In this case, the diffractive optics is modeled as an infinitely thin phase plate and the light's propagation is calculated using the appropriate scalar diffraction theory.

Using the scalar approach, the design of optical diffracting elements for parallel optical processing systems has become possible. There are now highly efficient space invariant spot array generators that provide a signal in the Fourier transform plane of a lens, and space variant lens arrays, which provide a signal in the device's focal plane. Diffractive beam samplers and beam shapers are also in use. Some detailed designs can be found in the work of Mait [2].

Common optimization techniques include phase retrieval [6,7] nonlinear optimization, and search methods or linear systems analysis.

8.2.1.3 Vector Diffraction Theory

8.2.1.3.1 The Resonant Domain

In recent years, more attention has been paid to elements that push against the validity of the scalar approximations or violate them completely. These works in the so-called resonance domain are characterized by diffracting elements structures, with size (w), that lie within the illuminating wavelength (λ) range

$$\frac{\lambda}{10} < w < 10\lambda. \tag{8.2}$$

The strongest resonance effects are produced when the size of the elemental features approach the wavelength of the illuminating light. In this case, polarization effects and multiple scattering must be taken into account by using the electromagnetic theory rigorously. Such DE when working at optical wavelengths can be fabricated using techniques, such as direct-write electron beam lithography.

In the resonant domain, the diffraction efficiency changes significantly with a slight change of grating parameters, such as period, depth, and refractive index.

Diffraction analysis of these situations requires a vector solution of Maxwell's equations that avoids the approximations present in scalar theories.

A relatively simple method for finding an exact solution of Maxwell's equations for the electromagnetic diffraction is by grating structures. It has been used successfully and accurately to analyze holographic and surface-relief grating structures, transmission and reflection planar dielectric/absorption holographic gratings, dielectric/metallic surfaces relief gratings, multiplexed holographic gratings, and so on. [8,9].

In this theory, the surface-relief periodic grating is approximated by a stack of lamellar gratings. The electromagnetic fields of each layer of the stack are split into spatial harmonics having the same periodicity as the grating. These spatial harmonics are determined by solving a set of coupled wave equations, one for each layer. The electromagnetic fields within each layer are matched to those in the two adjacent layers and to the fields associated with the backward and forward propagating waves or evanescent waves in two exterior regions. The amplitudes of the diffracted fields are obtained by solving this system of equations.

Rigorous methods have led to several approaches: the integral, differential, or variational [10,11]; analytic continuation [12]; and variational methods and others [13]. The integral approach covers a range of methods based on the solution of integral equations. In some methods, the wave equations are solved by numerical integration through the grating. Recently, a method has been proposed for the calculation or the diffraction efficiency that includes the response of photosensitive materials that have a non-uniform thickness variation or erosion of the emulsion surface due to the developing process [14a, 14b].

Another method used to model all diffractive effects rigorously is to solve Maxwell's equations by using the finite element method (FEM) that is based on a variational formulation of the scalar wave equation. The FEM is a tool in areas, such as geophysics, acoustics, aerodynamics, astrophysics, laser function, fluid dynamics and electromagnetics. It is also used to model complex structures. With this method, the analysis of complicated material structures can be calculated [15].

Some hybrid integral–variational methods have also been studied [11,16]. In this case, an FEM is used to solve the Helmholtz equation for the interior of a DE. A boundary element method, a Green's function approach, is used to determine the field exterior to the DE. These two methods are coupled at the surface of the DE by field continuity conditions. This work has been applied to the design of a subwavelength diffractive lens in which the phase is continuous and the analysis of diffraction from inhomogeneous scatterers in an unbounded region.

There is a similar method for the design of periodic subwavelength diffracting elements [17,18] that begins with an initial structure derived from scalar theory, which uses simulated annealing to improve its performance. However, the infinitely periodic nature of the structures allows the rigorous coupled wave theory (RCW) developed by Moharam and Gaylord [19] to be used for the diffraction model. These methods are flexible and have been applied to surface-relief gratings and to gradient-index structures.

This principle can be used to form different DEs, such as a diffractive optic beam deflector for high-power laser systems, where laser-induced damage limits the usefulness of conventional elements. Reflection gratings operating as a polarization-sensitive beamsplitter has also been proposed [20]. Some numerical simulation uses a single diffraction grating in the resonant domain for pulse compression [21]. These computer-synthesized diffractive elements can be made using the fabrication methods well known within integrated circuits technology [22]. This has been pushed toward compact small-size and low-cost elements by industrial requirements. An additional requirement is to incorporate several sophisticated functions into a single component.

Vector theory allows the analysis of the polarization properties of surface-relief gratings and diffracted beams whenever cross-coupling between the polarization states takes place. This is shown in the design and tolerance of components for magneto-optical heads [23]. Another example is a polarizing beam splitter (PBS) that combines the form birefringence of a spatial frequency grating, with the resonant refractivity of a multilayer structure [24]. The results demonstrate very high extinction ratios (1,000,000:1) when PBS is operated at the designed wavelength and angle of incidence, and good average extinction ratios (from 800:1 to 50:1) when the PBS is operated for waves of 20° angular bandwidth, with wavelength ranging from 1300 nm to 1500 nm, combining features, such as small size and negligible insertion losses. The design has been optimized using rigorous couple-wave analysis (RCWA).

8.2.1.4 Achromatic Diffractive Elements

Diffractive optical elements (DEs) operate at the wavelength for which they were designed. When operating at a different wavelength, chromatic aberrations arise. This is a characteristic feature of diffractive lenses. As described by Bennett [25], a complete analysis of the chromatic aberrations of holograms must take into account the lateral displacement of an image point (lateral dispersion), and its longitudinal dispersion, the change of magnification, third- and higher-order chromatic aberrations, and amplitude variation across the reconstructed wavefronts from thick holograms.

The Abbe value of a diffractive lens υ_{diff}, defined over the wavelength range from λ_{short} to λ_{long} [26]

$$\upsilon_{\text{diff}} = \frac{\lambda_0}{\left(\lambda_{\text{short}} - \lambda_{\text{long}}\right)} \tag{8.3}$$

will always be negative as long as $\lambda_{\text{long}} > \lambda_{\text{short}}$. The absolute value of υ_{diff} is much smaller than the Abbe value for a conventional refractive lens. For this lens, aberration coefficients can be derived.

Since many potential DE applications require the simultaneous use of more than one wavelength, correction of chromatic aberration is essential.

The unique properties of the DE can be used to correct the aberration of the optical systems that consists of conventional optical elements and DEs by combining this with refractive elements to produce achromatic diffractive/refractive hybrid lenses for use in optical systems. Figure 8.2 shows some hybrid eyepiece designs. Many of these elements have been designed for use with spectral bands ranging from the visible to mid-wave infrared and long-wave infrared regions. These show that a DE is very effective in the correction of primary chromatic aberrations in the infrared region and of primary and secondary chromatic aberrations for visible optical systems. Generally, a DE can improve optical system performance while reducing cost and weight. One can reduce the number of lens elements by approximately one-third; additional benefits may include reducing the sensitivity of the system to rigid body misalignments.

The advantages offered by hybrid refractive-diffractive elements are particularly attractive in infrared systems where the material used is a significant proportion of the overall cost. Hybrid elements allow, for example, passive athermalization of a lens in a simple aluminum amount with a minimum number of elements in which two elements do the athermalization while the dispersive properties of a diffractive

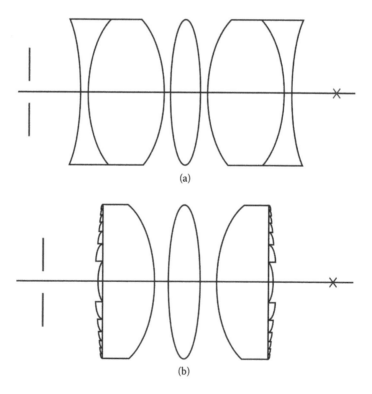

(a)

(b)

FIGURE 8.2 Eyepiece designs. (a) Erfle eyepiece and (b) an equivalent hybrid diffractive–refractive eyepiece. (From Missig MD, Morris GM. *Appl. Opt.*, 34, 2452–2461, 1995.)

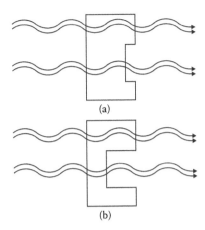

FIGURE 8.3 Effect of wavelength shift on (a) half-wave and (b) multiple-wave phase holograms. The path-length difference on wavelength shift is greater when the etch depth is optimum. With the correct etch depth, the phase delay at the second wavelength is zero, and there is no diffraction. (From Ford JE, et al., *Opt. Lett.*, 21, 80–82, 1996.)

surface are used to achromatize the system. In order to achieve their full effectivity, hybrid elements must include a conventional aspheric lens with a diffractive structure on one surface [27]. Diamond turning permits the aspheric profile and diffractive features to be machined on the same surface in a single process.

Another method of achromatic DE design is that of Ford et al. [28], where the DE acts differently for each of the two wavelengths. The phase-relief hologram can be transparent at one wavelength (λ) yet diffracting efficiently at another (λ') provided that the phase delay is an integral number of wavelengths at λ and a half-integer number of wavelengths at λ'. In other words, there is an integral-multiple phase retardation of 2π of one wavelength until the suitable phase retardation of the second wavelength is achieved (Figure 8.3). In another method the DE is corrected for chromatic aberration designed by combining two aligned DEs made of different materials [29].

8.3 Fabrication Techniques

The design of a diffractive optical element must include specifications for microstructure necessary to obtain the desired performance. With an appropriate fabrication technique, these microstructures will introduce a change in amplitude or phase that alters the incident wavefront.

A factor that has stimulated much of the recent interest in diffractive optics has been new manufacturing techniques that give the designer greater control over the phase function that introduces the diffracting element, resulting in reasonably high diffraction efficiency. In fact, a scalar diffraction theory analysis indicates that a properly designed surface profile can have a first-order diffraction efficiency of 100% at the design wavelength.

In this section, we discuss the main fabrication techniques. These are holographic recording, mask fabrication, and direct-writing techniques, as shown in Figure 8.4.

8.3.1 Holographic Recording

Holograms with predefined optical transfer functions are referred to as holographic optical elements (HOEs). The optical transfer function of an HOE is based on diffraction by the diffraction equation (Equation 8.1) and is recorded using a simple wavefront that satisfies, at least approximately, the Eikonal equation [30]. The wavelength dependence of this grating will depend on its structure, as shown in Figure 8.5.

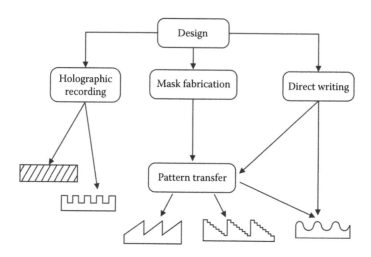

FIGURE 8.4 Main fabrication techniques of diffractive optical elements.

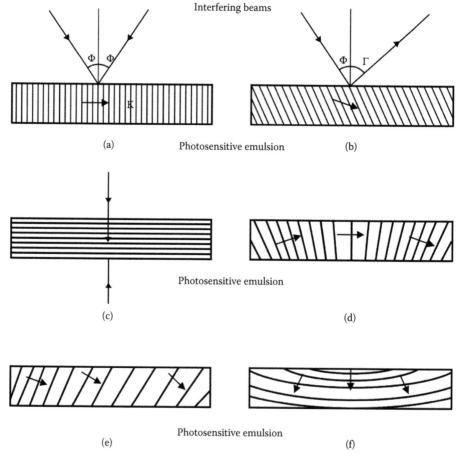

FIGURE 8.5 Different fringes forms of the holographic optical elements. (From Jamison T, et al., *Proc. SPIE*, 2132, 44–70, 1994.)

In principle, given an arbitrary input wavefront, an HOE can be designed to transform into a desired output wavefront. In such a situation, the required HOE recording beams would most likely be produced by computer-generated holograms in conjunction with conventional refractive and reflective optical elements. Using the recent advances in micromachining and microelectronic techniques, the fabrication of such structures with micron and submicron minimal details has become practical.

For both kinds of diffractive elements (HOEs and DOEs), the grating effect is dominant and defines their function and limitations. In general, grating dispersion is much stronger than prism dispersion. Thus, chromatic (wavelength) dispersion strongly influences (and limits) the imaging properties of HOEs and DOEs. Moreover, almost all applications of HOEs and DOEs are the result of effectively controlling chromatic dispersion and chromatic aberrations.

Although DOEs typically have a periodic (grating) structure that is always located at their surface as a surface relief pattern, HOEs also have a periodic structure that is located either on the surface or within the volume of the holographic material, Figure 8.6.

Amplitude or phase modulation at high spatial frequencies can be obtained from a holographic recording. Off-axis diffractive optical elements have grating-like structures with submicron carrier frequency and diffraction efficiencies as high as 90%. The holographic recording process is rather complicated and is extremely sensitive to vibration, which can be avoided by using an active fringe stabilization system. With this technique, it is possible to obtain positioning errors below $\lambda/40$.

Probably one of the best-known materials is dichromated gelatine, which can be used to produce elements that introduce a phase-index modulation either in its bulk or on its surface. This material is sensitive to the blue part of the spectrum, although there are some dyes that can be incorporated to make it sensitive to a different wavelength [31].

Other materials are photopolymers based on the photopolymerization of free-radical monomers, such as acrylate esters [32]. A relatively large number of free-radical monomers suitable for use in holographic recording have been developed. These allow the rapid polymerization of free-radical monomers with any of the common laser lines in the visible spectrum. Problems with these materials include the inhibition of free-radical polymerization due to the presence of dissolved oxygen. To compensate, a high initial exposure to oxygen is required, which causes a significant volume contraction, distorting the recorded fringe pattern. Reprocity failure, reduced diffraction efficiency at low spatial frequencies, and time-consuming post-exposure fixing are limitations that are overcome in a photopolymer based on cationic ring-opening polymerization [33]. Among those photopolymers with good stability and

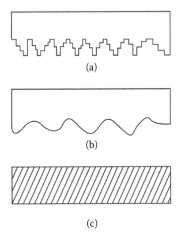

(a)

(b)

(c)

FIGURE 8.6 Illustration of diffractive optical elements (DOEs) and holographic optical elements (HOEs). (a) Multilevel DOE. (b) Surface (embossed) HOE. (c) Volume (Bragg) HOE; phase or amplitude modulation in the bulk of the material. (From Jannson T, et al., *Proc. SPIE*, 2152, 44–70, 1994.)

high-index modulation are those made by Dupont [34], the laboratory made with poly vinyl alcohol as a base and the ones containing acrylamide, and some dyes [35].

Surface-relief DE can be fabricated by holographic exposure in different materials, such as photoresists [36], chalkogenide glasses [37], semiconductor-doped glasses [38], and in liquid [39] and dry self-developing photopolymer materials [40–42], and so on. Two types can be distinguished: those that approximate a staircase (Fresnel lens) and those based on diffractive optical elements (Fresnel zone plates, gratings, etc.).

It has been shown [43] that almost any object intensity distribution can be interferometrically recorded and transferred to a binary surface relief using a strongly nonlinear development. As a result, the sinusoidal interference pattern is then transformed into a rectangular-shaped relief grating.

8.3.2 Mask Fabrication: Binary Optical Elements

Binary optical elements are staircase approximations of kinoforms, which have multiple levels created by a photolithographic process, Figure 8.7. The term binary optics comes from the binary mask lithography used to fabricate the multilevel structures.

The optical efficiency of a diffraction grating depends on the phase encoding technique. Binary optics is based in the creation of multilevel profiles, which requires exposure, development, and etching with different masks that are accurately aligned to each other. The number of phase levels realized through multiple binary masks depends on the specific encoding approach.

To explain the principle of these elements, assume that a blazed grating is to be written having the phase profile shown in Figure 8.8a. Here the total phase shift over the grating is 2π radians and the period of the grating is defined as d. This grating would yield 100% diffraction efficiency into the first order. To fabricate this grating using binary optics techniques, masks are designed having increasingly finer resolutions of $d/2$, $d/4$, $d/8$, and so on. Each mask is deposited sequentially onto a glass substrate. After the deposition of the first mask, the surface is etched in such a way that the phase difference between masked and unmasked areas is π radians, as shown in Figure 8.8b. However, the diffraction efficiency of this binary-phase-only mask is only 40.5%. To get higher diffraction efficiencies, increasingly finer masks are deposited one after the other and the substrate is etched in such a way as to produce increasingly smaller phase shifts. For the eighth-phase level grating of Figure 8.8c, the diffraction efficiency

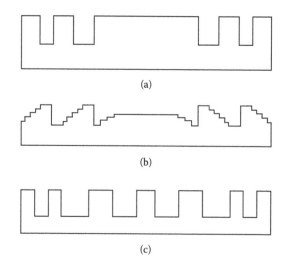

(a)

(b)

(c)

FIGURE 8.7 Binary optics fabricated by binary semiconductor mask lithography: (a) Fresnel zone, (b) Fresnel lenslet, and (c) Dammann grating.

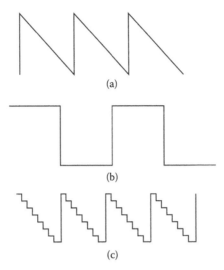

(a)

(b)

(c)

FIGURE 8.8 DE profiles: (a) phase profile for a grating with 100% diffraction efficiency; (b) binary phase grating profile; and (c) step phase grating profile. (From Davis JA, Cottrell DM. *Proc. SPIE*, 2152, 226–236, 1994.)

reached 95%. However, to reach higher diffraction efficiencies, the size of the elemental features must decrease in order to maintain the periodicity [44].

These DE are constrained by spatial and phase quantization [45]. The complexity and quality of the reconstructed image determines the spatial complexity and phase resolution of the DE. The important issues in using masks are the alignment between the successive steps and the linewidth errors. This limits the fabrication of multilevel phase elements to three or four masks, corresponding to eight- or 16-phase levels.

Masks can be generated with electron-beam or laser beam lithography. These are amplitude elements that have to be transformed into surface-relief structures by exposure, chemical processing, and etching of the photoresist. These processes permit the fabrication of sawtooth, multilevel, or continuous profiles. For more rugged elements with high optical quality, the photoresist profiles are then transferred into a quartz substrate by standard techniques, such as reactive ion etching.

With electron-beam lithography, one can write gratings with periods down to 100 nm, but beyond that they are limited by the proximity effect. Electron-beam lithography is a highly flexible means of generating arbitrary structures, even micro-lenses [46]. However, in the case of elementary feature sizes of the order of 50–100 nm, this approach is limited by the positioning accuracy during the writing process.

Binary and multilevel diffractive lenses with elementary feature sizes of the order of submicrometers have been produced on silicon and gallium phosphide wafers by using the CAD method, direct writing electron-beam lithography, reactive ion etching, antireflection coating, and wafer dicing. Measurements indicate that it is possible to obtain aberration-free imaging and maximum diffraction efficiencies of 70% for lenses with numerical apertures (NAs) as high as 0.5. This technique has been applied to off-axis arrays for 18-channel parallel receiver modules [23].

8.3.2.1 Photolithography

The interesting field of photolithography has developed as a result of the introduction of resist profiles to produce microoptical elements (Fresnel lenses, gratings, kinoforms, etc.). The thickness of the resist film that must be altered can be several micrometers thick to obtain the required profile depth of the optical element. The efficiency of those elements depends on the shape and quality of the resist profiles. Blazed and multilevel diffractive optical elements can reach a higher efficiency than binary optical elements.

Surface-relief lithography diffractive elements generated show promise for applications to magneto-optic heads for data storage due to their polarization selectivity, planar geometry, high diffractive efficiency, and manufacturability. Former applications of these elements had been limited due to the lack of information on their polarization properties.

The use of lithographic techniques opens the way to the development of optical elements, which are economical, with high-resolving power, and flexible design. These ideas are used in many systems at optical or near-infrared wavelengths.

8.3.2.2 Gray-Tone Masks

The gray-scale masks are an alternative approach to the multiple mask technique; it requires only one exposure and one etching process and yields a continuous profile. The gray levels are made by varying a number of transparent holes in a chromium mask that are so small that they are not resolved during the photolithographic step. Diffraction efficiencies reported are of the order of 75% for an element etched in fused silica, $\lambda = 633$ nm.

This process requires linearization of the photographic emulsion exposure as well as linearization of photoresist exposure: both are hard to reproduce [47].

8.3.3 Direct-Writing Techniques

High-intensity pulsed lasers can uniformly ablate material from the surface of a wide range of substrates [48]. Proper choice of laser wavelength allows a precise control of depth that can be applied in many materials that absorb in this region of the spectrum. These lasers have been used in lithographic processes. Direct-writing techniques yield higher-phase resolutions (of the order of 64–128 phase levels) than photolithographic methods, but at the expense of reduced spatial resolution.

Another alternative, however, is to use an excimer laser with an ultraviolet waveguide to etch diffractive structures directly into the substrate without masks or intermediate processing steps (Figure 8.9) [49]. This technique can be applied to a large spectrum of substrate materials, such as glass, diamond, semiconductors, and polymers, and can also reduce time and cost to produce a diffractive element.

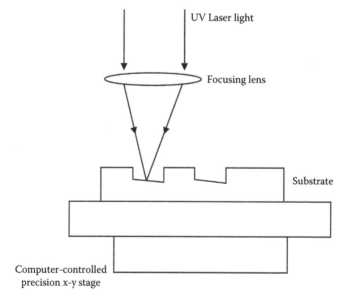

FIGURE 8.9 Direct write photoablation of the substrate. (From Duignan MT. *Technical Digest Series-Opt. Soc. Am.*, 11, 129, 1994.)

Figures 8.10 and 8.11 show the fabrication steps and a schematic of one of the systems: in this particular case, a He–Cd laser is used to fabricate the DE on a photoresist substrate [50].

Direct writing in photoresist, with accurate control of the process parameters, enables one to fabricate a complex continuous relief microstructure with a single exposure and development operation, which has been shown to produce excellent results [43]. Because writing times can be relatively long (many hours for typical microstructures of 1 cm²), latent image decay must be compensated. A number

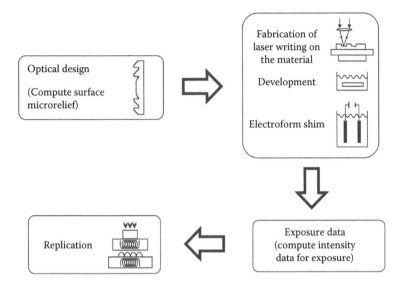

FIGURE 8.10 Fabrication steps for continuous-relief micro-optical elements. (From Gale MT, et al., *Opt. Eng.,* 33, 3556–3566, 1994.)

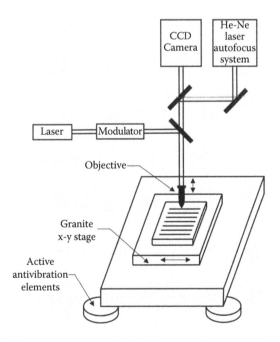

FIGURE 8.11 Schematic of laser-writing system for the fabrication of continuous relief micro-optical elements. (From Gale MT, et al. *Opt. Eng.,* 33, 3556–3566, 1994.)

of factors determine the fidelity of the developed microstructure. The dominant experimental errors in the writing process are surface structures of the coated and developed photoresist films, the profile of the focused laser spot, the accuracy of the exposure dose, the line straightness, and the accuracy of the interline distance of the raster scan on the substrate.

An example of elements fabricated by direct laser writing in photoresist [51] is a fanout element and diffractive microlens with NA = 0.5, which has been produced with a diffraction efficiency of 60%.

8.3.4 Replication Techniques

The main attraction of micro-optical elements lies in the possibility of mass production using replication technology. Replication technology is already well established for the production of diffractive foil, display holograms, and holographic security features in which the microrelief structures have a typical grating period of 0.5–1 μm with a maximum depth of about 1 μm. These are produced with a hot roller press applied to rolls of plastic up to 1 m wide and thousands of meters in length, [52]. For deeper microstructures, other replication techniques are required, such as hot-press, casting, or molding. In all cases, it is necessary first to fabricate a metal shim, usually of nickel (Ni), by electroplating the surface of the microstructure. Figure 8.12 [51] illustrates the steps involved in the fabrication of these shims. The recorded surface-relief microstructure in photoresist is first made conducting, either by the evaporation of a thin film of silver or gold of the order of 100 nm, or by using a commercial electronless Ni deposition bath. A Ni foil is then built up by electroplating this structure to a thickness of about 300 μm. Finally, the Ni is separated from the resist/substrate and cleaned to give the master (first-generation) replication shim. This master shim can be used directly for replication, by hot-embossing or casting. It can also be supplied to a commercial shim facility for recombination to produce a large-area production shim.

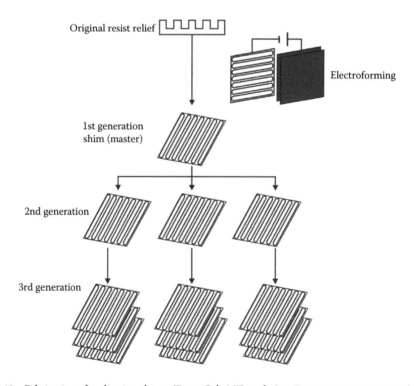

FIGURE 8.12 Fabrication of replication shims. (From Gale MT, et al. *Opt. Eng.,* 33, 3556–3566, 1994.)

The first-generation master can be used to generate multiple copies by electroplating further generations. The silver or nickel surface is first passivated by immersion in a dichromate solution or by O_2 plasma treatment, followed by further electroplating to form a copy that can readily be separated. In this way, numerous copies can be made from a single recorded microrelief.

Advances in the sol–gel process have made it possible to replicate fine-patterned surfaces in high-purity silica glass by a molding technique [53]. The different types of diffractive optics that have already been replicated include binary grating, blazed grating, hybrid diffractive/refractive optical element, and plano-kinoform. The requirements of the optics define the design of the molds to be used to produce the desired optical components. To manufacture the mold, a tool that contains the required relief pattern must be fabricated. The mold is fabricated and used in the sol–gel process to produce prototype parts. Quality control then provides the necessary input to determine what, if any, changes are necessary in either mold or procedure for the final DE [54].

There are three important advantages of the sol–gel replication process:

1. It provides a cost-effective way of producing optical elements with fine features. Although the mold surface is expensive, its cost can be amortized over a large volume of parts, thus making the unit cost relatively low.
2. The process can produce optical elements in silica glass, one of the best optical materials. The advantages of silica include a very high transmission over a broad wavelength range from 0.2 to 3.2 µm, excellent thermal stability, radiation hardness, and chemical treatment. Therefore, it can be used in harsh environments, such as space, or for military uses and in high-power systems.
3. In the sol–gel replication process, there is a substantial shrinkage that is controlled by adjusting the processing parameters. This shrinkage has been accurately quantified and has been found to be very uniform in all three dimensions, making it possible to fabricate parts with structures smaller than those made by other processing techniques. This reduces imperfections and tool marks by a factor of 2.5, reducing scattered light at the design wavelength.

Other techniques are compatible with the microfabrication techniques used in the semiconductor industry and require the generation of a gray-level mask, such as that fabricated in high-energy beam sensitive (HEBS)-glass by means of a single electron-beam direct-write step [55]. This mask was used in an optical contact aligner to print a multilevel DE in a single optical exposure. A chemically assisted ion-beam etching process has been used to transfer the DE structure from the resist into the substrate [47].

8.3.4.1 Plastic Optics

It is important to mention molded plastic DEs: they are low cost and can be mass-produced [56]. They can be produced in different shapes: rotationally symmetric, aspheric, and hybrid refractive/diffractive lenses. These are used in various applications, such as fixed focus and zoom camera lenses, camera viewfinders, diffractive achromatized laser-diode objectives, and asymmetric anamorphic diffractive concentrating and spectral filtering lenses for range-finding and autofocus applications.

Planar micro-optical elements can be found in an increasing number of applications in optical systems and are expected to play a major role in the future. Typical elements for application at visible and infrared wavelengths are surface-relief microstructures with a maximum depth of about 5 µm. These can be mass-produced using current replication techniques, such as hot-embossing, molding, and casting [50].

Polyamide lenses with periodic and aperiodic radial profiles have been reported by Furlan et al. [57], which made diffractive THz using a 3D printer 0.625 THz.

8.4 Dynamic Diffractive Elements

In recent years, there has been a great deal of interest in active, dynamic diffractive elements that can be electrically programmed or switched. These elements can be divided into two classes. The first class uses an element-by-element addressing structure to produce diffracting patterns as spatial light modulators.

The second class switches on a pre-patterned diffraction structure that has been configured during fabrication. These devices could expand the range of application of the DE through the real-time control of an element's optical function. Both these devices have a large range of designs and methods for dynamically generating the phase or amplitude modulation of a spatial pattern in response to an electrical signal.

8.4.1 Programmable Diffraction Elements

Binary optics can be programmed to produce patterns with a large dynamic range. These have two functions. First, the spatial light modulator (SLM) serves as a programmable low-cost test for more complicated nonprogrammable binary optical elements. Secondly, the programmability of this system allows real-time image processing in which the optical element can be changed rapidly.

One way to obtain such elements is by using electro-optic material, such as a liquid crystal (LC) layer. The LC materials exhibit a large field-induced birefringence effect, resulting in a local change in the index of refraction, polarization angle, or both. Another alternative is to use the diffractive optical elements written onto a spatial light modulator (SLM) [58]. In this case, each phase region will be encoded onto an individual pixel element whose size is limited by the resolution of the SLM. These phase regions are limited by the operating physics of the SLM.

The SLM has been used for the fabrication of micro-scale medical devices [59], magneto-optic spatial light modulators, optical interconnections, lens arrays [59], subdiffraction patterns, [60], derivative lenses [61], nondiffractive lenses, and Damman gratings.

8.4.2 Switched Optical Elements

Switched optical elements use transmitting or reflecting structures, which incorporate a material that exhibits an index of refraction that can be varied electrically. When an electric field is applied to the resulting composite structure, a range of predetermined optical characteristics emerge, which spatially modulate or redirect light in a controlled manner. The effect on an incident wavefront may be controlled by varying the applied electric field.

These devices are capable of producing diffraction effects when a drive signal is applied, or in some designs, when it is removed.

Typically, SLMs are restricted to relatively small pixel arrays on the order of 256×256 and with correspondingly low diffraction efficiency. Monolithic holograms, on the other hand, have extremely high resolution, high optical quality, and diffraction efficiency, with one million times the pixel density. Such elements can be used in devices that are significantly different, especially for SLMs, if the material is also of sufficiently high optical quality to permit series stacking [62]. Figure 8.13 shows a generic device made of stacks of switchable holograms.

Among the different approaches to these switching DE is the placing of electrodes over a layer of liquid crystals to respond to the localized fields with two-dimensional distribution birefringence. It is also possible to fill a surface-relief binary optical element or a sinusoidal relief grating etched on dichromated gelatin with a layer of liquid crystals [63,64]. Switching times ranged between 20 and 50 ms for an applied voltage of 20 $V_{rms,}$ have been reported. Some other work involves special materials, such as LC selective polymerization, or fabrication of holographic gratings by photopolymerization of liquid crystalline monomers that can be switched with the application of an electric field [65].

Most of the reported approaches involve liquid crystals in one way or another, although, in principle, semiconductor techniques could also be used [66].

One of the most popular materials is the polymer-dispersed liquid crystal (PDLC) formed *in situ* by a single-step photopolymer reaction [62]. These materials are composites of LC droplets imbedded in a transparent polymer host whose refractive index falls between the LC ordinary and extraordinary indices. By electrically modulating the index match between LC droplets and polymer host, the characteristics of the volume holographic diffraction may be reduced. Fine-grained PDLCs have recently

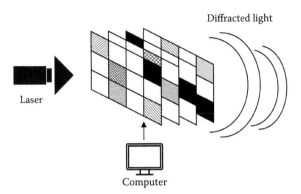

FIGURE 8.13 General active holographic interconnect device. Each plane contains an independent array of individually controlled patches of prerecorded holograms. By electrically addressing patches throughout the volume, many different diffractive functions can be programmed. (From Domash LH, et al. *Proc. SPIE,* 2152, 127–138, 1994.)

become available for electrically switchable holographic elements. The mechanism for the formation of the hologram grating is described by Sutherland, 1994b. They have high diffraction efficiency, narrow angular selectivity, low voltage switching, and microsecond switching speed in elements with good optical quality as well as for the storage of the holographic image. Electro-optical read-out can be used with this new material. Applications are for switchable notch filters for sensor application, reconfigurable optical interconnections in optical computing, fiber-optic switches, beam steering in laser radar, and tunable diffractive filters for color projection displays.

Dynamic-focus lenses that are controlled electrically are used in autofocusing devices for tracking in CD pickups, optical data storage components, and many other purposes. Some applications require continuous focusing; others call for switchable lenses with a discrete number of focal lengths. The basic concept is a diffractive lens material whose diffractive characteristics can be turned off by the application of an electric field. Using such a material, an electro-optic diffractive lens may be switched between two states—transparency (infinite focus) and finite focus.

A number of light-modulating SOE devices for display applications use structures that can be referred to as hybrid; that is, structures that combine a fixed array of individually switched electrodes with a pre-patterned diffractive structure.

8.5 Applications

8.5.1 Micro-Optical Diffracting Elements

Micro-optical devices, such as diffractive and refractive microlenses have received considerable attention in the optics R&D community.

Technological advances make it possible to micromachine optical and diffractive devices with structures that have the same order of magnitude as the wavelength of the incident light. Devices that were once considered impractical because of the limitations of bulk optics are now designed and easily fabricated with advanced microelectronics technology.

Micro-optical elements can be refractive, such as hemispherical lenslet and lenslet arrays, diffractive, such as kinoforms, grating structures, and so on, or a combination of both, such as Fresnel microlenses. They can be continuous surface-relief microstructures [51], binary or multilevel reliefs or made by direct laser writing [43,67]. A recent approach has been proposed by Xu et al. [68] preparing a lenticular microlens array (LMA) with large dynamic range using polyvinyl chloride (PVC)/dibutyl phthalate (DBP) gels using an electrostatic repulsive force, obtaining a reconfiguration on the membrane LMA with potential applications in imaging, information processing, biometrics, and displays.

The ability to combine various optical functions, for example, focusing and deflection, and the reduced thickness and weight of DE in comparison to their refractive counterparts essentially explains the concern of diffractive optics in micro-optics.

In processing optical materials, the two main classes of diffractive optics are higher-power lasers and their periphery (such as the interconnection of a high-power Nd:YAG laser with a fiber bundle) and the use of DE to shape the laser in order to provide the illumination beam required for the same application, such as the production of diffuse illumination with high-power CO_2 lasers [69].

On the other hand, such elements can be applied to holography for memory imaging, nondestructive testing in interferometry, wavefront shaping, and spatial filters. There have been many practical applications, such as diffraction gratings to shape the phase and polarization of incident fields, reflectors for microwave resonant heating systems, microwave lenses, mode converters in integrated optical devices, for dispersion compensation and pulse shaping in ultrafast optics, and so on [15].

Technology for making binary optics uses advanced sub micrometric processing tools and micromachining processes to create novel optical devices. One potential role of binary optics is to integrate very large scale integration (VLSI) of microelectronic devices with micro-optical elements [70]. Because small feature sizes and stringent process control have been two major considerations, attention has focused on microlithography during the past few years. The rapid growth of commercial devices, such as miniature compact disk heads, demands both higher accuracy and lower-cost microlenses' fabrication methods.

Binary optics microlenses arrays are typically fabricated from bulk material by multi-mask-level photoresist patterning and sequential reactive-ion etching to form multistep phase profiles that approximate a kinoform surface. To fabricate an efficient microlens' array, eight-phase-level zones are necessary. The main parameters involved in its design are the wavelength (λ), the microlens' diameter (d'), the focal length (f), $f_\# = f/d'$, and the smallest feature size or critical dimension (D). For a typical binary optic microlens with eight phase levels, the D value is $D = (\lambda f_\#)/4$. The minimum value of VLSI is of the order of 0.5–1 µm. This limits the speed of binary optic microlenses designed for wavelengths (diode laser) from 0.632 µm to 0.850 µm to $f/6$ and $f/3$, respectively. Nevertheless, higher-speed microlenses can be fabricated for infrared applications.

As already mentioned, the diffraction efficiency of the light diffracted to the first-order focus increases with the number of phase levels; in practice, it decreases with the number of processing factors. Values of 90% have been obtained for eight-phase-level microlenses. The extent to which this is acceptable will depend on the application.

The surface relief of these diffractive microlenses has a planar structure of the order of the design wavelength. In a typical system, this reduces the volume and weight of the component relative to an all-refractive design.

Along with these developments in micro-optics technology is the development of micro-electromechanical (MEM) technology, which is based on micromachining methods for processing 3D integrated circuits. MEM and micro-optics technologies have one critically important feature in common: both technologies are compatible with VSLI circuit processing. This feature means that the final device can be produced in volume at low cost. The standard VLSI process is generally confined to the surface of the wafer (Si or GaAs), extending only several micrometers under the surface. Multilayers of metal and dielectric are either deposited/grown on the surface or etched into the surface.

Some micro-optical DEs have been applied in optical choppers, optical switches, and scanners.

8.5.2 Optical Disk Memory Read–Write Heads

The optical head is an important component in optical disk storage. In it, a laser beam is focused on a 1 µm diameter spot on the surface of the disk. The conventional optical head usually contains several optical elements, such as a beamsplitter prism, a diffraction grating, a collimating lens, and a cylindrical lens, as shown in Figure 8.14 [71].

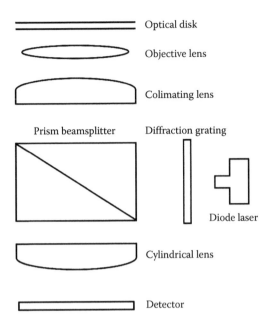

FIGURE 8.14 Configuration of a common optical head. (From Huang G, et al. *Proc. SPIE,* 2152, 30, 261–265, 1994.)

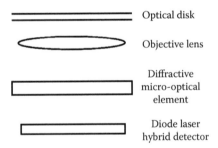

FIGURE 8.15 Optical head using a diffractive micro-optical element. (From Huang G, et al. *Proc. SPIE,* 2152, 30, 261–265, 1994.)

The disk moves under the optical axis of the head as it rotates. In this system, it is necessary to detect and correct focus error to an accuracy of about ± 1 μm. This focus error is determined from the total intensity of the light reflected by the optical disk.

Some systems have been suggested for replacing each of the optical elements with a diffractive micro-optical element performing the three optical functions required for an optical head: splitting the beam, focusing, and tracking the error signals (Figure 8.15) [71].

8.5.3 Optical Interconnects

Diffractive optics will play an important role in optical interconnects and optical interconnecting networks necessary in high-parallel-throughput processing. Diffractive optical interconnect elements provide several advantages over conventional bulk elements, such as spherical and cylindrical lenses [73].

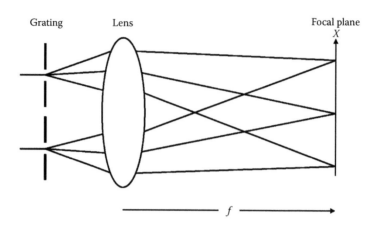

FIGURE 8.16   A grating is the simplest device for fanning out signals.

One of the simplest devices for fanning out signals in optical interconnecting systems is diffraction grating. A basic fan-out arrangement, consisting of a diffraction grating and a collecting lens, is shown in Figure 8.16. Diffraction by a periodic pattern, such as in a Damman grating, divides the incident wave into many beams that are then focused by the collimating lens onto the detector plane. The amount of light in each beam is determined by the specific pattern of the grating. The light focused in the different orders illuminates photodetectors or fibers, depending on the application of the system. These systems can compensate for wavelength dispersion and distortion that occur in diffractive fan-out elements [73].

The use of diffractive optics in interconnects is of considerable interest for several reasons. First, multiple diffractive optic elements can be cascaded on to planar substrates and more easily packaged with planar electronic substrates. To be effective, however, the diffractive optical system must separate and distribute optical signals in several dimensions. With diffractive optical interconnects for digital switching networks, current technology has the ability to form four-dimensional, free-space optical interconnects with boundary conditions [74]. The optics must be packaged with standard board substrates, and have alignment tolerances sufficient for board insertion, replacement, and changes of length caused by temperature variations.

Bidirectional information transfer is necessary at each information port. Parallel data transfer to increase information transfer rates is also important. It must be possible to broadcast greater data processing signal loads to multiple lateral and longitudinal locations.

In addition, the fabrication of microdiffractive optics using microlithographic and holographic methods can produce many optical functions, which can be used in both space variant and invariant systems. With these fabrication methods, units can be mass-produced, lowering overall system costs. Fabrication of diffractive optics uses computer-aided design (CAD) and microstructuring techniques. A reduction in the reflection losses has been achieved using common techniques used in microelectronics technology such as ion-beam-sputter deposition. To reduce crosstalk and feedback, antireflection (AR) coatings or AR-structured surfaces have been suggested [75].

DOEs have proven to be useful in optical interconnection and routing systems, especially where volume, weight, and design flexibility are important. Their characteristics can be increased by making them polarization-selective [76].

8.5.4 Polarizing Elements

As mentioned earlier, another important application for diffractive elements is their ability to polarize light. Polarization-selective computer-generated holograms (CGH) or birefringent CGH (BCGH) have been found to be useful for image processing, photonic switching, and the packaging of optoelectronic devices and systems. With BCGH, it is possible to perform two completely distinct functions using each of the two orthogonal polarizations.

Other applications are polarized beamsplitters (PBS) used for read–write magneto-optic disk heads [77], polarization-based imaging systems [78,79], and optical information processing, such as free-space optical switching networks [80]. These require the PBS to provide high extinction ratios, tolerate a wide angular bandwidth, a broad wavelength range of the incident light, and have a compact size for efficient packaging. For these applications, traditional birefringent structures, such as the Wollaston prism or multilayer structures do not meet these requirements.

Another interesting result is the bifocal polarization holographic lens [81] where two holographic modulations, bulk birefringence and surface relief, are induced in the medium at the same time. The resultant holographic element has two focal planes, and the polarization of light in the focal points depends on the polarization of the incident light. This kind of lens can be used to read–write information in two planes simultaneously or separately.

Diffractive optical elements (DOE) have proven to be useful components in optical interconnection and routing systems, especially where volume, weight, and design flexibility are important. Their usefulness has been increased by making them polarization-selective using two wet etched anisotropic calcite substrates, joined at their etched surfaces and with their optical axes mutually perpendicular. The gap was filled with an index-matching polymer [76]. This element is less sensitive to fabrication errors. This method has been used to obtain elements that change the form of the emerging wavefront, depending on the polarization of the incident light, and has been applied in Fresnel lenses, gratings, and holograms that generate different images in their Fourier plane.

8.5.5 Beam Shaping

In many applications one needs to reshape the laser beam intensity. The advantage of DE is that the beam energy is redistributed rather than blocked or removed, so that energy is preserved.

Some designs have been proposed using computer-generated holograms where the Gaussian beam has been converted into a ring distribution [82], or using a two-element holographic system to obtain a flat distribution [83]. Another proposed system is a Gaussian to top hat converter using a multilevel DE, which can be fabricated with standard VLSI manufacturing equipment [84]. This distribution is useful in material processing, laser radar, optical processing, and so on. An interesting application is the collimation of high power laser diodes [85].

Free-space digital optical systems require optical power supplies that generate two-dimensional arrays of uniform intensity [51]. The resultant spot arrays are used to illuminate optoelectronic logic devices' arrays to optically encode and transfer information. The favored method for creating these regularly spaced beam arrays is to illuminate a computer-designed Fourier-plane hologram using a collimated laser source. These surface-relief gratings, also referred to as multiple beamsplitters, are designed using scalar diffraction theory by means of a computer optimization process that creates an array of beams of uniform intensity [51]. The quality of the hologram is measured by its diffraction efficiency in coupling light into a set of designated orders and the relative deviation of the beam intensities from their targeted values.

Other beam shapers include the Laguerre–Gaussian beam, which has a phase singularity that propagates along its axis [86]. Work has also been done to convert a Gaussian-profile beam into a uniform-profile beam in a one-dimensional optical system as well as rotationally symmetric optical systems both for different fractional orders and different parameters of the beam [87]. Another important application is the pseudo-nondiffracting beam DE, characterized by an intensity distribution that is almost constant axially over a finite axial region and a long propagation distance along the optical axis [88] and the axicons. An axicon is an optical element that produces a constant light distribution over a long distance along the optical axis. A diffractive axicon with a discrete phase profile can be fabricated using lithographic fabrication techniques. Other elements can be fabricated with linear phase profiles [89].

Another beam-shaping procedure is the projection pattern that can be applied to change the physical or chemical state of a surface with visible light or ultraviolet radiation. Important applications in industrial production processes are microlithography and laser material processing.

In conventional methods of pattern projection, a real value (mostly binary) transmission mask pattern is reproduced on the target surface by imaging or shadow casting. This pattern is then formed by diffraction of the illuminating wave at the mask where the diffracted wave is transformed by propagation, either through a lens or through free space, to the target surface.

The use of DE allows us to add phase components to the mask, giving a complex transmission coefficient. This method is called phase masking. It can be used to improve the steepness of the edges in projected patterns by reduction of the spatial bandwidth. Also, the mask may be located at some distance from the target surface or its optical conjugate. The mask then contains the pattern to be projected in a coded form. When it is in the far field of the target surface, this code is essentially a Fourier transformation [90].

Acknowledgments

The author wishes to thank M. Sc. Alicia Torales and D.I. Raymundo Mendoza for their great help.

References

1. O'Shea DC, Suleski TJ, Kathman AD, Prather DW. *Diffractive Optics: Design, Fabrication and Test, Tutorial Texts in Optical Engineering*. SPIE Press, Bellingham, Washington USA, Vol. TT62; 2004.
2. Mait JN. Understanding diffractive optic design in the scalar domain, *J. Opt. Soc. Am. A*, **12**, 2145–2150; 1995.
3. Welford WT. *Aberrations of Optical System*. Boston, MA: Adam Hilger, 1996.
4. Sweat WC. Describing holographic optical elements as lenses, *J. Opt. Soc. Am.*, **67**, 803–808; 1977.
5. Sweat WC. Mathematical equivalence between a holographic optical element and an ultra-high index lens, *J. Opt. Soc. Am.*, **69**, 486–487; 1979.
6. Fienup JR. Reconstruction and synthesis applications for an iterative algorithm, *Proc. SPIE*, **373**, 147–160; 1981.
7. Mendes-Lopes J, Benítez P, Miñano J, Santamaría A. Simultaneous multiple surface design method for diffractive surfaces, *Opt. Express*, **24**, doi:10.1364/; 5584–5590 2016.
8. Moharam MG, Pommet DA, Grann EB. Implementation of the Rigorous Coupled-Wave Technique: Stability, Efficiency, and Convergence. *Diffractive Optics: Design, Fabrication, and Applications*, 1; 1994.
9. Maystre D. Rigorous vector theories of diffraction gratings. In *Progress in Optics*, Wolf, E. (ed.). Amsterdam: North-Holland, Vol. **21**, 1; 1984.
10. Noponen E, Saarinen J. Rigorous synthesis of diffractive optical elements, *Proc. SPIE*, **2689**, 6, 54–65; 1996.
11. Mirotznik MS, Prather DW, Mait J. Hybrid finite element-bounder element method for vector modeling diffractive optical elements, *Proc. SPIE*, **2689**, 2, 2–13; 1996.
12. Bruno OP, Reitich F. Diffractive optics in nonlinear media with periodic structure, *J. Opt. Soc. Am. A*, **12**, 3321–3325; 1995.
13. Prather DW, Mirotznik MS, Mait JN. Boundary element method for vector modeling diffractive optical elements, *Proc. SPIE*, **2404**, 28–39; 1995.
14a. Kamiya N. Rigorous couple-wave analysis for practical planar dielectric gratings: 1. Thickness-change holograms and some characteristics of diffraction efficiency, *Appl. Opt.*, **37**, 5843–5853; 1998a.
14b. Kamiya N. Rigorous couple-wave analysis for practical planar dielectric gratings: 2. Diffraction by a surface-eroded hologram layer, *Appl. Opt.*, **37**, 5854–5863; 1998b.
15. Lichtenberg B, Gallagher NC. Numerical modeling of diffractive devices using the finite element method, *Opt. Eng.*, **33**, 3518–3526; 1994.

16. Cotter NPK, Preist TW, Sambles JR. Scattering-matrix approach to multilayer diffraction, *J. Opt. Soc. Am. A*, **12**, 1097–1103; 1995.

17. Noponen E, Turunen J, Wyrowski F. Synthesis of paraxial domain diffractive elements by rigorous electromagentic theory, *J. Opt. Soc. Am. A*, **12**, 1128; 1995.

18. Zhou Z, Drabik TJ. Optimized binary, phase-only diffractive optical elements with sub-wavelength features for 1.55 μm, *J. Opt. Soc. Am. A*, **12**, 1104–1112; 1995.

19. Moharam MG, Gaylord TK. Rigorous coupled-wave analysis of surface-gratings with arbitrary profiles, *J. Opt. Soc. Am.*, **71**, 1573–1574; 1981.

20. Lightbody MTM, Layet B, Taghizadeh MR, Bett T. Design of novel resonance domain diffractive optical element, *Proc. SPIE*, **2404**, 96–107; 1995.

21. Ichikawa H. Analysis of femtosecond-order pulses diffracted by periodic structures, *J. Opt. Soc. Am. A*, **16**, 299–304; 1999.

22. Wei M, Anderson EH, Attwood DT. Fabrication of ultrahigh resolution gratings for X-ray spectroscopy, *Technical Digest Series-Optical Society of America, vol. 11*, 91; 1994.

23. Haggans CW, Kostuk R. Use of rigorous coupled-wave theory for designing and tolerancing surface-relief diffractive components for magneto-optical heads, *Proc. SPIE*, **1499**, 293–296; 1991.

24. Tyan RC, Sun PC, Fainman Y. Polarizing beam splitters constructed of form-birefringent multilayer grating, *Proc. SPIE*, **2689**, 8, 82–89; 1996.

25. Bennett J. Achromatic combinations of hologram optical elements, *Appl. Opt.*, **15**, 542–545; 1976.

26. Buralli DA. Using diffractive lenses in optical design, Technical Digest Series-Optical Society of America, vol. 11, 44; 1994.

27. McDowell AJ, Conway PB, Cox AC, Parker R, Slinger CW, Wood AP. Investigation of the defects introduced when diamond-tuning hybrid components for use in infrared optical systems. Technical Digest Series-Optical Society of America, vol. 11, 99–99; 1994.

28. Ford JE, Xu F, Fainman Y. Wavelength-selective planar holograms, *Opt. Lett.*, **21**, 80–82; 1996.

29. Arieli Y, Noach S, Ozeri S, Eisenberg N. Design of diffractive optical elements for multiple wavelengths, *Appl. Opt.*, **37**, 6174–6177; 1998.

30. Sommerfeld, A. *Optics*, New York, NY: Academic Press, Vol. 4 (1962).

31. Solano C, Roberge PC, Lessard RA. Methylene blue sensitized gelatinas a photosensitive medium for conventional and polarizing holography, *Appl. Opt.*, **26**, 1989–1977; 1987.

32. Lessard RA, Thompson BJ (eds.). *Selected Papers on Photopolymers,* Published by SPIE, The International Society for Optical Engineering, Bellingham, Washington, USA, Vol. MS114; 1995.

33. Close DH, Jacobsob AD, Margerun JD, Brault RG, McClung FJ. Hologram recording on photopoly mer holograms, *Appl. Phys. Lett*, **14**, 159–16; 1969.

34. Chen T, Chen Y. Graded-reflectivity mirror based on a volume phase hologram in a photopolymer film, *Appl. Opt.*, **37**, 6603–6608; 1998.

35. Pascual I, Marquez A, Belendez A, Fimia A, Campos J, Yzuel MJ. Fabrication of computer-generated phase holograms using photopolymers as holographic recoding material, *Proc. SPIE*, **3633**, 302–305, 1999.

36. Zaidi SH, Brueck SRJ. High aspect-ratio holographic photoresist gratings, *Appl. Opt.*, **27**, 2999–3002; 1988.

37. Tgran VG, Viens JF, Villeneuve A, Richardson K, Duguay MA. Photoinduced self-developing relief gratings in thin film chalcogenide As_2s Si_3 glasses, *J. Lightwave Technol.*, **15**, 1343–1345; 1997.

38. Amuk AY, Lawandy NM. Direct laser writing of diffractive optics in glass, *Opt. Lett.*, **22**, 1030–1032; 1997.

39. Boiko YB, Granchak VM, Dilung II, Solojev VS, Sisakian IN, Sojfer VA. Relief holograms recording on liquid photopolymerizable layers. In *Three-Dimensional Holography: Science, Culture, Education,* Jeong, T. H. and V. B. Markov (eds.), Published by SPIE, The International Society for Optical Engineering, Bellingham, Washington, USA, **1238**, 253–257; 1991.

40. Calixto S. Dry polymer for holographic recording, *Appl. Opt.*, **26**, 3904–3910; 1987.

41. Calixto S, Paez GP. Micromirrors and microlenses fabricated on polymer materials by means of infrared radiation, *Appl. Opt.*, **35**, 6126–6130; 1996.
42. Neuman J, Wieking KS, Detlef K. Direct laser writing surface relief in dry, self-developing photopolymer films, *Appl. Opt.*, **38**, 5418–5421; 1999.
43. Ehbets P, Herzig HP, Prongué D, Gale AT. High-efficiency continuous surface-relief gratings for two-dimensional array generation, *Opt. Lett.*, **17**, 908–910; 1992.
44. Beretta S, Cairoli MM, Viardi M. Optimum design of phase gratings for diffractive optical elements obtained by thin-film deposition, *Proc. SPIE*, **1544**, 2–9; 1991.
45. Arrizon V, Testorf M. Efficiency limit of spatially quantized Fourier array illuminators, *Opt. Lett.*, **22**, 197–199; 1997.
46. Oppliger Y, Sixt P, Stauffer JM, Mayor JM, Regnault P, Voirin G. One-step shaping using a gray-tone mask for optical and microelectronics applications, *Microelectron. Eng.*, **23**, 449–452; 1994.
47. Däschner W, Long P, Larson M, Lee SH. Fabrication of diffractive optical elements using a single optical exposure with a gray level mask, *J. Vac. Sci. Technol. B*, **13**, 6–9; 1995.
48. Braren B, Dubbowski JJ, Norton DP. (eds.). *Laser Ablation in Materials Processing: Fundamentals and Applications*. Material Research Society, Pittsburgh, PA, Vol. 285, 64–71; 1993.
49. Duignan MT. Micromachining of diffractive optics with excimer lasers, *Tech. Digest Series-Opt. Soc. Am.*, **11**, 129; 1994.
50. Gale MT, Rossi M, Pedersen J, Schütz H. Fabrication of continuous-relief micro-optical elements by direct laser writing in photoresists, *Opt. Eng.*, **33**, 3556–3566; 1994.
51. Gale MT, Rossi M, Schütz H, Ehbets P, Herzig HP, Prongué D. Continuous-relief diffractive optical elements for two-dimensional array generation, *Appl. Opt.*, **32**, 2526–2533; 1993.
52. Kluepfel B, Ross F. (eds.). *Holography Market Place*. Berkeley, CA: Ross Books, 1991.
53. Noguès JL, LaPaglia AJ. Processing properties and applications of sol–gel silica optics, *Proc. SPIE*, **1168**, 61–82; 1989.
54. Moreshead W, Nogues JL, Howell L, Zhu BF. Replication of diffractive optics in silica glass, *Proc. SPIE*, **2689**, 16, 142–152; 1996.
55. Wu C. Method of Making High Energy Beam Sensitive Glasses, U.S. Patent No. 5,078,771; 1992.
56. Meyers MM. Diffractive optics at Eastman Kodak Company, *Proc. SPIE*, **2689**, 31, 228–254; 1996.
57. Furlan W, Ferrando V, Monsoriu J, Zagrajek P, Czerwińska EB, Szustakowski M. 3D printed diffractive terahertz lenses, *Opt. Lett.*, **41**, 1748–1752; 2016.
58. Parker WP. Commercial applications of diffractive switched optical elements (SOE's), *Proc. SPIE*, **2689**, 27, 195–209; 1996.
59. Shaun DG, Nguyen A, Obata K, Koroleva A, Narayan RJ, Chichkov BN. Parallel phase-shifting digital holography with adaptive function using phase-mode spatial light modulator, *Biomed. Opt. Express*, **2**, 3167–3178; 2011.
60. Wei-Feng H, Yu-Wen C, Yuan-Hong S. Implementation of phase-shift patterns using a holographic projection system with phase-only diffractive optical elements, *Appl. Opt.*, **50**, 3646–3652; 2011.
61. Lin et al. 2012.
62. Sutherland RL, Natarajan LV, Tondiglia VP, Bunning TJ, Adams WW. Development of photopolymer-liquid crystal composite materials for dynamic hologram applications, *Proc. SPIE*, **2152**, 303–312; 1994a.
63. Sutherland RL, Natarajan LV, Tondiglia VP, Bunning TJ, Adams WW. Electrically switchable volume gratings in polymer-dispersed liquid crystal, *Appl. Phys. Lett.*, **64**, 1074–1076; 1994b.
64. Sainov S, Mazakova M, Pantcheva M, Tonchev D, Holographic Diffraction Grating Controlled by Means of Nematic Liquid Crystal, *Mol. Cryst. Liq. Cryst.*, **152**, 609–612; 1987.
65. Stalder M, Ehbets P. Electrically switchable diffractive optical element for image processing, *Opt. Lett.*, **19**, 1–3; 1994.
66. Zhang J, Sponolor MB. Switchable liquid crystalline photopolymer media for holography, *J. Am. Chem. Soc.*, **14**, 1516–1520; 1992.

67. Domash LH, Gozewski C, Nelson A. Application of switchable polaroid holograms, *Proc. SPIE*, **2152**, 13, 127–138; 1994.

68. Oakdale, J. S., J. Ye, W. L. Smith, and J. Biener. Post-print UV curing method for improving the mechanical properties of prototypes derived from two-photon lithography, *Opt. Express*, **24**, 27077–27086, (2016).

69. Xu M, Jim B., He R, Ren H. Adaptive lenticular microlens array based on voltage-induced waves at the surface of polyvinyl chloride/dibutyl phthalate gels, *Opt. Express*, **24**, 8142. doi:10.1364/OE.24.008142; 2016.

70. Wyroski F, Esdonk H, Zuidema R, Wadmann S, Notenboom G. Use of diffractive optics in material processing, *Proc. SPIE*, **2152**, 16, 139–144; 1994.

71. Montamedi ME. Micro-opto-electro-mechanical systems, *Opt. Eng.*, **33**, 3505–3517; 1994.

72. Huang G, Wu M, Jin G, Yan Y. Micro-optic element for optical disk memory read-write heads, *Proc. SPIE*, **2152**, 30, 261–265; 1994.

73. Herzig HP, Dändliker R. Diffractive components: Holographic optical elements. In *Perspectives for Parallel Interconnects*, Lalanne P. and P. Chavel (eds.), Berlin: Springer-Verlag, 1993.

74. Schwab M, Lindlein N, Schwider J, Amitai Y, Friesem AA, Reinhorn S. Achromatic diffractive fan-out systems, *Proc. SPIE*, **2152**, 12, 173–184; 1994.

75. Rakuljic GA, Leyva V, Yariv A. Optical data storage by using orthogonal wavelength-multiplexed volume holograms, *Opt. Lett.*, **15**, 1471–1473; 1992.

76. Pawlowski E, Kuhlow B. Antireflection-coated diffractive optical elements fabricated by thin-film deposition, *Opt. Eng.*, **33**, 3537–3546; 1994.

77. Nieuborg N, Kirk AG, Morlion B, Thienpont H, Veretennicoff LP. Highly polarization-selective diffractive optical elements in calcite with an index-matching gap material, *Proc. SPIE*, **3010**, 123–126; 1997.

78. Ojima M, Saito A, Kaku T, et al. Compact magnetooptical disk for coded data storage, *Appl. Opt.*, **25**, 483–489; 1986.

79. Kinnstatter K, Ojima K. Amplitude detection for the focus error in optical disks using a birefringent lens, *Appl. Opt.*, **29**, 4408–4413; 1990.

80. Kunstmann P, Spitschan HJ. General complex amplitude addition in a polarization interferometer in the detection of pattern differences, *Opt. Commun.*, **4**, 166–172; 1971.

81. McCormick FB, Tooley FAP, Cloonan TJ, et al. Experimental investigation of a free-space optical switching network by using symmetric self-electro-optic-effective devices, *Appl. Opt.*, **31**, 5431–5446; 1992.

82. Martinez-Ponce G, Petrova T, Tomova N, Dragostinova V, Todorov T, Nikolova L. Bifocal-polarization holographic lens, *Opt. Lett.*, **29**, 1001–1003; 2004.

83. Miler M, Koudela I, Aubrecht I. Zone areas of diffractive phase elements producing focal annuli, *Proc. SPIE*, **3320**, 210–213; 1998.

84. Aleksoff CC, Ellis KK, Neagle BG. Holographic conversion of Gaussian beam to near-field uniform beam, *Opt. Eng.*, **30**, 537–543; 1991.

85. Kosoburd T, Akivis M, Malkin Y, Kobrin B, Kedmi S. Beam shaping with multilevel elements, *Proc. SPIE*, **2152**, 48, 214–224; 1994.

86. Goering R, Hoefer B, Kraeplin A, Schreiber P, Kley E, Schmeisser V. Integration of high power laser diodes with microoptical components in a compact pumping source for visible fiber laser, *Proc. SPIE*, **3631**, 191–197; 1999.

87. Miyamoto Y, Masuda M, Wada A, Takeda M. Electron-beam lithography fabrication of phase holograms to generate Laguerre–Gaussian beams, *Proc. SPIE*, **3740**, 232; 1999.

88. Zhang Y, Gu B, Dong B, Yang G. Design of diffractive phase elements for beam shaping in the fractional Fourier transform domain, *Proc. SPIE*, **3291**, 58–67; 1998.

89. Liu B, Domg B, Gu B. Implementation of pseudo-nondiffracting beams by use of diffractive phase elements, *Appl. Opt.*, **37**, 8219–8223; 1998.
90. Lunitz B, Jahns J. Computer generated diffractive axicons, *Eur. Opt. Soc.*, **22**, 26–27; 1999.
91. Velzel CHF, Wyrowski F, Esdink H. Pattern projection by diffractive and conventional optics: A comparison, *Proc. SPIE*, **2152**, 24, 206–211; 1994.

References from Figures

Davis JA, Cottrell DM. Binary optical elements using spatial light modulators, *Proc. SPIE*, **2152**, 26, 226–236; 1994.
Jannson T, Aye TM, Savant G. Dispersion and aberration techniques in diffractive optics and holography, *Proc. SPIE*, **2152**, 5, 44–70; 1994.
Gray, S., F. Clube, and D. Struchen. The Holographic Mask Aligner, *Holographic Systems, Components and Applications*, Neuchâtel, CH Conference Publication No. 379. Institution of Electrical Engineers, London, 265 (1993).
Missig, M. D. and G. M. Morris. Diffractive optics applied to eyepiece design, *Appl. Opt.*, **34**, 2452–2461 (1995).

9

Polarization and Polarizing Optical Devices

Rafael Espinosa-Luna
and Qiwen Zhan

9.1 Introduction

Light has been, is, and sure will be one of the most common physical manifestations of energy in nature. To describe it, the comparison of its behavior with other known physical phenomena, like interference and propagation of waves in water, has been inspirational. The parallel and independent work done on the characteristics of light and in electricity and magnetism, has led on to understand that light is composed of a unique entity, formed by three elements: an electric field, a magnetic field, and a propagation direction. Light is a kinematic manifestation of energy propagation. Experience has shown the speed of light depends on the physical characteristics of the transmission medium, being the vacuum where the maximum speed can be reached and as denser the matter, the slower the propagation speed on that medium. The light is a very small part of the entire electromagnetic spectrum, where the visible light (VIS) usually is located between 380 nm (violet color) and 780 nm (red color). It is also common to define the VIS wavelength interval from 400 nm to 800 nm.

The basic equations that describe the behavior of the electromagnetic field and its interaction with matter are given by the Maxwell's equations, Equations. 9.1 through 9.4 [1]:

$$\nabla \cdot \vec{D} = \rho \tag{9.1}$$

$$\nabla \cdot \vec{B} = 0 \tag{9.2}$$

$$\nabla \times \vec{E} = -\frac{\partial \vec{B}}{\partial t} \tag{9.3}$$

$$\nabla \times \vec{H} = \vec{j} + \frac{\partial \vec{D}}{\partial t}, \tag{9.4}$$

where $\vec{E}$ is the electric field, $\vec{D}$ is the electric displacement, $\vec{B}$ is the magnetic field, $\vec{H}$ is the magnetic field intensity, $\vec{j}$ is the current density, ρ is the volume charge density, and the divergence and the rotational vector differential operations are given by $(\nabla \cdot)$ and $(\nabla \times)$, respectively. Equation 9.1 is the Gauss' law for the electric field; Equation 9.2 is the Gauss' law for the magnetic field; Equation 9.3 is the Faraday's law or the magnetic induction law (basis of the electric motor); and Equation 9.4 is the Ampère-Maxwell's equation.

The response of a medium to the electromagnetic field depends on its intensity, phase, and direction of incidence, but also of the own nature of matter. The corresponding relationships between the electric field components and the medium response, for a linear response to the electromagnetic field, are described by the material or constitutive equations which, for most optical media and considering only a linear response, are given by

$$\vec{D} = \varepsilon_0 \vec{E} + \vec{P} = \varepsilon_0 \vec{E} + \varepsilon_0 \chi \vec{E} = \varepsilon \vec{E} = \varepsilon_0 n^2 \vec{E} \Rightarrow n = \sqrt{\varepsilon / \varepsilon_0} \tag{9.5}$$

$$\vec{B} = \mu_0 \vec{H} + \vec{M} = \mu \vec{H} \tag{9.6}$$

$$\vec{j} = \sigma \vec{E}, \tag{9.7}$$

where $\vec{P}$ is the electric polarization and $\vec{M}$ is the magnetization of the optical medium (in general $\vec{M} = 0$, except for ferromagnetic materials), the permittivity or dielectric constant of free space and of medium are ε_0 and ε, respectively; μ_0 and μ are the permeability of the free space and the medium, respectively, σ is the electric conductivity, and n is named the refractive index of the medium.

A material medium is classified according to its response to some specific external stimulus, than can has an electric, magnetic, mechanical or thermal origin, among many other possible characteristics. In this sense, an optical medium is physically classified depending on his response to the electric or the magnetic field components, according to the following convention. Denoting the temperature by T, if $\sigma \approx 1/T$, it is a conductor; when $\sigma \approx T$, it is named semiconductor, but if $\sigma \ll 1$, it is a dielectric or non-conducting $(\vec{j} \approx 0)$. If $\mu \gg 1$, it is a non-magnetic material; when $\mu > 1$, it is a paramagnetic medium; but, if $\mu < 1$, the material is diamagnetic.

The Lorentz equation describes the force experienced by an electric charge q, moving with a velocity $\vec{v}$, subject to electric and magnetic fields, namely

$$\vec{F} = q\left(\vec{E} + \frac{\vec{v}}{c} \times \vec{B}\right) \tag{9.8}$$

where $\times$ represents the cross or vector product operation.

There are some very important and commonly used physical properties associated with an optical medium (usually $\mu \approx \mu_0$) termed by specific names, like the following. A medium is uniform or homogeneous if the values of (ε, μ) do not depend on their spatial position; they have the same values at any point. An isotropic medium signifies that the values of (ε, μ) are independent of the direction of propagation of the field. Non-dispersive medium means (ε, μ) do not depend of the frequency $\omega = 2\pi\nu$ associated to the field, and free of charge or current implies $\rho = 0$ or $\vec{j} = 0$, respectively.

The Maxwell's equations take on a very simple form when the electromagnetic field propagates into a medium with the previous physical characteristics

$$\nabla \cdot \vec{E} = 0, \ \nabla \cdot \vec{B} = 0, \ \nabla \times \vec{E} = -\frac{\partial \vec{B}}{\partial t}, \ \nabla \times \vec{B} = \varepsilon\mu\frac{\partial \vec{E}}{\partial t}. \tag{9.9}$$

Applying the curl to the Faraday's law and substituting the Ampère-Maxwell's law

$$\nabla \times \nabla \times \vec{E} = -\frac{\partial}{\partial t}\nabla \times \vec{B} = -\frac{\partial}{\partial t}\left(\varepsilon\mu\frac{\partial \vec{E}}{\partial t}\right) = -\varepsilon\mu\frac{\partial^2 \vec{E}}{\partial t^2}, \tag{9.10}$$

where

$$\nabla \times \nabla \times \vec{E} = \nabla\left(\nabla \cdot \vec{E}\right) - \nabla^2 \vec{E} = -\nabla^2 \vec{E} \tag{9.11}$$

then

$$\nabla^2 \vec{E}(\vec{r}, t) - \varepsilon\mu\frac{\partial^2 \vec{E}(\vec{r}, t)}{\partial t^2} = 0. \tag{9.12}$$

The fact that $\nabla \cdot \vec{E}(\vec{r}, t) = 0$ implies the electric field is spatially homogeneous, a consideration of great importance because it is part of the foundation of the conventional polarization.

A similar relationship can be obtained for the magnetic field by following a parallel procedure, where now the time derivative must be applied to the Ampère-Maxwell's law, as the first step. As a consequence, the wave equation describes the propagation of the electromagnetic field through this optical medium, accordingly to

$$\nabla^2 \vec{\psi}(\vec{r}, t) - \frac{1}{v^2}\frac{\partial^2 \vec{\psi}(\vec{r}, t)}{\partial t^2} = 0, \tag{9.13}$$

where $\vec{\psi}(\vec{r},t) = \vec{E}(\vec{r},t), \vec{B}(\vec{r},t), \nabla^2$ represents the Laplacian operation, and $v = 1/\sqrt{\varepsilon\mu}$ is the speed or phase velocity of the electromagnetic component in the optical medium. For most optical materials $v = c/n$, where $c = 2.998 \times 10^8 m/s$ is the speed of light in vacuum.

The simplest solution to the wave Equation 9.13 is given by the harmonic plane wave solution. This describes the propagation along the wave vector direction $\vec{k}$ of a wave with vector amplitude $\vec{E}_0$ varying harmonically in time and space

$$\vec{E}(\vec{r},t) = \vec{E}_0 \cos\left(\omega t - \vec{k} \cdot \vec{r}\right). \tag{9.14}$$

In order to easily handle the mathematical operations related with the electromagnetic field theory, it is common to use the complex notation, where the real part of the final result obtained contains the physically measurable quantity, able to be registered, directly or indirectly, as a physical response. Then, the simple harmonic plane wave solution takes on the form

$$\vec{E}(\vec{r},t) = \vec{E}_0 \exp\left[i\left(\omega t - \vec{k} \cdot \vec{r}\right)\right]. \tag{9.15}$$

As a matter of fact, Equation 9.15 can also be represented mathematically by its complex conjugate, equivalently, as

$$\vec{E}(\vec{r},t) = \vec{E}_0 \exp\left[i\left(\vec{k} \cdot \vec{r} - \omega t\right)\right], \tag{9.16}$$

because any of them represents a traveling plane harmonic wave and the real part is the same in both expressions. It is important to take into account that Equation 9.15 is associated to a plane wave traveling to the observer, while Equation 9.16 travels away from the observer.

The application of Equation 9.15 to the wave Equation, Equation 9.12, leads to the dispersion relation, which provides a linear response of the wave number k to the frequency

$$k = \omega\sqrt{\varepsilon\mu} = \frac{\omega}{v}, \quad \omega = vk = ck/n. \tag{9.17}$$

Other important relationships can be derived from the application of Equation 9.15 to Equation 9.9, like those obtained when it is applied to the Gauss' law for the electric field and to the Faraday's law, which provide the respective results

$$\nabla \cdot \vec{E} = 0 \Rightarrow \vec{k} \cdot \vec{E}_0 = 0 \tag{9.18}$$

and

$$\nabla \times \vec{E} = -\frac{\partial \vec{B}}{\partial t} \Rightarrow \vec{k} \times \vec{E} = \omega\vec{B}. \tag{9.19}$$

Equations 9.18 and 9.19 show that a harmonic plane wave is characterized by three quantities mutually orthogonal to one another and that $\vec{E}$ and $\vec{B}$ are in phase. Another interesting result derived from Equation 9.18 is the amplitude value between the electric and the magnetic field components

$$\left|\omega\vec{B}\right| = \left|\vec{k} \times \vec{E}\right| \Rightarrow \left|\vec{B}\right| = \frac{n}{c}\left|\vec{E}\right| \Rightarrow \left|\vec{E}\right| = \frac{c}{n}\left|\vec{B}\right|. \tag{9.20}$$

As a consequence of the previous relationships, the polarization properties associated to a magnetic (electric) field component can be completely determined once the electric (magnetic) field component has been totally characterized.

Note that the best register obtained by the detector system defines the more appropriate entity to describe the polarization of the electromagnetic field through the electric or the magnetic field components. For the most optical media, expression 9.20 shows the electric field amplitude is bigger than the magnetic field amplitude. This is the main reason why the electric field component is vastly employed in optics in general, and particularly in polarization-related areas, instead of the magnetic field. In some areas, like in geophysics, the characterization of the magnetic field component through magnetic induction devices is the usual way to work with the very low frequencies technique (VLF), the analogous conceptual scheme of the ellipsometry technique employed in optics.

Energy in nature is commonly identified when it is associated to classical mechanics (any reader surely has experimented its effects after an undesired and sudden visit on the floor); in this sense, it is easy to quantify the potential or the kinetic energy that any mass can possess under some specific conditions. The direct exposure to the sunlight has shown the light, an electromagnetic wave, also has associated energy. The total energy U stored in an electromagnetic (propagating) field, is given by

$$U = 1/2 \left(\vec{D} \cdot \vec{E} + \vec{B} \cdot \vec{H} \right), \tag{9.21}$$

where the dot represents the inner or scalar product. For a plane wave traveling in a homogeneous, isotropic, non-magnetic medium,

$$U = \frac{\varepsilon \vec{E} \cdot \vec{E} + \vec{B}. \dfrac{\vec{B}}{\mu}}{2} = \frac{\left(\varepsilon + \dfrac{n^2}{\mu c^2} \right) E^2}{2} = \frac{(2\varepsilon) E^2}{2} = \varepsilon E^2. \tag{9.22}$$

The total energy flows through the propagating medium and is described by the Poynting's vector $\vec{P}$ as the time rate of the flow of electromagnetic energy per unit area. In free space, it is given as

$$\vec{P} = \vec{E} \times \vec{H}. \tag{9.23}$$

Since a long time ago, one of the most appreciated goals of the scientific community has been to "trap" the light. Recently, there have been some inspirational results, where it has been to get slow light in various transmitting media [2].

A very common situation is the steady or time-independent solution to the wave Equation 9.13 when the electric field is time harmonic. Taking the Equation 9.16,

$$\vec{E}(\vec{r}, t) = \vec{E}(x, y, z) e^{-i\omega t}. \tag{9.24}$$

The wave equation at free space takes the form

$$\nabla^2 \vec{E}(x, y, z) e^{-i\omega t} - \varepsilon \mu \vec{E}(x, y, z) \frac{\partial^2}{\partial t^2} \left[e^{-i\omega t} \right] = 0, \tag{9.25}$$

where

$$\frac{\partial^2 e^{-i\omega t}}{\partial t^2} = -\omega^2 e^{-i\omega t}, \quad \varepsilon \mu \omega^2 = k^2. \tag{9.26}$$

The resultant equation, once the time factor is taken out, is best known as the Helmholtz equation:

$$\nabla^2 \vec{E}(x, y, z) + k^2 \vec{E}(x, y, z) = 0. \tag{9.27}$$

The invention of the laser has been a great stimulus to make optical engineering in order to provide different output beam shapes and the Helmholtz equation has been the mainframe to test different possibilities.

Consider a wave traveling along the z-direction, Equation 9.16, able to be expressed as the scalar expression

$$u(x, y, z)e^{ikz}. \tag{9.28}$$

Substituting into Equation 9.27 and taking off the exponential term, Equation 9.27 reduces to

$$\frac{\partial^2 u}{\partial x^2} + \frac{\partial^2 u}{\partial y^2} + \frac{\partial^2 u}{\partial z^2} + 2ik\frac{\partial u}{\partial z} = 0. \tag{9.29}$$

The need to generate low diverging beams has been the seed for the generation of Gaussian-like output wavefront solutions. This means $u(x,y,z)$ must fulfill the paraxial or slow varying envelope approximation.

The variation of propagation is slow on the scale of wavelength that in mathematical terms is interpreted as

$$\left|\frac{\partial^2 u(x,y,z)}{\partial z^2}\right| \ll 2k\left|\frac{\partial u(x,y,z)}{\partial z}\right| = \frac{4\pi}{\lambda}\left|\frac{\partial u(x,y,z)}{\partial z}\right|. \tag{9.30}$$

The variation of propagation is slow compared to the transverse extent of the wave

$$\left|\frac{\partial^2 u(x,y,z)}{\partial z^2}\right| \ll \left|\frac{\partial^2 u(x,y,z)}{\partial x^2}\right|, \left|\frac{\partial^2 u(x,y,z)}{\partial y^2}\right|. \tag{9.31}$$

Then the Equation 9.29 takes the form known as the paraxial or slow varying approximation wave equation

$$\frac{\partial^2 u}{\partial x^2} + \frac{\partial^2 u}{\partial y^2} + 2ik\frac{\partial u}{\partial z} = 0. \tag{9.32}$$

The possible solutions to this equation have been broadly studied and they depend on the symmetry associated to the electric field. For example, if Equation 9.32 is solved in the Cartesian coordinate system, that means the beam size and shape are not perfectly circular, and the Hermite-Gauss polynomials, with HG_{mn} modes, represent the solution [3,4]

$$u(x,y,z) = E_0 H_m\left(\sqrt{2}\frac{x}{w(z)}\right) H_n\left(\sqrt{2}\frac{y}{w(z)}\right)\frac{w_0}{w(z)}e^{-i\varphi_{mn}(z)}e^{-i\frac{k}{2q(z)}r^2}, \tag{9.33}$$

where E_0 is a constant electric field amplitude, $H_m(x)$, $H_n(y)$ are the Hermite polynomials, $w(z)$ is the transversal beam size evaluated at the distance z (also known as the radius at which the field amplitude drops to its $1/e$ value), w_0 is the narrowest beam size (beam waist), $q(z)=z-iz_0$ is the complex number beam parameter, $z_0 = \pi w_0^2 / \lambda$ is the Rayleigh range (a measure of how far the beam is collimated), and $\varphi_{mn}(z)=(m+n+1)\tan^{-1}(z/z_0)$ is the Gouy phase shift.

Recently, special attention has been given to the scalar general wave solution in the cylindrical coordinate system, where the Helmholtz equation, in the slow varying envelop approximation is reduced to [3,4]

$$\frac{1}{r}\frac{\partial}{\partial r}\left(r\frac{\partial u}{\partial r}\right)+\frac{1}{r^2}\frac{\partial^2 u}{\partial \phi^2}+2ik\frac{\partial u}{\partial z}=0. \tag{9.34}$$

One solution to Equation 9.34 is provided by the Laguerre-Gauss polynomials [3,4]:

$$u(r,\phi,z)= E_0\left(\sqrt{2}\frac{r}{\omega}\right)^l L_p^l\left(2\frac{r^2}{\omega^2}\right)\frac{w_0}{w(z)}\exp\left[-i\varphi_{pl}(z)\right]\exp\left[i\frac{k}{2q(z)}r^2\right]\exp(il\phi), \tag{9.35}$$

where $L_p^l(2r^2/\omega^2)$ are the associated Laguerre polynomials, $\varphi_{pl}(z)=(2p+l+1)\tan^{-1}(z/z_0)$ is the Gouy phase shift, and the last term, $\exp(il\phi)$, is a vortex term. Another solution to Equation 9.34 can be obtained if the wave presents rotational ϕ-symmetry or azimuthal symmetry [3,4]:

$$u(r,z)= E_0\frac{w_0}{w(z)}\exp\left[-i\varphi(z)\right]\exp\left[i\frac{k}{2q(z)}r^2\right]J_0\left(\frac{\beta r}{1+iz/z_0}\right)\exp\left[-\frac{\beta^2 z/(2k)}{1+iz/z_0}\right], \tag{9.36}$$

where $\varphi(z)$ is the Gouy phase shift, $J_0(\beta, r, z, z_0)$ is the zeroth-order Bessel function of the first kind, and β is a constant scale parameter with inverse longitude dimensions.

Equations 9.33, 9.35, and 9.36 reduce to the fundamental Gaussian solution when $m=n=0, l=p=0$, and $\beta=0$, respectively:

$$u(r,z)= E_0\frac{w_0}{w(z)}\exp\left[-i\varphi(z)\right]\exp\left[-\frac{k}{2q(z)}r^2\right]. \tag{9.37}$$

Observe that this solution is azimuthally symmetric, just as a paraxial Gaussian beam should be.

The solutions to the *scalar* Helmholtz paraxial equation, given by Equations 9.33 and 9.35 through 9.37, are related to uniform distributions of the electric field amplitude and intensity. They are associated to *spatially homogeneous polarization* or scalar beams, where the electric field oscillations do not depend on the localization of the observer within the cross-section of the beam of light.

9.1.1 Linearly Polarized and Unpolarized Light

From Equation 9.15, a special situation can be obtained if the vector amplitude $\vec{E}_0$ oscillates within a fixed plane defined by its amplitude direction and the wavevector propagation direction $\vec{k}$. This plane is named the polarization plane and is oriented at an angle $\theta = \tan^{-1}(E_y/E_x)$, measured from the horizontal x- (where $\theta=0°$) to the vertical y-axis ($\theta=90°$), in a right-handed coordinate Cartesian system ($\hat{x}\times\hat{y}=\hat{z}$), where (E_x, E_y) are the vector components of the electric field along the x- and y-axes, respectively (see Figure 9.1). In other words, the positive angles are measured in the counter-clockwise sense (*where an observer registers the polarization looking to the source*). This condition defines

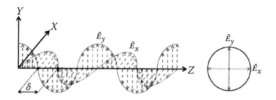

FIGURE 9.1 Propagation along the z-axis of two orthogonal linear polarized plane waves $\vec{E}_x$, $\vec{E}_y$ with phase difference δ.

a linearly polarized wave or state. Consider a monochromatic linearly polarized light propagating along the z-axis, with Equation 9.15 expressed as

$$\vec{E}(r,t) = \vec{E}_0(x_0, y_0)\exp\left[i(\omega t - kz)\right].\tag{9.38}$$

Depending on the azimuthal orientation θ of the amplitude $\vec{E}_0(x_0, y_0)$, any linear polarization state can be obtained. For example:

1. Linear horizontal polarization, E_x, $\theta = 0°$, the complex and real parts are given by

$$\vec{E}_x(z,t) = \hat{x}E_0 x \exp\left[i(\omega t - kz)\right];\ \vec{E}_x(z,t) = \hat{x}E_o x \cos(\omega t - kz).\tag{9.39}$$

2. Linear vertical polarization, E_y, $\theta = 90°$,

$$\vec{E}_y(z,t) = \hat{y}E_0 y \exp\left[i(\omega t - kz)\right];\ \vec{E}_y(z,t) = \hat{y}E_0 y \cos(\omega t - kz).\tag{9.40}$$

3. Linear $\pm 45°$ polarization, $E_{0x} = \pm E_{0y}$ where $|E_{0x}| = |E_{0y}| = E_0$ implies $\theta = \pm 45°$,

$$\vec{E}_\pm(z,t) = E_0(\hat{x} \pm \hat{y})\exp[i(\omega t - kz)];\ \vec{E}_\pm(z,t) = E_0(\hat{x} \pm \hat{y})\cos(\omega t - kz).\tag{9.41}$$

Note any of the linear $\pm 45°$ polarizations have the same amplitudes along the (x, y) plane. For simplicity, a zero phase value has been considered for the linear polarization, Equation 9.38. The linear $-45°$ polarization is also known, indistinctly, as linear $135°$ polarization.

The magnitude of Equation 9.38 is named the intensity of the wave and is given by

$$\vec{E} \cdot \vec{E}^* = \vec{E}_0 \cdot \vec{E}_0 = E_0^2 \equiv .\tag{9.42}$$

In general, a beam of light is named polarized when there exists a deterministic, well-behaved description of its electric or magnetic field components and has a well-defined, non-random, phase value.

On the opposite side, an unpolarized or nonpolarized beam of light has electromagnetic field amplitudes randomly oriented and valued, and also randomly valued phases. The Sun emits unpolarized light and generally the light generated under combustion also has this random nature. For some applications, it is necessary to use unpolarized light with controllable intensity. There are some devices, named depolarizers, whose principal function is to convert totally polarized states, mainly linear, into unpolarized or quasi-depolarized light. These devices can be natural, for example integrating spheres (highly reflecting rough surfaces), particles suspended in liquid solutions, where the depolarization is generated by multiple reflections that generate random phase shifts in the incident polarized beam. They can also be man-made designed, like the quartz ethalons, where a random phase shift is generated locally to a collimated totally polarized beam (the quartz-wedge achromatic design is one of the most successful configurations), or liquid crystal displays-based pseudo-depolarizers (see Section 9.4.6).

9.2 Interference of Polarized Light

Consider the addition of two linearly polarized plane waves, traveling along the same direction z:

$$\vec{E}(\vec{r},t) = \hat{p}E_{01}e^{i(\omega t - kz + \delta_1)} + \hat{q}E_{02}e^{i(\omega t - kz + \delta_2)} = \left[\hat{p}E_{01}e^{i\delta_1} + \hat{q}E_{02}e^{i\delta_2} \right]e^{i(\omega t - kz)}, \tag{9.43}$$

where $\hat{p}, \hat{q}$ are arbitrary unitary vectors defined in the plane (x, y) perpendicular to the propagating direction. The field obtained is a polarized plane wave, where the orientation of the polarization plane depends on the specific values assigned to the amplitudes E_{01}, E_{02}, the phase shifts δ_1, δ_2, and the relative orientation θ_{pq} between the unitary vectors.

The instant intensity of Equation 9.43 is given as

$$I = \vec{E} \cdot \vec{E}^* = E_{01}^2 + E_{02}^2 + \hat{p} \cdot \hat{q} 2E_{01}E_{02}\cos(\delta_1 - \delta_2). \tag{9.44}$$

The third term into Equation 9.44 is called the interference and its value depends on the amplitude of the waves, the phase difference $\delta = \delta_1 - \delta_2$, and the relative orientation between $\hat{p}$ and $\hat{q}$ ($\hat{p} \cdot \hat{q} = |\hat{p}||\hat{q}|\cos\theta_{pq} = \cos\theta_{pq}$),

$$I = E_{01}^2 + E_{02}^2 + 2E_{01}E_{02}\cos\theta_{pq}\cos\delta. \tag{9.45}$$

Equation 9.45 takes on the familiar form used in the interference area of optics when the unitary vectors are parallel ($\theta_{pq} = 0°$):

$$I = E_{01}^2 + E_{02}^2 + 2E_{01}E_{02}\cos\delta. \tag{9.46}$$

When the interference term in Equation 9.46 is positive ($-\pi/2 < \delta < \pi/2$), the superposition of the waves is named constructive; on the other hand, if the interference term is negative ($\pi/2 < \delta < 3\pi/2$), the superposition of the waves is named destructive. There does not exist interference when the waves are out-of-phase or un-phased ($\delta = \pm\pi/2, \pm3\pi/2$).

Observe that other possibilities exist if the unitary vectors are anti-parallel ($\theta_{pq} = 180°$),

$$I = E_{01}^2 + E_{02}^2 - 2E_{01}E_{02}\cos\delta, \tag{9.47}$$

where the previous conditions for constructive or destructive interference are now reversed.

A very special situation occurs when the unitary vectors are mutually orthogonal, where the interference term vanishes independent of the phase difference value and the resultant intensity is the addition of the intensity of each wave. This situation sets for one of the physical conditions for the optical polarization phenomena.

Finally, Equation 9.45 indicates that the interference will always be present in some degree, dependent of the relative angle θ_{pq} between the unitary vectors $\hat{p}, \hat{q}$, and will only be absent when they are orthogonal or the waves are out-of-phase.

An experimental setup is designed with the intention to extract some desired information from a system under study, by taking control on the experimental variables. In this sense, the experimental results

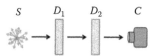

FIGURE 9.2   Series configuration.

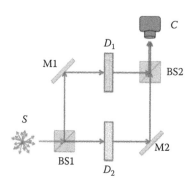

FIGURE 9.3   Parallel configuration.

must provide enough information in order to distinguish the response of the system under study from the characteristic behavior of its own experimental setup without any sample. This is done by following a procedure named calibration.

There are two basic optical experimental setups, defined according to [5]:

1. *Series configuration.* If the intensity is transmitted through the optical devices with linear responses defined by D_1 and D_2 in this order, from the source to the photo-detector or CCD camera C, the physical effect is undistinguishable from the other one where the intensity is transmitted through the optical device $D = D_2 D_1$ (see Figure 9.2).
2. *Parallel configuration.* If the intensity is divided, usually by a beamsplitter (50: 50), *BS1*, the two beams reflected by mirrors, *M1* and *M2*, transmitted through the optical devices D_1, D_2, respectively, and the emerging intensity is coherently superimposed by a second beamsplitter, *BS2*, the effect at the photo-detector or CCD camera C is the same as if the intensity had been transmitted through the single device $D = D_1 + D_2$ (see Figure 9.3).

9.2.1 Elliptical, Circular, and Linearly Polarized Light

Consider the propagation of a linearly polarized plane wave E_θ ($0° \leq \theta \leq 360°$), where the horizontal E_{0x} and the vertical E_{0y} components are affected by $-\delta/2$ and $+\delta/2$ arbitrary phase shifts, respectively, where $\theta = \tan^{-1}\left(E_{0y}/E_{0x}\right)$ is its instant slope. The resultant electric field can be conceptualized as the linear superposition of two orthogonal vector components [6]:

$$\vec{E}(r,t) = \hat{x}E_{0x}\exp\left[i\left(\omega t - kz - \delta/2\right)\right] + \hat{y}E_{0y}\exp\left[i\left(\omega t - kz + \delta/2\right)\right] \tag{9.48}$$

$$\vec{E}(r,t) = \left(\hat{x}E_{0x}e^{-i\delta/2} + \hat{y}E_{0y}e^{i\delta/2}\right)\exp\left[i\left(\omega t - kz\right)\right]. \tag{9.49}$$

Taking into account that $e^{i\delta/2} = e^{-i\delta/2}e^{i\delta}$, Equation 9.49 can be reduced to

$$\vec{E}(r,t) = \left(\hat{x}E_{0x} + \hat{y}E_{0y}e^{i\delta}\right)\exp\left[i\left(\omega t - kz - \delta/2\right)\right]. \tag{9.50}$$

The resulting polarization state depends on the electric field amplitude components and the δ phase shift value. The real part of Equation 9.49 is

$$\vec{E}_e(r,t) = \hat{x}E_{0x}\cos(\omega t - kz - \delta/2) + \hat{y}E_{0y}\cos(\omega t - kz + \delta/2). \tag{9.51}$$

The behavior described by the resultant electric field can be easily visualized if, at an observing fixed plane $(x_0, y_0, z_0 = 0)$, it is analyzed during a temporal period $\Delta t = \tau$ or $0 \le \omega t = 2\pi v t \le 2\pi$. If $\delta > 0°$, Equation 9.51 describes an ellipse whose rotation sense is given according to the period when it is completed (increasing time); in this case, it rotates in the clockwise direction. It will be considered as positive and the observer takes on register lectures looking to the source. It is named an elliptical right-hand polarization. If $\delta < 0°$, the ellipse begins its rotation in the counter-clockwise sense, generating an elliptical left-hand polarization state. Also by fixing the time $t = 0$, the analysis can be realized as a function of the electric field spatial behavior into a wavelength interval, $0 \le k_z z = \left(\dfrac{2\pi}{\lambda}\right)\lambda \le 2\pi$, by taking equally spaced intervals.

Due to its very common use, the following are very special cases of the general elliptical polarization state, Equation 9.50:

1. Suppose $E_{0x} = E_{0y} = E_0$ and $\delta = \pi/2$

$$\vec{E}_r(r,t) = E_0\left(\hat{x}e^{-i\pi/4} + \hat{y}e^{i\pi/4}\right)e^{i(\omega t - kz)} = E_0\left(\hat{x} + \hat{y}e^{i\pi/2}\right)e^{i\left(\omega t - kz - \frac{\pi}{4}\right)} \tag{9.52}$$

$$\vec{E}_r(r,t) = E_0\left[\hat{x}\cos(\omega t - kz - \pi/4) + \hat{y}\cos(\omega t - kz + \pi/4)\right]. \tag{9.53}$$

e intensity of Equation 9.53 can be directly calculated, according to

$$I_r = \vec{E}_r \cdot \vec{E}_r^* = E_0^2. \tag{9.54}$$

Analyzing the propagation of Equation 9.53 during a time period τ, at a fixed point $z = z_0 = 0$, the electric field describes a circle rotating in the clockwise sense, with a constant amplitude. It is named circular right-hand polarization, E_r. Table 9.1 contains some characteristic values obtained when increasingly intervals $\Delta \omega t = \dfrac{\pi}{4}$, $\Delta t = 0.125\tau$ are taken within a full period.

For an observer located at a fixed position along the propagation direction z, increasing times means the registering lectures are realized looking to the source.

The type of polarization can also be easily identified if the analysis is realized at a fixed time $t = t_0 = 0$, as a function of the spatial behavior associated to the Equation 9.53, taking equally

TABLE 9.1 At a Fixed Point along the Propagation Direction $z = z_0 = 0$, the Wave Given by Equation 9.53 Describes a Circle in the Perpendicular Plane (x_0, y_0). The Circle Follows the Clockwise Rotation Sense, Associated to a Circular Right-Hand Polarization State, E_r.

t	0τ	0.125τ	0.250τ	0.375τ	0.500τ	0.625τ	0.750τ	0.875τ	1.000τ
ωt	0	$\pi/4$	$2\pi/4$	$3\pi/4$	$4\pi/4$	$5\pi/4$	$6\pi/4$	$7\pi/4$	$8\pi/4$
$\vec{E}_r(t)$	$\dfrac{(\hat{x}-\hat{y})E_0}{\sqrt{2}}$	$\hat{x}E_0$	$\dfrac{(\hat{x}-\hat{y})E_0}{\sqrt{2}}$	$-\hat{y}E_0$	$\dfrac{(\hat{x}+\hat{y})E_0}{\sqrt{2}}$	$-\hat{x}E_0$	$\dfrac{(-\hat{x}+y)E_0}{\sqrt{2}}$	$\hat{y}E_0$	$\dfrac{(\hat{x}+\hat{y})E_0}{\sqrt{2}}$

TABLE 9.2 At a Fixed Time $t = t_0 = 0$ and Choosing Increasingly, Equally Spaced Points $\Delta z = 0.125\lambda$ along the Direction of Propagation kz, the Wave Given by Equation 9.53 Describes a Circle in the Perpendicular Plane (x_0, y_0). The Circle Follows the Clockwise Rotation Sense for an Observer Looking to the Source, Associated to a Circular Right-Hand Polarization State, E_r

z	0λ	0.125λ	0.250λ	0.375λ	0.500λ	0.625λ	0.75λ	0.875λ	1.000λ
$k_z z$	0	$\pi/4$	$2\pi/4$	$3\pi/4$	$4\pi/4$	$5\pi/4$	$6\pi/4$	$7\pi/4$	$8\pi/4$
$\vec{E}_r(z)$	$\dfrac{(\hat{x}+\hat{y})E_0}{\sqrt{2}}$	$\hat{y}E_0$	$\dfrac{(-\hat{x}+\hat{y})E_0}{\sqrt{2}}$	$-\hat{x}E_0$	$\dfrac{-(\hat{x}+\hat{y})E_0}{\sqrt{2}}$	$-\hat{y}E_0$	$\dfrac{(\hat{x}-y)E_0}{\sqrt{2}}$	$\hat{x}E_0$	$\dfrac{(\hat{x}+\hat{y})E_0}{\sqrt{2}}$

spaced intervals $k\Delta z = \dfrac{\pi}{4}$, $\Delta z = 0.125\lambda$ in a full wavelength spatial interval. Table 9.2 shows the result obtained.

As depicted in Table 9.2, the electric field has different orientations in each plane (x, y), at different positions z, always maintaining a same magnitude or intensity

$$I_r = \vec{E}_r \bullet \vec{E}_r = \frac{(\hat{x}+\hat{y})E_0}{\sqrt{2}} \cdot \frac{(\hat{x}+\hat{y})E_0}{\sqrt{2}} = \hat{x}E_0 \cdot \hat{x}E_0 = \frac{(\hat{x}-\hat{y})E_0}{\sqrt{2}} \cdot \frac{(\hat{x}-\hat{y})E_0}{\sqrt{2}} = \ldots = \hat{y}E_0 \cdot \hat{y}E_0 = E_0^2. \quad (9.55)$$

Table 9.2 is some kind of photographic representation of the wave field, Equation 9.53, at different spatial positions inside a complete interval equivalent to a wavelength period. Care must be paid to the correct interpretation of this table. If the observer takes the initial point of her/his registration, moving from $z = \lambda$ to $z = 0$ (looking to the source), it can be deduced that the electric field rotates in the clockwise sense, which means Equation 9.53 represents a circular right-hand polarization state. However, if the observer takes the register in the opposite sense, moving from $z = 0$ to $z = \lambda$ (looking to the propagation direction), it can be deduced that the electric field rotates in the counter-clockwise sense, consequently, for an observer that registers in the propagation direction, Equation 9.53 describes a circular left-handed polarization state.

2. When $E_{0x} = E_{0y} = E_0$ and $\delta = -\pi/2$, Equation 9.50 takes the form

$$\vec{E}(r, t) = \left(\hat{x}E_{ox} + \hat{y}E_{0y}e^{-\frac{i\pi}{2}} \right) \exp\left[i(\omega t - kz + \pi/4) \right]. \quad (9.56)$$

$$\vec{E}(r, t) = E_0 \left[\hat{x}\cos(\omega t - kz + \pi/4) + \hat{y}\cos(\omega t - kz - \pi/4) \right] \equiv \vec{E}_l. \quad (9.57)$$

By following a similar procedure to the previous case, it can be shown that Equation 9.57 describes a circle that rotates according to the counter-clockwise sense for an observer looking to the source. Under these conditions, a circular left-hand polarization state, E_l, is obtained and described by Equation 9.57.

3. When $\delta = 0°$, Equation 9.50 takes on the form

$$\vec{E}(r, t) = \left(\hat{x}E_{ox} + \hat{y}E_{0y} \right) \exp\left[i(\omega t - kz) \right], \quad (9.58)$$

where the real part takes the form

$$\vec{E}(r, t) = \left[\hat{x}E_{0x} + \hat{y}E_{0y} \right] \cos(\omega t - kz). \quad (9.59)$$

The resultant electric field has a constant amplitude oriented along $\theta = \tan^{-1}(E_{0y}/E_{0x})$, modulated by a traveling, harmonic wave. This is a linear polarized state. For example, if

4. $E_{0y} = 0$, $\delta = 0°$, the linear horizontal E_x polarization is obtained

$$\vec{E}(r,t) \equiv \vec{E}_x(z,t) = \hat{x}E_{0x}\exp[i(\omega t - kz)], \tag{9.60}$$

where the real part is

$$\vec{E}_x(z,t) = \hat{x}E_{0x}\cos(\omega t - kz). \tag{9.61}$$

5. $E_{0x} = 0$, $\delta = 0°$, the linear vertical E_y polarization is obtained

$$\vec{E}(r,t) \equiv \vec{E}_y(z,t) = \hat{y}E_{0y}\exp[i(\omega t - kz)] \tag{9.62}$$

where the physically realizable part is

$$\vec{E}_y(z,t) = \hat{y}E_{0y}\cos(\omega t - kz). \tag{9.63}$$

6. When $E_{0y} = E_{0x} = E_0$, $\delta = 0°$, the linear +45° polarization state is generated, E_+

$$\vec{E}(r,t) \equiv \vec{E}_+(z,t) = E_0(\hat{x}+\hat{y})\exp[i(\omega t - kz)] \tag{9.64}$$

or

$$\vec{E}_+(z,t) = E_0(\hat{x}+\hat{y})\cos(\omega t - kz). \tag{9.65}$$

7. When $E_{0y} = E_{0x} = E_0$, $\delta = 180°$, the linear −45° polarized state is obtained, E_-

$$\vec{E}(r,t) \equiv \vec{E}_-(z,t) = E_0(\hat{x}-\hat{y})\exp[i(\omega t - kz - \pi/2)], \tag{9.66}$$

where the real part is

$$\vec{E}(r,t) \equiv \vec{E}_-(z,t) = E_0(\hat{x}-\hat{y})\cos(\omega t - kz - \pi/2). \tag{9.67}$$

9.2.2 Jones Parameters

Any polarized state can also be ordered as a 2×1 column matrix of complex elements, according to the following notation convention, known as the Jones parameters or the Jones vectors [7,8]:

$$\vec{E}(z,t) = \begin{pmatrix} E_x \\ E_y \end{pmatrix} = \begin{pmatrix} E_{0x}\exp[i(\omega t - kz + i\delta_x)] \\ E_{0y}\exp[i(\omega t - kz + i\delta_y)] \end{pmatrix} = e^{i(\omega t - kz)}\begin{pmatrix} E_{0x}\exp(i\delta_x) \\ E_{0y}\exp(i\delta_y) \end{pmatrix}, \tag{9.68}$$

where the total intensity I of the optical field has been defined as

$$I = I_x + I_y = E_x E_x^* + E_y E_y^* = E_{ox}^2 + E_{oy}^2 = |E_x|^2 + |E_y|^2. \tag{9.69}$$

Note the intensity can also be obtained by the matrix multiplication:

$$I = \begin{pmatrix} E_x^* & E_y^* \end{pmatrix} \begin{pmatrix} E_x \\ E_y \end{pmatrix}. \tag{9.70}$$

It is usual to use the Jones vector at its minimal expression, which is obtained once the propagating term $\exp\left[i(\omega t - kz)\right]$ is taken out, keeping only its amplitude and phase. The Jones vector can be represented in its full or non-normalized form or as a normalized matrix with respect to its intensity $(I = 1)$. In this way, the Jones vector for the basic completely polarized states are given as normalized and non-normalized, respectively:

1. Linear horizontally E_x polarized light, $E_{0y} = 0$, then $E_{0x}^2 = 1$

$$E_x = \begin{pmatrix} 1 \\ 0 \end{pmatrix}, \quad E_x = \begin{pmatrix} E_{0x} \exp(i\delta_x) \\ 0 \end{pmatrix}. \tag{9.71}$$

2. Linear vertically E_y polarized light, $E_{0x} = 0$, then $E_{0y}^2 = 1$

$$E_y = \begin{pmatrix} 0 \\ 1 \end{pmatrix}, \quad E_y = \begin{pmatrix} 0 \\ E_{0y} \exp(i\delta_y) \end{pmatrix}. \tag{9.72}$$

3. Linear $+45°$ polarized light, $E_{0x} = E_{0y}, \delta = \delta_y - \delta_x = 0°$, then $2E_{0x}^2 = 1 \Rightarrow E_{0x} = E_{0y} = 1/\sqrt{2}$

$$E_+ = \frac{1}{\sqrt{2}} \begin{pmatrix} 1 \\ 1 \end{pmatrix}, \quad E_+ = \begin{pmatrix} E_{0x} \exp(i\delta_x) \\ E_{0x} \exp(i\delta_x) \end{pmatrix}. \tag{9.73}$$

The same result can be obtained by superposition: $E_+ = E_x + E_y$, where $E_{0x} = E_{0y}$ and $\delta_x = \delta_y$.

4. Linear$-45°$ polarized light, $E_{0x} = -E_{0y}$ or $E_{0x} = E_{0y}$ and $\delta_y - \delta_x = \pi$, then $2E_{0x}^2 = 1$,

$$E_- = \frac{1}{\sqrt{2}} \begin{pmatrix} 1 \\ -1 \end{pmatrix}, \quad E_- = \begin{pmatrix} E_{0x} \exp(i\delta_x) \\ E_{0x} \exp\left[i(\delta_x + \pi)\right] \end{pmatrix} = \begin{pmatrix} E_{0x} \exp(i\delta_x) \\ -E_{0x} \exp(i\delta_x) \end{pmatrix}. \tag{9.74}$$

observe that $E_- = E_x + E_y$ where $E_{0x} = -E_{0y}$ and $\delta_x = \delta_y$ or equivalently, $E_{0x} = E_{0y}$ and $\delta_y = \delta_x + \pi$.

5. Circular right-hand E_r polarized light, $E_{0x} = E_{0y}$ and $\delta = \delta_y - \delta_x = \pi/2$, then $2E_{0x}^2 = 1$,

$$E_r = \frac{1}{\sqrt{2}} \begin{pmatrix} 1 \\ i \end{pmatrix}, \quad E_r = \begin{pmatrix} E_{0x} \exp(i\delta_x) \\ E_{0x} \exp\left[i(\delta_x + \pi/2)\right] \end{pmatrix}. \tag{9.75}$$

Note that $E_r = E_x + E_y$, where $E_{0x} = E_{0y}$ and $\delta_y = \delta_x + \pi/2$.

6. Circular left-hand E_l polarized light, $E_{0x} = E_{0y}$ and $\delta = \delta_y - \delta_x = -\pi/2$, then $2E_{0x}^2 = 1$

$$E_l = \frac{1}{\sqrt{2}} \begin{pmatrix} 1 \\ -i \end{pmatrix}, \quad E_l = \begin{pmatrix} E_{0x} \exp(i\delta_x) \\ E_{0x} \exp\left[i(\delta_x - \pi/2)\right] \end{pmatrix}. \tag{9.76}$$

7. Elliptical E_{el} polarized light, $E_{el} = E_x + E_y$, where in general $E_{0x} \neq E_{0y}$ and $\delta = \delta_y - \delta_x \neq \pm\dfrac{\pi}{2}, m\pi$, for m-integer.

$$E_{el} = \begin{pmatrix} E_{0x} \exp(i\delta_x) \\ E_{0y} \exp\left[i\left(\delta_y\right)\right] \end{pmatrix} \tag{9.77}$$

Depending on the phase difference, δ, it can be right- or left-handed rotating.

In general, the Jones parameters can also be superposed to generate any totally polarized state. This is possible because the addition implies superposition of *coherent* polarization states. Note that the Jones matrix obeys the "principle of optical equivalence" [9], which can be adapted to this situation as: two beams that have the same Jones vectors are indistinguishable with respect to intensity, degree of polarization, and polarization form.

9.2.3 Stokes Parameters

Polarization states can be represented directly by measurable quantities, like the intensity. In 1852, G. G. Stokes provided an experimental procedure to measure the polarization state associated to any beam of light, independently of its polarization (totally polarized, partially polarized, or totally non-polarized) or spectral nature (monochromatic or not), in terms of intensities [9]:

$$s_0 = I_{0°} + I_{90°} \tag{9.78}$$

$$s_1 = I_{0°} - I_{90°} \tag{9.79}$$

$$s_2 = I_{+45°} - I_{135°} \tag{9.80}$$

$$s_3 = I_r - I_l. \tag{9.81}$$

Where the first three parameters are measured using a linear polarizer (see Section 9.2.6) with transmission axis along $\theta = 0°$, $45°$, $90°$, and $135°$ (or equivalently $-45°$), respectively, measured from the horizontal x-axis to the vertical y-axis, and the circular right- (r) and left-handed (l) polarizations are measured using a polarization analyzer formed by a quarter-wave plate, with fast axis horizontal, and a linear polarizer with transmission axis located at $45°$ and $135°$, respectively (see Section 9.2.6).

Observe that the first Stokes parameter, Equation (9.78), represents the total intensity present in the beam of light under study, a zero value means the absence of light. The second parameter, Equation (9.79), represents the tendency to linear horizontal polarization, s_2 represents the predominance or tendency to linear $+45°$ polarization, and s_3 is interpreted as the tendency to rotation in the clockwise sense (considering positive values). For the natural or unpolarized light, s_0 is the unique non-zero Stokes parameter. A beam where only the last Stokes parameter provides a null value means the absence of elliptical or circular polarization. Other clearly general results can be obtained by analyzing the physical meaning of Equation 9.78 through 9.81.

For totally polarized light and a plane wave solution, the Stokes parameters are represented in terms of the orthogonal component amplitudes (E_x, E_y) and their phase difference, $\delta = \delta_y - \delta_x$. Interpreting the Stokes procedure in terms of the complex electric field or in terms of amplitudes and phase difference, the following specific relationships are obtained:

$$s_0 = E_x E_x^* + E_y E_y^* = E_{0x}^2 + E_{0y}^2 \tag{9.82}$$

$$s_1 = E_x E_x^* - E_y E_y^* = E_{0x}^2 - E_{0y}^2 \tag{9.83}$$

$$s_2 = E_x E_y^* + E_y E_x^* = 2 E_{0x} E_{0y} \cos\delta \tag{9.84}$$

$$s_3 = i(E_x E_y^* - E_y E_x^*) = 2 E_{0x} E_{0y} \sin\delta. \tag{9.85}$$

The Stokes parameters can also be represented as ensemble or temporal averages on the orthogonal complex components of the electric field (this is the way a partially or totally unpolarized beam of light can be represented formally):

$$s_0 = \left\langle E_x E_x^* + E_y E_y^* \right\rangle = \left\langle E_x E_x^* \right\rangle + \left\langle E_y E_y^* \right\rangle \tag{9.86}$$

$$s_1 = \left\langle E_x E_x^* - E_y E_y^* \right\rangle = \left\langle E_x E_x^* \right\rangle - \left\langle E_y E_y^* \right\rangle \tag{9.87}$$

$$s_2 = \left\langle E_x E_y^* + E_y E_x^* \right\rangle = \left\langle E_x E_y^* \right\rangle + \left\langle E_y E_x^* \right\rangle = 2 Re \left\langle E_x E_y^* \right\rangle \tag{9.88}$$

$$s_3 = i \left\langle E_x E_y^* - E_y E_x^* \right\rangle = i \left(\left\langle E_x E_y^* \right\rangle - \left\langle E_y E_x^* \right\rangle \right) = -2 Im \left\langle E_x E_y^* \right\rangle. \tag{9.89}$$

The Stokes parameters are better known as Stokes vectors when they are ordered as a column matrix 4×1:

$$S = \begin{pmatrix} s_0 \\ s_1 \\ s_2 \\ s_3 \end{pmatrix} = \begin{pmatrix} s_0 & s_1 & s_2 & s_3 \end{pmatrix}^T, \tag{9.90}$$

where T denotes the transposition operation. It is notorious to know that scientists and engineers accept names like the Stokes vector, even when they know its name is incorrect because it does not have direction, it only has magnitude. Even the name of matrix is not completely correct, because the Stokes vectors do not always fulfill the principle of matrix equivalence, which ensures two matrices are equivalent if their elements have the same values. For example, suppose two beams, both with the same orthogonal amplitudes and the same phase difference. Indeed, they can have the same phase difference δ, even when their respective absolute phases could be different $\delta = \delta_{ax} - \delta_{by} = \delta_{Ax} - \delta_{By}$, where $\delta_{ax} \neq \delta_{Ax}$ and $\delta_{bx} \neq \delta_{By}$. This fact makes one beam different from the other [10].

The superposition of two incoherent beams can be expressed as the sums of their respective Stokes parameters:

$$S^a + S^b = \begin{pmatrix} s_0^a \\ s_1^a \\ s_2^a \\ s_3^a \end{pmatrix} + \begin{pmatrix} s_0^b \\ s_1^b \\ s_2^b \\ s_3^b \end{pmatrix} = \begin{pmatrix} s_0^a + s_0^b \\ s_1^a + s_1^b \\ s_2^a + s_2^b \\ s_3^a + s_3^b \end{pmatrix} = S. \tag{9.91}$$

This rule is applied to any number of beams of light. On the other hand, it is possible, under restricted conditions, to deal with the superposition of coherent beams of light, even when they are represented by their Stokes parameters [11].

Equations 9.82 through 9.85 fulfill the condition:

$$s_1^2 + s_2^2 + s_3^2 = s_0^2, \tag{9.92}$$

In general, the Stokes parameters associated to any homogeneous beam of light, obey the relationship

$$s_1^2 + s_2^2 + s_3^2 \leq s_0^2 . \tag{9.93}$$

In terms of the normalized Stokes parameters, the following special polarizations are easily identified:

1. Unpolarized light: $S_{un} = (1\ 0\ 0\ 0)^T$.
2. Linear horizontal polarization: $S_x = (1\ 1\ 0\ 0)^T$.
3. Linear vertical polarization: $S_y = (1 -1\ 0\ 0)^T$.
4. Linear polarization at 45° with respect to the horizontal: $S_{45°} = (1\ 0\ 1\ 0)^T$.
5. Linear polarization at 135° (−45°) with respect to the horizontal: $S_{135°} = (1\ 0 -1\ 0)^T$.
6. Circular right-hand polarization (looking to the source): $S_r = (1\ 0\ 0\ 1)^T$.
7. Circular left-hand polarization (looking to the source): $S_l = (1\ 0\ 0 -1)^T$.
8. Elliptical right-hand polarization r, if c > 0, (left-hand l, if c < 0), (observing to the source): $S_e = (1\ a\ b\ c)^T$, where $a^2 + b^2 + c^2 = 1$.
9. Absence of light: $S_{dark} = (0\ 0\ 0\ 0)^T$.

where T denotes the transpose matrix operation.

9.2.4 Poincarè Sphere

The Poincarè sphere is a geometrical representation of the polarization states, used to represent all the possible conventional polarization states, where the amplitude and the phase difference of the orthogonal components of the electric field are the required information to represent them. In this order, the totally polarized states are represented as points on the surface of a sphere with unitary radius, while the polarized part of partially polarized states is represented as points inside the surface. The origin represents unpolarized states and points outside its surface have no physical meaning. It can be shown that any polarization state can be obtained as the superposition of a right- and a left-handed polarization state.

Any totally polarized state can be represented as a point $P2\chi, 2E$ on the Poincarè sphere, whose spherical coordinates are the azimuth angle 2χ ($0 \leq 2\chi \leq 2\pi$) and the ellipticity angle 2ε ($-\pi/2 \leq 2\varepsilon \leq \pi/2$), see Figure 9.4.

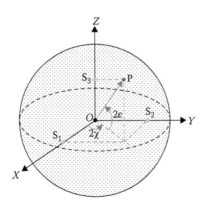

FIGURE 9.4 Representation of polarization states by the Poincarè sphere.

Some typical polarization states are represented according to [6, 8]:

1. Linear horizontal polarization, $2\chi = 0°$, $2\varepsilon = 0°$.
2. Linear vertical polarization, $2\chi = 180°$, $2\varepsilon = 0°$.
3. Linear +45° polarization, $2\chi = 90°$, $2\varepsilon = 0°$.
4. Linear −45° or 135°, $2\chi = 270°$, $2\varepsilon = 0°$.
5. Any other linear polarization state is defined in the equator, 2χ, $2\varepsilon = 0°$.
6. Circular right-handed polarization, $2\chi = 0°$, $2\varepsilon = 90°$.
7. Circular left-handed polarization, $2\chi = 0°$, $2\varepsilon = -90°$.
8. Any elliptical polarization state, 2χ, $2\varepsilon \neq 0°$, where 2e > 0 means right-handed and 2e < 0 means left-handed..

When Equation 9.92 is normalized with respect to the total intensity, the equality takes on the form of the sphere equation with unitary radius, with main axes along the normalized values of the last three normalized parameters of the Stokes vector:

$$\left(s_1^2 + s_2^2 + s_3^2\right)/s_0^2 = 1. \tag{9.94}$$

This representation is named the Poincarè sphere. Considering the normalized Poincarè sphere, with unitary radius $s_0 = 1$, three different possibilities deserve special attention.

1. If $s_3^2 = 0$, this means the beam is not circular right- or left-handed polarized, an infinite number of polarization states within the unitary radius circular cross-section exist (the equator):

$$s_1^2 + s_2^2 \leq 1. \tag{9.95}$$

   All of them are associated to linear polarizations, the linear horizontal totally polarized ($s_1 = 1$, $s_2 = 0$), linear vertical totally polarized ($s_1 = -1$, $s_2 = 0$), linear +45° totally polarized ($s_1 = 0$, $s_2 = 1$), and linear −45° totally polarized ($s_1 = 0$, $s_2 = -1$) being special cases. The solutions associated to the inequality correspond to linear partial polarized states, the remaining to linear total polarized states. The solution $s_1 = s_2 = 0$ corresponds to totally unpolarized light. The Stokes parameters are defined as real quantities; consequently, it does not make sense to consider a complex solution to Equation 9.94.

2. If $s_2^2 = 0$, an infinity number of polarization states are also possible within the circular section:

$$s_1^2 + s_3^2 \leq 1. \tag{9.96}$$

   The linear horizontal ($s_1 = 1$, $s_3 = 0$), the linear vertical ($s_1 = -1$, $s_3 = 0$), the circular right-hand ($s_1 = 0$, $s_3 = 1$), and the circular left-hand ($s_1 = 0$, $s_3 = -1$) totally polarized states are special solutions. With exception to the solutions along the $|s_1| < 1$ axis and the $|s_3| < 1$ axis, the remaining solutions correspond to elliptical polarization states.

3. If $s_1^2 = 0$, all the possible polarizations have the same orthogonal amplitude components $|E_x| = |E_y|$, and the unitary radius circular cross-section is defined by

$$s_2^2 + s_3^2 \leq 1. \tag{9.97}$$

   Once again, with the exception to the pure linear ±45° and circular r- and l-handed polarization states, the remaining solutions correspond to elliptical polarization states.

Note that the Poincarè sphere can be easily related with the Stokes parameters. Considering the normalized Stokes parameters

$$s_1 = r\cos(2\varepsilon)\cos(2\chi) = \cos(2\varepsilon)\cos(2\chi) \tag{9.98}$$

$$s_2 = r\cos(2\varepsilon)\sin(2\chi) = \cos(2\varepsilon)\sin(2\chi) \tag{9.99}$$

$$s_3 = r\sin(2\varepsilon) = \sin(2\varepsilon), \tag{9.100}$$

where $r = 1$ for totally polarized light (states on the surface of the Poincarè sphere).
Conversely [6],

$$\chi = \frac{1}{2}\tan^{-1}\left(\frac{s_2}{s_1}\right) \tag{9.101}$$

$$\varepsilon = \frac{1}{2}\sin^{-1}(s_3) \tag{9.102}$$

$$r = \left(s_1^2 + s_2^2 + s_3^2\right)^{\frac{1}{2}}. \tag{9.103}$$

The degree of polarization, DoP, is a very useful scalar metric used to identify the percentage of polarized light present into a beam of light; it is defined as

$$0 \le \text{DoP} = \left(s_1^2 + s_2^2 + s_3^2\right)^{\frac{1}{2}} / s_0 \le 1, \tag{9.104}$$

where the lower limit means totally unpolarized light, the intermediate values are interpreted as partial polarization, and the upper limit is associated to totally polarized light.

Note that the degree of polarization can also be defined with respect to the Poincarè sphere coordinates (r, χ, ε). From Equations 9.98 through 9.100, considering all the possible polarization states $(0 \le r \le 1)$:

$$0 \le \text{DoP} = r / s_0 \le 1. \tag{9.105}$$

It is a basic assumption to take into account that s_0 is the total intensity present within the beam (polarized plus unpolarized intensities). The lower limit $(r = 0)$ means the beam of light is totally unpolarized, the intermediate values $(0 < r < s_0)$ mean the beam is partially polarized, and the upper limit $(r = s_0)$ implies the beam is totally polarized. Values up to $r > s_0$ do not have physical meaning; when these values are present into an experimental measurement, the experimental setup has probably not been calibrated properly and if these values are present within theoretical results, surely some basic considerations are missing.

Table 9.3 summarizes the different representation of the most common states of polarization discussed here.

9.2.5 Anisotropy as the Origin of the Polarization Phenomena

The propagation of light into an optical medium can be affected by several conditions. If the system depends on the values associated to the wavelength, $n = n(\lambda)$, the medium is chromatic; otherwise, it is achromatic. If the response of the system depends of the amplitude of the electric or magnetic field of the light, it is non-linear; otherwise, it is linear. A medium can be considered basically as homogeneous or non-homogeneous, a spatial characteristic associated to the degree of uniformity. If the optical medium is homogeneous, then n has a constant value at any point; otherwise, it is non-homogeneous. The medium is isotropic if the value of n does not depend on the beam's propagation direction (for example, the vitreous materials, the cubic system or the isomeric crystals); otherwise, if $n = n(\vec{r})$

TABLE 9.3   Equivalent representation of the most common states of polarization

SOP	Algebraic	Jones Vector	Stokes Vector	Poincaré Sphere	Ellipse Polarization (During a Period)
Linear, X	$\hat{x}E_0 e^{i(\omega t - kz)}$	$\begin{pmatrix} 1 \\ 0 \end{pmatrix}$	$I_0 \begin{pmatrix} 1 \\ 1 \\ 0 \\ 0 \end{pmatrix}$	$\chi = 0°, \varepsilon = 0°$	$\longleftrightarrow \bar{E}_x$
Linear, Y	$\hat{y}E_0 e^{i(\omega t - kz)}$	$\begin{pmatrix} 0 \\ 1 \end{pmatrix}$	$I_0 \begin{pmatrix} 1 \\ -1 \\ 0 \\ 0 \end{pmatrix}$	$\chi = 90°, \varepsilon = 0°$	$\downarrow \bar{E}_y$
Linear, +45°	$E_0(\hat{x} + \hat{y})e^{i(\omega t - kz)}$	$\dfrac{1}{\sqrt{2}}\begin{pmatrix} 1 \\ 1 \end{pmatrix}$	$I_0 \begin{pmatrix} 1 \\ 0 \\ 1 \\ 0 \end{pmatrix}$	$\chi = 45°, \varepsilon = 0°$	$\nearrow \bar{E}_{45°}$
Linear, −45°	$E_0(-\hat{x} + \hat{y})e^{i(\omega t - kz)}$	$\dfrac{1}{\sqrt{2}}\begin{pmatrix} 1 \\ -1 \end{pmatrix}$	$I_0 \begin{pmatrix} 1 \\ 0 \\ -1 \\ 0 \end{pmatrix}$	$\chi = 135°, \varepsilon = 0°$	$\searrow \bar{E}_{-45°}$
Circular Right-hand	$E_0(\hat{x} + \hat{y}e^{\frac{i\pi}{2}})e^{i(\omega t - kz)}$	$\dfrac{1}{\sqrt{2}}\begin{pmatrix} 1 \\ i \end{pmatrix}$	$I_0 \begin{pmatrix} 1 \\ 0 \\ 0 \\ 1 \end{pmatrix}$	$\chi = 0°, \varepsilon = 45°$	$\bigcirc \bar{E}_r$
Circular Left-hand	$E_0(\hat{x} + \hat{y}e^{\frac{-i\pi}{2}})e^{i(\omega t - kz)}$	$\dfrac{1}{\sqrt{2}}\begin{pmatrix} 1 \\ -i \end{pmatrix}$	$I_0 \begin{pmatrix} 1 \\ 0 \\ 0 \\ -1 \end{pmatrix}$	$\chi = 0°, \varepsilon = -45°$	$\bigcirc \bar{E}_\varepsilon$
Any elliptical Pol. state	$(\hat{x}E_{0x}e^{i\delta_x} + \hat{y}E_{0x}e^{i\delta_y})e^{i(\omega t - kz)}$	$\begin{pmatrix} \cos\chi \\ \sin\chi^{e^{i\delta}} \end{pmatrix}$	$I_0 \begin{pmatrix} 1 \\ a \\ b \\ c \end{pmatrix}$	$\chi, \varepsilon \neq 0°$	$\bar{E}_{\varepsilon\varepsilon}$
			$a^2 + b^2 + c^2 = 1$		
Un-polarized Light	$\langle \bar{E}_0 e^{i(\omega t - kz + \delta)} \rangle$		$I_0 \begin{pmatrix} 1 \\ 0 \\ 0 \\ 0 \end{pmatrix}$	$r = 0$	$\bar{E}_{UN}$

depends of the propagation direction, the medium is anisotropic or non-isotropic. Two opposite situations are usually the most commonly present in nature: molecules follow a well-defined, geometrical order (crystals) or are located at random positions (amorphous). A fractal is a molecular structure random locally, but it is a crystal when observed from large distances.

Optical media are constituted by matter and matter by molecules and atoms. In crystallography sciences, all the known substances or compounds can be found forming seven crystal lattice systems and 14 Bravais lattices (see Figure 9.5). A crystal lattice is a three-dimensional structure, defined by three axes (a, b, c) and three angles (α, β, γ), according to [12]:

1. Cubic (*s,b,fc*): $a = b = c$, $\alpha = \beta = \gamma = 90°$. Examples: *NaCl*, *Cu*, Zinc-blended structures.
2. Tetragonal (*s, bc, fc*): $a = b \neq c$, $\alpha = \beta = \gamma = 90°$. Examples: SnO_2, TiO_2.
3. Orthorombic (*s, bc, fc, bbc*): $a \neq b \neq c$, $\alpha = \beta = \gamma = 90°$. Examples: KNO_3, $BaSO_4$.
4. Monoclinic (*s, bc*): $a \neq b \neq c$, $\alpha = \gamma = 90°$, $\beta \neq 90°$. Example: Monoclinic sulphur.
5. Rhombohedral (*s*): $a = b = c$, $\alpha = \beta = \gamma \neq 90°$. Example: Calcite $CaCO_3$.
6. Hexagonal (*s*): $a = b \neq c$, $\alpha = \beta = 90°$, $\gamma = 120°$. Examples: *ZnO*, *CdS*, Graphite.
7. Triclinic (*s*): $a \neq b \neq c$, $\alpha \neq \beta \neq \gamma \neq 90°$. Examples: $K_2Cr_2O_7$, H_3BO_3.

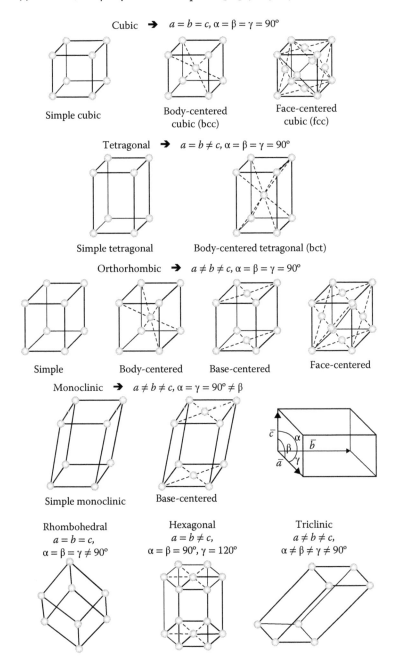

FIGURE 9.5 The 14 Bravais lattice crystalline system.

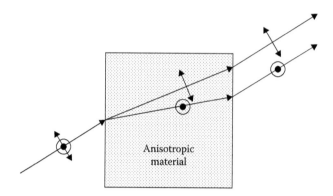

FIGURE 9.6   Light transmitted through an anisotropic material.

Where the Bravais lattices are indicated within the parenthesis as simple or primitive s (atoms or molecules located only at each of the eight vertex of the structure), body-centered bc (located at vertex and center of the lattice), face-centered fc (located at vertex and at center of each face), and base-centered bbc (at vertex and center of the base face).

Of fundamental importance in polarization optics are the non-isotropic optical media, which can be divided into uniaxial or biaxial crystals. The uniaxial crystals have two different refractive indices, originated by the difference in phase velocity along two orthogonal directions. The biaxial crystals have three different refractive indices. In general, a higher phase velocity along an axis (a, b, c) provides the smallest refractive index and this axis is called the fast axis and conversely, in the slow axis the phase velocity is lowest and has the highest refractive index. The birefringence is defined as the difference value between the refractive index along the slow axis and the refractive index along the fast axis:

$$\text{birefringence} \equiv n_{\text{slow}} - n_{\text{fast}} . \tag{9.106}$$

The optical axis into a crystal is defined as the direction along which a transmitted beam does not suffer birefringence. Inside a uniaxial crystal, it is the direction perpendicular to the isotropic plane. All the polarizing optical devices are based in the anisotropic response to the electric or the magnetic field associated to the light (see Figure 9.6).

9.2.6 Jones and Mueller Matrices

The most common and simple interaction of matter and polarized light is provided by the linear response. Mathematically, it is represented as a number or as a matrix, depending if its response is of scalar, vector, or tensorial nature, which corresponds usually to a medium that is spatially homogeneous and isotropic, spatially homogeneous and non-isotropic, and spatially non-homogeneous and non-isotropic, respectively. If the optical system does not depolarize the light, its linear response can be described by the Jones matricial formalism. On the other hand, the Mueller matrix formalism [13–20] is the most powerful tool to describe any non-polarizing, partially depolarizing or totally depolarizing optical media. The Mueller matrix formalism is applied to describe the linear interaction of intensity with matter from a phenomenological point of view. These formulations opened the possibility to operate numerically and computationally between the theoretical foundations and the physically measurable quantities, giving rise to an impressive development in optics, electro-optics, photonics, classical electrodynamics, classical statistical mechanics, and quantum electrodynamics, among other relevant areas. The Mueller-Jones formalism is a natural bridge between the theoretical, microscopic entities and the phenomenological world. It could be asserted that the Jones and the Mueller matrix formalism are connected one with the other in the same way as classical statistical mechanics are connected with thermodynamics.

Consider the linear response of an optical system to an incident beam of light E, with orthogonal complex components E_x, E_y:

$$E_x^o = j_{xx} E_x + j_{xy} E_y \tag{9.107}$$

$$E_y^o = j_{yx} E_x + j_{yy} E_y, \tag{9.108}$$

where E_x^o, E_y^o are the output orthogonal complex components of the electric field E^o and the $j's$ are the Jones complex coefficients (the linear response of the medium). Equations 9.107 and 9.108 can be expressed directly in terms of matrix products, accordingly to

$$\begin{pmatrix} E_x^o \\ E_y^o \end{pmatrix} = \begin{pmatrix} j_{xx} & j_{xy} \\ j_{yx} & j_{yy} \end{pmatrix} \begin{pmatrix} E_x \\ E_y \end{pmatrix} \Rightarrow E^0 = JE, \tag{9.109}$$

where J is the Jones matrix and E, E^o are represented by their Jones parameters. Note the Jones formalism requires the existence of a well-defined electric field (amplitude and phase); the unpolarized light does not fulfill this physical condition. This is the reason why the Jones formalism applies only to non-depolarizing process. It works fine with or without absorption or attenuation in the linear response.

Not all the complex 2×2 matrices are Jones matrices. A deterministic Jones matrix must fulfill the following conditions, in order to be physically realizable [8]:

$$0 \leq Tr\left[J\left(J^T\right)^* \right] \leq 2 \tag{9.110}$$

$$|Det\ J|^2 \leq 1, \tag{9.111}$$

where Tr and Det denotes the trace and the determinant operations, respectively. The Jones matrix can also be written as a linear combination of the Pauli spin σ matrices [13]:

$$J = \sum_{m=0}^{3} c_m \sigma_m, \tag{9.112}$$

where

$$\sigma_0 = \begin{pmatrix} 1 & 0 \\ 0 & 1 \end{pmatrix}, \sigma_1 = \begin{pmatrix} 1 & 0 \\ 0 & -1 \end{pmatrix}, \sigma_2 = \begin{pmatrix} 0 & 1 \\ 1 & 0 \end{pmatrix}, \sigma_3 = \begin{pmatrix} 0 & -i \\ i & 0 \end{pmatrix}; c_m = \frac{Tr\left(J\sigma_m\right)}{2}. \tag{9.113}$$

A note of caution: there are different Pauli spin basis, where their physical meaning is related with the reference system employed to describe the polarization of light.

The Mueller matrix M is the linear response of a given optical medium to the intensity associated to polarization, represented by a Stokes vector S. It is expressed as a 4×4 matrix of real elements [5]:

$$S^o = MS \Rightarrow \begin{pmatrix} s_0^o \\ s_1^o \\ s_2^o \\ s_3^o \end{pmatrix} = \begin{bmatrix} m_{00} & m_{01} & m_{02} & m_{03} \\ m_{10} & m_{11} & m_{12} & m_{13} \\ m_{20} & m_{21} & m_{22} & m_{23} \\ m_{30} & m_{31} & m_{32} & m_{33} \end{bmatrix} \begin{pmatrix} s_0 \\ s_1 \\ s_2 \\ s_3 \end{pmatrix}. \tag{9.114}$$

Each m_{ij} Mueller element represents the specific response to certain incident polarizations [14] and each element is associated to some specific optical characteristic as non-depolarization, depolarization, linear or circular birefringence, linear or circular dichroism, linear or circular attenuation, among other polarimetric parameters [15–20]. A Mueller matrix contains all the polarimetric information associated to the optical medium under study. Table 9.4 summarizes some of the usual polarimetric parameters derived from the Mueller matrix.

Observe that the Jones matrix is formulated in terms of theoretical quantities, derived either from classical electrodynamics and classical statistical mechanics or can be formulated also from quantum statistical electrodynamics. On the other hand, the Mueller matrix is formulated in terms of phenomenological, experimentally measurable quantities [5]. There exists one connection between both matricial formulations, named the Mueller-Jones matrix.

TABLE 9.4   Polarization-Dependent Parameters Derived from the Mueller Matrix. Where M_r Is the Mueller Matrix Associated to the Retardance Contribution, Obtained from the Polar Decomposition Method [17].

Depolarization index, DI(M)	$0 \leq DI(M) = \left\{ \sum_{j,k=0}^{3} m_{jk}^2 - m_{00}^2 \right\}^{1/2} / \sqrt{3} m_{00} \leq 1$
Degree of polarization, DoP(M,S)	$0 \leq DoP(M,S) = \dfrac{\sqrt{(s_1^o)^2 + (s_2^o)^2 + (s_3^o)^2}}{s_0^o} = \left[\dfrac{\sum_{j=1}^{3}(m_{j0}s_0^i + m_{j1}s_1^i + m_{j2}s_2^i + m_{j3}s_3^i)^2}{m_{00}s_0^i + m_{01}s_1^i + m_{02}s_2^i + m_{03}s_3^i} \right]^{\frac{1}{2}} \leq 1$
Anisotropic degree of depolarization, Add	$0 \leq Add = \dfrac{(DoP)_{Max}^o - (DoP)_{min}^o}{(DoP)_{Max}^o + (DoP)_{min}^o} \leq 1$
Total diattenuation, D(M)	$0 \leq D(M) = \sqrt{m_{01}^2 + m_{02}^2 + m_{03}^2} / m_{00} \leq 1$
Linear diattenuation, LD	$0 \leq LD(M) = \sqrt{m_{01}^2 + m_{02}^2 + m_{03}^2} / m_{00} \leq 1$
Circular diattenuation, CD	$0 \leq CD(M) = \sqrt{m_{03}^2} / m_{00} \leq 1$
Total polarizance, P(M)	$0 \leq P(M) = \sqrt{m_{10}^2 + m_{20}^2 + m_{30}^2} / m_{00} \leq 1$
Linear polarizance, LP	$0 \leq LP(M) = \sqrt{m_{10}^2 + m_{20}^2 + m_{30}^2} / m_{00} \leq 1$
Circular polarizance, CP	$0 \leq CP(M) = \sqrt{m_{30}^2} / m_{00} \leq 1$
Q(M) metric, Q(M)	$0 \leq Q(M) = \dfrac{\sum_{j,k=0}^{3} m_{jk}^2}{\sum_{k=0}^{3} m_{0k}^2} = \dfrac{3[DI(M)]^2 - [D(M)]^2}{1 + [D(M)]^2} = \dfrac{\left\{ \sum_{j,k=1}^{3} m_{jk}^2 \right\} / m_{00}^2 + [P(M)]^2}{1 + [D(M)]^2} \leq 3$
Gil-Bernabeu theorem, TGB	$Tr(M^t M) = 4m_{00}^2$
Polarization dependent loss, PDL	$PDL = 10\log(T\max / T\min) = 10\log \left[\dfrac{m_{00} + \left(m_{01}^2 + m_{02}^2 + m_{03}^2\right)^{1/2}}{m_{00} - \left(m_{01}^2 + m_{02}^2 + m_{03}^2\right)^{1/2}} \right]$
Gain, G	$0 \leq g = \dfrac{s_0^o}{s_0^i} = \dfrac{m_{00}s_0^i + m_{01}s_1^i + m_{02}s_2^i + m_{03}s_3^i}{s_0^i} \leq 1$
Total retardance, R	$R = \cos^{-1}\left(\dfrac{Tr(M_R - 1)}{2} \right)$
Linear retardance, δ	$\delta = \left(\sqrt{\left[M_R(1,1) + M_R(2,2)\right]^2 + \left[M_R(2,1) - M_R(1,2)\right]^2} - 1 \right)$
Circular retardance, Cr	$Cr = \dfrac{1}{2}\tan^{-1}\left\{ \dfrac{M_R(2,1) - M_R(1,2)}{M_R(1,1) + M_R(2,2)} \right\}$

If a system does not depolarize, its linear response to the polarization state of the light can be described either by the Jones (J) or by the Mueller matrix (M), which takes on the name of Mueller-Jones matrix M_J. Given a J matrix, its M_J can be derived from [21,22]

$$M_J = A\left(J \otimes J^*\right)A^{-1}, \ A^{-1} = \frac{1}{2}\left(A^T\right)^*, \tag{9.115}$$

where $\otimes$ indicates a Kronecker matricial or direct product and

$$A = \begin{bmatrix} 1 & 0 & 0 & 1 \\ 1 & 0 & 0 & -1 \\ 0 & 1 & 1 & 0 \\ 0 & i & -i & 0 \end{bmatrix}. \tag{9.116}$$

A is the transformation matrix when the observer describes the polarization looking to the source. If the observer looks to the propagating direction, then A^* is the appropriate transformation matrix to be used into Equation 9.115.

Note the definition of the Stokes parameters can be expressed in terms of the matrix A:

$$S = \begin{pmatrix} s_0 \\ s_1 \\ s_2 \\ s_3 \end{pmatrix} = \begin{bmatrix} 1 & 0 & 0 & 1 \\ 1 & 0 & 0 & -1 \\ 0 & 1 & 1 & 0 \\ 0 & i & -i & 0 \end{bmatrix} \begin{pmatrix} E_x E_x^* \\ E_x E_y^* \\ E_y E_x^* \\ E_y E_y^* \end{pmatrix}, \tag{9.117}$$

where the column matrix contains the elements of the Φ coherence matrix, which is employed to describe partially polarized optical fields [8]:

$$\Phi = \begin{bmatrix} E_x E_x^* & E_x E_y^* \\ E_y E_x^* & E_y E_y^* \end{bmatrix}. \tag{9.118}$$

Note the degree of polarization can also be defined as [8]

$$0 \leq \mathrm{DoP} = \frac{I_{\mathrm{Pol}}}{I_{\mathrm{Unpol}}} = \left[1 - \frac{4\mathrm{Det}\,\Phi}{\left(Tr\,\Phi\right)^{\frac{1}{2}}}\right]^{\frac{1}{2}} \leq 1. \tag{9.119}$$

The operation of the optical polarization devices like linear polarizers, wave-phase or wave plate retarders, rotators, among many other optical instruments, can be represented by Jones or by Mueller matrices. The propagation of incident polarized light (E, S) through a given experimental arrangement composed of n optical linear devices, can be easily handled by using the Jones or the Mueller matrix formalisms, according to its configuration [5]

$$E^0 = J_n J_{n-1} \ldots J_2 J_1 E = J_{\mathrm{equivalent}} E = JE \tag{9.120}$$

$$S^0 = M_n M_{n-1} \ldots M_2 M_1 S = M_{\mathrm{equivalent}} S = MS, \tag{9.121}$$

and the successive changes in polarization states through the cascade optical system can be represented as points that describe paths on the Poincarè sphere [19], whenever a *series configuration* is employed. Observe that a series configuration means the incident radiation (measured as intensity or also as irradiance) goes through all the optical devices between the source and the photo-detector. The Equations 9.120 and 9.121 define the matrix multiplication for the series configuration (Figure 9.2).

On the other hand, if the incident radiation is divided to generate two equal beams E_i (usually employing a 50:50 beamsplitter), transmitted through devices J_1 and J_2, respectively, and if the emerging radiation E^o can be added coherently, the effect is the same as if the radiation had been transmitted through the single optical device J, defined by (see the parallel configuration, Figure 9.3)

$$E_1^o = J_1 E_i,\ E_2^o = J_2 E_i,\ if\ E^o = E_1^o + E_2^o \Rightarrow E^o = J_1 E_i + J_2 E_i = (J_1 + J_2)E_i = JE_i\ \therefore\ J = J_1 + J_2. \tag{9.122}$$

Observe that this rule defines the matrix addition and also applies for the Mueller matrix addition, even when the interfering beams, represented as Stokes vectors, were able to become from independent sources (as a consequence, an incoherent superposition should be generated)

$$S^o = S_1^o + S_2^o;\ S_1^o = M_1 S_i,\ S_2^o = M_2 S_i \Rightarrow S^o = M_1 S_i + M_2 S_i = (M_1 + M_2)S_i \equiv MS_i\ \therefore\ M = M_1 + M_2. \tag{9.123}$$

The following are examples of typical optical polarization devices represented by Jones and by Mueller matrices.

1. Linear polarizer. It is an anisotropic optical element that attenuates non-uniformly the orthogonal components of an optical beam. Consider there is total transmission along the x-axis (defined as the transmission axis) and total attenuation is along the y-axis (the extinction axis). An ideal linear polarizer can be formed by dichroic materials (natural as the tourmaline gemstone or artificial like those based in large metallic absorbing molecular chains or cylinders) or can be made from birefringent materials (natural as the calcite or artificial as a liquid crystal display, LCD). The Jones and the Mueller matrices of a linear polarizer with transmission axis along the x-axis, rotated by the angle θ are given, respectively, by [10,6]

$$J_P(\theta) = \begin{pmatrix} \cos^2\theta & \sin\theta\cos\theta \\ \sin\theta\cos\theta & \sin^2\theta \end{pmatrix} \tag{9.124}$$

$$M_P(2\theta) = \frac{1}{2} \begin{bmatrix} 1 & \cos 2\theta & \sin 2\theta & 0 \\ \cos 2\theta & \cos^2 2\theta & \sin 2\theta\cos 2\theta & 0 \\ \sin 2\theta & \sin 2\theta\cos 2\theta & \sin^2 2\theta & 0 \\ 0 & 0 & 0 & 0 \end{bmatrix}. \tag{9.125}$$

A linear polarizer with vertical transmission axis is obtained when $\theta = 90°$; if $\theta = \pm 45°$, a linear polarizer with transmission axis along the $\pm 45°$ is obtained. The linear horizontal polarizer is obtained when $\theta = 0°$

$$J_H(\theta = 0°) = \begin{pmatrix} 1 & 0 \\ 0 & 0 \end{pmatrix},\ M_H(2\theta = 0°) = \frac{1}{2} \begin{bmatrix} 1 & 1 & 0 & 0 \\ 1 & 1 & 0 & 0 \\ 0 & 0 & 0 & 0 \\ 0 & 0 & 0 & 0 \end{bmatrix}. \tag{9.126}$$

2. Phase shift retarder. A phase-shift device is also known as a retarder or compensator. It generates a phase shift δ between the orthogonal components E_x, E_y, the fast axis is located along the x-axis, rotated by the angle θ [10,6]:

$$
J_R(\theta, \delta) = \begin{pmatrix} \cos\dfrac{\delta}{2} + i\sin\dfrac{\delta}{2}\cos 2\theta & i\sin\dfrac{\delta}{2}\sin 2\theta \\[2mm] i\sin\dfrac{\delta}{2}\sin 2\theta & \cos\dfrac{\delta}{2} - i\sin\dfrac{\delta}{2}\cos 2\theta \end{pmatrix} \tag{9.127}
$$

$$
M_R(\theta, \delta) = \begin{pmatrix} 1 & 0 & 0 & 0 \\ 0 & \cos^2 2\theta + \cos\delta\sin^2 2\theta & (1-\cos\delta)\sin 2\theta\cos 2\theta & -\sin\delta\sin 2\theta \\ 0 & (1-\cos\delta)\sin 2\theta\cos 2\theta & \sin^2 2\theta + \cos\delta\cos^2 2\theta & \sin\delta\cos 2\theta \\ 0 & \sin\delta\sin 2\theta & -\sin\delta\cos 2\theta & \cos\delta \end{pmatrix}. \tag{9.128}
$$

When $\delta = \pi$, a phase shift is associated to a half-wave plate ($\lambda/2$), and a quarter-wave plate ($\lambda/4$) is obtained when $\delta = \pi/2$, both with x-fast axis rotated by an angle θ. If $\theta = 0°$ (fast axis horizontal), the Jones $J_{\lambda/2}$, $J_{\lambda/4}$ and the Mueller $M_{\lambda/2}$, $M_{\lambda/4}$ matrices are obtained as

$$
J_{\lambda/2}(\theta=0°, \delta=\pi) = \begin{pmatrix} i & 0 \\ 0 & -i \end{pmatrix}, \quad M_{\lambda/2}(\theta=0°, \delta=\pi) = \begin{pmatrix} 1 & 0 & 0 & 0 \\ 0 & 1 & 0 & 0 \\ 0 & 0 & -1 & 0 \\ 0 & 0 & 0 & -1 \end{pmatrix} \tag{9.129}
$$

$$
J_{\lambda/4}(\theta=0°, \delta=\pi/2) = \frac{1}{\sqrt{2}}\begin{pmatrix} 1+i & 0 \\ 0 & 1-i \end{pmatrix}, \quad M_{\lambda/4}(\theta=0°, \delta=\pi/2) = \begin{pmatrix} 1 & 0 & 0 & 0 \\ 0 & 1 & 0 & 0 \\ 0 & 0 & 0 & 1 \\ 0 & 0 & -1 & 0 \end{pmatrix}. \tag{9.130}
$$

A simple way to generate a phase shift to a given incident polarization state can be realized by properly using a metallic cylinder, generating with this method an angularly-resolved wave-retarder [23].

3. Rotator. It is an optical device that generates a rotation on the orthogonal components E_x, E_y through an angle γ

$$
J_{ROT}(\gamma) = \begin{pmatrix} \cos\gamma & \sin\gamma \\ -\sin\gamma & \cos\gamma \end{pmatrix}, \quad M_{ROT}(2\gamma) = \begin{pmatrix} 1 & 0 & 0 & 0 \\ 0 & \cos 2\gamma & \sin 2\gamma & 0 \\ 0 & -\sin 2\gamma & \cos 2\gamma & 0 \\ 0 & 0 & 0 & 1 \end{pmatrix}. \tag{9.131}
$$

4. Circular polarizers. A circular polarizer is a physical device formed by a horizontal linear polarizer ($\theta=0°$), located between a pair of quarter-wave retarders with fast axis oriented at $\theta = \pm 45°$,

forming a relative angle of 90° one with the other. Converts incident light into circular right-
(J_r, M_r) or left-handed (J_l, M_l) polarization states, respectively, accordingly to the following:

$$J_r \equiv J_R\left(\theta=45°,\delta=\frac{\pi}{2}\right)J_H\left(\theta=0°\right)J_R\left(\theta=-45°,\delta=\frac{\pi}{2}\right)=\frac{1}{2}\begin{bmatrix} 1 & i \\ -i & 1 \end{bmatrix} \tag{9.132}$$

$$M_r \equiv M_R\left(\theta=45°,\delta=\frac{\pi}{2}\right)M_H\left(\theta=0°\right)M_R\left(\theta=-45°,\delta=\frac{\pi}{2}\right)=\begin{bmatrix} 1 & 0 & 0 & 1 \\ 0 & 0 & 0 & 0 \\ 0 & 0 & 0 & 0 \\ 1 & 0 & 0 & 1 \end{bmatrix} \tag{9.133}$$

$$J_l \equiv J_R\left(\theta=-45°,\delta=\frac{\pi}{2}\right)J_H\left(\theta=0°\right)J_R\left(\theta=45°,\delta=\frac{\pi}{2}\right)=\frac{1}{2}\begin{bmatrix} 1 & -i \\ i & 1 \end{bmatrix} \tag{9.134}$$

$$M_l \equiv M_R\left(\theta=-45°,\delta=\frac{\pi}{2}\right)M_H\left(\theta=0°\right)M_R\left(\theta=45°,\delta=\frac{\pi}{2}\right)=\begin{bmatrix} 1 & 0 & 0 & -1 \\ 0 & 0 & 0 & 0 \\ 0 & 0 & 0 & 0 \\ -1 & 0 & 0 & 1 \end{bmatrix}. \tag{9.135}$$

It can be shown that the knowledge of the Jones matrix of a given optical device and the application
of the operational definition of the Mueller-Jones matrix, Equation 9.115, leads to the corresponding
Mueller matrix. It is a nice exercise to reproduce the Mueller matrices shown previously, Equations 9.124
through 9.135.

Observe that not all the real 4×4 matrices are Mueller or Mueller-Jones matrices; they must fulfill
several mathematical criteria in order to get physical sense [8, 15, 18–20, 24–26].

9.2.7 Natural and Partially Polarized Light

Light emitted from the Sun is named natural and it is unpolarized light. Light generated under chemical
combustion is unpolarized. In general, the light emitted by biological tissues under fluorescence
phenomena is usually linearly polarized (as a matter of fact, this effect is used as an auxiliary tool to
determine the molecular orientation of proteins under ultraviolet illumination). Man-made lamps are
unpolarized, partially polarized or totally polarized light sources. The partially polarized light is made
of a combination of unpolarized and polarized light; in fact, almost all the light around us is partially
polarized, even when the main source is the sunlight. There are two main reasons responsible of this
behavior: (1) the light scattered by particles in the atmosphere at a plane perpendicular to the sun's
direction becomes partly polarized linearly and (2) the light reflected by the horizontal surfaces around
us is linear horizontally polarized.

A simple way to determine the degree of linear polarization in a beam can be realized with the help
of a linear polarizer and a photo-detector, applying the relationship

$$0 \leq \mathrm{DoP}_L = \frac{I_{\mathrm{Max}}-I_{\min}}{I_{\mathrm{Max}}+I_{\min}} \leq 1, \tag{9.136}$$

where I_{Max}, $I_{\min}$ are the maximum and the minimum values detected when the linear polarizer is rotated
azimuthally. The minimum value ($\mathrm{DoP}_L = 0$) means the light is natural or unpolarized (the light transmitted through a linear polarizer is reduced uniformly in intensity with respect to its incident value

due to absorption by a dichroic linear polarizer), the intermediate values are associated to partial linear polarization, and the maximum value ($\text{DoP}_L = 1$) implies the light is totally linearly polarized. The degree of linear polarization from the skylight has $15\% \leq \text{DoP}_L \leq 75\%$ while the man-made constructions have $80\% \leq \text{DoP}_L \leq 95\%$, depending on the weather conditions [27].

But not all the polarized light is linear. This is why the previous relation applies only to test the presence of linear polarizations. Note that a circular polarization is undistinguishable from natural or unpolarized light if the Equation 9.136 is used to test it (its value would be zero because the intensity transmitted through a linear polarizer is always constant, $I_{\text{Max}} = I_{\text{min}} = I_0$); instead, Equation 9.104 must be applied to analyze the light considering the existence of partial and generalized polarization states properly. In the following sections, some general methods are described to measure or analyze any polarization state.

In general, the total intensity I_T present within a beam of light partially polarized is made of a part of polarized light (any polarization state, I_{pol}) and a complementary part of unpolarized light I_{un},

$$I_T = I_{\text{pol}} + I_{\text{un}} \Rightarrow S_T = S_{\text{pol}} + S_{\text{un}}, \tag{9.137}$$

where

$$S_{\text{pol}} = \begin{pmatrix} \sqrt{s_1^2 + s_2^2 + s_3^2} \\ s_1 \\ s_2 \\ s_3 \end{pmatrix}, \ S_{\text{un}} = \begin{pmatrix} s_0 - \sqrt{s_1^2 + s_2^2 + s_3^2} \\ 0 \\ 0 \\ 0 \end{pmatrix}. \tag{9.138}$$

This relation automatically defines the percentage of unpolarized light present into the beam.

Other logical options can be proposed. Observe the partially polarized light s_0^2 can also be written as a normalized relationship relating the polarized s_p^2 and the unpolarized s_u^2 components, according to

$$s_0^2 \geq s_1^2 + s_2^2 + s_3^2 \equiv s_p^2 \Rightarrow s_0^2 = s_p^2 + s_u^2, \ s_p = \sqrt{s_1^2 + s_2^2 + s_3^2}, \tag{9.139}$$

where the normalized components make the physically realizable limits clear,

$$0 \leq s_0^2 \leq 1, \begin{cases} s_0^2 = 0 \Rightarrow \text{Null condition} \\ s_0^2 = 1 = s_u^2 \ or \ s_p^2 = 0 \Rightarrow \text{Total unpolarization} \\ s_0^2 = 1 = s_p^2 \ or \ s_u^2 = 0 \Rightarrow \text{Total polarization} \\ 0 < s_0^2 = s_p^2 + s_u^2 < 1, \ s_p^2 \neq 0 \neq s_u^2 \Rightarrow \text{Partial polarization} \end{cases} \tag{9.140}$$

9.3 Animal and Human Sensitivity to Polarized Light

The polarized light has always been present before the human presence on Earth, being that the sunlight scattered from the atmosphere and after reflected by smooth surfaces the main, natural source of linear horizontal polarization. It has been employed by many animal species as localization, feeding, reproduction, defense, and by navigation purposes. The sensitivity to colors and polarization in animal vision in general is a theme broadly studied and recent notable reports are available [28].

Some of the animals that see and use the linear polarized light are bees, dragonflies, horseflies, fruit flies, houseflies, deerflies, tabanid flies, mayflies, libellulas, crickets, ants, wasps, desert locusts, Monarch butterflies, and also aquatic insects and amphibians. The circular polarization is used by scarab beetles, by crustaceans and cephalopods for defense and for reproductive selection as well [28]. The linear horizontal polarization is the most common in the surface of the Earth and also in the water bodies and oceans.

On the other hand, there exists evidence that some pre-columbian cultures, such as the Inca in Perú and probably the Olmecas in México, had directly observed the "Haidinger brush" or the "Malta cross" polarization effect when the reflection of light close to the Brewster angle (60° measured from the vertical direction) occurred on mirrors or in quiet lakes water [29].

The first study related with the human vision response to linear polarized light was reported by Wilhem Karl von Haidinger, in 1844, which is known as the "Haidinger brush" [28] or the "Maltese Cross" effect [30], which can be observed if a source of white light is observed through a linear polarizer. For example, observe a white region on a liquid crystal display (which provides linear polarization along the screen's diagonal) and rotate it with respect to the line of vision; you will observe an image formed by a blue line and a yellow line, at 90 degrees one from the other, forming the effect mentioned previously. This is an induced false color effect originated in the human macula, which has the properties of a radial analyzer, with peak absorption at 460 nm (blue color, complimentary of the yellow color). As a matter of fact, the form and size of this image could be used to test the macula or the retine health of the human eye because experience has shown, different eyes register different quality and image forms. A Mueller matrix model has been reported to describe the Haindinger's brushes effect [31]. The formation of the Maltese cross occurs only under illumination at the visible range (400–800 nm) and disappears for longer wavelengths.

Gundo von Boehm, in 1940, found that the human eye is also able to detect elliptical polarization and also linear rotating polarization, under certain conditions. The effect, known as "Boehm's brush" effect, is most visible when a small source $(1° − 2°)$ emitting polarized light is observed against a dark background in the peripheral visual field $(15° − 20°)$ and is only perceived if the electric vector of the source is rotating from 1 to 2 Hz [28]. The brush formed is perpendicular to the orientation of the electric field and has the same color as the polarized light source; it is weaker (narrower and shorter) at longer wavelengths. The Rayleigh scattering within the retina gives rise to this phenomenon.

9.4 Generation, Detection, and Identification of the Different Polarization States

The possibility to generate any polarized state is as important as the capability to analyze them (detection and identification). There are two basic methods to obtain polarized light: designing a polarizing emitting source like a polarized gas laser, or by polarization converters, where any polarization state can be generated by properly modifying the polarization associated to any kind of light. The experimental setup employed to generate polarization is named Polarization State Generator (PSG). *A desirable PSG should provide polarization states either not correlated or correlated with the source nature, depending on the user requirements.*

In this sense, a Polarization State Analyzer (PSA) is an experimental setup used to measure the intensity and the phase associated to polarized light; in other words, the PSA is an optical instrument that provides all the required information to know with a high degree of confidence to the polarization state associated to a beam of light. In practice, all of this is reduced to the total conversion from any polarization state to total linear or circular polarization, measured by a photo-sensitive device ideally insensitive to polarization. The exception occurs when interferometric techniques are used, which able the correct detection of amplitude and phase [32].

Both, the PSG and PSA employ anisotropic optical elements, like linear polarizers and phase wave retarders. The number and types of optical elements, their configuration and operation, have given rise to different methods and polarimetric techniques [6]. There exist a parallel and symbiotic route followed by the theoretical fundamentals and the development of experimental techniques.

In practice, the basic tools employed to both generate, analyze or detect any polarization state are a couple of linear polarizers, half-wave plates, and quarter-wave plates, in addition to a light source and a photo-detector.

9.4.1 Production and Detection of Linearly Polarized Light: The Malus' Law

The process to convert any unpolarized, partially polarized, or totally polarized light into linear polarization can be realized basically by two different physical mechanisms: selective absorption (dichroism) or selective birefringence. The first mechanism implies some loss of the original energy emerging from the source and an energy dissipation process. The second one is related with polarization selective propagation along two different directions, originating two different linear polarization states, with no energy loss present. To the knowledge of the authors, there does not exist yet a perfect mechanism through which all the incident unpolarized energy can be transformed into a single, purely linear polarized beam.

Tourmaline is a natural gemstone with a high degree of wavelength-dependent absorption along a well-defined direction or axis. This property is named dichroism. It is some of the gemstones with the broader color spectrum (the complimentary colors are the absorbed ones). A black color means the tourmaline sample completely absorbs the visible wavelengths. In practice, usually both, dichroism and birefringence, are being present in some degree. Laboratory experience has demonstrated that any mechanical stress exempted un-properly within optical devices generates spurious birefringence if they are not being mounted adequately.

A linear polarizer is a device, natural or artificially created, which can be used as a generator or as an analyzer of linear polarization. In nature, calcite is a mineral that polarizes linearly the light under transmission along a defined molecular direction, the polarization axis. The first commercially available linear polarizers were built by Edwin Land, in 1931 [33]. They were constituted by a parallel arrangement of electrical conductive molecular chains, where the electric field incident was absorbed preferably along them, but permitting the transmission along its transversal direction, where a lower loss by Joule conduction was generated.

A very important figure of merit to evaluate the quality of a linear polarizer is the Polarization or Transmission Extinction Ratio, ER, which is defined as the transmission ratio of optical powers of polarization parallel I_p to the transmission axis with respect to the polarization perpendicular I_s to the transmission axis (named extinction axis), $ER = I_p / I_s$. Similarly, it is defined as the ratio of the maximum to the minimum intensity, $ER = I_{max} / I_{min}$. The ER determination is of a particular importance in optical fibers [34]. Commercial available linear polarizers can be found as general purpose sheets with transmission extinction ratio (ER) > 4,000:1, or constructed by laminating a polymer polarizer thin film between two high-precision fused silica windows, with high efficiencies from UV-VIS-NIR and antireflection coating options; there are polarizers based in borosilicate glass with aligned particles, and ER > 10,000:1. Calcite has an ER of 100,000:1 for wavelengths inside the interval (200 nm to 2200 nm). Recently, a linear polarizer based in the use of nano-particles with the highest reported ER >1,000,000:1 is available. There are thin film exists with high quality BK7 substrates, high-energy linear polarizers, with coating designs to be used by Nd: YAG lasers, with ER > 100:1.

The Malus' law is usually employed to define a pure linear polarization present in a beam of light. It can be implemented under several equivalent procedures. Some of them can be easily tested.

1. Experimentally. If the source is unpolarized with emission intensity I_0, then use two linear polarizers, P_1, P_2, keeping one of them fixed, while rotating azimuthally the second one (see Figures 9.7 and 9.8). Find the relative angle θ at which the intensity is minimum (ideally zero). This is the cross-condition, which means $\theta = 90°$. Rotate to $\theta = 0°$ and register this value as $I_0 / 2$ (this should be the maximum intensity transmitted because the first linear polarizer ideally reduces by 50% the incident intensity from the unpolarized source). Take equally spaced step angular intervals,

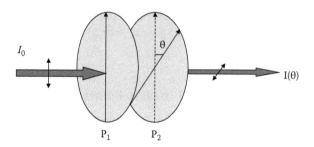

FIGURE 9.7   Experimental determination of the Malus' law.

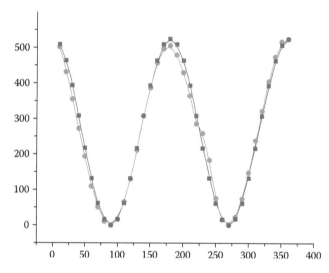

FIGURE 9.8   Typical experimental (circles) and theoretical (squares) results associated to the Malus' law, plotted as power (arbitrary units) vs rotation of transmission axis of polarizer P_2 (measured in degrees).

for example $\Delta\theta = 10°$ from $0° \le \Delta\theta \le 360°$ (also a half-round rotation provides enough information). Order and plot the results (intensity versus θ). The Malus' law should be obtained:

$$I(\theta) = \frac{1}{2} I_0 \cos^2\theta \equiv I \cos^2\theta . \tag{9.141}$$

2. Experimentally. If the source is linearly polarized, use only one polarizer, as analyzer. Find the minimum intensity and repeat the previous procedure (the intensity increases twice because there is not a first linear polarizer that reduces the intensity from the source by a half).

3. Theoretically. Using the Mueller-Stokes formalism. Consider the source emits an arbitrarily polarization with Stokes vector S, a first linear polarizer (generator) fixed with transmission axis horizontal, and a second linear polarizer (analyzer) with transmission axis forming an azimuthal angle θ with the horizontal. Using the Equation 9.125, the output Stokes vector S^o is

$$S^o = M_P(2\theta) M_P(2\theta = 0°) S = \frac{1}{4} \begin{pmatrix} (s_0 + s_1)(1 + \cos 2\theta) \\ (s_0 + s_1)(\cos 2\theta + \cos^2 2\theta) \\ (s_0 + s_1)(\sin 2\theta + \sin 2\theta \cos 2\theta) \\ 0 \end{pmatrix}. \tag{9.142}$$

a. If the source emits unpolarized light with intensity $(s_0 = I_0, s_1 = s_2 = s_3 = 0)$, the intensity registered at the photo-detector is given by the Malus' law

$$I(\theta) \equiv s_0^o = \frac{I_0}{4}(1 + \cos 2\theta) = \frac{I_0}{2}\cos^2\theta. \tag{9.143}$$

b. If the source emits linearly polarized light oriented along the transmission axis of the first ideal linear polarizer, no absorption occurs and the Stokes vector transmitted keeps unchanged $(s_0 = I_0 = s_1, s_2 = s_3 = 0)$. The intensity registered now is twice that registered by Equation 9.143

$$I(\theta) \equiv s_0^o = \frac{I_0}{4}\left[2(1 + \cos 2\theta)\right] = I_0\cos^2\theta. \tag{9.144}$$

4. Theoretically. Malus' law can also be easily derived for an ideal analyzer (a linear polarizer) and a linearly polarized beam of light (source polarized). Suppose a linear polarization is incident with its electric field E oriented at an angle θ with respect to the transmission axis of the analyzer. This means the part of the electric field transmitted $E^o(\theta)$ is given by the projection of the electric field along the transmission axis

$$E^o(\theta) = E\cos\theta. \tag{9.145}$$

Equation 9.145 is also an equivalent of the Malus law for the amplitude of the electric field. In terms on intensity, Equation 9.145 becomes the Malus' law, Equation 9.144.

9.4.2 Polarization by Absorption through Transmission: Types of Polarizers

Color is a wavelength-selective light absorption process characteristic of matter. Absorption can occur along specific directions in matter, which means the existence of spatial anisotropy. Tourmaline is a natural gemstone that exhibits selective absorption along some specific direction, allowing the transmission of specific wavelengths along a well-defined direction; this effect is named selective absorption or dichroism. A human-made linear polarizer based in the dichroic mechanism is an arrangement of a long chain of metallic molecules (separated at distances shorter than the incident wavelength, to avoid diffraction effects) sandwiched between two transparent thin films, generating absorption of the components of the electric field parallel to the chains and allowing the transmission only of the components perpendicular to them. This invention by Edwin Land [33] has given rise to other optical devices, opening the possibility to the generation and use of polarized light into broad areas, such as food and medicine characterization, drug control, and image display, among many other applications.

The left-hand side image in Figure 9.9 shows a schematic representation of the operation of a linear polarizer, P, based in the dichroic effect or selective absorption by transmission (the transmission axis is located, in this case, at angle θ with respect to the vertical reference axis). On the other hand, the right-hand side image in Figure 9.9 shows a dichroic linear polarizer totally obstructing a portion of a liquid crystal display (LCD). This image shows that the LCD emits light linearly polarized (in this case, at +45° with respect to the horizontal). Hint: you now have access to a linear polarizer source, using any LCD-based device!

9.4.3 Polarization by Absorption under Reflection: Polarizing Prism

Light is always partly polarized under reflection by a plane surface; as a matter of fact, the partially linearly horizontal polarization is the most common form of polarization generated by surfaces on the Earth. Light

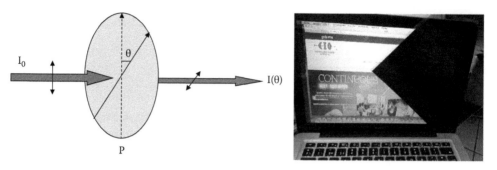

FIGURE 9.9   Dichroic linear polarizer.

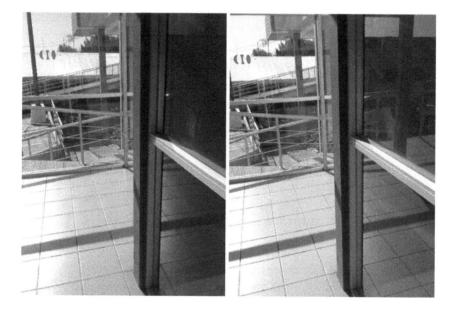

FIGURE 9.10   Images taken through a linear polarizer with transmission axis set horizontally (left-hand side) and vertically (right-hand side), respectively.

can be completely polarized with the electric field oscillating perpendicularly to the incidence plane if the incidence occurs at the polarization or the Brewster angle, which is a geometric condition where the axis of the radiative dipole of the excited molecules coincides with the reflection angle. Phenomenologically, the polarization or Brewster angle θ_B is reached when the following relationship is fulfilled:

$$n_i \sin\theta_i = n_t \sin\theta_t ; \; \theta_i + \theta_t = 90° \text{ and } n_i < n_t \Rightarrow \theta_i \equiv \theta_B \therefore n_t = n_i \tan\theta_B, \tag{9.146}$$

where $n_{i,t}$ represent the refractive indices of the plane interface between two arbitrary media and $\theta_{i,t}$ represent the angles of incidence and transmission, respectively. The knowledge of three parameters allows the determination of the remaining one. If a linear polarization parallel to the incidence plane incides at the Brewster angle, all the energy is transmitted (this condition gives rise to a filter function). Figure 9.10 shows the image obtained when the observation is made through a dichroic, linear polarizer with transmission axis set horizontally (left-hand side) and vertically (right-hand side), respectively. This figure shows that the intensity suppressed is associated to reflected linear vertical polarization (the glass windows polarize the light vertically) and horizontal polarization (the floor polarizes the light horizontally), respectively.

Using the principle of the Brewster angle [35, 36], the spatially resolved refractive index of dielectric samples can be easily determined by using unconventional polarization [37], see Section 9.7. From the refraction or Snell's law, given in the first part of Equation 9.146, another phenomenological extreme behavior can be obtained, giving rise to the total internal reflection effect, which is presented when the incidence occurs at the critical angle θ_t:

$$n_i \sin\theta_i = n_t \sin\theta_t; \; \theta_t = 90° \text{ and } n_i > n_t \Rightarrow \theta_i \equiv \theta_c \therefore n_t = n_i \sin\theta_i . \tag{9.147}$$

At the critical angle, the reflected light is clearly distinguished from all the transmitted light that goes through the interface, but when $\theta_i > \theta_c$, all the transmitted light is reflected in the same direction as the reflected light. The reflected light contains information that permits to distinguish the part normally reflected from the reflected due to the critical condition. This condition generates a phase difference between the components parallel and perpendicular to the plane of incidence.

Based in the previous phenomena the Fresnel rhombs were designed, which consist of transparent rhombs where the light is internally reflected twice at the critical angle associated to the refractive index of the material composition. They are designed usually to induce a phase difference of 45° degrees between both components (E_p, E_s) under each of the two internal reflections, originating a total phase difference of 90°. If a linearly polarized light is incident at an azimuth angle of 45° to the plane of incidence (which means both, parallel and perpendicular components have the same amplitude and phase difference), after two internal reflections, the light will emerge circularly polarized. The relations for the phase change $\delta = \delta_p - \delta_s$ between the orthogonal components of the electric field associated to the incident linear polarization, considering the rhomb is surrounded by air, are given as [38]

$$\tan\left(\delta_p / 2\right) = \left[n\left(n^2 \sin^2\theta - 1\right)^{1/2} \right] / \cos\theta \tag{9.148}$$

$$\tan\left(\delta_s / 2\right) = \left(n^2 \sin^2\theta - 1\right)^{1/2} / n\cos\theta , \tag{9.149}$$

when $\delta = 45°$ implies $n = 1.4966$, a value close to the bosorilicate crown glass [38]. The number of reflections provides the final phase difference desired. A quarter-wave plate is obtained when two internal reflections occur, but four internal reflections generate a total phase difference of 180°, that is, a half-wave plate. One of the most relevant results provided for the Fresnel rhombs is just that the dispersion relation, usually given by Sellmeier relationships for the glasses, opens the possibility to use them almost as achromatic dephasers.

9.4.4 Polarization by Double Refraction

Calcite ($CaCO_3$) is a natural crystal, with a high transparency within the 200 nm to 2200 nm wavelength interval. It has the outstanding possibility to generate total linear polarization from a given totally unpolarized incident beam of light. This effect is manifested as a double image (Figure 9.11, left-hand side), where each one of them has associated a mutually orthogonal linear polarization. Using a linear polarizer as an analyzer, the ordinary image (Figure 9.11, center) and the extraordinary image (Figure 9.11, right-hand side), can be directly obtained. The horizontal line has been depicted as the reference.

There are different commercially available calcite-based linear polarizers, all with an ER > 100,000:1, transmittances up to 95% within the wavelength range from 350—2300 nm uncoated, such as the following designs:

1. Wollaston prism polarizer. Transforms an incident unpolarized beam into two mutually orthogonal linearly polarized beams, separated by angles around 20°. Can be used in lasers with low and medium output power.

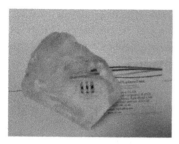
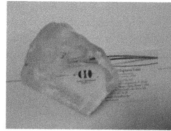

FIGURE 9.11   Double image generated by transmission through natural calcite crystals and the separation of the ordinary and the extraordinary images after a dichroic polarizer is used as an analyzer, respectively.

2. Rochon prism polarizer. Works like a Wollaston prism, with the difference that the output beams are slightly separated from one another.
3. Glan-Thompson polarizer. It is constructed of two optical cemented calcite prisms, where one polarized beam is side-deviated by total internal reflection, while the other is transmitted along basically following the normally incident illuminating beam.
4. Glan-Laser prism polarizers. They are designed similarly to the Glan-Thompson polarizers, but with air-spaced calcite prism pair interface angles, optimized for minimum insertion loss. One or two exit ports allow the rejected polarization component to leave the system, avoiding the absorption of the usually employed high-power lasers.

9.4.5 Polarization by Scattering

When unpolarized light interacts with particles and molecules with a size smaller than the incident wavelengths, light is re-emitted linearly polarized into a plane perpendicular to the incident direction of the illuminating source. As a matter of fact, this is not just a plane, it is just a region defined by the set of perpendicular planes to the direction of incidence. The Earth's atmosphere contains particles, those with a size smaller that the incident light from the Sun and also from the Moon, scatter light generating light totally unpolarized along the illuminating propagating direction $\hat{k}_i$, totally linear polarized within the plane perpendicular to $\hat{k}_i$, and partly polarized in all the remaining directions.

This effect can be simulated easily by transmitting light from a diode or pointing laser through a transparent fishbowl with some whole milk drops mixed into water. The first two images in Figure 9.12 have been taken when a dichroic linear polarizer with the transmission axis set vertically, which implies that the scattered light is linearly polarized horizontally (the first two images are darker than the two remaining ones, which were obtained by using a linear polarizer with transmission axis set horizontally).

9.4.6 Depolarizers

Sometimes it is desirable to use unpolarized light for different applications. There exist devices able to transform a partially or totally polarized beam to quasi-randomly polarized light. This can be done by using reflection, transmission, or scattered properties of polarized light under certain conditions that allow the induction of spatially or temporally quasi-random phase shifts within the polarized beam's cross-section. There are commercially available depolarizer devices based in quartz-wedge designs and also based in liquid crystal displays. Depolarizers usually exhibit some of the following characteristics.

FIGURE 9.12 Light scattered by the atmosphere can be simulated using milk mixed with water. A linear polarizer with transmission axis vertically (left-side) and horizontally (right-side), respectively, showing the light scattered is linearly polarized horizontally along the perpendicular direction to the incident unpolarized light.

- Are optical devices that decrease and even annul the degree of polarization of light.
- Depolarizers induce some degrees of disorder and couple polarized light into unpolarized light. Retardation and/or diattenuation vary rapidly in space.
- Depolarization usually is associated with scattering. For example: rough surfaces such as ground glass; small particles suspended in liquid, as whole milk mixed with water; clouds, optical fibers, biological tissues, etc. Random scattering leads to wide angular spectrum and occurs loss of power per unit angle solid.

Most depolarizers are actually so-called pseudo-depolarizers or polarization mixers.

Typical pseudo-depolarizers work for linear (known) incident polarization and for given detectors; for example, if the detector integrates radiation over certain bandwidth, a depolarizer can be made of crystals that have spectral dependent retardation. Each linearly polarized component of the spectrum is converted into elliptical polarization, but with different elevation angle and ellipticity. When the detector integrates the power over the whole spectrum, the polarization dependence vanishes. Thus, the device plus the detector act as a depolarizing system. Light from depolarizing devices is not natural light. The output state can be reversed to obtain the incident state of polarization (this cannot be done for natural light). Pseudo-depolarizers are often useful when isotropy is desired or polarization sensitivity needs to be removed.

9.5 Phase Retarders

A phase retarder, dephaser, or compensator is a birefringent device that extends a phase delay between the two orthogonal electric field components of an incident beam of light (see Figure 9.13). The birefringence anisotropy implies the existence of two different directions (the fast and the slow axis, respectively), usually one of them perpendicular to the other, where the light travels at two different velocities. In one direction, the original phase suffers a greater phase reduction, while in the other, it is lower with respect to the initial phase of the incident wave. After the beam travels, a given optical path, both components interfere coherently to generate a new polarization state, being the phase difference the mechanism responsible of this transformation (visualized in terms of the Poincarè sphere, the phase change means a longitude transformation).

9.5.1 Fixed Phase Retarders: Half-Wave and Quarter-Wave Retarders

A fixed phase retarder is a device designed, naturally or artificially, to impinge a fixed phase delay among the orthogonal electric field components of a transmitted beam of light. The dephaser can be

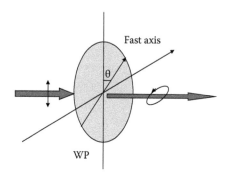

FIGURE 9.13   Wave-plate retarder, with fast axis oriented at angle θ with respect to the vertical axis.

conceptualized due to two basic different mechanisms: one at which one component is delayed and the other advanced and the other in which only one component is delayed or advanced (the resultant polarization is achieved once the affected and the non-disturbed components are added).

There are commercially available fixed wave retarders based in mica, quartz, MgF2, among many other compounds. Those based in quartz can be of zero-order, multiple-order, or achromatic type are designed to generate phase delays of 90°, 180°, 360° and can operate from 250 nm to 1550 nm, according to the desired application with or without antireflection coatings. The zero-order wave retarders are temperature insensitive phase retarders for moderate bandwidth laser diode or tunable laser applications with a temperature coefficient of $10^{-4}\lambda/°C$. The multiple-order wave plates generate a differentially retard on the phase of polarized beams, producing multiple $\lambda/2.\lambda/$ retardation, being more sensitive to temperature, with a temperature coefficient of $1.5\times10^{-3}\lambda/°C$. For instance, a multiple-order wave retarder designed for a wavelength of 532 nm produces the same retardation at 1064 nm. The achromatic wave retarders generate a fixed phase retard to waves within a broad wavelength interval; they are very stable and reliable passive devices.

9.5.2 Variable Phase Retarders

A variable phase retarder is a device capable to induce a desired phase change. Variable retarders have been constructed from quartz wedges (Soleil-Babinet compensator), liquid-crystal displays (nematic molecules oriented through an external DC voltage), quartz or fused silica anisotropy generated by electro-mechanical induced stress (photo-elastic modulators).

The Soleil-Babinet compensator. It is designed with a pair of relatively bulky quartz wedges, where one of them is displaced with respect to the other, generating a variable optical path that induces a variable phase difference between the orthogonal components of a transmitted polarized beam of light and operates within a broad wavelength range (usually from 0 to 1000 nm or from 1000 to 2000 nm).

The Berek's variable wave plate retarder design is based in the principle of total reflection at the critical angle. Depending on the incident polarization, it generates different phase retarders with minimal beam displacement, at lower prices than the Soleil-Babinet compensators.

The Liquid Crystal Display (LCD)-based variable retarders use nematic liquid crystal molecules contained in cells with parallel faces made of high quality transparent glass with Indium Tin Oxide thin films. Applying an external direct voltage changes the orientation of the nematic molecules, generating the desired birefringence along a preferred direction within the cell. The birefringence induces a phase difference between the orthogonal components of a transmitted beam of light, proportional to the DC voltage applied. The Liquid Crystal on Silicon (LCoS) Displays are the basis for the Spatial Light Modulators (SLM), which are some of the most successful devices employed to manipulate the polarized light with control at a pixel-microscale.

Photoelastic modulators operate based on the photoelastic effect, where a mechanical stress induces the desired birefringence. The stress is induced electronically usually in a suitable transparent material like fused silica, through an attached piezoelectric transducer. The bar vibrates at the speed of a longitudinal sound wave, at a fixed, low frequency ultrasound range (20 kHz to 100 kHz). A mechanically characterized soft transparent material could be employed to induce any spatial phase distribution, by mechanical stress (try it with different geometries, the polarization distributions could be amazing and you could get useful dephasing tools).

9.6 Uses of Polarizers, Retarders, and Polarized Light

There are so many applications given to the optical polarization devices and they are so commonly known that any interested reader can find them easily by using any Internet research engine. In this section, only some few of their many applications are mentioned.

9.6.1 Photo-Elastic Analysis

In general, an isotropic substance able to show an anisotropic or birefringent behavior due to external mechanical stress or compression is termed photo-elastic. When the mechanical stress is related linearly with the change in the refractive index of the medium, the polarization techniques can be applied to determine it through the photo-elastic analysis [39].

Consider an initially homogeneous and isotropic substance with refractive index n, measured at a given wavelength λ, and distributed spatially filling a width d. Suppose a mechanical external action generate stresses σ_1, σ_2 along the axis 1, 2, respectively and that as a consequence, the initial refractive index is modified, taking values n_1, n_2 along each axis, respectively. The following relationships can be obtained [40]:

$$n_1 - n = c_1\sigma_1 - c_2\sigma_2 \tag{9.150}$$

$$n_2 - n = c_1\sigma_2 - c_2\sigma_1, \tag{9.151}$$

where c_1, c_2 are the coefficients of the corresponding optical stresses. Resting Equation 9.151 from Equation 9.150,

$$n_1 - n_2 = (c_1 + c_2)(\sigma_1 - \sigma_2) \equiv C(\sigma_1 - \sigma_2), \tag{9.152}$$

where the left-hand side of this Equation represents the birefringence induced into the medium and the right-hand side, the linear relationship with respect to the induced mechanical stress. Equation 9.152 deserves special attention just because it connects an optical response with a mechanical response experimentally. On the other hand, a monochromatic beam of light with wavelength λ, traveling a distance d through a birefringent medium, experiences a relative dephase δ between the waves propagating along each axis, given by [39]

$$\delta = \frac{2\pi d}{\lambda}(n_1 - n_2) = \frac{2\pi d}{\lambda}C(\sigma_1 - \sigma_2). \tag{9.153}$$

From physical optics, it is known the number N of interference fringes is related directly with the relative dephase δ within a full 2π cycle or a wavelength spatial period,

$$N = \delta / 2\pi. \tag{9.154}$$

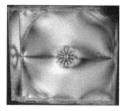

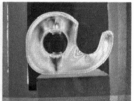

FIGURE 9.14 Photoelastic effect shown by transparent, common objects.

FIGURE 9.15 Authentication of bee honey using polarimetric techniques (GIPYS Lab-CIO).

Substituting Equation 9.154 into Equation 9.153, a directly measurable entity is obtained by using a proper polarimetric arrangement [39]:

$$\sigma_1 - \sigma_2 = N\lambda / dC. \tag{9.155}$$

An easy way to identify the quality of transparent materials, based on the fixed mechanical stress can be done by illuminating the sample under a pair of crossed dichroic linear polarizers, as depicted in Figure 9.14.

9.6.2 Sacharimetry

Optical activity is the name assigned to the effect of rotation of the electric field induced by cholesteric molecules. As a consequence, the presence and concentration of cholesteric molecules in matter can be determined by polarization techniques. The sacharimetry is a commonly used apparatus to measure the concentration of glucose in liquid samples. This technique measures the rotation of the linear polarization by an angle θ after its transmission through the sample under study. At a well-accepted approximation, the concentration C of glucose of sugar in a liquid solution (usually distilled water) is given by $\theta = \alpha LC$, where $\alpha = 11.2° / cm$ and L is the path of the light traveled inside the solution.

Polarimetric techniques have also been employed to authenticate bee honey from adultered or fake samples [41]. Figure 9.15 shows an experimental arrangement to determine the Mueller matrix associated to bee honey.

9.6.3 Determination of the Complex Refractive Index by Ellipsometry

The first attempts to measure the polarization state of light were based on the null intensity condition, where the task was to determine the angular, azimuthal values (ψ, θ) associated to the combination of a quarter-wave plate and a linear polarizer, respectively. One of the pioneering techniques was the ellipsometry, which measures the amplitude ratio and phase difference between the orthogonal linear p- and s-polarization states, respectively. In other words, the ellipsometric techniques measure changes in polarization states under reflection or transmission through an interface that separates the incident (n_i, θ_i) from the transmitted (n_t, θ_t) medium. It is usual to define the linear polarizations with respect to the plane of incidence, which is perpendicular to the boundary that separates both media. The information of interest is contained in the ellipsometric parameter ρ [42]:

$$\rho = \tan \Psi \exp(i\Delta) \equiv \frac{r_p}{r_s} \equiv \frac{\left(\dfrac{E_r}{E_i}\right)_p}{\left(\dfrac{E_r}{E_i}\right)_s}, \tag{9.156}$$

where

$$\tan \Psi = \frac{|r_p|}{|r_s|}, \Delta = \delta_{rp} - \delta_{rs}. \tag{9.157}$$

The lower index r, i indicates reflection and incidence, respectively, and r_p, r_s are the amplitude Fresnel parameters for reflection parallel and perpendicular to the plane of incidence, respectively

$$r_p \equiv \frac{E_{rp}}{E_{ip}} = \frac{n_t \cos\theta_i - n_i \cos\theta_t}{n_t \cos\theta_i + n_i \cos\theta_t} = |r_p| \exp(i\delta_{rp}) \tag{9.158}$$

$$r_s \equiv \frac{E_{rs}}{E_{is}} = \frac{n_i \cos\theta_i - n_t \cos\theta_t}{n_i \cos\theta_i + n_t \cos\theta_t} = |r_s| \exp(i\delta_{rs}). \tag{9.159}$$

A similar approach exists for the case of transmission, where the amplitude Fresnel coefficients must be used as the ellipsometric parameter ρ,

$$t_p \equiv \left(\frac{E_t}{E_i}\right)_p = \frac{2n_i \cos\theta_i}{n_t \cos\theta_i + n_i \cos\theta_t} \tag{9.160}$$

$$t_s \equiv \left(\frac{E_t}{E_i}\right)_s = \frac{2n_i \cos\theta_i}{n_i \cos\theta_i + n_t \cos\theta_t}. \tag{9.161}$$

The amplitude Fresnel coefficients for reflection and transmission or the reflectance and the transmittance fulfills a closed condition when there is nor absorption or scattering:

$$R + T = 1, R = |r|^2, T = |t|^2. \tag{9.162}$$

Once the ellipsometric parameter ρ is determined experimentally, the complex refractive index (n, κ) can be determined, for example for a transparent-to-absorbing interface [42]:

$$n = \mathrm{Re}\left\{\tan\theta_i \left[1 - \frac{4\rho}{(1+\rho)^2} \sin^2\theta_i\right]^{1/2}\right\}, \kappa = \mathrm{Im}\left\{\tan\theta_i \left[1 - \frac{4\rho}{(1+\rho)^2} \sin^2\theta_i\right]^{1/2}\right\}, \tag{9.163}$$

where the real part is associated to the refractive index who provides information about the relative velocity of light in vacuum and the medium, and the imaginary part is associated to the absorption of light by the medium (this absorption can be due to Joule heat if the medium is metallic or to elastic or non-elastic scattering, among many other physical mechanisms associated with dispersion in general).

Ellipsometry is also employed to determine the thickness of parallel film stacks. For example, consider a two-interface system, where the incident medium is air ($n_0 = 1$), the second medium (n_1) has an unknown thickness d, and the third medium (n_2) has an infinite thickness. The reflection on the first interface is named primary, and the main reflection generated from the second interface is the secondary beam reflection. The thickness d can be determined once the following condition is reached:

$$2dn_1 \cos\theta_1 = m\lambda , \qquad (9.164)$$

where m is an integer number.

9.6.4 Determination of the Mueller Matrix by the Ideal Polarizer Arrangement

The Mueller matrix can be determined by different methods [6,14,43–45]. An easy and accessible method to determine the Mueller matrix of a given static system is the Ideal Polarimeter Arrangement, IPA, where the polarization optical elements required are a light source (polarized or unpolarized), two linear polarizers, two half-wave retarders, two quarter-wave retarders, and a photo-detector (see Figure 9.16).

A very easy and experimentally intuitive way to generate and analyze any polarization state is given by the following procedure:

1. Define the light source. For the most general case, suppose it emits unpolarized light.
2. Configure the polarization state generator, PSG, using a linear polarizer (LP) with transmission axis defined by the azimuthal angle θ_1, a half-wave plate ($\lambda / 2$) with fast axis defined by the azimuthal angle ϕ_1, and a quarter-wave plate ($\lambda / 4$) with fast axis defined by the azimuthal angle φ_1.
3. Illumine the sample under study.
4. Configure the polarization state analyzer, PSA, using a quarter-wave plate (φ_2), a half-wave plate (ϕ_2), and a linear polarizer (θ_2).
5. Using a non-sensitive to polarization photodetector, register the intensity.

The Polarization State Generator (PSG) and the Polarization State Analyzer (PSA) can be modeled algebraically through the Mueller-Stokes formulism,

$$S^d = M_P(\theta_2) M_{\frac{\lambda}{2}}(\phi_2) M_{\frac{\lambda}{4}}(\varphi_2) M_{\text{sample}} M_{\frac{\lambda}{4}}(\varphi_1) M_{\frac{\lambda}{2}}(\phi_1) M_P(\theta_1) S^s, \qquad (9.165)$$

where $S^{d,s}$ denote the Stokes vectors detected and generated by the source, respectively, M_{sample} is the Mueller matrix associated to the sample under study.

The generation and analysis of linear horizontal polarization S_x is registered by the photo-detector as I_{xx} when $\theta_{1,2} = \phi_{1,2} = \varphi_{1,2} = 0°$ in both PSG and PSA, respectively (if the source emits linear horizontal

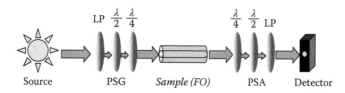

FIGURE 9.16 Experimental configuration to generate and to analyze a complete polarization state with the Ideal Polarimetric Arrangement.

polarized light with a high degree of polarization, the first linear polarizer can be omitted from the experimental arrangement). Follow this procedure:

1. With the generation of linear horizontal polarization, S_x, an analysis of vertical polarization, S_y is obtained I_{xy} when $\theta_1 = 0°$, $\phi_1 = 0°$, $\varphi_1 = 0°$ in the PSG and $\varphi_2 = \phi_2 = \theta_2 = 90°$ in the PSA (in the absence of affectation from the sample, it should be registered as a null intensity).
2. The generation of circular right-handed polarization, S_r, an analysis of $135°$ $or - 45°$ linear polarization, S_{-45} is registered as I_{r-} when $\theta_1 = \phi_1 = 0°$, $\varphi_1 = 45°$ in the PSG, and $\varphi_2 = \phi_2 = \theta_2 = -45°$ in the PSA, respectively.
3. To generate a $+45°$ linear polarization, S_{45}, and to analyze circular left-hand polarization, S_l, I_{+l} is registered when $\theta_1 = \phi_1 = 22.5°$, $\varphi_1 = 45°$ in the PSA and $\varphi_2 = -45°, \phi_2 = \theta_2 = 0°$.
4. Follow a similar procedure in order to generate and analyze each of the six basic polarization states.

The angles are measured as positive in the counter-clockwise sense (rotating from the x-axis to the y-axis, observing to the source) and the right (left) sense of rotation occurs when the electric field component rotates in the clockwise (counter-clockwise) direction. The generation and analysis of the six basic polarized states (S_x, S_y, S_{45}, S_{135}, S_r, S_l) imply 36 intensity measurements, which can be used to obtain a generally stable, well-behaved Mueller matrix for any static optical element [15,44]:

$$m_{00} = \left(I_{xx} + I_{xy} + I_{yy} + I_{yx}\right)/2$$
$$m_{01} = \left(I_{xx} + I_{xy} - I_{yy} - I_{yx}\right)/2$$
$$m_{02} = \left(I_{+x} + I_{+y} - I_{-x} - I_{-y}\right)/2$$
$$m_{03} = \left(I_{rx} + I_{ry} - I_{lx} - I_{ly}\right)/2$$
$$m_{10} = \left(I_{xx} - I_{xy} + I_{yx} - I_{yy}\right)/2$$
$$m_{11} = \left(I_{xx} - I_{xy} - I_{yx} + I_{yy}\right)/2$$
$$m_{12} = \left(I_{+x} - I_{ry} - I_{lx} + I_{ly}\right)/2$$
$$m_{13} = \left(I_{rx} - I_{ry} - I_{lx} + I_{ly}\right)/2 \quad .$$
$$m_{20} = \left(I_{x+} - I_{x-} + I_{y+} - I_{y-}\right)/2$$
$$m_{21} = \left(I_{x+} - I_{r-} - I_{y+} + I_{y-}\right)/2$$
$$m_{22} = \left(I_{++} - I_{+-} - I_{-+} + I_{--}\right)/2$$
$$m_{23} = \left(I_{r+} - I_{r-} - I_{l+} + I_{l-}\right)/2$$
$$m_{30} = \left(I_{xr} - I_{xl} - I_{yl} + I_{yr}\right)/2$$
$$m_{31} = \left(I_{xr} - I_{xl} - I_{yr} + I_{yl}\right)/2$$
$$m_{32} = \left(I_{+r} - I_{+l} - I_{-r} + I_{-l}\right)/2$$
$$m_{33} = \left(I_{rr} - I_{rl} - I_{lr} + I_{ll}\right)/2$$

$$(9.166)$$

It is important to point out that better results can be obtained if the transmission axis of the first linear polarizer is maintained always fixed.

9.6.5 Determination of the Mueller Matrix by an Incomplete, Commercial Polarimeter

There are both PSG and PSA equipments commercially available that can be operated semi-automatically for the complete determination of the Mueller matrix, at a relatively accessible cost (Thorlabs, DPC5000,

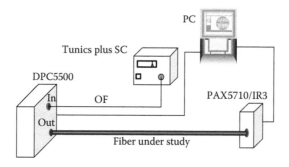

FIGURE 9.17 Diagram of the experimental setup employed for the determination of the Mueller matrix of optical fibers under a fiber-to-fiber configuration. (From Salas-Alcántara, K. M. et al., *Appl. Opt.*, 53, 269–277, 2014. [45])

PAX 5710 IR3). A specific methodology has been published for the handling of the experimental data provided by this configuration (see Figure 9.17) [45]:

$$m_{00} = \frac{1}{2}\left(s_0^{pd} + s_0^{sd}\right), m_{01} = \frac{1}{2}\left(s_0^{pd} - s_0^{sd}\right), m_{02} = s_0^{+d} - m_{00}, m_{03} = s_0^{rd} - m_{00}$$

$$m_{10} = \frac{1}{2}\left(s_1^{pd} s_0^{pd} + s_1^{sd} s_0^{sd}\right), m_{11} = \frac{1}{2}\left(s_1^{pd} s_0^{pd} - s_1^{sd} s_0^{sd}\right), m_{12} = s_1^{+d} s_0^{+d} - m_{10}, m_{13} = s_1^{rd} s_0^{rd} - m_{10}$$

$$m_{20} = \frac{1}{2}\left(s_{21}^{pd} s_0^{pd} + s_2^{sd} s_0^{sd}\right), m_{21} = \frac{1}{2}\left(s_2^{pd} s_0^{pd} - s_2^{sd} s_0^{sd}\right), m_{22} = s_2^{+d} s_0^{+d} - m_{20}, m_{23} = s_2^{rd} s_0^{rd} - m_{20}$$

$$m_{30} = \frac{1}{2}\left(s_3^{pd} s_0^{pd} + s_3^{sd} s_0^{sd}\right), m_{31} = \frac{1}{2}\left(s_3^{pd} s_0^{pd} - s_3^{sd} s_0^{sd}\right), m_{32} = s_3^{+d} s_0^{+d} - m_{30}, m_{33} = s_3^{rd} s_0^{rd} - m_{30}$$

$$\text{(9.167)}$$

where s^{jd} is the detected Stokes vector for the incident polarization states $j = x, y, +, -, r, l,$ and d indicates the state detected by the equipment's software [45].

Why does the application of different methods, even on a same static system, provide slightly different Mueller matrices? The answer depends on several factors, for example, the type of statistics of the noise present within the intensity measurements [46], the optimal number of intensity measurements [47], the number of likelihood estimation [48], the generation and analysis of certain polarization states [49], among others.

9.6.6 Determination of the Polarized Light Pollution, PLP

The use of artificial light by night and the human-made constructions have an unfortunate impact on the reproduction and the increasing mortality in animal populations. The term "Ecological Light Pollution, ELP" has been used since the year 2004 to describe the effects related with the disruption of the natural patterns of light experienced by organisms in ecosystems [50]. Their effects have been documented and have attracted the attention of the scientific community to a specific kind of light generated artificially: the polarized light. This is the reason why the term of "Polarized Light Pollution, PLP" has been adopted recently to call the attention of the society, government, biologists, physiologists, ecologists, physicists, architects, engineers, and to resource managers also [27]. The polarized light pollution, PLP, can be summarized with the following characteristics and effects in the environment [27]:

1. "The linear polarization generated by reflection from smooth human-made objects (buildings and windows, solar cell panels, asphalt roads, among others), or by the light scattered in the atmosphere or hydrosphere at unnatural times or locations, making rise to artificial polarizers. It is necessary to re-design them to reduce their effects in the environment."

2. "The artificial polarizers act like ecological traps that threaten populations of polarization-sensitive species. This characteristic also offers an opportunity for the creation of ecological traps against some dangerous species for human or agriculture".

3. "Artificial polarized light has modified and negatively affected the predatory relationships between species maintained by naturally occurring patterns of polarized light, altering the community structure, diversity and animal dynamics".

9.6.7 Non-Polarizing Optical Devices

Not all of the material media are sensitive to the polarization nature of light. The so-named non-polarizing optical devices are optical elements whose responses do not depend on the incident polarization and do not change the state of polarization of the incident light. Some of them are isotropic materials, vacuum/air interfaces, and special optics/coatings. There exists an open discussion related with the own asymmetry of the universe, which is supposed to be related with the presence of circular polarization associated to the dark ground energy.

9.7 Introduction to Spatially Inhomogeneous Polarization States (Unconventional Polarization)

Recently, the generation of special shapes of beam fronts and phases has also originated special polarization states, which have associated spatially non-homogeneous polarization states [3,4]. Due to its growing interest, accelerated by their increasing number of applications and scientific research needs, this area has received the name of unconventional polarization. Strictly speaking, the non-homogeneous spatial distributions of electric and magnetic field components have been present into any laboratory where highly focused linear polarization has been generated. As a matter of fact, the unconventional polarization has been present within the scattering of light by one-dimensional rough surfaces under conical configuration studies [51–56], where the radial and the azimuthal polarization behavior have been reported experimentally [52, 54]. Consider an incident spatially homogeneous linear vertical electric field distribution within the beam cross-section; at the focus, the orientation of the electric field oscillations is no longer spatially homogeneous, the field has components whose distribution depends on the numerical aperture of the convergent lens, *even along the propagating direction*. As a matter of fact, this is one method employed to generate the z-polarization, a type of unconventional polarization, where the electric field is parallel to the direction of propagation.

From the different types of spatially non-homogeneous electromagnetic field distributions, the cylindrical vector beams deserve special importance because they can be generated not only by transmission in optical fibers [4], modifying the inner cavities in lasers [4], and also can be generated and transport information in free space [53]. The solutions to the vector or general wave Helmholtz equation within the paraxial and the slow varying approximation are the mathematical origin of the cylindrical vector beams [4].

The wave equation for the electric field component, considering the plane wave solution Equation 9.16, is reduced to

$$\nabla \times \nabla \times \left[\vec{E}(\vec{r})e^{-i\omega t} \right] + \varepsilon\mu \frac{\partial^2}{\partial t^2}\left[\vec{E}(\vec{r})e^{-i\omega t} \right] = 0 \tag{9.168}$$

$$\nabla \times \nabla \times \vec{E}(\vec{r}) - k^2\vec{E}(\vec{r}) = 0. \tag{9.169}$$

The Equation 9.169 is the vector Helmholtz wave equation; its different solutions represent spatially non-homogeneous distributed electromagnetic fields that give rise to the unconventional polarization of light [4].

Equation 9.169 has many solutions, where the one associated to a high cylindrical axially symmetry has the form in cylindrical coordinates [4]:

$$\vec{E}(r, z) = U(r, z)\exp\left[ikz\right]\hat{e}_\varnothing, \tag{9.170}$$

where $\hat{e}_\phi$ is a unitary vector in the azimuthal direction. Applying the paraxial approximation to Equation 9.169, with the form considered in Equation 9.170, leads to

$$\frac{1}{r}\frac{\partial}{\partial r}\left(r\frac{\partial U}{\partial r}\right)-\frac{U}{r^2}+2ik\frac{\partial U}{\partial z}=0 . \tag{9.171}$$

Observe that the second term is different in Equation 9.171 and Equation 9.165. Now the Helmholtz equation does not depend on the azimuthal coordinate ϕ, which means the solution $U(r,z)$ is, in effect, azimuthally symmetric. A trial solution is given by

$$U(r,z)=E_0 J_1\left(\frac{\beta r}{1+iz/z_0}\right)\exp\left[-\frac{i\beta^2 z/(2k)}{1+iz/z_0}\right]u(r,z), \tag{9.172}$$

where $u(r,z)$ is the fundamental Gaussian solution, Equation 9.39, $J_1(r, \beta, z, z_0)$ is the Bessel function of first order and first kind. The general solution to the vector wave equation under the paraxial approximation in the cylindrical coordinate system is named *azimuthally polarized vector Bessel-Gauss beam solution* [4]:

$$\vec{E}_\phi(r,z,t)=E_0 J_1\left(\frac{\beta r}{1+iz/z_0}\right)\exp\left[-\frac{i\beta^2 z/(2k)}{1+iz/z_0}\right]u(r,z)\exp\left[i(kz-\omega t)\right]\hat{e}_\phi . \tag{9.173}$$

The general vector wave equation has the same form for the electric or the magnetic field, considering a plane wave solution

$$\nabla\times\nabla\times\vec{H}(\vec{r})-k^2\vec{H}(\vec{r})=0 . \tag{9.174}$$

Following a similar procedure, the solution for a beam with *magnetic field azimuthally polarized*, under the paraxial approximation, can be found to be

$$\vec{H}_\phi(r,z,t)=-H_0 J_1\left(\frac{\beta r}{1+iz/z_0}\right)\exp\left[-\frac{i\beta^2 z/(2k)}{1+iz/z_0}\right]u(r,z)\exp\left[i(kz-\omega t)\right]\hat{e}_\phi . \tag{9.175}$$

Applying the relationship $\vec{k}_z\times\vec{E}_r=\omega\vec{B}_\phi=\mu\omega\vec{H}_\phi$ to Equation 9.175, the radial non-conventional polarized electric field can be obtained.

An experimental setup employed to generate cylindrical vector optical beams is depicted in Figure 9.18.

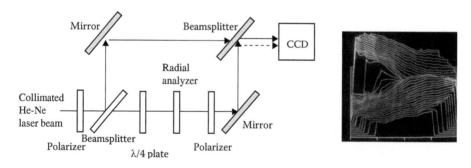

FIGURE 9.18 Experimental setup employed to generate cylindrical vector optical fields.

In general, a cylindrical vector beam can be generated as the superposition of the linearly polarized Hermite-Gauss and the Laguerre-Gauss modes. Note that the cylindrical vector beams have a singularity point at the beam's center.

A particular and direct application can be derived as a simplified solution to describe the amplitude profile of cylindrical vector beams if they have relatively large cross-sections. Such beams have a notorious symmetry, azimuthal or radial, and for small values of the parameter β, the vector Bessel-Gauss beam at his narrowest cross-section (waist) takes the form

$$\vec{E}_j(r, z) = E_0 r \exp\left(-\frac{r^2}{w^2}\right)\hat{e}_j, \ j = r, \phi \ . \tag{9.176}$$

This approximation represents the Laguerre-Gauss, LG_{01}, modes without the vortex term $\exp(i\varphi)$.

9.7.1 Generation of Radial and Azimuthal Polarizations: Active and Passive Methods

Cylindrical vector beams can be expressed as the linear superposition of orthogonally polarized Hermite-Gauss HG_{01} and HG_{10} modes [4]:

$$\vec{E}_r = HG_{10}\hat{x} + HG_{01}\hat{y} \tag{9.177}$$

$$\vec{E}_\phi = HG_{01}\hat{x} + HG_{10}\hat{y}, \tag{9.178}$$

where $\vec{E}_r$, $\vec{E}_\phi$ are the radial (the instant electric field oscillates radially diverging from the center within a ring-shaped pattern of equally distributed intensity) and the azimuthal (the instant electric field oscillates azimuthally or tangentially within a ring-shaped pattern of equally distributed intensity) polarizations, respectively. Figure 9.19 shows (from left-to-right) radial, azimuthal and generalized cylindrical non-conventional polarization states.

There are many other possible unconventional polarization states derived from the superposition of polarized Hermite-Gauss, Laguerre-Gauss, or Bessel-Gauss modes.

The unconventional polarization states can be created fundamentally from two different methods, active or passive.

Active methods for the generation of cylindrical vector beams. If the generation involves amplifying media, like the laser intracavity resonators, the method is named active. Usually they include some form of birefringence or dichroic, non-isotropic, affectation on the fundamental mode. These mode discriminations include naturally birefringence material like the calcite crystal or induced birefringence, quarts crystals, Brewster angle reflectors, axial intracavity dichroism with conical axicon, among many others [3,4].

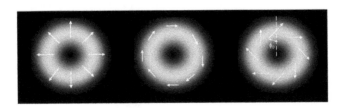

FIGURE 9.19 (From left-to-right) Radial, azimuthal, and generalized cylindrical unconventional polarization states.

Passive methods for the generation of unconventional polarization states in free space. The passive methods are based on the modification of conventional polarization states (usually linear or circular) to generate unconventional polarization states, where the most common are the radial, azimuthal, and z-polarization states. Maybe one of the simplest ways to build a radial polarizer is the following. Cut small triangular pieces from a dichroic linear polarizer where the transmission axis is located along the high of each regular triangle. Join all the pieces in order to get a closed geometrical figure (4, 5, 6, …, pieces), make a circular cut with the external planar basis of each piece.

Another possibility to generate radial polarization is by illuminating a radial polarizer with a circular polarization state, according to Figure 9.20.

Other simple method to generate both, radial and azimuthal polarizations is provided by the incidence of a linearly polarized beam in a thin metallic cylinder at an oblique or conical angle α and a lens doublet to collimate the annularly scattered beam of light. If the linear polarization is parallel to the plane defined by the cylinder axis and the conical angle of incidence α (σ-polarized), a conical scattered beam with radial polarization is generated [57]. On the other hand, if the linear polarization is perpendicular to the plane defined by the cylinder axis and the conical angle of incidence (π-polarized), an azimuthal polarization is generated. Both situations are depicted schematically in Figures 9.21a and b, respectively (taken from Saucedo-Orozco et al., 2014 [57]).

In the conventional polarization, a linear polarization (birefringent, like calcite, or dichroic, as the sheet polarizers) is also an analyzer, where usually the cross configuration is used to identify the presence of linear polarization. In the unconventional polarization, a radial polarizer serves as an azimuthal analyzer and, conversely, an azimuthal polarizer serves also as a radial analyzer.

In general, a radial or an azimuthal polarizer/analyzer device is made of birefringent or dichroic materials, where its local axis of transmission is aligned along the radial/azimuthal directions. There

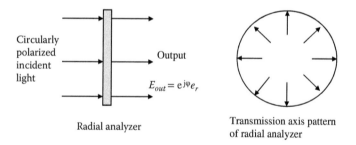

FIGURE 9.20 Conversion of circular polarization to radial polarization using a passive method.

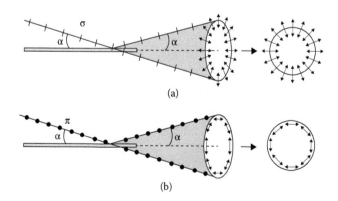

FIGURE 9.21 Generation of unconventional (a) radial polarization and (b) azimuthal polarization.

are commercially available radial, azimuthal and z-polarizers, ranging from single birefringent lens [58] to liquid crystal-based optomechanical devices [59]. All of them have as requirement, the incidence of linear or circular polarizations, and some of them in addition, a collimated beam incidence [4,37,58–60].

A note of caution must be taken into account. If the incidence in a radial converter or polarizer is circularly polarized, the Berry's phase factor $e^{i\phi}$ is present within the beam; it has been measured using interferometric methods [61]. In Cartesian coordinates, an incident circular right-hand polarization is written as

$$\vec{E}_{r,in}(\vec{r},t) = E_0\left(\hat{x} + \hat{y}e^{i\pi/2}\right)e^{i\left(kz-\omega t+\frac{\pi}{4}\right)},\tag{9.179}$$

where this polarization is expressed in cylindrical coordinates as

$$\vec{E}_{r,in}(r,\phi,z,t) = E_0\left[\left(\hat{e}_r\cos\phi - \hat{e}_\phi\sin\phi\right) + i\left(\hat{e}_r\sin\phi + \hat{e}_\phi\cos\phi\right)\right]e^{i\left(kz-\omega t+\frac{\pi}{4}\right)}\tag{9.180}$$

or

$$\vec{E}_{r,in}(r,\phi,z,t) = e^{i\phi}E_0\left(\hat{e}_r + i\hat{e}_\phi\right)e^{i\left(kz-\omega t+\frac{\pi}{4}\right)},\tag{9.181}$$

The function of the radial polarizer is to take off the azimuthal polarized part present in the circularly polarized beam of light, allowing the transmission only of the radially polarized component (indeed, it depends on the own nature of the radial polarizer, because it can serve as a filter if it is constructed from dichroic materials or as a phase converter if the materials employed are birefringent).

The use of spiral phase elements, SPE, makes possible to compensate the Berry's phase from a given cylindrical beam. Figure 9.22 shows a typical experimental arrangement used to compensate the Berry's phase.

Consider a radial polarizer with radially symmetric transmission axis. The output beam becomes

$$\vec{E}_{out}(r,\phi,z,t) = e^{i\phi}E_0\left(\hat{e}_r\right)e^{i\left(kz-\omega t+\frac{\pi}{4}\right)}.\tag{9.182}$$

Note this is not a pure radial polarization, just because the Berry's phase is present azimuthally. The Berry's phase geometric factor can be compensated with a spiral phase element with an opposite helicity, in order to get a pure radial polarization. There are available commercially spiral phase elements or plates, with a broad phase possibilities [62].

One of the most versatile, popular and powerful passive methods used to generate any desired unconventional, vectorial optical polarization state uses reflective liquid crystal spatial light modulators (LCoS-SLM). This generator is reported as capable of generating and controlling the phase, amplitude and polarization (ellipticity and orientation) on a pixel-by-pixel basis [63].

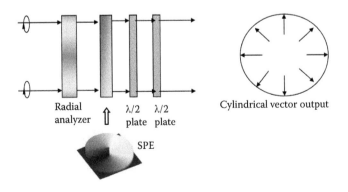

FIGURE 9.22 Experimental setup employed to compensate the Berry's phase from a radially polarized beam.

The applications of cylindrical vector beams to particle trapping, machining, imaging, data storage, remote sensing, among many other potential uses, makes of the unconventional polarization, a must-be for any people who work into the fundamentals and applications of polarized light nowadays.

9.7.2 Generalized Poincarè Sphere: Hybrid Poincarè Sphere

The Poincarè sphere representation has also been employed to represent cylindrical polarized states of light, like the radial and the azimuthal unconventional polarization states [64,65]. Even when it has been known from a long time ago that the electromagnetic field has linear momentum, the angular momentum has only been associated with the sense of rotation or circularly polarized light. The cylindrical vector beams are represented by Laguerre-Gaussian functions, denoted as LG_p^l, where the first order correspond to the mode values $N = 2p + |l| = 1$ [65]. It has been demonstrated that cylindrical beams have associated an orbital angular momentum (OAM) of $l\hbar$ per photon, which depends on the mode order present in the azimuthal phase term $e^{\mp i\phi}$ [66], where the negative (positive) sign is associated to right-handed (left-handed) Laguerre-Gaussian mode. In this sense, a sphere can be constructed from the superposition of a right- (north pole point) and a left-handed (south-pole point) Laguerre-Gaussian mode, respectively, where the equator is defined by the Hermite-Gaussian modes, denoted as $HG_{m,n}$. By following a similar development as the employed for the (conventional) Poincarè sphere, the modified normalized Stokes parameters can be defined in terms of observable quantities like the intensities, as [67]

$$o_0 = \left(I_{HG_{1,0}^{0°}} + I_{HG_{1,0}^{90°}} \right) / \left(I_{HG_{1,0}^{0°}} + I_{HG_{1,0}^{90°}} \right) \tag{9.183}$$

$$o_1 = \left(I_{HG_{1,0}^{0°}} - I_{HG_{1,0}^{90°}} \right) / \left(I_{HG_{1,0}^{0°}} + I_{HG_{1,0}^{90°}} \right) \tag{9.184}$$

$$o_2 = \left(I_{HG_{1,0}^{45°}} - I_{HG_{1,0}^{135°}} \right) / \left(I_{HG_{1,0}^{45°}} + I_{HG_{1,0}^{135°}} \right) \tag{9.185}$$

$$o_3 = \left(I_{LG_0^1} - I_{LG_0^{-1}} \right) / \left(I_{LG_0^1} + I_{LG_0^{-1}} \right), \tag{9.186}$$

where $I_{HG_{m,n}^{\propto}}$ represents the intensity registered when a cylindrical lens converter or a spiral phase plate is rotated at an azimuthal angle $\propto$ when a Hermite-Gaussian beam is transmitted, $I_{LG_p^l}$ is the intensity associated to a Laguerre-Gaussian beam with modes $l = \mp 1, p = 0$.

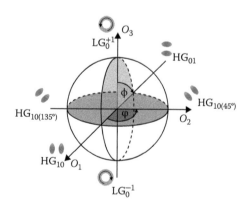

FIGURE 9.23 First-order Poincarè sphere for the representation of the orbital angular momentum (OAM) beams of light. (From Lassen, M. et al., *Phys. Rev. Lett.*, 102, 163602, 2009. [68])

Figure 9.23 shows a first-order Poincarè sphere, where the orbital angular momentums (OAM) are being plotted. In general, points on the surface of the sphere represent the superposition of OAMs states (the axial and azimuthal angles are indicated by φ and ϕ, respectively).

Acknowledgments

R. E. L. acknowledges to CONACYT-México (project 232956) for the financial support to spend his Sabbatical Year at UD and to Prof. Q. Z. and his Group for their hospitality and for sharing fruitful ideas related with optical polarization issues. Authors give thanks to Karla María Salas-Alcántara, Guadalupe López-Morales, Jacqueline Isamar Muro-Ríos, and to the CIO-students of the Optomechatronics Polarization Laboratory practice for their help with several figures presented here and with some corrections to this manuscript. R. E. L. dedicates this chapter to his beloved wife Jeannette Marvella and sons Rafmar and Juan Diego and his maternal grandfather Juan and great uncle Francisco Luna Pérez.

References

1. Jackson J.D. *Classical Electrodynamics*, 3rd Ed., New York, NY: John Wiley and Sons, 1998.
2. Khurgin JB. Slow light in various media: A tutorial, *Adv. Opt. Photonics.* **2**, 287–310; 2010.
3. Zhan Q. Cylindrical vector beams: From mathematical concepts to applications, *Adv. Opt. Photonics.* **1**, 1–57; 2009.
4. Zhan Q. *Vectorial Optical Fields, Fundamentals and Applications*, Hackensack, NJ: World Scientific; 2014.
5. Parke III GN. *Matrix Optics*, PhD Thesis, MIT, 1948.
6. Goldstein D. *Polarized Light*, New York, NY, 2nd Ed., Marcel Decker, 2003.
7. Jones RC. A new calculus for the treatment of optical systems, I. Description and discussion of the calculus, *J. Opt. Soc. Am.* **31**, 488–493; 1941.
8. Brosseau C. *Fundamentals of Polarized Light: A Statistical Optics Approach*, New York, NY: Wiley, 1998.
9. Stokes GG. On the composition and resolution of streams of polarized light from different sources, *Trans. Cambridge Phil. Soc.* **9**, 399; 1952.
10. Shurcliff WA. *Polarized Light, Production and Use*, Cambridge, MA: Harvard University Press, 1962.
11. Pancharatnam S. Generalized theory of interference and its applications, Proceedings of the Indian Academy of Sciences. **44** (6), 398–417; 1956.
12. Kittel C. *Introduction to Solid State Physics*, 7th Ed., New York, NY: John Wiley and Sons, 1996.
13. Mandel L, Wolf E. *Optical Coherence and Quantum Optics*, Cambridge: Cambridge University Press, 1995.
14. Bickel WS, Bailey WM. Stokes vectors, Mueller matrices, and polarized scattered light, *Am. J. Phys.* **53** (5), 468–478; 1985.
15. Gil JJ, Ossikovski R. *Polarized Light and the Mueller Matrix Approach*, New York, NY: CRC Press, 2016.
16. Savenkov SN. Jones and Mueller matrices: structure, symmetry relations and information content, Ch. 3, 71-119, in *Light Scattering Reviews 4, Single Light Scattering and Radiative Transfer*, Ed. Kokhanovsky AA, New York, NY: Springer; 2009.
17. Lu SY, Chipman RA. Interpretation of Mueller matrices based on polar decomposition, *J. Opt. Soc. Am. A.* **13** (5) 1106–1113; 1996.
18. Gil JJ, Bernabeu E. A depolarization criterion in Mueller matrices, *Opt. Acta.* **32** (3), 259–261; 1985.
19. Gil JJ, Bernabeu E. Depolarization and polarization indexes of an optical system, *Opt. Acta.* **33** (3), 185–189; 1986.
20. Espinosa-Luna R, Bernabeu E. On the Q(M) depolarization metric, *Opt. Commun.* **277** (2), 256–258; 2007.
21. O´Neill EL. Introduction to Statistical Optics, Reading, MA:Addison-Wesley, 1963.
22. Espinosa-Luna R, Rodríguez-Carrera D, Bernabeu E, Hinojosa-Ruíz S. Transformation matrices for the Mueller-Jones formalism, *Optik* **119** (16), 757–765; 2008.

23. Saucedo-Orozco I, Espinosa-Luna R, Zhan Q. Angularly-resolved variable wave-retarder using light scattering from a thin metallic cylinder, *Opt. Commun.* **352**, 135–139; 2015.

24. Espinosa-Luna R, Atondo-Rubio G, Bernabeu E, Hinojosa-Ruíz S. Dealing depolarization of light in Mueller matrices with scalar metrics, *Optik* **121** (12), 1058–1068; 2010.

25. Anderson DGM, Barakat R. Necessary and sufficient conditions for a Mueller matrix to be derivable from a Jones matrix, *J. Opt. Soc. Am. A* **11** (8), 2305–2319; 1994.

26. Espinosa-Luna R. The degree of polarization as a criterion to obtain the nine bilinear constraints between the Mueller-Jones matrix elements, *Appl. Opt.* **46** (24), 6047–6054, 2007.

27. Horváth G, Kriska G, Malik P, Robertson B. Polarized light pollution: A new kind of ecological photopollution, *Front. Ecol. Environ.* **7** (6): 317–325; 2009.

28. Horváth G. (Ed.), *Polarized Light and Polarization Vision in Animal Sciences*, 2nd Ed., Berlin: Springer, 2014.

29. Lunazzi JJ. Óptica precolombina del Perú, *Rev. Cub. de Física.* **24** (2), 170–174; 2007.

30. Wood E. *Crystals and Light: An Introduction to Optical Crystallography*, 2nd. Ed., New York, NY: Dover, 1977.

31. Misson GP. A Mueller matrix model for Haidinger´s brushes, *Ophthal. Physiol. Opt.* **23**, 441–447; 2003.

32. Malacara D, Servin M, Malacara Z. *Interferogram Analysis for Optical Testing*, 2nd Ed., New York, NY: Marcel Dekker, Inc, 2008.

33. Land EH. Some aspects of the development of sheet polarizers, *J. Opt. Soc. Am.* **41** (12) 957–963; 1951.

34. Penninckx D, Beck N. Definition, meaning, and measurement of the polarization extinction ratio of fiber-based devices, *Appl. Opt.* **44** (36), 7773–7779; 2005.

35. Brewster D. On the laws which regulate the polarisation of light by reflection from transparent bodies, *Philosophical Transactions of the Royal Society of London* **105**, 125–159; 1815.

36. Maradudin AA, Luna RE, Méndez ER. The Brewster effect for a one-dimensional random surface, *Waves in Random Media* **3** (1), 51–60; 1993.

37. López-Morales G, Rico-Botero VM, Espinosa-Luna R, Zhan Q. Refractive index measurement of dielectric samples using highly focused radially polarized light, *Chin. Opt. Lett.* **15** (3), 030004, 1–4; 2017.

38. Bennett JM. A critical evaluation of rhomb-type quarterwave retarders, *Appl. Opt.* **9** (9), 2123–2129; 1970.

39. Sirohi RS, Chau FS. *Optical Methods of measurement*, Ch. 7, New York, NY: Marcel Dekker, 1999.

40. Jessop HT. *Photoelasticity: Principles and Methods*, New York, NY: Dover, 1949.

41. Espinosa-Luna R, Saucedo-Orozco I, Santiago-Lona CV, Francso-Sánchez JM, Magallanes-Luján A. Polarimetric applications to identify bee honey, *Proc. SPIE.* **8287**, PP82870F1–PP82870F6; 2011.

42. Fujiwara H. *Spectroscopic Ellipsometry, Principles and Applications*, Tokyo: John Wiley & Sons, 2007.

43. Luna, 1998.

44. López-Téllez JM, Bruce NC. Mueller-matrix polarimeter using analysis of the nonlinear voltaje-retardance relationships for liquid-crysta variable retarders, *Appl. Opt.* **53** (24), 5359–5366; 2014.

45. Salas-Alcántara KM, Espinosa-Luna R, Torres-Gómez I, Barmenkov YO. Determination of the Mueller matrix of UV-inscribed long-period fiber grating, *Appl. Opt.* **53** (2), 269–277; 2014.

46. Li X, Hu H, Liu T, Huang B, Song Z. Optimal distribution of integration time for intensity measurements in degree of linear polarization polarimetry, *Opt. Express* **24** (7), 7191–7200; 2016.

47. Goudail F, Bnire A. Estimation precision of the degree of linear polarization and the angle of polarization in the presence of diferente sources of noise, *Appl. Opt.* **49** (4), 683–693; 2010.

48. Hu H, Anna G, Goudail F. On the performance of the physicality-constrained maximum-likelihood estimation of Stokes vector, *Appl. Opt.* **52** (27), 6636–6644; 2013.

49. Tyo JS. Design of optimal polarimeters: maximization of signal-to-noise ratio and minimization of systematic error, *Appl. Opt.* **41** (4), 619–630; 2002.

50. Longcore T, Rich C. Ecological light pollution, *Front. Ecol. Environ.* **2** (4), 191–198; 2004.

51. Depine RA, Antispecular enhancement in s- and p-polarized electromagnetic waves scattered from random gratings, *Opt. Lett.* **16** (19), 1457–1459; 1991.

52. Luna RE, Méndez ER, Escamilla H, Conical Scattering by a one-dimensional randomly rough metallic surfaces, *Proc. SPIE* **1983**: 313–314; 1993.

53. Li L, Chan CH, Tsang L, Numerical simulation of conical diffraction of tapered electromagnetic waves from random rough surfaces and applications to passive remote sensing, *Radio Sci.* **29** (3): 587–598; 1994.

54. Luna RE, Méndez ER, Scattering by one-dimensional random rough metallic surfaces in a conical configuration, *Opt. Lett.* **20** (7): 657–659; 1995.

55. Luna RE, Acosta-Ortíz SE, Zou LF. Mueller matrix for characterization of one-dimensional rough perfectly reflecting surfaces in a conical configuration, *Opt. Lett.* **23** (14), 1075–1077; 1998.

56. Novikov IV, Maradudin AA, The Stokes matrix in conical scattering from a one-dimensional perfectly conducting randomly rough surface, *Radio Science* **34** (3): 599–614; 1999.

57. Saucedo-Orozco I, López-Morales G, Espinosa-Luna R. Generation of unconventional polarization from light scattered by metallic cylinders under conical incidence, *Opt. Lett.* **39** (18), 5341–5344; 2014.

58. S-waveplate (Radial Polarization Converter), Altechna, http://www.altechna.com.

59. Radial-Polarization converter, Arcoptix, http://www.arcoptix.com.

60. Espinosa-Luna R, López-Morales G, Rico-Botero VM, Aguilar-Fernández E. Spatial average symmetry associated to unconventional polarization, *Rev. Mex. Fis.* **63** (2), 205–210; 2017.

61. Zhan Q, Leger JR. Interferometric measurement of Berry's phase in space-variant polarization manipulations, *Opt. Commun.* **213** (4-6), 241–245 (2002).

62. Zero-Order Vortex Half-Wave retarders, Thorlabs, http://www.thorlabs.com.

63. Spatial Light Modulators, Holoeye, http://www.holoeye.com.

64. Padgett MJ, Courtial J. Poincarè-sphere equivalent for light beams containing orbital angular momentum, *Opt. Lett.* **24** (7), 430–432; 1999.

65. Milione G, Sztul HI, Nolan DA, Alfano RR. High-Order Poincaré Sphere, Stokes Parameters, and the Angular Momentum of Light, *Phys. Rev. Lett.* **053601**, 1–4; 2011.

66. Allen L, Beijersbergen MW, Spreeuw RJC, Woerdman JP. Orbital angular momentum of light and the transformation of Laguerre-Gaussian laser modes, *Phys. Rev. A* **45** (11), 8145–8189; 1992.

67. Padgett MJ, Courtial J. Poincaré-sphere equivalent for light beams containing orbital angular momentum, *Opt. Lett.* **24** (7), 430–432; 1999.

68. Lassen M, Leuchs G, Andersen UL. Continuous variable entanglement and squeezing of orbital angular momentum states, *Phys. Rev. Lett.* **102**, 163602 (2009).

Appendix 9.1 Conventional Polarizers and Analyzers

A linear polarizer cannot be defined as a device that converts *any* kind of light into linear polarization because it does not generate *any* polarization state when the incident beam is linear and perpendicular to its transmission axis. The proper way to define a linear polarizer is just as an optical device that attenuates the incident orthogonal components E_x, E_y of the electric field non-uniformly.

A linear polarizer, Equation 9.125, converts light totally unpolarized and with a broad-wavelength spectrum (white light) to totally polarized light. Consider that the white light is described by the un-normalized Stokes vector S_{un}, which is supposed to have both amplitude and phase electric field components E_x, E_y changing randomly with time. The amplitudes are independent of one another, $E_{0x}^2 + E_{0y}^2 = E_{0x}^2 + E_{0y}^2$, and there is no preference between the two, $E_{0x}^2 = E_{0y}^2$. Similarly, the phase difference between the orthogonal components fluctuate randomly, independently of the sign and magnitude of the product $E_{0x}E_{0y}$, consequently $2E_{0x}E_{0y}\cos\delta = 0 = 2E_{0x}E_{0y}\sin\delta$,

$$S = \begin{pmatrix} E_{0x}^2 + E_{0y}^2 \\ E_{0x}^2 - E_{0y}^2 \\ 2E_{0x}E_{0y}\cos\delta \\ 2E_{0x}E_{0y}\sin\delta \end{pmatrix} = \begin{pmatrix} 2E_{0x}^2 \\ 0 \\ 0 \\ 0 \end{pmatrix} \equiv S_{un}. \tag{9.187}$$

Consider a linear polarizer with transmission axis rotated at the angle θ, Equation 9.125, and unpolarized light incident,

$$M_P(2\theta)S_{un} = \frac{1}{2}\begin{bmatrix} 1 & \cos 2\theta & \sin 2\theta & 0 \\ \cos 2\theta & \cos^2 2\theta & \sin 2\theta\cos 2\theta & 0 \\ \sin 2\theta & \sin 2\theta\cos 2\theta & \sin^2 2\theta & 0 \\ 0 & 0 & 0 & 0 \end{bmatrix}\begin{pmatrix} 2E_{0x}^2 \\ 0 \\ 0 \\ 0 \end{pmatrix} = \begin{pmatrix} E_{0x}^2 \\ E_{0x}^2\cos 2\theta \\ E_{0x}^2\sin 2\theta \\ 0 \end{pmatrix} \equiv S^o = \begin{pmatrix} s_0^o \\ s_1^o \\ s_2^o \\ s_3^o \end{pmatrix}. \tag{9.188}$$

Depending on the orientation angle θ, all the possible totally polarized linear states can be obtained.

A linear polarizer operates properly as a polarization state generator or as a polarization state analyzer only for linear polarization. This means that the same device can generate linear polarization or analyze it, being the crossed or null condition a common set up employed to test the presence of pure linear polarization. Also, it behaves as a "transparent" device for linear polarization incident parallel to the transmission axis. Consequently, a linear polarizer is also a linear analyzer once it is operated properly.

The circular polarizer must not be defined as a device that converts *any* kind of light into circular polarization. There exists an exception: a circular right-hand polarizer can not convert left-hand polarization into circular polarization and conversely, a circular left-hand polarizer can not convert circular right-hand polarization into circular left-hand polarization; in both cases, the beam of light is totally attenuated. In practice, this null condition is used to analyze the presence of pure circular polarization into a beam of light. Also, a circular polarizer behaves as a "transparent" device for an incident beam with the same kind of rotation sense. This means a circular right-hand polarizer must allow the total transmission of incident circular right-hand polarization, and a similar criterion must exist for a circular left-hand polarizer. Consequently, a circular polarizer is also a circular analyzer once it is operated properly.

Observe the Mueller matrix associated to a true circular polarizer is given as Equation 9.32 through 9.135:

$$M_r = \frac{1}{2}\begin{bmatrix} 1 & 0 & 0 & 1 \\ 0 & 0 & 0 & 0 \\ 0 & 0 & 0 & 0 \\ 1 & 0 & 0 & 1 \end{bmatrix}, \quad M_l = \frac{1}{2}\begin{bmatrix} 1 & 0 & 0 & -1 \\ 0 & 0 & 0 & 0 \\ 0 & 0 & 0 & 0 \\ -1 & 0 & 0 & 1 \end{bmatrix}. \tag{9.189}$$

Both matrices convert light into circular right- and left-handed polarized, respectively

$$S_r = \frac{1}{2}\begin{bmatrix} 1 & 0 & 0 & 1 \\ 0 & 0 & 0 & 0 \\ 0 & 0 & 0 & 0 \\ 1 & 0 & 0 & 1 \end{bmatrix}\begin{pmatrix} s_0 \\ s_1 \\ s_2 \\ s_3 \end{pmatrix} = \frac{1}{2}\begin{pmatrix} s_0 + s_3 \\ 0 \\ 0 \\ s_0 + s_3 \end{pmatrix} \tag{9.190}$$

$$S_l = \frac{1}{2}\begin{bmatrix} 1 & 0 & 0 & -1 \\ 0 & 0 & 0 & 0 \\ 0 & 0 & 0 & 0 \\ -1 & 0 & 0 & 1 \end{bmatrix}\begin{pmatrix} s_0 \\ s_1 \\ s_2 \\ s_3 \end{pmatrix} = \frac{1}{2}\begin{pmatrix} s_0 - s_3 \\ 0 \\ 0 \\ -s_0 + s_3 \end{pmatrix} \tag{9.191}$$

where $s_0^2 \geq s_1^2 + s_2^2 + s_3^2$ opens the possibility to deal with partially or unpolarized light.

However, in practice, a linear polarizer with transmission axis horizontal and a quarter-wave plate with fast axis at $\theta = +45°, -45°$ are employed jointly as circular right- and left-hand polarizers, respectively. The equivalent Mueller matrices for both arrangements are given by

$$\tilde{M}_r = M_R(\theta = 45°, \delta = 90°)M_P(2\theta = 0°) = \frac{1}{2}\begin{bmatrix} 1 & 1 & 0 & 0 \\ 0 & 0 & 0 & 0 \\ 0 & 0 & 0 & 0 \\ 1 & 1 & 0 & 0 \end{bmatrix} \tag{9.192}$$

$$\tilde{M}_l = M_R(\theta = -45°, \delta = 90°)M_P(2\theta = 0°) = \frac{1}{2}\begin{bmatrix} 1 & 1 & 0 & 0 \\ 0 & 0 & 0 & 0 \\ 0 & 0 & 0 & 0 \\ -1 & -1 & 0 & 0 \end{bmatrix}. \tag{9.193}$$

Note that the circular Mueller matrices generated experimentally are not the same as their respective true Mueller circular polarizers:

$$M_r \neq \tilde{M}_r, \quad M_l \neq \tilde{M}_l \tag{9.194}$$

The effect generated into an arbitrary beam of light by matrices 9.192 and 9.193, respectively, is given by

$$S^o = \tilde{M}_r S = \frac{1}{2}\begin{bmatrix} 1 & 1 & 0 & 0 \\ 0 & 0 & 0 & 0 \\ 0 & 0 & 0 & 0 \\ 1 & 1 & 0 & 0 \end{bmatrix}\begin{pmatrix} s_0 \\ s_1 \\ s_2 \\ s_3 \end{pmatrix} = \frac{1}{2}\begin{pmatrix} s_0 + s_1 \\ 0 \\ 0 \\ s_0 + s_1 \end{pmatrix} \equiv \tilde{S}_r \tag{9.195}$$

$$S^o = \tilde{M}_l S = \frac{1}{2}\begin{bmatrix} 1 & 1 & 0 & 0 \\ 0 & 0 & 0 & 0 \\ 0 & 0 & 0 & 0 \\ -1 & -1 & 0 & 0 \end{bmatrix}\begin{pmatrix} s_0 \\ s_1 \\ s_2 \\ s_3 \end{pmatrix} = \frac{1}{2}\begin{pmatrix} s_0 + s_1 \\ 0 \\ 0 \\ -(s_0 + s_1) \end{pmatrix} = \tilde{S}_l. \tag{9.196}$$

Observe that the circular polarization states generated by Equations 9.190 and 9.191 and Equations 9.195 and 9.196 depend strongly on the nature of the illuminating source. Properties related to the coherency of the incident light could provide different physical responses associated to the materials under study if the user employs as polarization state generators Equation 9.189, or Equations 9.192 and 9.193.

Do all the previous circular polarizers fulfill the crossed or null condition and the "transparency" condition? Considering the incident Stokes vector is associated to circular right-, $S_r = (1\ 0\ 0\ 1)^T$, and left-hand polarization, $S_l = (1\ 0\ 0-1)^T$, respectively,

$$S^o = M_r S_r = S_r, \quad S^o = M_r S_l = S_{Null} \tag{9.197}$$

$$S^o = M_l S_l = S_l, \quad S^o = M_l S_r = S_{Null} \tag{9.198}$$

$$S^o = \tilde{M}_r S_r = S_r / 2, \quad S^o = \tilde{M}_r S_l \neq S_{Null} \tag{9.199}$$

$$S^o = \tilde{M}_l S_l = S_l / 2, \quad S^o = \tilde{M}_l S_r \neq S_{Null}. \tag{9.200}$$

Results show that the true circular polarizers fulfill both desired conditions and consequently can be used also as analyzers. On the other hand, the traditional circular polarizers reduce by a half the incident circular polarization intensity and do not fulfill the desired "transparency" condition, so they cannot be used as circular analyzers (in the following paragraphs, it will be shown that reversing the position of the linear and the quarter-wave plate, gives rise to a traditional circular analyzer).

Even when the traditional experimental circular polarizers generate circular polarization, the output Stokes vectors generated, $\tilde{S}_r$, $\tilde{S}_r$, do not contain the same information as the theoretical circular polarizers provide, as has been discussed some time ago [14]. This observation deserves special attention because it gives rise to different possibilities.

1.a Employing the true circular polarizers and applying Equations 9.190 and 9.191, respectively, following the Bickel and Bailey observation,

$$S_r = M_r S = \frac{1}{2} \begin{pmatrix} s_0 + s_3 \\ 0 \\ 0 \\ s_0 + s_3 \end{pmatrix} = \frac{1}{2} \begin{pmatrix} E_{0x}^2 + E_{0y}^2 + 2E_{0x}E_{0y}\sin(\delta_y - \delta_x) \\ 0 \\ 0 \\ E_{0x}^2 + E_{0y}^2 + 2E_{0x}E_{0y}\sin(\delta_y - \delta_x) \end{pmatrix} \tag{9.201}$$

$$S_l = M_l S = \frac{1}{2} \begin{pmatrix} s_0 - s_3 \\ 0 \\ 0 \\ s_3 - s_0 \end{pmatrix} = \frac{1}{2} \begin{pmatrix} E_{0x}^2 + E_{0y}^2 - 2E_{0x}E_{0y}\sin(\delta_y - \delta_x) \\ 0 \\ 0 \\ 2E_{0x}E_{0y}\sin(\delta_y - \delta_x) - E_{0x}^2 - E_{0y}^2 \end{pmatrix}. \tag{9.202}$$

1.b Using now the traditional experimental circular polarizers and applying Equations 9.192 and 9.193, respectively,

$$\tilde{S}_r = \tilde{M}_r S = \frac{1}{2} \begin{pmatrix} s_0 + s_1 \\ 0 \\ 0 \\ s_0 + s_1 \end{pmatrix} = \frac{1}{2} \begin{pmatrix} E_{0x}^2 + E_{0y}^2 + E_{0x}^2 - E_{0y}^2 \\ 0 \\ 0 \\ E_{0x}^2 + E_{0y}^2 + E_{0x}^2 - E_{0y}^2 \end{pmatrix} = \begin{pmatrix} E_{0x}^2 \\ 0 \\ 0 \\ E_{0x}^2 \end{pmatrix} \tag{9.203}$$

$$\tilde{S}_l = \tilde{M}_l S = \frac{1}{2} \begin{pmatrix} s_0 + s_1 \\ 0 \\ 0 \\ -s_0 - s_1 \end{pmatrix} = \frac{1}{2} \begin{pmatrix} E_{0x}^2 + E_{0y}^2 + E_{0x}^2 - E_{0y}^2 \\ 0 \\ 0 \\ -E_{0x}^2 - E_{0y}^2 - E_{0x}^2 + E_{0y}^2 \end{pmatrix} = \begin{pmatrix} E_{0x}^2 \\ 0 \\ 0 \\ -E_{0x}^2 \end{pmatrix}. \tag{9.204}$$

Under the observation of Bickel and Bailey, the differences are evident: the phase information is missing from the total intensity element when the traditional experimental circular polarizers are used. Applying this conception to the generation of linear polarization from an unpolarized source, the following result is obtained:

$$S^o = M_P(2\theta)S_{un} = \frac{1}{2}\begin{bmatrix} 1 & 1 & 0 & 0 \\ 1 & 1 & 0 & 0 \\ 0 & 0 & 0 & 0 \\ 0 & 0 & 0 & 0 \end{bmatrix}\begin{pmatrix} E_{0x}^2 + E_{0y}^2 \\ 0 \\ 0 \\ 0 \end{pmatrix} = \frac{1}{2}\begin{pmatrix} E_{0x}^2 + E_{0y}^2 \\ E_{0x}^2 + E_{0y}^2 \\ 0 \\ 0 \end{pmatrix} \equiv S_x. \tag{9.205}$$

However, the Mueller-Stokes formalism implies the use of normalized entities to operate the propagation of electric field components (represented by normalized Stokes vectors) through optical devices (represented by normalized Mueller matrices). The anisotropic effect generated by the medium on the transmitted electric field is present within the form of the matrix that represents the optical device under consideration, while the kind of light is present within the form of the Stokes vectors. This can be directly verified by observing the electric field orthogonal components, their amplitudes and phases, and also the linear effect induced by the medium, according to Equations 9.107 through 9.109. Otherwise, how could it be explained the possibility to generate polarized light from an unpolarized light source after the transmission through a polarization state generator, for example a traditional circular polarizer, if no memory-removal is originated under such conversion? Consider once again the Equations 9.201 through 9.204 under this scope,

$$M_r S = \frac{1}{2}\begin{pmatrix} s_0 + s_3 \\ 0 \\ 0 \\ s_0 + s_3 \end{pmatrix} = S^{o1} \equiv \begin{pmatrix} s_0^{o1} \\ 0 \\ 0 \\ s_3^{o1} \end{pmatrix} \equiv \begin{pmatrix} \left(E_{ox}^2 + E_{0y}^2\right)^{o1} \\ 0 \\ 0 \\ \left(2E_{0x}E_{0y}\sin\delta\right)^{o1} \end{pmatrix} \tag{9.206}$$

$$M_l S = \frac{1}{2}\begin{pmatrix} s_0 - s_3 \\ 0 \\ 0 \\ s_3 - s_0 \end{pmatrix} = S^{o2} \equiv \begin{pmatrix} s_0^{o2} \\ 0 \\ 0 \\ s_3^{o2} \end{pmatrix} \equiv \begin{pmatrix} \left(E_{ox}^2 + E_{0y}^2\right)^{o2} \\ 0 \\ 0 \\ \left(2E_{0x}E_{0y}\sin\delta\right)^{o2} \end{pmatrix} \tag{9.207}$$

$$\tilde{M}_r S = \frac{1}{2}\begin{pmatrix} s_0 + s_1 \\ 0 \\ 0 \\ s_0 + s_1 \end{pmatrix} = S^{o3} \equiv \begin{pmatrix} s_0^{o3} \\ 0 \\ 0 \\ s_3^{o3} \end{pmatrix} \equiv \begin{pmatrix} \left(E_{ox}^2 + E_{0y}^2\right)^{o3} \\ 0 \\ 0 \\ \left(2E_{0x}E_{0y}\sin\delta\right)^{o3} \end{pmatrix} \tag{9.208}$$

$$\tilde{M}_l S = \frac{1}{2}\begin{pmatrix} s_0 + s_1 \\ 0 \\ 0 \\ -s_0 - s_1 \end{pmatrix} = S^{o4} \equiv \begin{pmatrix} s_0^{o4} \\ 0 \\ 0 \\ s_3^{o4} \end{pmatrix} \equiv \begin{pmatrix} \left(E_{ox}^2 + E_{0y}^2\right)^{o4} \\ 0 \\ 0 \\ \left(2E_{0x}E_{0y}\sin\delta\right)^{o4} \end{pmatrix}. \tag{9.209}$$

Note that depending on the numerical values assigned to the incident polarization state S, the circular output polarization is *always* obtained, even when different output intensity values s_0^{oj}, $j = 1, 2, 3, 4$

could be obtained. For example, if $S = S_{un} = (1\ 0\ 0\ 0)^T$, circular right- and left-hand polarizations are generated with the same normalized intensity,

$$S_r = \tilde{S}_r, \quad S_l = \tilde{S}_l. \tag{9.210}$$

The experimental arrangement employed to generate the true circular Mueller matrix, Equations 9.189, can be obtained by the following procedure [14]:

$$M_r = M_R(\theta = 45°, \delta = \pi/2) M_P(2\theta = 0°) M_R(\theta = -45°, \delta = \pi/2) \tag{9.211}$$

$$M_l = M_R(\theta = -45°, \delta = \pi/2) M_P(2\theta = 0°) M_R(\theta = 45°, \delta = \pi/2). \tag{9.212}$$

Note that Equation 9.211 can be obtained from Equation 9.212 just by a face-to-face rotation (rotation as a unity of the complete cascade arrangement by 180° either around the x- or the y-axis, considering the propagation along the z-axis and a right-hand Cartesian coordinate system, $\hat{x} \times \hat{y} = \hat{z}$). Note that they are invariant under azimuthal rotation respect to any incident polarization,

$$M_r = M_{Rot}(-2\theta) M_r M_{Rot}(2\theta); M_l = M_{Rot}(-2\theta) M_l M_{Rot}(2\theta), \tag{9.213}$$

where the matrix of rotation M_{Rot}, is given by

$$M_{Rot}(2\theta) = \begin{bmatrix} 1 & 0 & 0 & 0 \\ 0 & \cos 2\theta & \sin 2\theta & 0 \\ 0 & -\sin 2\theta & \cos 2\theta & 0 \\ 0 & 0 & 0 & 1 \end{bmatrix}. \tag{9.214}$$

A point against the use of the traditional method employed to generate circular polarization using only a linear polarizer and a quarter-wave plate is its strong dependence on the linear polarization nature of the source. Usually, this fact has been passed unnoticed because the typical examples have considered only unpolarized light as the source of light. However, note the incident light interacts firstly with the linear polarizer; at an extreme case, observe that if the light is linearly polarized with its electric field perpendicular to the transmission axis, no more light is transmitted to the quarter-wave plate and a null intensity is obtained. From Equations 9.192 and 9.193, respectively,

$$S^o = \tilde{M}_r S_y = \frac{1}{2} \begin{pmatrix} s_0 + s_1 \\ 0 \\ 0 \\ s_0 + s_1 \end{pmatrix} = \frac{1}{2} \begin{pmatrix} 1 - 1 \\ 0 \\ 0 \\ 1 - 1 \end{pmatrix} = S_{Null} \tag{9.215}$$

$$S^o = \tilde{M}_l S_y = \frac{1}{2} \begin{pmatrix} s_0 + s_1 \\ 0 \\ 0 \\ -(s_0 + s_1) \end{pmatrix} = \frac{1}{2} \begin{pmatrix} 1 - 1 \\ 0 \\ 0 \\ -1 + 1 \end{pmatrix} = S_{Null}. \tag{9.216}$$

On the other hand, this characteristic offers a usually ignored possibility. The traditional conversion to circular polarization has the advantage that it can generate *attenuated circular polarization* if the source emits light linearly polarized, where the attenuation could be controlled by the relative angular orientation θ_{rel} between the incident electric field and the transmission axis of the linear polarizer. This kind of attenuation follows the Malus' law, Equation 9.143; consequently, the circular attenuation is reduced in intensity as the $\cos^2 \theta_{rel}$, where obviously, the maximum intensity is obtained when $\theta_{rel} = 0°$. This can save the use of neutral density filters, beamsplitters, linear polarizers, or another kind of attenuators within an experimental setup employed for the generation of circular polarization. Also, the linear polarizer can be taken off from the traditional circular polarizer arrangement if the source emits total linear polarization and the quarter-wave plate is oriented properly. If the linear polarization is oriented horizontally (with Stokes vector S_x), the quarter-wave plate retarder must be oriented at $\theta = +45°, -45°$, in order to get circular right- and left-hand polarizations, respectively, with the maximum available intensity. Using Equation 9.128,

$$\tilde{S}_r = M_R\left(\theta = 45°, \delta = \pi/2\right)S_x = \begin{bmatrix} 1 & 0 & 0 & 0 \\ 0 & 0 & 0 & -1 \\ 0 & 0 & 1 & 0 \\ 0 & 1 & 0 & 0 \end{bmatrix}\begin{pmatrix} 1 \\ 1 \\ 0 \\ 0 \end{pmatrix} = \begin{pmatrix} 1 \\ 0 \\ 0 \\ 1 \end{pmatrix} \tag{9.217}$$

$$\tilde{S}_l = M_R\left(\theta = -45°, \delta = \pi/2\right)S_x = \begin{bmatrix} 1 & 0 & 0 & 0 \\ 0 & 0 & 0 & 1 \\ 0 & 0 & 1 & 0 \\ 0 & -1 & 0 & 0 \end{bmatrix}\begin{pmatrix} 1 \\ 1 \\ 0 \\ 0 \end{pmatrix} = \begin{pmatrix} 1 \\ 0 \\ 0 \\ -1 \end{pmatrix}. \tag{9.218}$$

Equations 9.217 and 9.218 can be named the *optimized traditional circular polarization generator*, which provides twice the energy because no linear polarizer is present; however, it is not invariant under azimuthal rotations.

The traditional circular analyzer is constructed with the same optical elements as the circular polarizer, but located in reverse order,

$$\tilde{A}_r \equiv M_H M_{\frac{\lambda}{4}}\left(\theta = -45°\right) = \frac{1}{2}\begin{bmatrix} 1 & 0 & 0 & 1 \\ 1 & 0 & 0 & 1 \\ 0 & 0 & 0 & 0 \\ 0 & 0 & 0 & 0 \end{bmatrix} = \left(\tilde{M}_r\right)^T \tag{9.219}$$

$$\tilde{A}_l \equiv M_H M_{\frac{\lambda}{4}}\left(\theta = 45°\right) = \frac{1}{2}\begin{bmatrix} 1 & 0 & 0 & -1 \\ 1 & 0 & 0 & -1 \\ 0 & 0 & 0 & 0 \\ 0 & 0 & 0 & 0 \end{bmatrix} = \left(\tilde{M}_l\right)^T. \tag{9.220}$$

Note that a traditional circular left-hand analyzer is reached from face-to-face zenithal or horizontal rotation on the traditional circular right-hand polarizer and, conversely, a similar rotation on the traditional circular left-hand polarizer generates a traditional circular right-hand analyzer. Consider the analysis of a general elliptically polarized state S_e,

$$\tilde{A}_r S_e = \frac{1}{2}\begin{bmatrix} 1 & 0 & 0 & 1 \\ 1 & 0 & 0 & 1 \\ 0 & 0 & 0 & 0 \\ 0 & 0 & 0 & 0 \end{bmatrix}\begin{pmatrix} s_0 \\ s_1 \\ s_2 \\ s_3 \end{pmatrix} = \frac{1}{2}\begin{pmatrix} s_0 + s_3 \\ s_0 + s_3 \\ 0 \\ 0 \end{pmatrix} = \begin{cases} s_0^{det} = 1 \text{ if } S_e = S_r \\ s_0^{det} = 0 \text{ if } S_e = S_l \end{cases} \tag{9.221}$$

$$\tilde{A}_l S_e = \frac{1}{2}\begin{bmatrix} 1 & 0 & 0 & -1 \\ 1 & 0 & 0 & -1 \\ 0 & 0 & 0 & 0 \\ 0 & 0 & 0 & 0 \end{bmatrix}\begin{pmatrix} s_0 \\ s_1 \\ s_2 \\ s_3 \end{pmatrix} = \frac{1}{2}\begin{pmatrix} s_0 - s_3 \\ s_0 - s_3 \\ 0 \\ 0 \end{pmatrix} = \begin{cases} s_0^{det} = 1 \ if \ S_e = S_l. \\ s_0^{det} = 0 \ if \ S_e = S_r \end{cases} \qquad (9.222)$$

A desirable circular analyzer is one that allows a perfect conversion of all the incident circular polarization into total energy. In this sense, the traditional circular analyzers fulfill this condition and they can also be employed under the crossed or null condition. This means a traditional circular analyzer can be used to test whether a beam of light is composed of totally circular polarized light.

10

Some Lens Optical Devices

Daniel Malacara-Doblado and Alejandro Téllez-Quiñones

10.1 Introduction

In this chapter, some of the most important optical systems using lenses will be described [1–4]. However, telescopes and microscopes are not described here, since they are the subjects of other chapters in this book. Since optical instruments cannot be studied without a previous background on the definitions of pupils, principal ray and skew and meridional rays, we will begin with a brief review of these concepts.

10.2 Principal and Meridional Rays

In any optical system with several lenses, two important meridional rays can be traced through the system, as shown in Figure 10.1 [5]. A meridional ray is in a common plane with the optical axis. A skew ray, on the other hand, is not.

1. A ray from the object plane, on the optical axis, to the image plane, also on the optical axis, is called an on-axis meridional ray.
2. A meridional ray from the edge (or any other off-axis point) on the object to the corresponding point on the image is called the chief or principal ray.

All the lenses in this optical system must have a minimum diameter to allow these two rays to pass through the whole system. The planes on which the on-axis meridional ray crosses the optical axis are conjugates to the object and image planes. Any thin lens at these positions does not affect the path of the on-axis meridional ray but affects the path of the principal ray. This lens is called a field lens. The diameter of the field lenses determines the diameter of the image (field). If a field lens limits the field diameter more than desired by stopping a principal ray with a certain height, we have an effect called vignetting. The image does not have a sharp boundary, but its luminosity decreases very rapidly toward

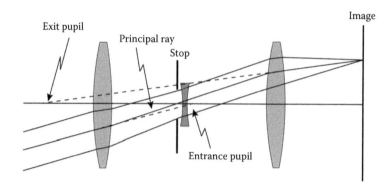

FIGURE 10.1   An optical system illustrating the concepts of stop, entrance pupil, and exit pupil.

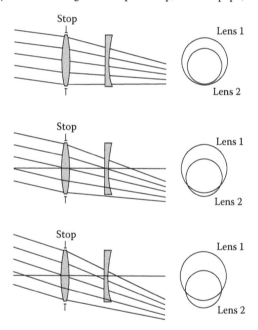

FIGURE 10.2   Vignetting in an optical system.

the outside of the field when maximum image height has been reached, that is, when vignetting begins, because the effective pupil size decreases very fast, as shown in Figure 10.2.

The planes on which the principal ray crosses the optical axis are said to be pupils or stop planes. At the first crossing of the ray, if it occurs before the first surface of the system, or at its straight extrapolation after leaving, the ray from the object, is the entrance pupil. At the last crossing of the ray, if it occurs after the last surface of the system, or at its straight extrapolation before arriving, the principal ray to the image is the exit pupil. Any lens located at the pupil planes affects the path of the meridional ray but not the path of the principal ray. This lens is an imaging or relay lens. The diameter of the relay lenses determines the aperture of the system.

10.3 Magnifiers

The most common and traditional use of a lens is as a magnifier. The stop of the system is the pupil of the observing eye. Since with good illumination this pupil is small (3 4 mm or even smaller), the aberrations that increase with the diameter of the pupil, such as spherical aberration, do not present any problem. On the other hand, the field of view is not small; hence, field curvature, astigmatism, and distortion are the most important aberrations to be corrected.

The angular magnification M is defined as the ratio of the angular diameter β of the virtual image observed through the magnifier to the angular diameter α of the object as observed from a distance of 250 mm (defined as the minimum observing distance for young adults). If the lens is placed at a distance from the observed object so that the virtual image is at distance l' from the eye, the magnification M is given by

$$M = \frac{\beta}{\alpha} = \frac{250}{(l'+d)}\left(\frac{l'}{f}+1\right), \tag{10.1}$$

where d is the distance from the lens to the eye, f is the focal length, and all distances are in millimeters.

The maximum magnification, obtained when the virtual image is as close as 250 mm in front of the observing eye, and the lens is close to the eye ($d = 0$) is

$$M = \frac{250}{f}+1. \tag{10.2}$$

If the virtual image is placed at infinity to avoid the need for eye accommodation, the magnification becomes independent of the distance d and has a value

$$M = \frac{250}{f}. \tag{10.3}$$

Here, we have to remember that eye accommodation occurs when the eye lens (crystalline) modifies its curvature to focus on close objects.

We can thus see that for small focal lengths f, the magnification is nearly the same for all lens' positions with respect to the eye and the observed object, as long as the virtual image is not closer than 250 mm from the eye.

It has been shown that if the magnifier is a single plano-convex lens, the optimum orientation to produce the best possible image with the minimum aberrations is

1. With the plane on the eye's side if the lens is closer to the eye than to the object, as shown in Figure 10.3a, and

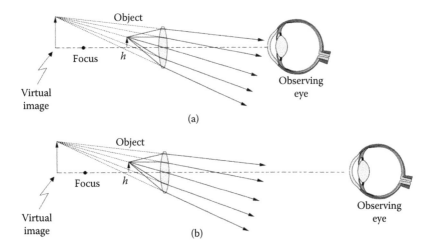

FIGURE 10.3 A simple magnifier (a) with the observing eye close to the lens and (b) with the observing eye far from the lens.

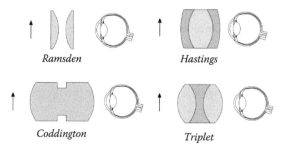

FIGURE 10.4 Some common magnifiers.

2. With the plane on the side of the object if the lens is closer to the object than to the eye, as shown in Figure 10.3b.

The disadvantage of the plano-convex magnifier is that if one does not know these rules one may use it with the wrong lens orientation. A safer magnifier configuration would be a double convex lens, but then the image is not the best. For this reason, most high-quality magnifiers are symmetrical. To reduce the aberrations and have the best possible image, more complicated designs can be used, as illustrated in Figure 10.4.

10.4 Ophthalmic Lenses

A human eye may have refractive defects that produce a defocused or aberrated image in the retina. The most common defects are as follows:

1. Myopia, when the image of an object at infinity falls in front of the retina. Eye accommodation cannot compensate this defocusing error. A myopic eye sees in focus only close objects (see Figure 10.5b).
2. Hypermetropia, when the image of an object at infinity falls behind the retina. This error can be compensated in young people by eye accommodation. A hypermetropic eye feels the problem if it cannot accommodate either due to age or to the high magnitude of the defect (see Figure 10.5c).
3. Astigmatism, when the rays in two planes passing through two perpendicular diameters on the pupil of the eye have different focus positions along the optical axis (see Figure 10.5d).

These refractive defects are corrected by means of a single lens in front of the eye. The geometry used to design an ophthalmic lens is shown in Figure 10.6. The eye rotates in its skull socket to observe objects at different locations away from the lens' optical axis. Thus, its effective stop is not at the pupil of the eye but at the center of rotation of the eye.

The ophthalmic lens is not in contact with the eye, but at a distance d_v of about 14 mm in front of the cornea. An image magnification is produced because of the lens separation from the cornea. The focus of the ophthalmic lens has to be located at the point in space that is conjugate to the retina. This point is in front of the eye for myopic eyes and behind the eye for hypermetropic eyes.

Since the distance d_v is a fixed constant parameter, the important quantity in the ophthalmic lens is the back (or vertex) focal length F_v. The inverse of this vertex focal length is the vertex power

$$P_v - \frac{1}{F_v}, \tag{10.4}$$

where P_v is in diopters if F_v is expressed in meters.

When an eye is corrected with an ophthalmic lens, the apparent image size changes. The magnification M produced by this lens is

$$M = \left(\frac{1}{1 - dP} - 1 \right) \times 100\%. \tag{10.5}$$

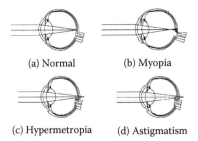

(a) Normal   (b) Myopia

(c) Hypermetropia   (d) Astigmatism

FIGURE 10.5 Geometry of vision refractive errors.

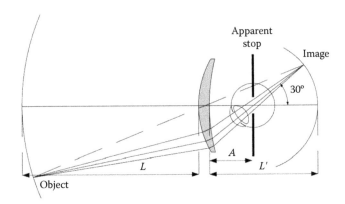

FIGURE 10.6 Geometry used to design eyeglasses.

The eye with refractive defects has a different image size than the normal (emetropic) eye. Myopia is due to abnormal elongation image size of the eye globe. Hypermetropia is due to an abnormal shortening of the eye globe. Astigmatism arises when the curvature of the cornea (or one of the surfaces of the eye lens) has two different values in mutually perpendicular diameters, like in a toroid. The diameter of the eye globe, with a refractive defect of power P, is given by the empirical relation

$$D = \frac{P_v}{6} + 14.5, \tag{10.6}$$

where P_v is the vertex power in diopters of the required lens and D is in millimeters.

When designing an ophthalmic lens, the lens surface has to be spherical, concentric with the eye globe.

Since the pupil of the eye has a small diameter, the shape of the lens has to be chosen so that the field curvature and the off-axis astigmatism are minimized, with an equilibrium that can be selected by the designer. Two solutions are found, as shown in the Tscherning ellipses in Figure 10.7. In these ellipses, we can observe the following:

1. Two possible solutions exist: one is the Ostwald lens and the other is the more curved Wollaston lens.
2. The solutions for zero-field curvature (off-axis power error) and for off-axis astigmatism are close to each other.
3. There are no solutions for lenses with a vertex power larger than about 7–10 diopters. If an aspheric surface is introduced, the range of powers with solutions is greatly extended.

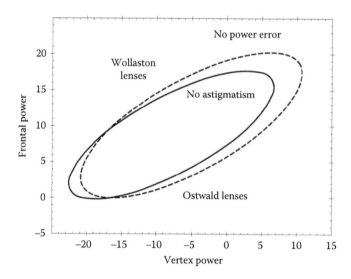

FIGURE 10.7   Ellipses for zero astigmatism and no power error in eyeglasses.

10.5 Achromatic Doublets and Collimators

A single lens has axial chromatic aberration. Its spherical aberration can be minimized with the proper shape (bending), but it can never be zero. The advantage of a doublet made by joining together two lenses with different glasses is that a good correction of both axial chromatic and spherical aberrations can be achieved.

Considering a doublet of two thin lenses with focal lengths f_1 and f_2, the focal length of the combination for red light (C) and blue light (F) can be made the same to eliminate the axial chromatic aberration if

$$\left(n_{1C}-n_{1F}\right)K_1 = \left(n_{2C}-n_{2F}\right)K_2, \tag{10.7}$$

where K_i has been defined by the lens maker's equation:

$$\frac{1}{f_i}-\left(n_i-1\right)\left(\frac{1}{r_{1i}}-\frac{1}{r_{2i}}\right)=\left(n_i-1\right)K_i. \tag{10.8}$$

From Equation 10.7 we can find

$$f_1 V_1 = -f_2 V_2, \tag{10.9}$$

where the Abbe number V_1 of the glass i has been defined as

$$V_i = \frac{\left(n_{iD}-1\right)}{\left(n_{iC}-n_{iF}\right)} \tag{10.10}$$

and from these expressions we finally obtain

$$f_1 = F\left(1 - \frac{V_2}{V_1}\right) \tag{10.11}$$

And

$$f_2 = F\left(1 - \frac{V_1}{V_2}\right). \tag{10.12}$$

We see that any two glasses with different values of the Abbe number V can be used to design an achromatic doublet. However, a small difference in the V values would produce lenses with low power, which are desirable. This means that large differences in the values of V are appropriate. Typically, the positive (convergent) lens is made with crown glass ($V > 50$) and the negative (divergent) lens with flint glass ($V < 50$). Of course, these formulas are thin-lenses approximations, but they produce a reasonable close solution to perform ray tracing in order to find an exact solution.

Another advantage of a doublet made with two different glasses is that the primary spherical aberration can be completely corrected if the proper shape of the lenses is used, as in Figure 10.8; the design data are given in Table 10.1 [6].

A disadvantage of the cemented doublet is that the primary spherical aberration is well corrected but not the high-order spherical aberration. A better correction is obtained if the two lens components are separated, as in Figure 10.9; the design data are given in Table 10.2 [6].

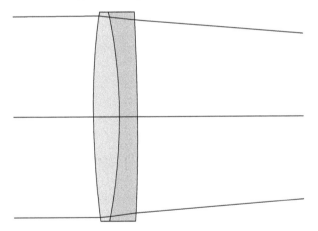

FIGURE 10.8 A cemented doublet.

TABLE 10.1 Achromatic Doublet

Radius of Curvature (mm)	Diameter (mm)	Thickness (mm)	Material
12.8018	3.41	0.3434	BaK-1
−9.0623	3.41	0.321	SF-8
−37.6563	3.41	19.631	Air
Focal ratio: 5.86632			
Effective focal length (mm):	20.0042		
Back focal length (mm):	19.6355		
Front focal length (mm):	−19.9043		

Source: From Malacara, D. and Z. Malacara, *Handbook of Lens Design*, Marcel Dekker, New York, 1994.

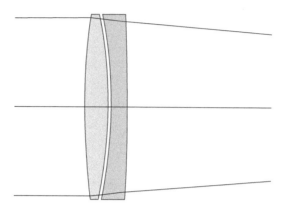

FIGURE 10.9   An air-spaced doublet.

TABLE 10.2   Broken-Contact Aplanatic Achromatic Doublet

Radius of Curvature (mm)	Diameter (mm)	Thickness (mm)	Material
58.393	20.0	4.0	BK7
−34.382	20.0	0.15	Air
−34.677	20.0	2.0	F2
Focal ratio: 4.99942			Air
Effective focal length (mm):	99.9884		
Back focal length (mm):	96.5450		
Front focal length (mm):	−99.3828		

Source: From Malacara, D. and Z. Malacara, *Handbook of Lens Design*, Marcel Dekker, New York, 1994.

Achromatic lenses are used as telescope objectives or as collimators. When used as laser collimators, the off-axis aberrations are not as important as in the telescope objective.

10.6 Afocal Systems

An afocal system by definition has an infinite effective focal length. Thus, the image of an object at an infinite distance is also at an infinite distance. However, if the object is at a finite distance in front of the system, a virtual image is formed, or vice versa. Let us consider, as in Figure 10.10, a real object with height H equal to the semi-diameter of the entrance pupil and the principal ray (dotted line) entering with an angle α. A marginal ray (solid line) enters parallel to the principal ray. Since the system is afocal and these two rays are parallel to each other, they will also be parallel to each other when they exit the system. Thus, we can write

$$M = \frac{\tan\beta}{\tan\alpha} = \frac{D_2}{D_1}\frac{X}{X'} = \frac{1}{M}\frac{X}{X'}.$$ (10.13)

The distance X is positive if the object is after the entrance pupil (virtual object) and the distance X' is positive if the image is after the exit pupil (real image). These two quantities always have the same sign. The total distance L between the object and the image is given by

$$L = L_p - X\left(1 - \frac{1}{M^2}\right),$$ (10.14)

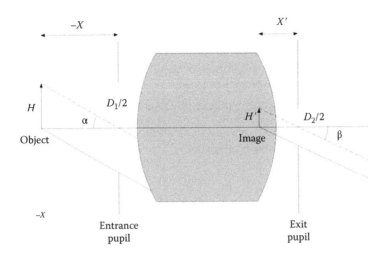

FIGURE 10.10   An afocal system forming a real image of a virtual object.

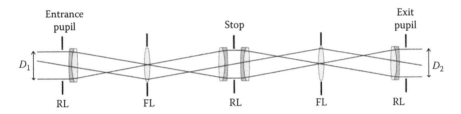

FIGURE 10.11   Optical schematics of a periscope.

where L_p is the separation between the entrance and exit pupils. We notice then that the lateral magnification H'/H is a constant equal to the angular magnification of the system.

Another interesting property of afocal systems is that, if the object and image planes are fixed in space, the image can be focused by axially moving the afocal system, without modifying the lateral magnification.

These properties find applications in microlithography.

10.7 Relay Systems and Periscopes

A periscope is an afocal instrument designed to observe through a tube or long hole, as shown in Figure 10.11.

An alternate series of imaging or relay lenses (RL) and field lenses (FL) form the system. The first imaging lens is at the entrance pupil and the last imaging lens is near the exit pupil, at the pupil of the observing eye. The field lenses are used to maintain the principal ray propagating to the system. Thin field lenses introduce almost no chromatic aberrations and distortion; hence, these are frequently thin single lenses. Imaging lenses, on the other hand, produce axial chromatic aberration. For this reason, they are represented here by doublets.

A disadvantage of this system is that all lenses have positive power, producing a large Petzval sum. Thus, the only way of controlling the field curvature is by the introduction of a large amount of astigmatism.

It must be pointed out that the ideal field lenses are at the intermediate images positions: i.e., where the meridional ray crosses the optical axis. Then, the images will be located on the plane of the field lenses and any dirt or imperfection on these lenses will appear sharply focused on top of the image.

This is not convenient. For this reason, field lenses are axially displaced a little from the intermediate images' positions.

Imaging lenses are located at the pupils' positions, where the principal ray crosses the optical axis. This location avoids the introduction of distortion by these lenses. However, the overall aberration balancing of the system may call for a different lens location.

As in any afocal system, if the object is located at infinity and the image is also located at infinity, the meridional ray is parallel to the optical axis in the object space as well as in the image space. Under this condition, the ratio of the slope β of the principal ray in the image space to the slope a in the object space is the angular magnification of the system, given by

$$M = \frac{\tan\beta}{\tan\alpha} = \frac{D_1}{D_2} \tag{10.15}$$

as in any afocal system, where D_1 is the diameter of the entrance pupil and D_2 is the diameter of the exit pupil.

10.8 Indirect Ophthalmoscopes and Fundus Camera

These ophthalmic instruments are periscopic afocal systems, designed to observe the retina of the eye. They have the following important characteristics:

1. The entrance pupil must have the same position and diameter as the eye pupil of the observed patient.
2. To have a good observed field of the retina of the patient, the angular magnification should be below 1. Thus, the exit pupil should be larger than the entrance pupil.

A version of this system is illustrated in Figure 10.12. The first lens in front of the eye is an imaging lens. The image of the retina is formed at the back focal plane of this lens. The height of the meridional ray (solid line) is the semi-diameter of the pupil of the observed eye. This small aperture makes the on-axis aberrations of this lens very small. This principal ray (dotted lines) arrives far from the center of the imaging lens, marking its diameter quite large if a good field angular diameter is desired. In order to form a good image of the pupil of the observed eye, this imaging lens has to be aspheric.

At the image of the pupil of the observed eye, a stop is located, which has three small windows. Two windows provide a stereoscopic view of the retina, sending the light from each window to a different observing eye. The third window is used to illuminate the retina of the observed eye.

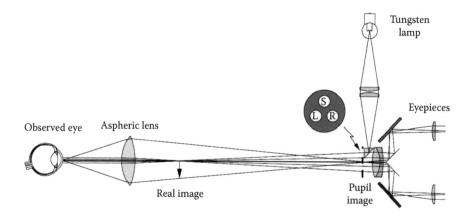

FIGURE 10.12   Portable stereoscopic indirect ophthalmoscope.

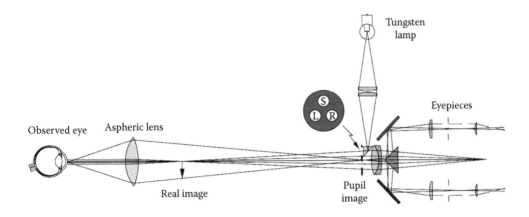

FIGURE 10.13   Schematics of a stereoscopic indirect ophthalmoscope.

The angular magnification M of the ophthalmoscope, as pointed out before, should be smaller than 1. Thus, the focal length of the eyepieces should be larger than the focal length of the aspheric lens (about five times larger).

A slightly different indirect ophthalmoscope is illustrated in Figure 10.13, using a second imaging lens at the pupil of the observed eye, thus producing an erected image at the focal planes of the eyepieces. This is an achromatic lens, since the image of the pupil has been magnified. The aperture of this lens has a stop in contact with three small apertures: one for illumination and two for stereoscopic observation. The final image is observed with a pair of Huygens' eyepieces.

The angular magnification M of this ophthalmoscope is given by

$$M = \frac{\tan\beta}{\tan\alpha} = m\frac{f_a}{f_e}. \tag{10.16}$$

where f_a is the effective focal length of the aspheric lens, f_e is the effective focal length of the eyepiece, and m is the lateral magnification of the achromatic lens located at the image of the pupil of the observed eye.

To effectively use all the field width provided by the aspheric lens aperture, the tangent of the angular field semi-diameter α_e of the eyepiece should be equal to the tangent of the angular field semi-diameter β of the aspheric lens multiplied by the angular magnification M, as follows:

$$\tan\beta = M \tan\alpha_e. \tag{10.17}$$

10.9 Projectors

Projectors are designed to form the amplified real image of a generally flat object on a flat screen. There are several kinds of projectors, but their optical layout is basically the same.

10.9.1 Slide and Movie Projectors

The basic optical arrangement in a slide projector is shown in Figure 10.14. The light from the lamp must reach the image screen as much as possible, after illuminating the slide in the most homogeneous manner possible. If a lamp without reflector is used, a concave mirror with the center of curvature at the lamp is employed.

In order to achieve this, the illuminating system must concentrate the maximum possible light energy at the entrance pupil of the objective, after passing through the slide. The light must have a homogeneous

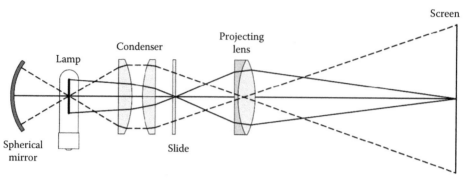

FIGURE 10.14   A classical slide projector.

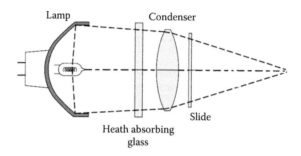

FIGURE 10.15   Illumination in a modern slide projector.

distribution over the slide. There are two basic illuminating systems. The classical configuration is shown in Figure 10.14, where the condenser can have several different configurations. The spherical mirror on the back of the lamp can be integrated in the lamp bulb.

Another more recent illumination configuration is shown in Figure 10.15. The lamp has a paraboloidal or elliptical reflector with a large collecting solid angle. This lamp produces a more or less uniform distribution of the light on a plane about 142 mm from the rim of the reflector. The condensing lens is aspheric.

10.9.2 Overhead Projectors

An overhead projector is shown in Figure 10.16. The light source is a lamp with a relatively small filament. The condenser is at the transparency plane, formed by a sandwich of two Fresnel lenses. This condenser forms, at the objective, an image of the small incandescent filament. Thus, the small image of the light source makes the effective aperture of the objective also small. This minimizes the need for spherical aberration and axial chromatic aberration corrections. Off-axis aberrations are nearly corrected with an almost symmetrical configuration of the objective, formed by two identical meniscus lenses.

10.9.3 Television Projectors

A basic difference between slide projectors and projectors used for television is that the object and the light source are the same. Thus, the objective has to have a large collecting solid angle. However, this large aperture increases the requirements for aberration corrections. A common system uses an inverted Schmidt configuration, as shown in Figure 10.17, but many others have been devised.

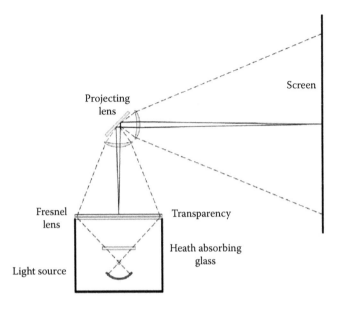

FIGURE 10.16 An overhead projector.

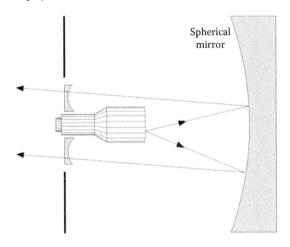

FIGURE 10.17 A television projector.

References

1. Kingslake R. Basic geometrical optics. In *Applied Optics and Optical Engineering*, Vol. 1, Kingslake, R. (ed.). San Diego, CA: Academic Press, 1965.
2. Kingslake R. Lens design. In *Applied Optics and Optical Engineering*, Vol. 3, Kingslake, R. (ed.). San Diego, CA: Academic Press, 1965.
3. Hopkins RR. Malacara D. Optics and optical methods. In *Methods of Experimental Physics, Geometrical and Instrumental Optics*, Vol. 25, Malacara D. (ed.). San Diego, CA: Academic Press, 1988.
4. Welford WT. Aplanatism and isoplonatism. In *Progress in Optics*, Vol. 13, Wolf, E. (ed.). Amsterdam: North Holland, 1976.
5. Hopkins RE. Geometrical optics. In *Methods of Experimental Physics, Geometrical and Instrumental Optics*, Vol. 25, Malacara D. (cd.). San Diego, CA: Academic Press, 1988.
6. Malacara D., Malacara Z. *Handbook of Lens Design*. New York, NY: Marcel Dekker, 1994.

11

Telescopes

Marija Strojnik and
Maureen S. Kirk

11.1 Traditional Transmissive Telescopes

11.1.1 Introduction

We may divide telescopes into three categories according to their primary function of application: the terrestrial telescope, the astronomical telescope, and the space telescope. While they all meet the same basic functional requirement of increasing the angle subtended by an object at the detector location, each of them satisfies different additional objectives.

The layout of the simplest basic telescope is shown in Figure 11.1. Its function is to increase the angle α that the object subtends at the viewer. By using the correctly curved optical surfaces of the two lenses or mirrors that comprise the basic telescope, the rays from the distant object are incident into the observer's eye with a much larger angle of incidence, β. The operation of a telescope is most easily understood with reference to Figure 11.2. The radiation originating at a distant object subtends an angle α when viewed with an unaided (naked) eye. When seen through the telescope, the object is magnified transversely to subtend an angle β at the viewer.

The function of a telescope is quite similar to that of a microscope, but for a small difference. They both magnify the angular extent of an object that subtends a small angle at the human eye. However, in the case of a telescope, the small angle subtended by the object arises as a consequence of the (very) large, or infinite, object distance. In a microscope, on the other hand, the small angle is a consequence of the very small size (height) of the object and the minimum accommodation distance of the human eye.

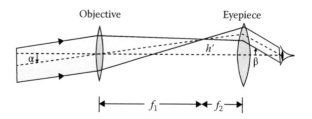

FIGURE 11.1   Basic (Keplerian) telescope with two converging lenses separated by the sum of their focal distances. The eyepiece lens, located closer to the viewer, is a strongly convergent lens. The collimated rays incident at an angle α are incident into the observer's eye on the right with a higher angle, β.

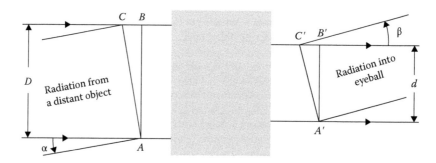

FIGURE 11.2   The function of a telescope is to increase the angle that the distant object subtends at the observer.

We may, therefore, summarize that we design and build the instruments, such as microscopes and telescopes, so that we can adapt the world around us to our limited visual opto-neural system [1]. Here we remember that the distribution of the rods and cones on the retina defines the limiting resolution of a normal human eye. The accommodation distance of the lens defines the minimum comfortable viewing distance of the human eye. The design procedures for classical telescopes have been described in the literature by many authors (see, e.g., Hopkins [2], Kingslake [3], Malacara and Malacara [4], Fisher et al. [5], and Smith [6]).

11.1.2 Resolution

The resolution of a telescope may be defined in a number of ways. When a telescope with a circular aperture is diffraction-limited, the image of a point source is not a point, but rather a bright disk surrounded by a series of bright rings of much smaller and decreasing amplitude. Figure 11.3 shows the normalized irradiance pattern observed in the focal plane of a traditional telescope with a circular aperture. Theoretically, there is zero irradiance in the first minimum, counting from the central peak, defining the first dark ring. About 84% of the optical energy is enclosed inside the first dark ring. According to the diffraction theory of imaging, the resulting irradiance distribution is proportional to the square of the Fourier transform of a disk aperture, that is, a Bessel function of the order 1 over its argument $[J_1(\rho)/\rho]$, in the case of a circular aperture; see Goodman [7], Born and Wolf [8], for more examples. The bright area enclosed by the first dark ring of the diffraction pattern is called the Airy disk. The insert in Figure 11.3 shows the corresponding gray-scale irradiance, assuming low exposure as expected for most telescopes applications. The first bright ring is still barely visible.

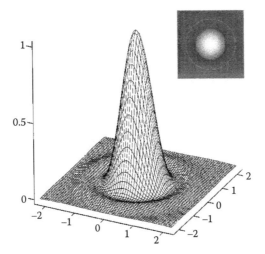

FIGURE 11.3 The normalized intensity pattern obtained in the focal plane of a traditional telescope with circular aperture. Theoretically, there is zero intensity in the first minimum, from the central peak, defining the first dark ring, or the radius of the resolution spot. The inset shows the corresponding gray-scale intensity, assuming a low exposure as expected for most applications of the telescopes. The first bright ring is barely visible.

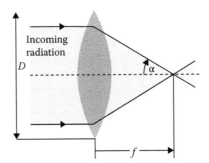

FIGURE 11.4 Any optical system, including a telescope, may be replaced by an equivalent optical system with limiting stop of diameter D and focal length f.

The radius of the bright disk is defined by the radius where the Bessel function becomes zero, in the expression for the first zero irradiance ring,

$$r = 1.22\lambda f/D\,[\text{m}]. \tag{11.1}$$

Here λ is wavelength [m], f is the focal length of the imaging system [m], and D is the diameter of the aperture stop [m]. These quantities are illustrated in Figure 11.4 for a general optical system with the equivalent focal distance f and the stop or aperture of diameter D. Here angle α is defined as $D/2f$ and represents the angular resolution of the telescope.

The larger the aperture is, the higher the angular resolution (the smaller the angle) of the telescope that may theoretically be achieved. For the same magnitude of the aberrations, the figuring and the alignment errors, the diameter of the diffraction spot increases with wavelength. Viewed from a different perspective, the same telescope may give a diffraction-limited performance in the infrared (IR) spectral region, but not in the visible [9]. A telescope is said to be diffraction-limited when the size of the diffraction spot determines the size of the image of a point source. Even though the stars come in different (huge) sizes, they are considered point sources due to their large (object) distances. The size of

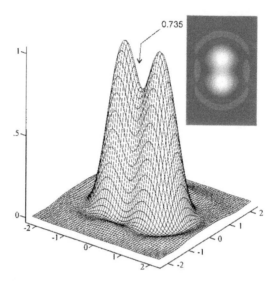

FIGURE 11.5   The normalized intensity patterns of two barely separated point objects obtained in the focal plane of a traditional telescope with a circular aperture. Theoretically, the intensity minimum between the two peaks reaches 0.735. The first dark ring is barely seen, surrounding both spots. The inset shows the corresponding gray-scale intensity distribution, assuming a low exposure as expected for most applications of the telescopes. The separate low exposure images are clearly identifiable as arising from two individual point sources. The first bright ring is nearly invisible.

their image depends on the telescope resolution or the seeing radius (when located on the Earth's surface), depending on the telescope diameter (see Section 11.2.4.2), rather than the actual star dimension. Astronomers identify the stars by the shape and the size of the diffraction spot diameter [10,11].

According to the generally accepted Rayleigh resolution criterion, two point sources may just be distinguishable when zero in the image-irradiance distribution of the first one coincides with the peak of the second one. In the image space, their separation is the distance given in Equation 11.1. Figure 11.5 presents the normalized irradiance pattern of two barely separated point objects, exhibited in the focal plane of a traditional telescope with a circular aperture. Theoretically, the irradiance minimum between the two peaks reaches 0.735. The first dark ring surrounding both spots is barely visible. The inset illustrates the corresponding gray-scale irradiance distribution, assuming a low exposure as expected for most telescopes' applications. The separate low exposure images are clearly identifiable as arising from two individual point sources. The first bright ring is barely visible.

11.1.3   Geometrical Optics

In addition to the diffraction effects, any optical instrument, including the telescope, also suffers from performance degradation due to aberrations, fabrication imperfections, and alignment errors. All these effects contribute to the spreading of the point image into a bright fuzzy disk. Aberrations are the variations by which an optical system forms an image of a point source that is spread out into a spot, even in the absence of diffraction effects. They are caused by the shape of the optical components and are a phenomenon purely of geometrical optics. In the paraxial optics approximation, imaging is seen as merely a conformal transformation: the image of a point is a point, the image of a straight line is a deformed line, and angles are preserved. Strictly speaking, spreading of the object point into an image spot is a consequence of aberrations, while deformation of a line into a curve is interpreted as an actual deformation. The true third-order aberrations are spherical aberration, astigmatism, and coma. A telescope or, for that matter, any optical system corrected for spherical aberration and coma, is referred to as aplanatic to the third order. When an optical system is additionally free of astigmatism, it is called aplanatic and anastigmatic. The shape of the surface on which the best image is located, the so-called Petzval surface, is also somewhat controllable within the third-order aberration theory.

Additionally, a tilt and defocus may also contribute to image deterioration. Either of them may be compensated for by the proper image displacement and image plane orientation.

When the geometrical spot size due to the effects of aberrations and misalignments is smaller than the disk inside the first dark ring of the diffraction pattern, the resolution is still given by Equation 11.1. When this condition is met, the optical system is said to be limited in its performance by the diffraction effects; or stated more succinctly, it is a diffraction-limited optical system.

The geometrical resolution of a moderately aberrated optical system is defined as the radius of the spot within which 90% of rays, originating at a point source, cross the image surface. This is often estimated visually, upon the review of the spot diagram. Figure 11.6 presents the spot diagrams for three image heights of a diffraction-limited Ritchey–Chretien telescope configuration with a two-element

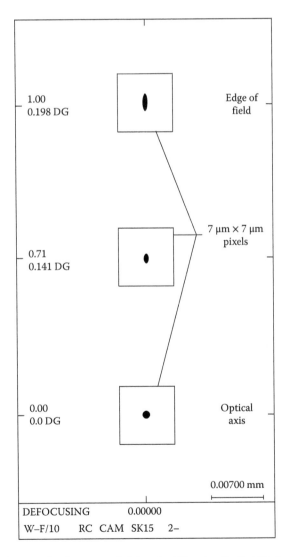

FIGURE 11.6 The spot diagrams for three image heights of a diffraction-limited Ritchey–Chretien telescope configuration with a two-element field corrector. (From Scholl MS. *Proc. SPIE*, **2019**, 407–412; 1993b.)

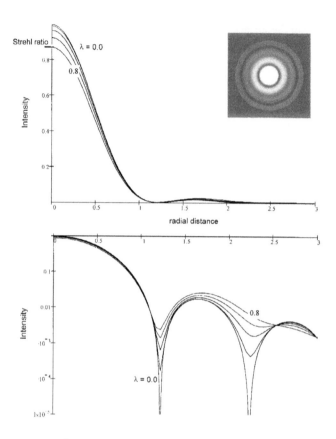

FIGURE 11.7 The intensity as a function of radial distance for an aberrated and diffracted basic telescope (see Figure 11.1) in the best focus location. The parameter is the wavefront error due to the spherical aberration at the aperture edge, increasing from zero to 0.8λ in increments of 0.2λ.

field corrector (see Section 11.2.2.3) [12]. The squares indicate the CCD pixel size, whose side has been chosen to be equal five times the resolution in Equation 11.1.

Figure 11.7 displays the irradiance as a function of radial distance for an aberrated and diffracted traditional telescope at the best focus location. The parameter is the wavefront error due to the spherical aberration at the aperture edge, increasing from zero to 0.8λ in increments of 0.2λ. The central peak (Figure 11.7, top) decreases as the aberration is increased, while the irradiance amplitude in the annulus increases. For increasing distances from the origin in the image space, the irradiance patterns become similar. The similarity to the diffraction limited pattern depends on the amount of the aberration. In the lower part, the same information is shown on a logarithmic scale.

Not surprisingly, in the paraxial focus, the effects of the aberrations rapidly overshadow the diffraction effects. The first few dark rings are drowned within the bright spot whose diameter is increased due to the aberrations. This is illustrated in Figure 11.8, where no irradiance zero is observed for low-order dark rings. The number of dark rings that disappear is related to the amount of aberrations.

The aperture diameter is a concern for space applications, because the weight of the primary mirror, and with it the whole telescope weight increases rapidly with the diameter. A multispectral spaceborne Cassegrain telescope features a primary mirror with 450 mm clear aperture, is made of zerodur, and is light-weighted by 50%. Its most prominent aberration is astigmatism, caused by gravity effects, deformation arising from the bonding processes and mounting of the primary mirror on the main telescope structure. Lin et al. [13] report that mechanical ground-support equipment, designed for the alignment

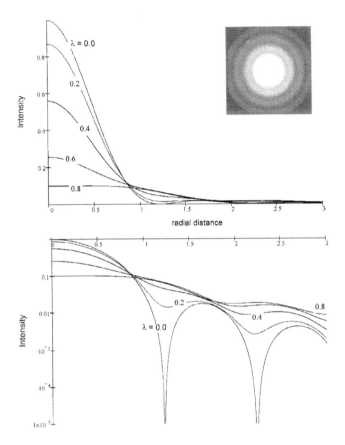

FIGURE 11.8 The intensity as a function of radial distance for an aberrated and diffracted basic telescope in the paraxial focus location. The parameter is the wavefront error due to the spherical aberration at the aperture edge, increasing from zero to 0.11λ in increments of 0.2λ.

and assembly processes represent the critical requirements imposed on the primary mirror assembly. Free-form surfaces enable design and implementation of novel optical systems by providing additional degrees of freedom to an optical designer. Gautam et al. [14] present an optical design of an off-axis Cassegarin telescope incorporating a free-form surface as the secondary mirror.

11.1.4 Modulation Transfer Function

The modern ray trace programs tend to summarize the performance of an optical system in terms of its capacity to image the individual spatial frequency components of the object faithfully. This presentation of results assumes that the object emittance (see the radiometry chapter elsewhere in this volume) is decomposed into its Fourier components. Then, the optical system may be considered as a black box that progressively decreases the modulation of the increasingly higher spatial frequencies. Furthermore, it usually modifies their phases as well. Imaging analysis theory then says that Fourier frequencies of the object multiplied by the optical transfer function, interpreted as a black box, provide Fourier frequencies of the image. The magnitude of an optical transfer function (OTF) is a modulation transfer function (MTF), often the quantity of the primary concern when the phase is not of interest. The significance of the modulation transfer function to modify the amplitudes of the spatial frequency components is illustrated in Figure 11.9.

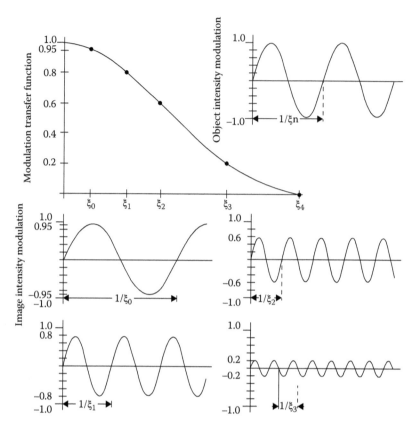

FIGURE 11.9 The significance of the modulation transfer function. It is zero for spatial frequencies higher than e4. No image components at frequencies higher than e4 are transmitted.

Figure 11.10 presents the best (theoretical) modulation transfer function as a function of position in the image plane of a traditional telescope with a circular aperture. Figure 11.10a features the three-dimensional view to illustrate MTF dependence on two spatial frequency coordinates. Figure 11.10b presents one cross-section of this rotationally symmetric function. There exists a minimum spatial frequency in the object space whose amplitude is zero in the image space. This spatial frequency is referred to as the cutoff frequency, because the optical system images no spatial frequencies larger than this one. Its value is $D/f\lambda$ for an optical system with focal length F and a circular aperture of diameter D.

The effects of the aberrations, component misalignment, gravity effects, and the imperfect surface figuring result in a deteriorated MTF. It does not achieve the values of the theoretical MTF. This behavior may be seen in Figure 11.11, showing the OTF as a function of the normalized spatial frequency coordinates for the case of 0.2–0.8λ of spherical aberration at the aperture edge. The theoretical MTF is shown for reference. For the MTF value of 0.4, as is sometimes required for high-resolution imaging, we see that the incremental increase in spherical aberration of 0.2λ results in a decrease in the corresponding maximum acceptable spatial frequency by about 15%. For smaller required values of MTF, the situation is similarly worsened. Assuming that one is interested in imaging a specific normalized spatial frequency, let us say the one at 0.5, the increment in spherical aberration by 0.2λ results in about 20% reduction in the modulation. The negative OTF arising from 0.8λ of spherical aberration corresponds to the phase reversal in addition to an insignificant amount of modulation. Warren Smith describes the MTF, the aberration effects, and image evaluation further in his classical textbook on *Modern Optical Engineering* [15].

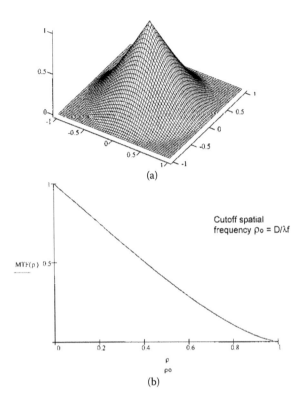

(a)

(b)

FIGURE 11.10 The best (theoretical) modulation transfer function as a function of position in the image plane of a traditional telescope with a circular aperture: (a) the three-dimensional view to illustrate the dependence on two coordinates; (b) one cross-section of this rotationally symmetric function. No object frequencies are imaged larger than the system cutoff frequency.

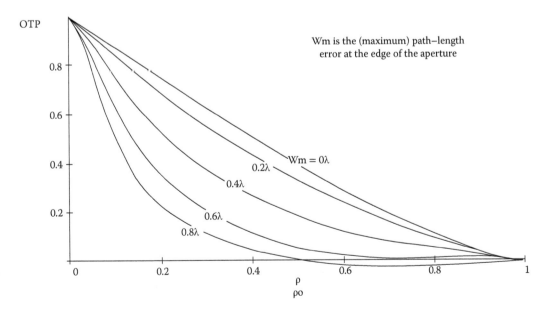

FIGURE 11.11 The optical transfer function (OTF) as a function of the normalized spatial frequency for the cases of 0–0.8λ of spherical aberration at the edge of the aperture in increments of 0.2λ.

The publicized case of the spherical aberration on the primary mirror of the original Hubble space telescope has been estimated to be about 0.5λ.

The performance of a designed (potential) telescope for the imaging of individual spatial frequencies is easily assessed when it is presented in the form of the MTF. Its predicted shape is compared with the optimum theoretically possible one for the same aperture diameter. The MTF of a high-resolution telescope originally designed to survey the Martian surface with the engineering goal of constructing topographic maps (for requirements, see [16]) is exhibited in Figure 11.12. The nearly perfect theoretically achievable MTF is obtained for the three field positions. The MTF value at the design spatial frequency of 71.7 cycles/mm is about 0.47. Figure 11.12b indicates that very small MTF degradation of a few percent is encountered for the range of displacements out of the best image surface for the potential CCD placement.

Generally, reduced amplitude of the spatial frequencies in the image space will contribute to an unfaithful and non-conformal imaging by the optical system. The amplitudes of other sources of optical and electronic noise may also surpass the magnitudes of the image amplitudes. Thus, there exists a minimum value of the MTF, and its corresponding spatial frequency, such that the spatial frequencies larger than it are not recoverable from the background noise. This value of MTF_{min} depends additionally on the amount of post-processing and image enhancement that is expected after the image capture. It is further limited by the degree and quality of the quantitative results that will be extracted from the measured irradiance distribution of the detected images. Therefore, the MTF in Figure 11.12 is shown only for the values of the MTF larger than 0.3 due to the high requirement for the recovery of quantitative data.

11.1.5 Refractory or Transmissive Telescopes

The simplest and most common telescope used in the refractory mode is the basic Keplerian telescope, illustrated in Figure 11.1. Two positive lenses with focal distances f_1 and f_2 are combined in such a way that their focal points coincide in the common space between them. The magnification of this system is

$$M = -f_1/f_2 = D_1/D_2. \tag{11.2}$$

The ratio of their focal lengths is equal to the required angular magnification. This means that the diameter of the first lens has to be much larger than the second one in order to provide requisite magnification.

A Keplerian telescope may be used as an image erector, or as a beam expander in the relatively low-power laser system. Internal focus allows the incorporation of a pinhole for beam centering and spatial cleaning. However, the Keplerian beam expander is not recommended for high-power laser applications due to excessive air heating in the focal volume, which may introduce thermal lensing effects, air breakdown, or even generation of the second-order electric field effects.

The ever-present need for telescope compactness favors a Galilean telescope configuration, depicted in Figure 11.13. Here, the objective is a strong positive lens, while the eyepiece is a negative lens, collocated in such a way that the focal points once again coincide, this time in the image space. The total telescope length is reduced by twice the focal distance of the eyepiece. Additionally, the focal point is not real, making it an ideal candidate for a beam expander in high-power laser beam applications.

The simplest Galilean telescope is presented in Figure 11.14. It is an example of an afocal system, both in the object and in the image space. The object space encompasses all the volume where the object may be located on the left side of the telescope (optical system), corresponding to the object location. We adhere to the convention that the light is incident from the left in this monogram. The image space includes all the space where the image may be found, usually on the right of the telescope, within the convention stated. An afocal system is the one that has an effective focal distance equal to infinity. The effective focal distance is the distance between the stop and the point where the ray incident parallel to

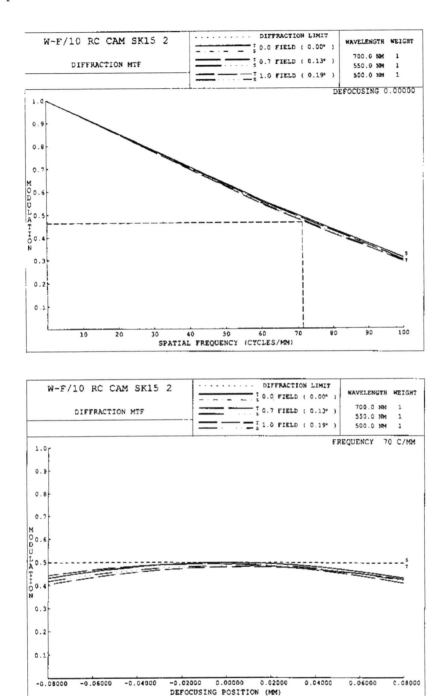

FIGURE 11.12 The modulation transfer function (MTF) of a high-resolution telescope designed to survey the Martian surface with a long-term goal of constructing topographic maps. (a) The nearly perfect theoretically achievable MTF is obtained for the three field positions. The MTF with the values larger than 0.3 is shown only due to the requirements for the recovery of quantitative data. (b) A very small MTF degradation of a few percent is encountered for the range of displacements out of the best image surface. (From Scholl MS. *Proc. SPIE*, **2019**, 407–412; 1993b.)

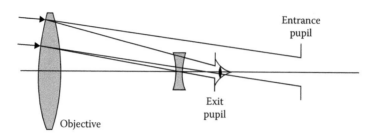

FIGURE 11.13   A Galilean telescope configuration.

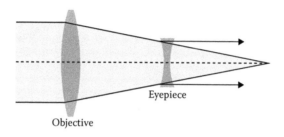

FIGURE 11.14   A Galilean telescope in afocal configuration.

the optical axis in the object space crosses the optical axis in the image space. In an afocal system, this point is at infinity.

11.1.6 Terrestrial Telescope

The terrestrial telescope is used by hunters and explorers for outdoor applications or as an "opera glass" to observe new clothes of fine ladies in the audience. For these purposes, it is absolutely mandatory that the image be erect, color-corrected, and a wide half-field-of-view is highly desirable. The terrestrial telescopes may be used in the visible spectral region and also in the very near infrared (NIR) for the nightvision, with the nightglow as the source of near IR illumination. The object distance for these telescopes ranges from about 50 m to 1000 m. For some applications like a vision scope, a telephoto arrangement may also be incorporated. The terrestrial telescopes tend to perform like surveillance cameras, for large object distances. They may have half fields-of-view of up to about 35 degrees.

The first lens, the one close to the object, is called the objective for its proximity to the object. The second lens, the one close to the observer, is referred to as the eyepiece because it is designed for direct (human) viewing of the terrestrial scene. The eyeball and the eyelashes require clearance, sufficient space for the easy unobstructed rotation around the eyeball axis, and a relatively large half-field of view. The eyepiece is an optical subsystem designed to meet these specifications. Eye relief is the distance between the last surface of the eyepiece and the first surface of the human eyeball (i.e., the cornea, the contact lens, or glasses). Eye relief has to be long enough to accommodate (long) eyelashes and/or spectacle lenses for those with astigmatism. The eyepiece is often a compound lens incorporating a field lens. The Huygens, Ramsden, and Kellner eyepieces represent successive improvements in the order of increasing degree of complexity, larger half-field of view (20 degrees), and consequently, the user's comfort. A review of a number of eyepieces and objectives is described by Smith [17] and Fisher [5].

The terrestrial telescope must, by necessity, incorporate an image erector system. A lens relay system makes the system quite long. A prism pair is considered more compact, but it is heavier.

Either Keplerian or Galilean telescope may be modified to meet the requirements of a terrestrial telescope. In practice, both the objective and the eyepiece are thick lenses, or a combination of lenses in order to minimize the chromatic and other aberrations that are annoying to a human observer, especially for large field angles.

Low vision individuals usually possess some ability to see objects, especially if those are made to appear close to the eyeball. A telescopic system is ideal for increasing field-of-view subtended at the eyeball. However, it often requires hands for support and switching between operational configurations. Arianpour et al. [18] describe a 1.6 mm thick scleral contact lens with two paths. The F/9.7 telescopic vision path uses 8.2 mm diameter annular entrance pupil and four internal reflections. The un-magnified F/4.1 vision path is mostly through the central lens aperture. Schuster et al. [19] describe a follow-up device, incorporating wink-controlled hands-free switching system. This eye-borne telescopic vision is embedded within scleral contact lenses.

11.2 Reflecting Telescopes

11.2.1 Increasing the Light-Collecting Area

An object is seen or detected by the telescope detector only if enough energy is collected to produce the requisite signal-to-noise ratio in the focal plane. The light sensitive device may be a photographic plate, or, for most modern applications, a CCD (charge-coupled device). Its additional advantages are easy readout and storage for further processing, reusability without the need for refurbishment, and high response linearity. CMOS devices are even better.

The best image produced by the telescope optics is located on a curved surface, requiring image field correctors to accommodate the planar surface of the CCD or other semiconductor (solid-state) detectors [20]. The CCD is preferably placed in a plane such that the pixel distance from the surface of the best focus is minimized for all pixels. This is illustrated in Figure 11.15 where the best image surface is displaced only slightly from the CCD flat surface, due to the incorporation of a two-element corrector system.

The photographic plate still remains the medium of choice for precise spectroscopic Earth-based measurements. Its resolution, environmental stability, and a faithfulness of recording have not yet been surpassed by any other (electronic) recording medium.

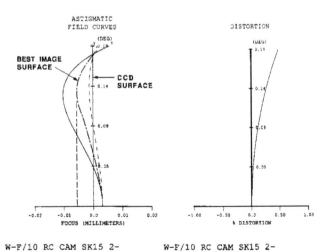

FIGURE 11.15 The CCD surface is placed in a plane such that the pixel distances from the surface of the best focus are minimized for all pixels. The best image surface is displaced only slightly from the CCD surface, due to the incorporation of a two-element corrector system (from Scholl [11]).

For a great majority of astronomical applications, it is desirable to have a large-diameter primary mirror to collect as much radiation as possible from faint and distant celestial sources. A number of fabrication and operational issues arise for a large diameter of the primary lens. First, these kinds of telescopes are not easily transportable; second, they require stability and placement in a controlled environment, such as that offered by a removable dome in most observatories.

There are basically three significant, fundamental challenges when incorporating a large-diameter primary lens into a transmissive telescope. The first one is the support and fabrication of two sides of an optical surface; the second one is mounting the large primary lens in the appropriate final structure. It is much easier to support the back of a single reflecting surface while it lays on the ground during the fabrication phase. This applies equally well to the fabrication phase as to the operational period. Finally, large optical quality glass blanks with homogeneous indices of refraction are difficult to produce.

Mirrors have other additional advantages over lenses. The transmissive components generally display chromatic aberration due to spectral dependence of the index of refraction, making them well cor-rected only over narrow spectral intervals. Broadband reflective coatings are now available over large spectral intervals with high reflectivity, permitting the use of a single design for wide spectral regions. Furthermore, metals perform well as reflectors, whether uncoated like aluminum or gold, or coated with multilayer dielectric film or gold. Finally, a mirror exhibits a lesser amount of aberrations than a lens of the same power.

All-reflective optical systems are known as *catoptric*. Mirror telescopes may be divided into one-, two-, and three-mirror systems. Flat-beam turning mirrors are not considered in the mirror count, as they have no power. Many well-corrected reflective systems employ a transmissive lens or a corrector plate for additional correction. When an optical system incorporates reflective as well as refractive com-ponents, it is known as a *catadioptric* system.

11.2.2 Catadioptric Telescope

Many nominally mirror telescopes additionally incorporate transmissive components for specific aber-ration correction most often field correction; they are known as catadioptric field correctors. The field corrector is a lens (or a combination of lenses) used to flatten the surface of the best image location, as commented with respect to Figure 11.15. An excellent corrector to the Ritchey–Chretien telescope design has been described by Rosin [21], with the corrector placed in the hole of the primary mirror. The Schmidt camera is probably the best-known system where the aberrations are corrected with a large transmissive element in the incoming beam.

11.2.2.1 One-Mirror Telescope

The simplest one-mirror system is a parabolic telescope, shown in Figure 11.16a, consisting of a simple parabola. When the stop is placed at the front focal point, this telescope has no spherical aberration and astigmatism, with the image located on a spherical surface. In this telescope, or a simple collimator, the focal plane instrument is located on optical axis, obstructing and interfering with the passage of the incoming beam.

In a Newtonian telescope, shown in Figure 11.16b, a small 45-degree flat mirror is placed in the beam path, deflecting the light out of the incoming beam and onto the workspace. More instruments may be located there without interfering with the beam passage. The great disadvantage of the reflecting telescopes is seen clearly in the case of the Newtonian telescope. Not all the area of the primary mirror is available to collect the radiation. The part of the beam obstructed by the beam-deflecting 45-degree mirror is reflected back to the object space. Also, its field of view is corresponding limited.

By using only the part of the parabola above its mechanical axis of symmetry, the need for the beam-turning mirror disappears. This configuration is known as the Herschelian telescope, as seen in Figure 11.16c. It represents one of the first concepts in the off-axis configurations now so popular for stray-light sensitive applications (see Section 11.2.6). The Herschelian telescope and the simple parabola

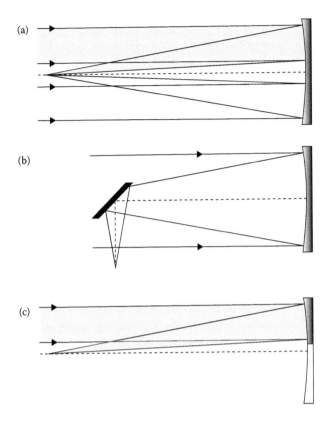

FIGURE 11.16 The development of a simple one-mirror telescope from (a) a simple parabola, to (b) a Newtonian telescope with an on-axis beam obstruction, to (c) a Herschelian telescope with an off-axis optical configuration that takes full advantage of the aperture of the primary mirror.

are the simplest collimator or telescope systems, depending on the direction of propagation of the incoming light.

The elementary implementation of the Schmidt telescope incorporates a spherical primary with a stop and a corrector plate at the center of the curvature of the primary mirror. The freedom from coma and astigmatism results from monocentric construction with the stop at the mirror's center of curvature. For the object at infinity, the image is free of spherical aberration, coma, astigmatism, and distortion. The focal surface has a radius equal to its focal length. Rosete-Aguilar and Maxwell [22] thoroughly discussed the design of the corrector plates.

The fabrication of the corrector plate in the Schmidt telescope is a complex task. One surface of the plate is a second-order polynomial in two variables, while the other side is a flat surface. Furthermore, this telescope is popular because its inventor came up with an ingenious way of fabricating the corrective surface. It is one of the first cases of stress polishing, also used much later on the Keck segments. First, the mirror is polished under vacuum conditions; then it is released to assume its stress-free form. This surface has been simplified in the Bouwers camera, on the basis of the work Schwarzschild had carried out on concentric surfaces [23]. In this case, the stop is located at the center of the curvature, and the corrector plate is a meniscus lens placed just before the focal point, so the rays transverse it only once. Its surfaces are concentric with the mirror surface, eliminating all off-axis aberrations. Maksutov designed his camera at about the same time, with the added improvement that the radii of curvature of the meniscus lens are related so as to correct the axial chromatic aberration.

11.2.2.2 Optical Performance of Telescopes with Central Obstruction

The central obscuration reduces the effective light-collection area of the primary mirror. The light-collection area then becomes

$$A = \pi(R^2 - r^2) = \pi R^2 (1 - \varepsilon^2)[m^2] \tag{11.3}$$

Here R is the radius of the primary, and r is the radius of the central obstruction, usually the secondary assembly, which includes the mirror, the mounts, and any baffling and support structures that interfere with the passage of the incident beam. The obscuration ratio $\varepsilon = r/R$ is a parameter with a value of zero in the absence of obscuration.

The irradiance of the image of a point object at infinity in the telescope with the central obscuration is displayed as a function of radial distance in Figure 11.17. The obscuration ratio ε is varied from 0 to 0.8 in increments of 0.2. The case corresponding to $\varepsilon = 0$ has been shown in Figure 11.3. The curves here are

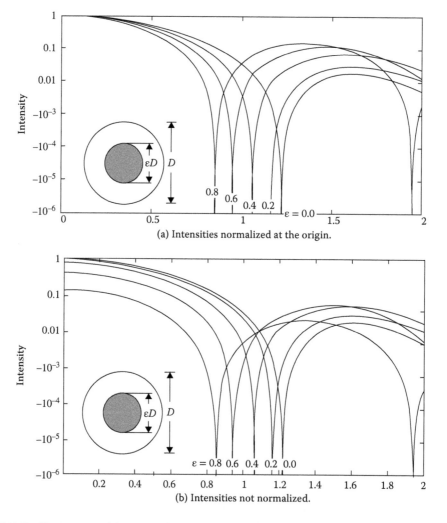

FIGURE 11.17 The intensity of the image of a point source at infinity in the telescope with a central obscuration as a function of radial distance with the obscuration ratio ε as a parameter, varying from 0 (no obscuration) to 0.8 in increments of 0.2.

presented in a logarithmic scale in order to demonstrate the trends clearly. As the obscuration radius increases, the first zero of the irradiance pattern moves to shorter radial distances, thus apparently increasing the resolution of the optical system [10]. This effect has also been utilized in microscopy to achieve the so-called *ultra-resolution*. Also, the successive zeros in the irradiance pattern are moved to shorter radial distances from the central peak.

In Figure 11.17a, where the irradiances of all curves are normalized to the value at the origin, we see that the increase in the beam obstruction ratio results in the increase in the irradiance height of the first bright annulus. The relative irradiance peak of the first annulus with respect to the central peak decreases with the increasing value of the obscuration ratio ε. In Figure 11.17b, we observe that the actual peak of the irradiance in the first annulus achieves the highest value for some intermediate value of ε, of about 0.5. This corresponds to the area ratio of 25%, the value considered highly desirable in the telescope designs. It represents a reasonable tradeoff between light collection area and size of the secondary mirror.

Figure 11.18 presents the integrated irradiance as a function of radial position for the obscuration ratio ε as a parameter, varying from 0 (no obscuration) to 0.8 in increments of 0.2. We identify the location of the dark rings when the integrated irradiance does not change with increasing radial distance, i.e., the curves come momentarily to a plateau. For all curves, the higher curve always corresponds to the case of lesser obscuration. All curves are normalized to the size of the light-collecting area: they approach asymptotically to 1 from below (become 1 at infinity). Here we can read off the curve that 84% of the energy is indeed contained within the first dark ring, for the cease of no obscuration. The radially integrated energy is a particularly convenient way of quantifying an aperture configuration when the spot sizes exhibit radial symmetry.

While the compactness of the optical spot (the image of a point source) is measured most effectively from the irradiance distributions in the focal plane, the MTF often provides the important complimentary information about the ability of the telescope to image a specific spatial frequency.

Thus, in Figure 11.19, we display the MTF as a function of radial spatial frequency coordinate, with the obscuration ratio as a parameter, varying from 0 (no obscuration) to 0.8 in increments of 0.2.

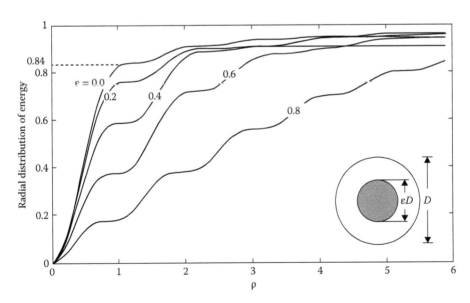

FIGURE 11.18 The integrated intensity as a function of radial position, with the obscuration ratio ε as a parameter, varying from 0 (no obscuration) to 0.8 in increments of 0.2. The compactness of the central spot may be recognized from the slope of this curve. There is 84% of energy enclosed within the first dark ring in an optical system without the obscuration.

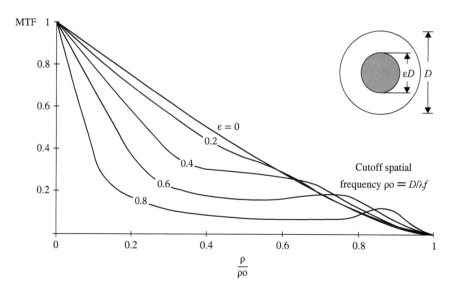

FIGURE 11.19 Modulation transfer function (MTF) as a function of radial spatial frequency coordinate, with the obscuration ratio ε as a parameter, varying from 0 (no obscuration) to 0.8 in increments of 0.2.

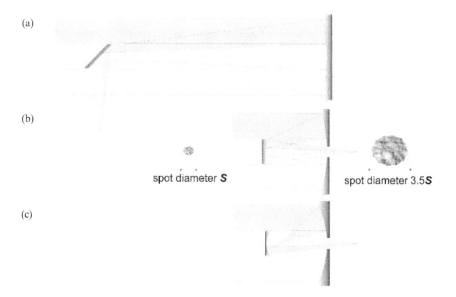

FIGURE 11.20 Comparison of (a) Newtonian, (b) Cassegrain, and (c) Gregorian telescopes under the conditions of the same f/number (F/#) and effective focal length. The image formed by the secondary mirror is magnified compared to that by the primary mirror.

Within the framework of imaging theory, this figure is equivalent to Figure 11.17. However, the performance degradation upon the inclusion of the central obscuration is potentially made much clearer in this presentation of results. For the obscuration ratio of 0.5 (area obscuration of 0.25), the imaging of the intermediate spatial frequencies (0.25–0.55) is decreased by approximately 40% with respect to the highest theoretically achievable value, with the MTF achieving only 0.25. For many imaging applications, this decreased value is considered too low for image reconstruction from under the noise.

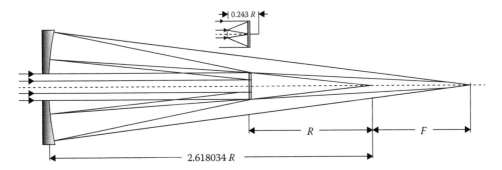

FIGURE 11.21 A Schwarzschild configuration representing aberration-free design on a flat field is appreciably larger than a Ritchey–Chretien telescope with the same F/# and effective focal length.

TABLE 11.1 Primary/Secondary Mirror Combinations in Telescopes

Schmidt plate	Primary	Secondary	Spherical aberration	Coma	Astigmatism	Field curvature
Aspheric	Spherical	Spherical	0	Yes	Yes	Yes
Aspheric[a]	Spherical	Hyperbolic	0	0	Yes	Yes
Aspheric[b]	Oblate spheroid	Oblate spheroid	0	0	0	Yes

[a] Classical form of the shortened Schmidt–Cassegrain.
[b] Ritchey–Chretien form.

11.2.2.3 Two-Mirror Telescopes

The evolution of two-mirror telescopes may be most easily appreciated by examining Figures 11.20 and 11.21, where four different telescope configurations are displayed with the constraint of the same aperture size, obstruction area, and the f-number. Figure 11.20 exhibits the following layouts: (a) Newtonian, (b) Cassegrain, and (c) Gregorian telescope. In Figure 11.21, a Schwarzschild configuration is presented separately due to its overall size, compared only with a Cassegrain telescope.

Strictly speaking, the simplest on-axis two-mirror telescope is a Newtonian, using just a beam-turning mirror to bring the focal plane and associated instruments out of the incoming beam. The two basic refractive telescopes may likewise be implemented as two-mirror systems. Figure 11.20b illustrates the case when the secondary mirror is divergent, resulting in an appreciably shorter instrument length and a smaller central obscuration, generally adopted for astronomical telescopes. This layout is known as a Cassegrain telescope when the primary mirror is a paraboloid and the secondary mirror is a hyperboloid. This configuration is necessary for the coincidence of the focal lengths behind the secondary mirror. The image formed by the primary mirror alone is smaller than the final image formed by the Cassegrain telescope because the secondary mirror actually magnifies it. When the secondary mirror is concave, there is a common focal point between the mirrors and the system is somewhat longer than the sum of the focal lengths. This layout is exhibited in Figure 11.20c, featuring the advantage that the final image just outside of the primary mirror is erect (head-up). This layout may be used for terrestrial applications. The configuration known as a Gregorian telescope incorporates an on-axis section of a paraboloid for the primary mirror, and an on-axis ellipsoid for the secondary mirror. In accordance with the established terminology, we will use shorter terminology and leave out *on-axis*, when not explicitly referring to an off-axis configuration, and *section* though this term is always implied.

Table 11.1 lists the possible combinations of mirror shapes in the telescope systems employing the corrector plate, and the aberrations that remain in the image. The corrector plate may be used in the incoming beam only for small telescopes where weight is not a significant problem. In other configurations, it is placed just before the image plane. This refractory component has low power, so that its surface

TABLE 11.2   Different Forms of Cassegrain Telescopes

	Primary Mirror	Secondary Mirror	Limiting Aberrations
Classical	Parabolic	Hyperbolic	Coma Astigmatism Field curvature
Ritchey–Chretien	Hyperbolic	Hyperbolic	Astigmatism Field curvature
Dall–Kirkham	Elliptical	Spherical	Coma Astigmatism Field curvature

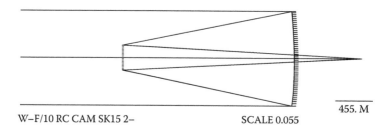

W–F/10 RC CAM SK15 2–   SCALE 0.055   455. M

FIGURE 11.22   Ritchey–Chretien layout of a 1 m diameter F-10 telescope proposed for the survey of the Martian surface with resolution of 0.25 m from a stable orbit. (From Scholl MS. *Proc. SPIE*, **2019**, 407–412; 1993b.)

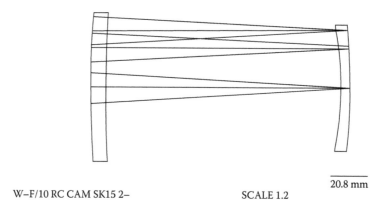

W–F/10 RC CAM SK15 2–   SCALE 1.2   20.8 mm

FIGURE 11.23   A two-element field corrector used to flatten the field of the telescope shown in Figure 11.22 and to permit good-quality imaging in the infrared spectral region in addition to that in the visible. (From Scholl MS. *Proc. SPIE*, **2019**, 407–412; 1993b.)

may be shaped to correct for the aberrations generated by the primary and the secondary mirrors. The Questar optical system uses a Schmidt-type corrector in the entering beam to correct the spherical aberration of two spherical mirrors.

Table 11.2 lists the different forms of Cassegrain telescopes. The Ritchey–Chretien configuration (see Figure 11.22) is used nowadays for nearly all large telescopes for astronomical applications, including a space application to survey the Martian surface from orbit prior to the identification of a potential landing site. Its residual aberration, astigmatism, may be corrected with the introduction of an additional component. The field curvature cannot be eliminated, as it is the consequence of the curvature of individual components; however, it may be decreased by the introduction of a small field correcting element, illustrated in Figure 11.23. C. G. Wynne wrote an excellent review of these telescope configurations [24].

Figure 11.21 exhibits the famous Schwarzschild layout that appears much like an inverted Cassegrain. This system is free of spherical aberration, coma, and astigmatism, when the stop is placed at the center of the curvature because the primary and the secondary mirrors are concentric. The radius of the image surface is equal to the focal length. This telescope is very large in comparison with other two-mirror layouts for the same diameter of the light-collecting aperture and the F/#.

An interesting review article points out how a secondary mirror might be fabricated from the central section of primary mirror that is cut out. Blakley [25] has proposed to call it Caesarian in honor of the first famous baby boy born by this method, while the mother was sacrificed for the good of the Roman Empire. The idea of using both parts of the blank actually appears to favor the joint survival of the mother and the child.

Afocal mirror combinations work very well, often including a beam-turning mirror to bring the focal plane instruments out of the main beam. A Mersenne telescope includes a parabola–parabola configuration. A very good review of afocal telescopes was presented by Puryayev [26] and Wetherell [27].

11.2.2.4 Unobscured Catadioptric Telescopes

There are three ways of implementing a mirror system without an obscuration. In the first, and the simplest one, the object, the image, and the mechanical axes of the parent components from which the off-axis segments are generated lie on same optical axis (see, for example, Scholl [28]). The Herschelian telescope (see Figure 11.16c) indicates that an off-axis segment may be cut out form an on-axis component, traditionally referred to as the parent component. Its optical axis (the mechanical axis of the parent component) then does not coincide with the physical center of the off-axis component. The beam incident parallel to the optical axis reflects into a focal point outside the incident beam path. Here, the object and the image remain on the optical axis, while the stop displaces the beam position on the optical element to an off-axis location. The term *off-axis* refers to the fact that the on-axis part of the parent component is not used for the imaging.

In Figures 11.20 and 11.21, the part of the beam incident above the mechanical axis has been shaded. This presentation illustrates how most common telescopes may be implemented in an off-axis layout, without any beam obstruction. It is interesting that the Gregorian configuration results in a better use of the parent component, as a larger diameter off-axis section may be used than in the other two-mirror configurations. Dials [29] describes the design of the high-resolution dynamic limb sounder (HIRDLS) instrument, which includes an off-axis Gregorian telescope developed to meet 0.7 arc seconds pointing and the 1% radiometric accuracy requirements.

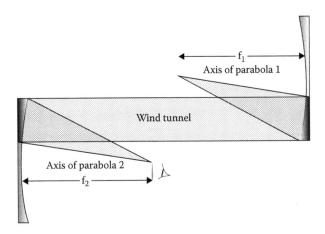

FIGURE 11.24 Two off-axis parabolas may be used to generate a collimated beam, needed in schlieren systems. The asymmetrical *z*-configuration results in a smaller amount of coma critical in the monochromators.

When implementing reflecting systems with finite conjugates (image and object distances), the use of a *z-configuration* is favored. The field-dependent aberration coma is canceled when imaging with two off-axis identical sections of a parabola with parallel optical axis above and below the collimated beam. This is recommended for source reimaging in the monochromators, for the placement of a wind tunnel in the collimated light, or in a *schlieren* system. The latter is illustrated in Figure 11.24. The less-compact layout is acceptable as it is frequently set up in the laboratory where space is not at a premium.

Off-axis parabolic telescopes are used often in astronomy and laser optics because they are practically aberration-free along the direction of the parabola axis. A detailed method to align high magnification off-axis afocal parabolic telescopes includes an initial pre-alignment using autocollimators. This is followed by fine tuning with a laser beam that travels through the telescope, reflects at a flat mirror in the focal plane and returns on itself following the same path. Tacca et al. [30] demonstrate this performance in experimental work on the gravitational wave interferometry detector, advanced Virgo.

The optical axis of the parent component may also be broken (may have a discontinuity), as in the second incorporation of the unobstructed optical systems [31]. The optical component may actually be tilted. Another way of incorporating off-axis configurations is when the optical axis is completely broken at each component, such as when three mirrors are employed for further aberration cancellation and field flattening. A combination of field-bias and aperture offset is also employed, as the third option for off-axis configuration of optical system [32].

11.2.2.5 Three-Mirror Telescopes

The number of aberrations that may be corrected in an optical system increases with the number of surfaces and their shapes. With three mirrors, the flat image surface is more easily achieved without using any refractory components. Generally, using an extra surface allows the correction of one more aberration than in a well-designed two-mirror system. The three-mirror systems are designed nowadays in an off-axis configuration.

Some three-mirror systems achieve the necessary performance by incorporating spherical mirrors. These systems, in general, tend to be optomechanically quite complex; however, the simplicity of fabrication, testing, and alignment of spherical (primary) mirrors makes them highly desirable.

M. Paul makes the first reference to the advantages of a three-mirror telescope, depicted in Figure 11.25 [33]. It is well known that two spherical surfaces may be replaced by an aspherical surface to achieve the same performance. He proposes replacing the hyperboloid in the classical Cassegrain with two spherical mirrors, with equal, but opposite signs of radii of curvature. The first mirror, with a very large radius of curvature is the secondary. Its image would be very far away if it were not for the spherical tertiary located inside the hole of the primary mirror. This forms an aplanatic and anastigmatic image in the space between the primary and the secondary mirrors. The secondary mirror, whose diameter

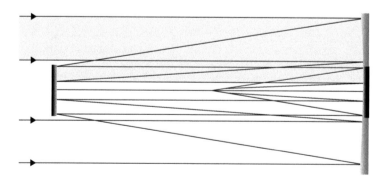

FIGURE 11.25 A three-mirror telescope proposed by Paul in which he replaces the hyperboloid in the classical Cassegrain with two spherical mirrors, with equal, but opposite radii of curvature. This telescope becomes an aplanatic and anastigmatic optical system. There is little space for the focal plane instruments, making this a good candidate for off-axis configuration.

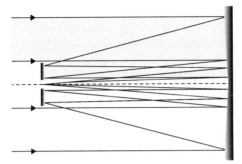

FIGURE 11.26 The primary and the tertiary form the same mirror surface, a hyperboloid. In this design with a central obscuration, the image is conveniently located outside the telescope volume, behind the hyperboloidal secondary. This system could be considered a version of a Ritchey–Chretien telescope with a decreased amount of aberrations.

is equal to that of the tertiary mirror, is in the path of the incoming beam. There is little space for the focal plane instruments, making this layout a good candidate for implementation in an off-axis layout.

The idea of using a single primary mirror figured as two distinct optical surfaces started to become feasible when numerically controlled (including diamond-turned) optical surfaces started to be applied in production. This procedure may be particularly appropriate for metal mirrors and/or for the applications in infrared. Thus, the delicate step of cutting a hole in the primary is avoided, no physical aperture is introduced, and the optical blank for the tertiary mirror comes integrated with the primary mirror [34]. A significant improvement on this idea follows with a design 20 years later [35]. Two reflections, one from the primary mirror and another from the tertiary mirror, are off the same mirror surface. Thus, the primary and the tertiary have the same form, a hyperboloid, and are figured on the same blank. As shown in Figure 11.26, the focal plane in this aplanatic design with a central obscuration is conveniently located outside the telescope volume, behind the hyperboloidal secondary. This system could be considered an improved version of a Ritchey–Chretien telescope, exhibiting even fewer aberrations.

Finally, three-mirror systems became popular with the advent of photolithography and laser beam handling systems. An unobscured configuration incorporating spherical mirrors with off-centered field and unit magnification may be corrected for all five Seidel aberrations. The Offner system for microlithography consists of two spherical mirrors that are approximately (exactly, in the third-order version) concentric. The primary and tertiary are the same mirror, and the secondary has a radius equal to half that of the primary-tertiary. In the third-order analysis, the aberrations are corrected in the concentric arrangement. The mirror spacing must be changed by a very small but critical amount to balance out the higher-order astigmatism. Additionally, the useful field is restricted to a ring or annulus, centered on the optical axis [36,37]. Afocal three-mirror systems are used for (high energy) laser beam handling and conditioning systems. They incorporate two paraboloids and an ellipsoid or a hyperboloid, resulting only in the Petzval curvature for the image location.

11.2.3 Astronomical Telescope

The primary function of the astronomical telescope is to collect as much energy (photons) as possible from a distant celestial source, often feeding the energy to a secondary set of instruments. This requires the incorporation of a beam-turning mirror in a so-called Coudé or Nasmyth arrangement. These layouts generally require the incorporation of the type of mounts that follow the stars to prolong the exposure time without image smearing.

In the astronomical telescope, the objective is known as the primary mirror because it has the primary or critical function of collecting the light. The eyepiece becomes the secondary mirror, because it no longer functions as an eyepiece. Its secondary function, in terms of its importance, is to present the image in a suitable form for signal integration, recording, and data processing.

In astronomical applications, the objective lenses in the Galilean and Keplerian telescope configuration are replaced with the mirrors of the same power in order to incorporate a large aperture without the associated weight and fabrication challenges.

In a mirror system completely equivalent to the refractive version, both the primary and the secondary mirrors are located on the optical axis: the secondary mirror obstructs the incident beam and projects the image inside the central section of the primary mirror. This procedure requires the construction of a physical hole in the primary mirror that represents an added risk in the fabrication and polishing step.

Astronomical telescopes generally have a very small field of view compared to that of a terrestrial telescope. They tend to be designed for a specific spectral region, from ultraviolet (UV) and visible to the far-infrared and millimeter applications. A most informative review of the requirements for the imaging in UV and some proposed solutions is offered by Courtes [38]. The term "camera" is sometimes used when referring to the telescope with wide-angle capabilities, such as Schmidt camera, with adequate half-field performance of up to more than five degrees. This term has arisen due to the similar performance expectations for the photographic camera [39].

The majority of the telescope configurations built in recent years takes advantage of the excellent performance of the Ritchey–Chretien design. Its design resolution is usually much higher than that actually achieved at the detector due to aberration introduced by the atmosphere for a larger diameter primary.

The importance of the coupling of the telescope's optics and the instrument's optics is that the combined optical system has to be designed for the optimum performance of the overall system. This was considered fortunate when the Hubble telescope was identified as having a certain amount of spherical aberration [40–42]. The smaller imaging camera was re-designed to compensate for the component longitudinal displacement, producing an aberration-corrected image in the camera image plane.

As the microwave detector technology continues to improve with the evolution of the semiconductor processes, future generations of cosmic microwave background telescopes are being designed for increasingly large field-of-view and diffraction-limited performance. Niemack [43] describes a new crossed Dragone telescope and receiver optics design with 10 times field-of-view of the current systems. This performance comes at a cost of reflection losses, due to the incorporation of many surfaces. Interestingly, each hexagonal detector includes its own optical system in the receiver, somewhat like a fly-eye.

The focal reducer is an optical relay system used to reduce the field-of-view of the telescope to that of the detecting focal plane array. The optical system is constrained to fit within the available space. Buisset et al. [44] describe a simple, cost effective focal reducer incorporating only spherical surfaces to implement on the 450 mm Thai National Telescope.

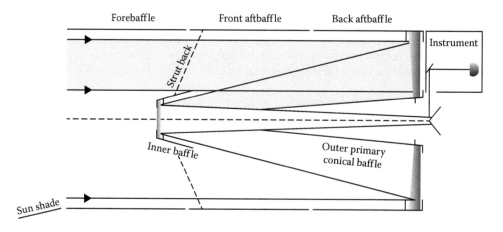

FIGURE 11.27 An astronomical telescope is used to collect the radiation and deliver it to the focal plane instruments. This is illustrated for one instrument and the pick-off mirror to deliver a portion of the beam.

In the space observational facilities, such as the Hubble, and the ISO telescopes, the focal plane remains fixed relative to the observatory housing. However, its remote operation and lack of accessibility requires that several instruments share the focal plane concurrently. These might include a spectrometer, a monochromator, a photometer, a radiometer, a spectroradiometer, and a wide- and a narrow-field imaging camera. A set of tertiary mirrors may have to be used to pick the beams off the main beam and send them into the individual instruments. This is illustrated for one instrument for the preliminary version of the SIRTF telescope in Figure 11.27 (Scholl, 1994b).

The need to incorporate a large number of pixels in the telescope focal plane necessitates the use of a pyramidal mirror that sends four beam segments onto four CCD detectors along four perpendicular directions. The Hubble telescope engineers used this elaborate technique to increase the effective number of pixels in the CCD by beam combining four separate devices. A similar architecture in reverse, using beam splitters, may also divide the available focal plane for a concurrent observation with several instruments.

The distant star is considered a point object whose image size is defined by the resolution of the telescope. In astronomical applications, we are interested in the angular position of the objects on the celestial sphere in general.

The half-angle corresponding to the diffraction blur radius is

$$\theta = r/f = 1.22\lambda/D \, [\text{sr}], \tag{11.4}$$

where θ is the semi-angle of resolution [sr] and r is the radius of the first zero in the Bessel function [m]. This resolution criterion applies only to small-aperture telescopes, or space-based instruments where the affects of the atmosphere do not destroy the local *seeing* conditions.

11.2.4 Atmospheric Effects

The atmosphere is a collection of gases that surround the Earth. It has two detrimental effects on the image quality of the telescope. The first one is that rays experience a different phase delay when incident on the primary mirror upon traversing the atmosphere having different amounts of air turbulence. The second one is that the absorption characteristics of the gases that constitute the atmosphere do not transmit the radiation of all spectral intervals of interest to astronomers—in particular, the atmosphere is quite opaque in several infrared bands and short wavelength UV regions. The latter is highly beneficial to protection of life on Earth.

11.2.4.1 Seeing

Earth-based telescopes have their resolution limited by turbulence in the atmosphere at their specific sites referred to as "seeing." The diameter of a "cell" with approximately the same conditions is 26 cm maximum for the best observational sites under the most favorable conditions. Lower values of 16–20 cm are more common. The seeing problems of large-diameter ground-based telescopes were discussed after the successful completion of multiple-mirror telescopes [46–48].

For large-aperture, Earth-based telescopes, the air movement in the atmosphere provides the limiting resolution, especially in the visible region. Generally, the "seeing" parameter of 10–20 cm is due to the stronger air turbulence within the first several meters from the ground. Furthermore, the stratospheric layer limits the isoplanatic patch angle. There, different incident rays are subject to distinct conditions due to fluctuations in air density, temperature, and humidity. In terms of resolving power, turbulence-induced wavefront distortions limit the telescope's aperture to an effective diameter of r_o, the coherence diameter of the atmosphere [49]. The seeing increases with the temperature stability, the presence of large water mass with a high thermal constant, the absence of human population and its different forms of polluting and perturbing the environment, and with the longer wavelengths.

Active systems are employed to improve the performance of the Earth-based astronomical telescopes. First, we measure the aberration introduced by the atmosphere in order to adjust the shape of the thin

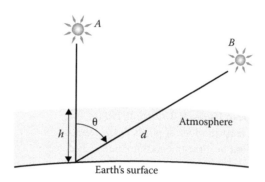

FIGURE 11.28 The air mass is the ratio of the apparent thickness of the atmosphere traversed by a ray coming from a star into the telescope at an angle θ divided by the thickness of the atmosphere experienced by a ray traversing the atmosphere from a point at a local zenith.

surface of the telescope primary mirror to compensate for this aberration. This intentional deformation of optical surface is referred to as "active optics" and requires the employment of deformable optical surfaces figured with the selective application of actuators [50].

The resolution of the Earth-based astronomical telescope is not as important as its light-gathering capability because of the effective resolution limitation imposed by the atmosphere. Fainter and more distant celestial objects may be detected by increasing the telescope diameter or by lengthening the signal-integration time.

Nian and Wen-Li [51] describe performance analysis of temperature, pressure, and air speed surrounding the retractable dome of 4 m Chinese Large Telescope. The results of their analysis show that the fluctuations in the refractive index of air are primarily caused by the inhomogeneous distribution of air temperature and speed.

11.2.4.2 Air Mass

Increased light-gathering capability may also be achieved by prolonging the signal collection and integration time. One way of accomplishing this objective in the surveillance cameras is by time delay and integration [52]. In the astronomical facility, this increased integration time is routinely achieved by employing tracking mounts in the observatory. Thus, the same object may be viewed for hours, limited only by the air mass and the sky background light. The alt-azimuth mounts are incorporated to compensate for the Earth's rotation.

This observational mode is limited by the facility's physical construction and the different air mass between the early/late and optimum (zenith) observation periods. The air mass is the ratio of the thickness of the atmosphere that is traversed by a ray from an object at an angle θ over that in a zenith (see Figure 11.28). In the case of star position B, the ray passes through the air thickness d, given by

$$d = h/\cos\theta[\mathrm{km}]. \tag{11.5}$$

Here h is the height of the air column in the zenith. One air mass corresponds to the observation of the star in zenith. The air mass increases quickly with the angle θ. For example, when the angle to the line of view to the star is 60 degrees, the light traverses two air masses. Under this condition, the air turbulence is increased appreciably with respect to the single air mass; and furthermore, the image of the star is displaced due to refraction in the thicker atmosphere.

The atmospheric deleterious effects may be avoided by placing a telescope in space or on the moon. Placing it on the dark side of the moon further avoids straight light noise from the Earth [53],

11.2.4.3 Active and Adaptive Optics

The atmosphere contributes phase error to the light signal originating at an extra-atmospheric object. There exist additional sources of phase error: the fabrication error of an optical component, alignment of an optical system, and mount jitter. All these phase errors add to the image deterioration. They are not known in advance, so a fixed component, such as a corrector plate or a corrective system, may not be built to compensate for their detrimental effects.

An adaptive optical component includes an optical element that adjusts or adapts its performance to the specific conditions and the associated electro-optical control system that makes it perform the requisite corrections. The most famous and the most used adaptive optics component is the lens inside the human eyeball. Its surfaces flatten in a young eye to focus from about 5 cm (with an effort) to infinity (5 m). It starts to perform with less flexibility around middle age (45-50 years).

An active system uses a feedback loop to minimize the undesirable residual phase aberration. A phase measurement is performed to obtain the information about phase error experienced by the system. The active system uses this information to send the instructions about the requisite changes on the surface shape of the deformable optical component to bring the aberration to zero.

The active optical system in the case of the human includes the complete visual neuro-motor system. When one cannot read the small print, the lymphatic neural system automatically sends an order to contract the muscles that control the lens shape (without the human even being aware of the process). Similarly, in a human-made active system, complex electromechanical control is used to change the shape of an optical component after the wavefront has been sensed as deformed and the degree of deformation has been quantified. Thus, a deformable component is just a subsystem of an active system. In the telescopes, a wavefront sensor is used to assure optimal alignment and configuration. Wavefront curvature sensing requires knowledge of two irradiance distributions, at equal distance before and after the focus. (Ref. C) Wu et al. [54] describe an algorithm where Zernike coefficients are found from a single irradiance distribution. They apply it to the optical design of the Hubble telescope and the modified Paul-Baker telescope. The well-performing algorithm outperforms existing algorithms in structural simplicity (two defocused images or a Hartmann sensor), making it suitable for extreme environments.

The spherical aberration discovered on the Hubble telescope after it had been launched in orbit might be an example of a phase error correctable with a deformable component. It would generally be expected that a large structure, such as a grand observatory would change its optomechanical characteristics after having been lifted into space. This includes experiences of the high accelerations and the associated variations, and then after adjustment to the gravity-free environment of space, and settling to the new temperature distribution, due to the absence of air. Indeed, the Hubble telescope incorporated a small number of actuators with a limited range to correct for the changes due to the different settled conditions of the space environment. However, this rudimentary system has not been found sufficient to perform the requisite corrections due to a large amount of spherical aberration. Due to the good fortune that the primary mirror was reimaged on the secondary mirrors of the wide-field and the planetary cameras (WFPC I), its surface shape was redesigned and adjusted on the second trial with the placement of new WFPC II [55]. This correction procedure is a form of an active system with a very long period (several years) between the identification of the phase error and the adjustment of optomechanical surfaces to correct it.

A study [56] was performed on the feasibility of correcting with an adaptive optic component the spherical aberration on the primary mirror of a large-diameter telescope with the following parameters: primary mirror diameter 2.4 m, mirror obscuration diameter 0.1056 m, mirror radius of curvature 4.8 m, design wavelength 0.5 μm, exhibiting 6.5 λ of spherical aberration.

Large space telescopes incorporating deployable and lightweight structures are reasonably expected to exhibit aberrations due to thermal deformation, gravitational release, and misalignment. Rausch et al. [57] developed a unimorph deformable mirror based on piezoelectric actuation. The space mirror is designed to correct for large aberrations of low order with high surface fidelity.

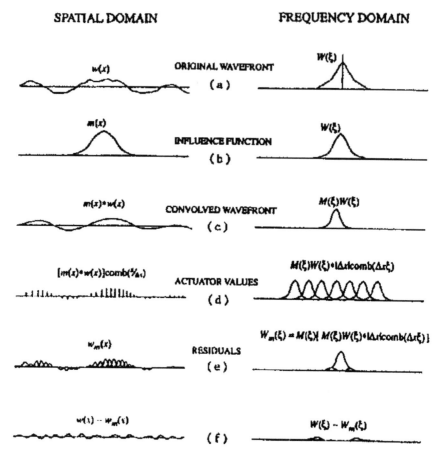

FIGURE 11.29   In an adaptive mirror, the surface is deformed to produce an equal but opposite phase error to that which is to be corrected in order to cancel the aberration. Its performance is best understood in spatial and spatial frequency domains. (a) The original aberrated wavefront is to be duplicated by applying a deformation to a thin faceplate using actuators. (b) The effect of an actuator force exerted at a specific position is to deform the surface in a shape similar to a Gaussian. (c) The amount of force applied at actuator locations reproduces the general shape of the wavefront error. (d) The specific values of the actuator amplitudes are calculated from the expected surface shape. (e) With the optimal choice of the width of the actuator influence function, controlled by the faceplate stiffness, the actuator separation may be set to determine the highest spatial frequency corrected in the phase. (f) The resulting wavefront error, given as the difference between the initial phase and the phase due to the surface deformation, contains only the high-frequency, low-amplitude phase components. (After Scholl MS, Lawrence GN. *Optical Engineering*, **29**(3), 271–278; 1990.)

Solar adaptive optics systems rely on wavefront sensors to produce a tomographic reconstruction of the turbulent phase. Marino and Wöger [58] evaluated a layer oriented cross-correlating Shack-Hartmann wavefront sensor theoretically and experimentally. They concluded that this device was not a practical instrument for atmospheric characterization.

In an adaptive mirror, the surface is deformed to produce an equal but opposite phase error to cancel out the original error. The combined effect of the aberrated phase and the component intentionally deformed with the opposite algebraic sign is expected to be close to zero. Only those spatial frequencies in the phase distribution that are smaller than the maximum spatial frequency in the actuator distribution may be corrected. Thus, some high frequencies remain in the corrected phase function. Figure 11.29 shows how the correction process actually works. Each actuator in a rectangular array exerts a force

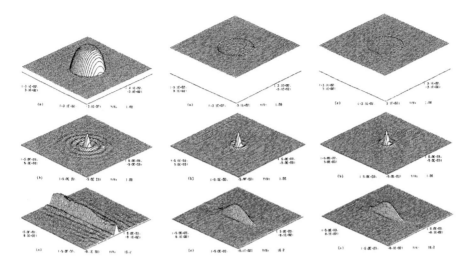

FIGURE 11.30 The simulation results of the correction of a wavefront with 6.5λ of spherical aberration with a deformable mirror demonstrate the high degree of correction possible with a deformable mirror of 2.4 m diameter. In each of the three cases, we show the wavefront shape (a) then the far-field image (b) and finally a transverse slice along the optical axis (c). Case 1: the energy in the point-spread function corresponding to this aberrated wavefront in the Cassegrain focus is spread out into the higher annuli (compare with Figure 11.7). Case 2: the significant improvement in the wavefront is observed upon correction with actuators placed on a square grid, 9.2 cm apart, producing a rather compact spot with a Strehl ratio of 0.86. Case 3: the quadrupling of the number of actuators, the corrected phase exhibits a nearly perfectly flat profile, and the far-field pattern closely resembles the diffraction-limited shape of Figure 11.3, with the Strehl ratio of 0.95. (After Scholl MS, Lawrence GN. *Optical Engineering*, **29**(3), 271–278; 1990.)

on the thin surface of the deformable component. Depending on the deformation characteristics of the thin faceplate, it is deformed in a controlled fashion with some coupling between the neighboring actuators. A zonal model of the actuator phase-plate interaction was used to model the possible range of phase-sheet deformations [59]. Obviously, a large-diameter mirror is more amenable to faithful re-shaping because more actuators and their control mechanisms can be fitted behind a large area. The physical space occupied by the actuators behind the mirrors makes them suitable correction candidates. When only few frequencies are to be corrected, placement of the actuators behind a smaller secondary is the preferred choice to conserve space.

Circular bimorph piezoelectric actuators may be good candidate for large adaptive telescopes. Wang and Yang [60] describe numerical modeling and analysis of the thermal effects in such actuators both under static and dynamic conditions. They find the effects of the thickness of the metallic layer especially important.

Figure 11.30a shows the aberrated wavefront and the resulting irradiance distribution in the telescope focal plane (see Figure 11.30b). There, the presence of many irradiance rings confirms a highly aberrated wavefront (see Figure 11.8). Figure 11.30c shows the irradiance slices along the optical axis, illustrated additionally in Figure 11.31. The image spot is significantly spread out along the optical axis. Using the deformable mirror model with the actuator spacing of 4.6 cm, the corrected spot (see Figure 11.30, right) has a Strehl ratio of 0.95. When the number of actuators is reduced by a factor of two in each direction, the Strehl ratio decreases to 0.86. With the application of an adaptive component, the irradiance distribution in the focal plane significantly improves, exhibiting only a single ring around the central bright spot. Also, the spot extent along the optical axis tightens appreciably. The corrected phase is seen to be nearly constant. A quantitative measure, such as the Strehl ratio, allows us to appreciate the real improvement of about 10% upon quadrupling the number of actuators. Such improvement confirms the diffraction-limited performance upon applying active correction.

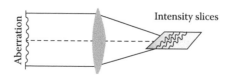

FIGURE 11.31 The wavefront with a (compensated) aberration is incident on a perfect mirror, which forms the far-field image in the (Cassegrain) focal plane. Customarily, the point-spread function is shown as a function of the distance from the optical axis. Here the intensity slices are given as a function of one transverse coordinate for a number of positions along the optical axis (also known as inside, in, and outside focus positions). In Figure 11.30c, we show the intensity slices along the optical axis, in order to assess the three-dimensional extent of the bright spot. (From Scholl MS. *Proc. SPIE*, **2019**, 407–412; 1993b.)

The incorporation of a fully-fledged deformable primary mirror into a space telescope would unquestionably improve its performance under most adverse conditions. The inclusion of the obligatory robust feedback loop and control system adds complexities and increases the potential for risk of a single-point failure. These considerations prohibit its implementation for the applications to space astronomy at this time.

Even if an adaptive component in a space system were technologically feasible, the correction implementation would not necessarily include a complete active system. The quality of the images and their review by the scientific community currently provide the wavefront sensing and evaluation feedback to generate the control commands. This is basically a quasi-static correction system with phase error evaluation performed infrequently (once a month).

Atmospheric changes occur frequently, as a matter of fact constantly, as we can ascertain when we observe the twinkling of a distant light at night. For this reason, many modern telescopes envision some active correction (see, for example, Hardy [61] and Beckers et al. [62]). Any correction of the atmospheric effects would have to be monitored and corrected with a frequency between 0.10 (for calm conditions) and 100 Hertz (for conditions of turbulence). Real-time compensation of atmospheric phase distortions is generally referred to as an active optics system [63,64]. Such a temporally dependent system includes by necessity an active system of sensing and a wavefront correction feedback loop.

Only some phase error may be corrected during each sampling period to avoid instabilities and transients. The phase error has to be sampled at a frequency at least five times higher than its highest significant temporal frequency to avoid system failure.

The resolution limitation imposed by the atmosphere a generally accepted phenomenon is on a large telescope. A great deal of effort has been expended in the last half a century to understand this behavior, in order to be able to correct for it with an active optics system. Atmospheric aberration may be described by the Kolmogorov spectral distribution of the wavefront as a function of height above the Earth surface [65],

$$W^2(\rho, \lambda, n) = \frac{0.38}{\lambda^2 \cdot \rho^{11/13}} \int_{h_{min}}^{h_{max}} \left[C_n(h) \right]^2 dh, \tag{11.6}$$

where C_n^2 is the refractive index structure of the atmosphere as a function of the height layer n, between h_{min} and h_{max}. Upon integration of this equation over all layers of the atmosphere, characterized by their respective refractive index structure, we can see that the amount of atmospheric aberration is proportional to the inverse wavelength and the variation in the index of refraction of the air. These coefficients, of course, are dependent on the site in general, and on the specific climatic and environmental conditions, in particular.

In terms of the seeing parameter r_0 that is used to describe empirically the effect of the atmosphere on the phase error, this equation may also be written out. Here, once again h_{min} and h_{max} include the integration over the whole atmosphere [66],

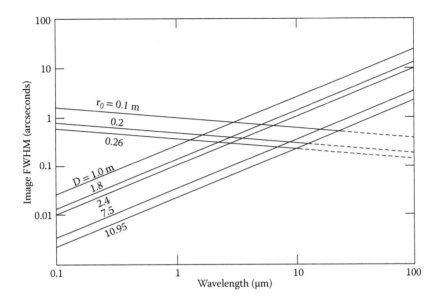

FIGURE 11.32 The angular resolution of the Earth-based telescope as a function of wavelength with two sets of parameters: the mirror diameter varies from 1 m to 10.95 m; the seeing radius includes the range from 0.1 m to 0.26 m, corresponding to good astronomical sites in the visible.

$$W^2(\rho, r_o) = \frac{0.023}{(r_o)^{5/3} \cdot \rho^{11/3}} \cdot \frac{e^{-\rho^2 \cdot (L_i)^2}}{\left[1 + \frac{1}{(L_o \cdot \rho)^2}\right]^{11/6}}, \qquad (11.7)$$

were L_i and L_o are the inner and outer scale, while ρ is the radial spatial frequency coordinate. When comparing Equations 11.6 and 11.7, we observe that the seeing radius increases with wavelength nearly linearly (with exponent of 1.2=6/5). (Ref. A)

$$r_o(\lambda) - \lambda^{1.2}\left[R_{o1}(\lambda_1)/\lambda_1^{1.2}\right] \qquad (11.8)$$

This equation may be rewritten for the visible region, for which the seeing parameter is usually known, generally falling between 0.1 and 0.2 m.

Using Equation 11.4 for the angular resolution, we obtain

$$\theta(\lambda) = [0.61/\lambda^{0.2}][\lambda_1/R_{o1}(\lambda_1)^{1.2}][\text{rad}]. \qquad (11.9)$$

Thus, the angular resolution θ of Earth-based telescopes without atmospheric correction increases with wavelength with an exponent of wavelength of −0.2. We recall that a small value for the angular resolution means a good high resolution. As the wavelength increases, θ in Equation 11.9 decreases slowly. The angular resolution of Earth-based telescopes as a function of wavelength is shown in Figure 11.32. This figure incorporates two resolution curves. The mirror diameter is varied from 1 m to 10.95 m (the diameter of the Keck telescope). Also, the seeing radius is varied to include (in the visible) the range that corresponds to good astronomical sites. The resolution limit represents the higher of the curves for the applicable parameters. The region of the expected resolution values has been shaded for emphasis. We note that in the visible and the near-infrared spectral regions the seeing limits the resolution of the telescope, independently of its diameter.

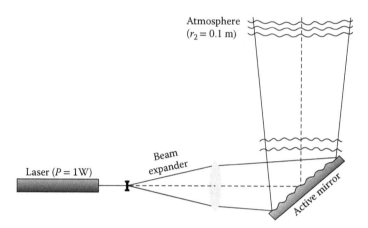

FIGURE 11.33 The optical model to assess the feasibility of correcting the atmospheric aberration with a deformable mirror includes the laser beam, first expanded to 0.5 m diameter, with a beam expander that introduces 0.2λ astigmatism at 0.48 µm; the phase error introduced by the atmosphere; and the actuator strengths, calculated so as to neutralize the beam aberration. (After Scholl MS. *Infrared Physics & Technology*, **38**, 87–92; 1997.)

Thus, large-diameter telescopes are well utilized on the Earth's surface only in the infrared. There, unfortunately, the selective atmospheric spectral transmission in combination with the Earth's self-emission generates a whole new set of problems.

The feasibility of correcting the atmospheric aberration using a deformable mirror has also been analyzed. A laser beam may first be expanded to 0.5 m diameter, with a beam expander that introduces 0.2-λ astigmatism at 0.48 µm. The effects of the atmospheric phase error are calculated, and the effects of the actuators to neutralize the beam aberration are determined, as shown in Figure 11.33. With the actuator separation on a square grid of 0.04 m, we may appreciate the sequence of irradiance and phase at each of the key locations, as shown in Figure 11.34. The initial irradiance and phase are depicted in Figure 11.34a, with the Strehl ratio of 1. The phase in Figure 11.34b presents some astigmatism after the beam passes through the aberrated beam expander, now having the Strehl ratio of 0.924. The phase is severely, randomly, and unpredictably aberrated upon transmission through the atmosphere in Figure 11.34c, with a significant Strehl ratio degradation to 0.344. Finally, the corrected phase in Figure 11.34d exhibits a much-improved Strehl ratio of 0.865. The residual phase error is characterized by presence of only the high-frequency components whose effects could not be eliminated by the relatively small given number of actuators (small number per distance).

The phase error introduced by the atmosphere, as detected just before the mirror surface, is equal to the deformation that needs to be placed on the mirror surface to produce the corrected wavefront. Ruiz et al. [67] propose a new concept for push-pull active optics. The push force is provided by means of individual airbag actuators. The vacuum, applied to the mirror back, and gravity provide the pull force in an experiment on 2.12 m telescope in the Sierra San Pedro, Baja California, Mexico. This work demonstrates a novel, cost effective technique. It incorporates a reduced number of actuators, with decreasing overall complexity.

Due to the practical consideration of insufficient exposure, artificial sources have been developed using an active laser beam as illuminator. In principle, a bright star could be used as a radiation source to measure the atmospheric phase error to determine the requisite correction. When the atmosphere is rapidly changing, we may have to use an active illumination system, such as a laser beam directed to the upper atmospheric layers. There, the excited sodium atoms generate fluorescence, which functions as a bright point source. As an example, the Strehl ratio has been improved from 0.025 to 0.64 on a 1.5 m telescope using an active system [68] and an artificial laser beacon. Lack of sensitivity to tilt is one of the disadvantages of this otherwise effective tool.

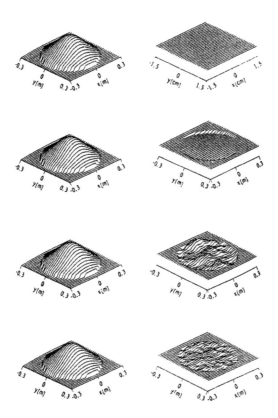

FIGURE 11.34 The feasibility of correcting the atmospheric aberration using a deformable mirror has been confirmed with the Strehl ratio increase from 0.344 due to atmospheric aberration to the much higher Strehl ratio of 0.865 upon the application of the active optics system to control the surface shape of the deformable mirror. While the beam and the phase are shown at each significant step in the active optics system, the atmospheric aberration affects only the phase. The laser beam intensity scattering within the atmosphere is not considered. (a) The initial phase at the output of the laser is constant, with the Strehl ratio of 1. (b) A small amount of astigmatism (0.2λ) is introduced at the aberrated beam expander, reducing the Strehl ratio slightly to 0.924. (c) The phase is severely, randomly, and unpredictably aberrated due to its propagation through the atmosphere, with a significant Strehl ratio degradation to 0.344. (d) Finally, the corrected phase exhibits only small-amplitude, high-frequency ripples after reflection off the deformable mirror with the actuator separation on a square grid of 0.04 m, with a much improved Strehl ratio of 0.865. The residual phase error is characterized by the high-frequency components whose presence could not be eliminated by the given small number of actuators. (After Scholl MS. *Infrared Physics & Technology*, **38**, 87–92; 1997.)

11.2.5 Space Telescopes

With the advent of larger and more powerful rockets, it became feasible in the late 1960s to send a satellite into space with communications and scientific remote-sensing payloads. This accomplishment was followed by telescope facilities, in the UV, visible, and IR. In the UV and IR spectral regions, the atmosphere is opaque, effectively making it impossible to obtain astronomical data from the ground observatories. The Far Infrared and Submillimeter Space Telescope (FIRST) is intended to open up for study [69] the far-infrared and entire submillimeter (85–900 μm) bands. When the seeing problem was first fully appreciated in the 1970s, a construction of a series of large observational telescope facilities incorporating large-diameter monolithic telescopes was initiated. These great observatories include the Hubble telescope in the visible, followed by the Infrared Space Observatory (ISO) in the infrared. Remote sensing is another important area of the use of space telescopes. For example, they collect radiation and perform imaging with instruments used to monitor ozone depletion in the stratosphere [70], ever improving ways of predicting weather, and to monitor the Earth's radiative balance to determine the extent of global warming.

The clear advantage of placing a telescope in space is that the detrimental effects of the atmosphere are avoided there. The equally obvious disadvantage is that the telescope orbiting Earth in space is not easily accessible for (routine) repairs and adjustments. This became obvious in the case of the incorrectly positioned primary mirror on the Hubble telescope resulting in approximately 0.5λ of spherical aberration [71,72]. One of the instruments that the telescope is feeding, the wide-field camera, was redesigned to compensate for this aberration and replaced in the observational facility several years later. In addition, COSTAR was added to allow the use of the non-imaging spectrographs and the faint object camera.

Telescopes may be classified according to their spectral intervals of observation: the visible, including blue and red; the UV; the near- and far-IR, and the millimeter and radio waves. The IR telescope facilities generally include the observations approaching a millimeter range where the detection methods change, and the telescope primary mirror functions just as a photon-collecting dish, even in space. Both the UV and the infrared regions are important to differentiate from the visible spectral regions. There exist different light transmission characteristics of the atmosphere and optical glasses in those two spectral regions. Additionally, the control of stray light in the infrared becomes of uttermost importance because it generates the thermal noise.

11.2.6 Infrared Telescopes

The IRAS was the first IR satellite to survey the infrared and millimeter skies, while the ISO was the first IR astronomical observatory in space [73] operating at wavelength bands from 2.5 to 200 μm. It was considered a success, having run longer than was originally planned due to its conservative design [74–77]. Its fruitful completion resulted in the plans for another large space observatory, the Space Infrared Telescope facility (SIRTF).

The material–transmission characteristics of infrared glasses favor the implementation of the all-reflective configurations in the IR spectral region. First of all, there is an increased amount of radiation absorption at the IR wavelength and glass has spectral transmission characteristics. Second, the index of refraction is larger, thus increasing the losses upon refraction at a glass-air interface. Such problems are eliminated in an all-reflective design.

Additionally, the reflectivities of aluminum, silver, and gold increase in the red and IR spectral region. This material characteristic makes it even more advantageous to incorporate all-metal reflecting telescopes for applications in these spectral regions.

There is one more important issue to address when dealing with telescopes operating in the longer-wavelength regions. It is the thermal noise, whose minimization is generally addressed during the early design stages in order to eliminate the so-called stray light. The importance of controlling the noise arising from sources outside the instrument field-of-view, whose radiation scatters inside the telescope barrel, has been analyzed and identified as important for assessing the performance of the emitter in the IR. This needs to be considered even for the sides not illuminated directly by the sun, due to their temperature. The temperature control of all parts of the telescope subsystems and components is of critical concern for the minimization of the internally generated stray light noise [78,79,45]. For these reasons, IR telescopes tend to be heavily baffled, as may also be appreciated upon studying Figure 11.27.

One detail that may be clearly noted upon studying this layout, with the exactly traced optical beam volume, is that the central part of the secondary mirror is not used for imaging. As this part of the mirror is seen directly by the detector, a small planar reflector obstructs it. It disperses the radiation out of beam. This small reflector misses its purpose to some degree in the telescopes where the secondary mirror is nodded with the intention of chopping the radiation, by looking at a different region of the sky as a reference and for calibration.

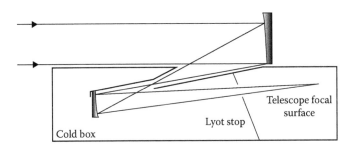

Cold box

Lyot stop

Telescope focal surface

FIGURE 11.35 A telescope designed with the objective of a decreased amount of stray light for an improved signal-to-noise ratio takes advantage of the off-axis configuration to fully employ the available light-collecting area of the primary mirror. To decrease the number of diffracting edges, a Gregorian layout incorporates the internal focus where a field stop may be located, and the Lyot stop for limiting the stray light. (After Scholl MS, Peez G. *Infrared Physics & Technology*, **38**, 25–30; 1997b.)

The secondary mirror often requires extra baffling during the design stage, resulting in a large obstruction to the incident radiation, much larger than the actual diameter of the secondary. With such an enlarged obscuration, the amount of light collected by an on-axis reflecting configuration is smaller, equal to $\pi(R^2 - r^2)$, where r is the radius of the obstruction, rather than that of the secondary mirror.

In the IR designs, the diameter of the secondary mirror tends to be rather small for two reasons. Due to its proximity to the detector, it must subtend a small angle at the detector in order to diminish the amount of stray light it generates and that the detector intercepts. Secondly, the small secondary mirror is easier to control while it oscillates between the source and the reference patch of the sky, while the primary mirror remains stationary due to its large mass. Of course, such fine control mechanisms add to the bulkiness of the secondary assembly. The primary mirror has to be slightly oversized to accommodate the extreme position of the nodding secondary mirror. The designers of IR telescopes have not yet come to an agreement whether it is better to have the stop at the primary or at the secondary mirror.

The telescopes in the large infrared facilities, such as the ISO, have been designed as on-axis configurations, due to the improved mechanical stability of centered systems, even though the shade introduces some mechanical asymmetry. In the last few decades, there has been an insurgence of small IR telescopes for dedicated missions to observe Earth, seas, shores and polar caps. These revolutionary designs incorporate the off-axis telescope layouts with the Lyot stop configuration to minimize the amount of stray light incident on the detector plane.

In this approach, the whole instrument is designed with the single overriding objective of maximizing the signal-to-noise ratio for detection of distant faint sources, as in the case of the wide-field infrared explorer. The same high signal-to-noise ratio may be achieved either with a large aperture for a large light-collecting area or a small amount of noise. In the design depicted in Figure 11.35, the telescope features an off-axis reflective layout in the Gregorian configuration. A long-wave IR imaging system for space exploration of faint targets is highly sensitive to stray-light radiation. Zhu et al. [80] present internal and external stray radiation suppression technique for long wave IR catadioptric telescope. The results show that the stray light suppression may be dramatically improved by iterative optimization of optomechanical configurations.

By choosing an off-axis reflecting mirror, the whole surface area of the primary is used as a light collection system. The secondary neither obstructs the radiation, nor does it modify the resolution (light distribution in the focal plane). The off-axis layout has the undesirable consequences of introducing field aberrations, such as astigmatism and coma. The image is located on a surface with an increased amount of (Petzval) curvature.

With ever increasing numbers of pixels in the microwave focal plane arrays, the optical imaging system must have diffraction-limited performance over the whole detector area in the new generation cosmic microwave background telescope. Niemack [43] describes the proposed crossed Dragone telescope and receiver optics designs featuring an increased field-of-view for the next generation systems.

Two-axes gimbaled telescope support systems have been considered for a long time the best that can be achieved for tracking a star for a stable image capture during a long exposure. A single point support provides repeatable thermal and gravitational deformation permitting the simplified control of the surface of the primary mirror with commands stored and provided by the look-up tables. Padin [81] reports that such design is most appropriate for survey telescopes.

Space surveillance telescope systems are used to detect space objects, including asteroids and artificial satellites in the Earth orbit. Hardy et al. [82] compare performance of three algorithms in their capacity to detect objects, using as a test case a geosynchronous Earth orbit satellite, ANIK-F1 entering the Earth eclipse. They find that unequal-cost probability function delivers improved performance over a large range of potential scenarios.

11.3 Filled-Aperture Telescope

The maximum diameter of constructed (Earth-based) astronomical telescopes is limited by the technological and engineering considerations of the day. First, there are fabrication issues, in terms of the uniformity of the blank, then its transport, finally the optomechanical forming and shaping of the surface, and the component testing to meet the specifications.

In the 1960s, the belief that the maximum feasible blank was limited to about 2.3 m led to the nearly simultaneous designs employing an arrangement of six mirrors, known as the multiple-mirror telescope (MMT); see, for example, [83,84]. The advantage of a large light-collecting area was traded off for the difficulty in phasing of the different segments. The segments must be aligned so well with respect to each other as to generate zero phase error between them, [83,85]. This successful segmented telescope configuration was followed by the first Keck mirror, with 356 hexagonal panels of 0.9 m on a side in a honeycomb arrangement. The first truly segmented mirror was about 11 m in diameter (maximum diameter of

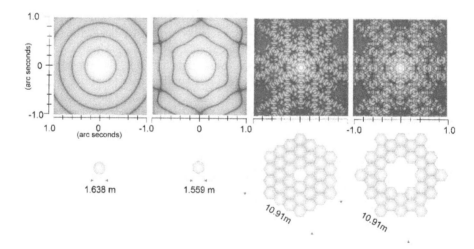

FIGURE 11.36 Diffraction patterns in a logarithmic scale illustrating the effect of the segments on the performance of the segmented array and the distribution of segments. (a) A single circular segment with diameter 1.638 m produces a set of circular rings, with the first-intensity zero at 0.33 arc seconds. (b) A single hexagonal segment with the diameter of the inscribed circle of 1.559 m, with the first zero ring a slightly deformed circle, and with resolution of 0.33 arc seconds. (c) Thirty-six hexagonal segments in a honeycomb configuration, with the central one missing, and with the first zero ring at 0.055 arc seconds. (d) Thirty-six hexagonal segments in a honeycomb configuration, with the central seven segments missing, and with the first zero ring at 0.045 arc seconds. The wavelength of 2.2 µm is taken for the resolution determination.

10.95 m). This telescope has been successfully fabricated and aligned. It has been used at visible as well as IR wavelengths since its use [86–88]. This system incorporates an active control to align and maintain in alignment the large number of segments [89]. Later, a second telescope was added to allow interferometry.

Figure 11.36 shows the diffraction patterns of two possible arrangements of 36 hexagonal segments with the diameter of the inscribed circle of 1.559 m. The diffraction pattern is exhibited on a logarithmic scale such that irradiance of 1 corresponds to gray level of 255, and the irradiance value of 10^{-6} and less is black, assigned the gray level of zero. The wavelength of 2.2 μm is assumed for the resolution calculations. In both layouts, the aperture area is the same; thus, the light-collecting efficiency does not change.

In Figure 11.36c, we see a layout with a single central segment missing, the actual Keck configuration, while in Figure 11.36d, the first ring of mirrors is additionally missing. The six remaining mirrors are placed on the outside of the configuration shown in Figure 11.36c. There are three potential benefits realized with this modification. In both figures, we observe the first zero ring that defines the resolution of the telescope, corresponding to a hexagonal shape of the aperture. The diameter of the first zero ring, representing the telescope resolution in Figure 11.36d is smaller than that in Figure 11.36c due to its larger outer diameter and larger inner hole. Additionally, the placement of six segments on the outside increases the diameter of the telescope. Therefore, the cutoff frequency is larger along these directions. If the segmented configuration is parabolic or hyperbolic in cross-section, then the curvature of the segment decreases with the distance from the apex of the primary mirror.

In Figure 11.36c and d we observe a strong dark ring at 0.33 arc seconds, which arises as a consequence of the diffraction pattern due to an individual segment, shown by itself in Figure 11.36b. Finally, we display for comparison the diffraction pattern of a circular aperture of diameter 1.638 m, producing a circular ring at the same location. While it may be difficult to appreciate it with the naked eye, (human) eyeball is not designed for such absolute measurements. We comment that the first zero ring in Figure 11.36b, c, and d is in fact approximately a rounded-off hexagon, roughly circular in shape. Looking at the irradiance distribution in Figure 11.36b, we may therefore conclude that the diffraction pattern of a hexagon generates a near-circular hexagonal ring as its first zero, followed by more clearly shaped hexagons. The first hexagon with a circle inscribed in the diffraction pattern of 36 segment configurations features a radius of 0.055 arc sec, representing the true resolution of the segmented configuration. This analysis assumes perfect phasing among segments.

In the 1970s, after its great success in the radio astronomy [90], the revolution dry techniques of aperture synthesis began to be studied in optics as well.

The funding was initially available for military research. Unfortunately, diluted aperture optical systems are limited to a narrow field-of-view, precluding their applications to surveillance [91]. Diluted apertures became of interest to the astronomical community for interferometric rather than imaging applications, for astrometric studies (exact star coordinates on the celestial sphere), and for such exotic research as the detection of planets outside our solar system (see, for example, Scholl, [92,93] and Strojnik and Peez [94]). It is believed that a single spatial frequency detected with an interferometric array could confirm the presence of such an ephemeral object as a dark, small planet orbiting a bright sun-like star.

The formation of the planets, solar systems, stars, galaxies, and the universe remains of great scientific and astronomical interest worldwide. Some of the planet detection projects have lost a bit of their appeal within the U.S. scientific community, although the Spitzer space telescope just reported finding the 1000th planet using the transit technique.

11.3.1 Increasing the Aperture Size

There are basically two ways of improving the radiometric sensitivity of a telescope. The first one that has been receiving a great deal of technological impetus, in parallel with the advances in the semiconductor technology due to the employment of similar materials, is the continued evolution of the detector technology [95] and the focal plane architecture. The second one belongs to the realm of the traditional optics: increasing the diameter of the light-collecting aperture to intercept more photons and decrease the diameter of the diffraction spot. Its challenges are often the limitations of the state of technology rather than the fundamental limits. The issues involved around the large-diameter telescopes started to be addressed at the successful completion and testing of the multiple-mirror telescope [96].

11.3.2 True Monolithic

When the Hubble telescope was first proposed, the 2.4 m diameter primary mirror was considered the largest mirror that could be figured and tested for use in space. Only a decade earlier, a 2 meter segment had been considered the best and the largest that could be fabricated. In order to build a larger telescope than that, a telescope had to be built up of several segments of smaller size. The technology limitation of the day is not to be taken lightly: a size limitation of a furnace to produce a quality blank is indeed technology limitation, requiring many improvements before it may be overcome. A great advantage of the technology development programs is that with time, efforts, funds, and human ingenuity, such limitations may be overcome. However, if the technology development stagnates, the progress might taper to a mere crawl.

11.3.3 Monolithic in Appearance

Today, there exist a number of monolithic mirrors under development or already incorporated in a telescope with diameters of 7–8.5 m that started to be developed as light-weighted mirrors [97–105]. This evolution includes even the replacement of the one in the housing originally prepared for the multiple-mirror telescope (see, for example, Olbert [106]). Actually, in most of these cases, it was human ingenuity rather than a breakthrough in technology or technology transfer from similar areas that prompted accelerated development. Possibly the most interesting case is the building of a primary from a set of hexagonal segments, whose surface is covered with a single thin sheet of glass—technology transfer from the adaptive optics [107].

11.3.4 Segmented for Any Size

The multiple-mirror telescope represented a truly novel way of achieving large diameters: if one can combine six mirrors, why not 12, or 18 or even more: 36 in the case of Keck, and 127 for the extremely large telescope (ELT). Figure 11.37 compares the sizes of the primary mirrors of the Hubble, the Keck, and the ELT. The anticipated growth of the segmented mirrors represents an extension of the theorem of natural numbers: if you can build a two-segment primary mirror, then assume that you can build an N-segment mirror, and see if you can build an $N + 1$ segment mirror. Of course you can, and you will, because the astronomical community wants to intercept photons from even fainter sources, and they just need to count them. This has been the design philosophy of the Keck telescope as well as the Spanish telescope in the Canaries [108]. An accurate calibration of the interaction matrix optimizes the performance of the adaptive optics system. The future European extremely large telescope is expected to exhibit a few thousand mirror modes. Meimon et al. [109] describe calibration strategy applicable to adaptive optics systems in a closed loop. They incorporate the slope-oriented Hadamard method to obtain 7-fold improvement (i.e., decrease) in the calibration time.

Likewise, the huge telescopes of the future will be constructed on the basis of the same principle. Consider the overwhelmingly large telescope (OWL) that incorporates 2000 segments of 2.3 m mirror diameter. In this telescope, the fabrication will be much simplified. It is contemplated that each segment be a sphere, fabricated at a rate of one per day. The primary mirror is estimated to weigh 1500 tons. The support structure is of the same size as that of the Eiffel tower, employing 4000 tubes of 2 m diameter. The secondary is planned to be larger than the Keck. Figure 11.38 shows the apertures of the OWL and the Keck telescope size for comparison.

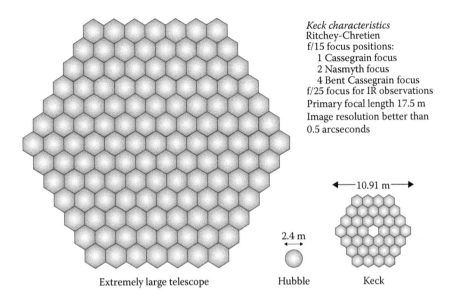

Keck characteristics
Ritchey-Chretien
f/15 focus positions:
  1 Cassegrain focus
  2 Nasmyth focus
  4 Bent Cassegrain focus
f/25 focus for IR observations
Primary focal length 17.5 m
Image resolution better than
0.5 arcseconds

Extremely large telescope   Hubble   Keck

FIGURE 11.37   The primary apertures of Hubble; Keck, with 36 segments; and the extremely large telescope, with 127 segments.

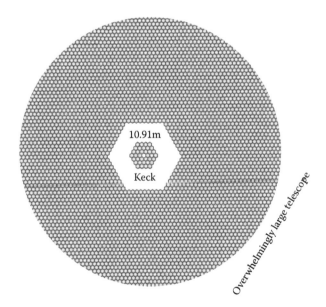

FIGURE 11.38   The primary apertures of the Keck telescope and the OWL, the overwhelmingly large telescope, incorporating 2000 spherical segments of 2.3 meter diameter.

At about the same time that we learned that mosaicking a mirror surface is beneficial to light collection but not necessarily to the resolution, we also discovered the detrimental effects of the atmosphere. The need to phase segmented mirrors brought into focus the application of interferometry for the detection of a specific feature in the spatial frequency domain. The spatial frequency domain corresponds to the space where a Fourier transform of the irradiance distribution is presented. Thus, it has been known for a long time that two separated apertures give information about one spatial frequency. This spatial frequency corresponds to the inverse separation between the apertures, along the line connecting the apertures. By changing the separation of the apertures, also referred to as the interferometer baseline, a continuous set of spatial frequencies could be sampled. The interferometry has been used since the 1950s in radio astronomy to synthesize the shape of radio sources, using a set of spatially separated radiation collecting mirror segments.

The telescope segments are often hexagonal in shape. However, the Zernike aberration theory has been developed for circular apertures. Ferreira et al. [110] developed a piecewise diffeo-morphism to transform a unit disk into a polygon. With this transformation, they deduced the Zernike coefficients of a polygon. Segmented primary mirrors represent potentially the simplest solution to the need for ever increasing telescope diameters. The optical path difference between neighboring segments must be less than a few nanometers in order to achieve diffraction-limited performance. Simar et al. [111] propose an inter-segment piston sensor upon measuring the coherence of a star image.

11.4 Diluted-Aperture Telescope

11.4.1 Increasing the Resolution, but Not the Area

A segmented mirror is assembled of segments whose shapes combine to form a continuous surface with a small (2 mm) separation between segments. While the surface may be made arbitrarily large by adding more segments, the shape of the individual segment dominates the diffraction pattern. When the segments are correctly phased, as in the Keck telescopes, the diffraction pattern is that of the full aperture (see Figure 11.36c). The established practice of interferometry is to measure a specific spatial frequency of a distant object through the visibility function. Additionally, the earlier experience of image synthesis employed in the radio frequency domain led the researchers to ask: why do we not separate the individual segments spatially, in order to sample different and higher spatial frequencies? The fundamental idea was to conserve the total light-collecting area of the segments, but collocate the segments at such position as to collect information about the diverse spatial frequencies of interest. This is the principle behind dilute aperture imaging. We hope to use an interferometric configuration to detect a planet around a nearby star. For example, if a planet is indeed orbiting the star, its presence is confirmed by one spatial frequency corresponding to the star–planet separation. The separation of the segments may be arranged to look for this spatial frequency. Thus, we may think of a diluted-aperture array as a configuration where the light-collecting area is of the same size as in the segmented telescope, the diffraction pattern or resolution is that of an individual mirror the extent of specific spatial frequencies that may be sampled is increased corresponding to the separation between individual apertures.

Conversely, we may achieve the same coverage in the spatial frequency domain by using a select set of apertures, but having a much smaller light-collecting area. One such distribution of small circular areas may replace the spatial frequency coverage even in the case of the Keck telescope. The advantage of incorporating the diluted-aperture configuration is that a much smaller area of segments is fabricated and constructed, while the arrays still sample the important, information-carrying spatial frequencies, according to the distribution of the segments. The disadvantages of diluted-aperture configurations are the difficulties of phasing the now physically distant and separated segments. In radio astronomy, due to significantly longer wavelengths, this task was successfully overcome. The challenges of image reconstruction with the unusual aperture distributions were tremendously helped with number-crunching abilities of the modern computers. Construction and control of the system of telescopes was aided with great advances in the control theory. While this is a complex task, there has been no fundamental limitation identified to prevent its accomplishment; it had already been implemented in the radio portion of the electromagnetic spectrum. Rather, the most obvious limitation has been the reduced applicability of an optical system with a narrow field-of-view. This makes it of decreased interest to the defense and surveillance communities, its original proponents. The small field-of-view is not considered a limitation in astronomical application where interest continues in measuring stellar diameters and star–planet separations, all subtending very small angles.

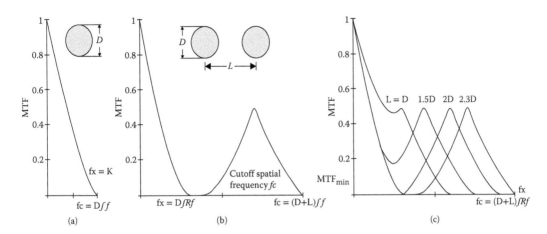

FIGURE 11.39 The modulation transfer function (MTF) of (a) the monolithic and (b, c) simplest diluted aperture composed of two apertures of the same diameter. Here f_c is the maximum spatial frequency imaged by the optical system; f_X is the maximum spatial frequency below which the MTF does not dip to zero; and f_F is the functional frequency, the maximum spatial frequency for which MTF remains higher than some MTF_{min}. As the latter depends on the specific application, the value of f_F also depends on the application rather than on the configuration of the diluted aperture. The f_c and f_X coincide in the case of a monolithic mirror and when the mirrors are close.

The primary function of the diluted-aperture configuration is to detect specific spatial frequencies rather than to form a faithful image of the object. Thus, those portions of the aperture that are involved in the imaging of the spatial frequencies of no interest may be safely eliminated. In Figure 11.39, we compare the MTF of the simplest diluted aperture with that for the filled one. One such ground-based configuration involves Keck I and Keck II [112] to perform ground-based interferometry. Two apertures of diameter D and center-to-center separation L may produce the MTF that exhibits zero values, as shown in Figure 11.39b. It is assumed that these spatial frequencies have no information of interest. When the aperture separation L is sufficiently small, the two MTF peaks start to overlap, and all the intermediate spatial frequencies are also imaged with some modulation (see Figure 11.39c). Here f_c is the maximum spatial frequency imaged by the monolithic optical system, as shown in Figure 11.39a; f_X is the maximum spatial frequency for which the MTF is higher than zero; and f_F is the functional frequency; that is, the maximum spatial frequency for which MTF remains higher than some minimally acceptable MTF_{min}. As the latter depends on the specific application, the value of the functional frequency f_F depends more on the application rather than on the configuration of the diluted aperture. For this reason, it is not a figure-of-merit for aperture optimization. The frequencies f_c and f_X coincide in the case of a monolithic mirror. Each of these frequencies is a significant figure of merit in different applications: high f_X and f_c are desirable for faithful imaging; high f_c is sought in interferometry, and high f_c, but not necessarily f_X, is required for imaging of select spatial frequencies.

Diluted imaging was analyzed in great depth for radiofrequencies. The first case of an optical system employing a diluted configuration for the visible and near-IR spectral regions was the multiple-mirror telescope, with six mirrors nearly in contact. An approximation to this layout is shown on the inset of Figure 11.40, described as a six-aperture redundant configuration with a dilution factor of 1.5. A redundant configuration is the one where each spatial frequency is sampled more than once, a highly desirable feature for an imaging system. A dilution ratio is the ratio of the area of the aperture of the monolithic mirror to the combined area of the sub-apertures with the same cutoff frequency.

The MTF of the six circular apertures in contact, depicted in Figure 11.40, illustrates the general features of diluted-aperture optical systems. Due to its relatively low dilution ratio, the MTF has a non-zero value for nearly the same spatial frequencies as a monolithic mirror. However, the amplitudes of

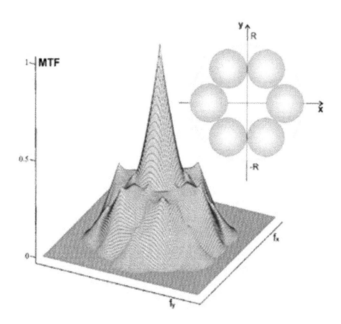

FIGURE 11.40 The modulation transfer function (MTF) of a six-aperture redundant configuration with a dilution factor of 1.5 illustrates the general features of the diluted aperture systems. Due to a low dilution factor, the MTF for this system covers about the same spatial frequencies as that of a monolithic aperture. The MTF exhibits a large plateau for moderate values of the radial spatial frequency.

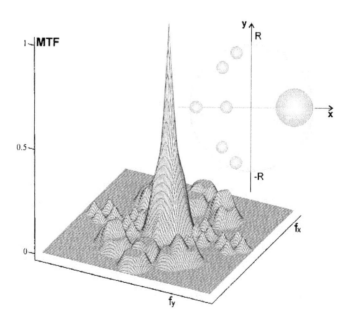

FIGURE 11.41 The polar stratospheric telescope (POST) unsymmetrical aperture configuration with a large mirror, flanked on one side by several smaller ones, with the purpose of using the large mirror for imaging up to its cutoff frequency and the smaller ones for the detection of a specific phenomenon at a given spatial frequency. (After Ford H. *Proc. SPIE*, **2199**, 298–314; 1994.)

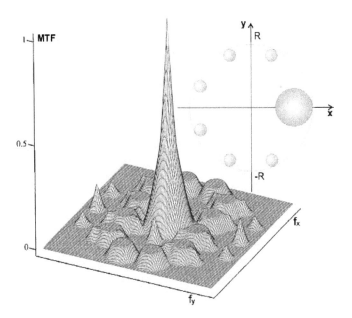

FIGURE 11.42 The diameters of the side mirrors in the POST layout are increased and their centers are redistributed in a non-redundant manner to obtain a complete coverage, even if with a low amplitude, of the spatial frequency plane up to the cutoff frequency.

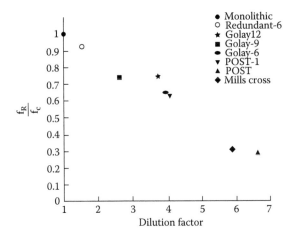

FIGURE 11.43 The increase in the dilution factor results in the decrease of the quality of imaging as measured by the maximum spatial frequency for which the modulation transfer function (MTF) is not zero, f_R, normalized to the cutoff frequency of the monolithic mirror f_c.

the spatial frequencies are smaller: for the first 35% of the covered radial spatial frequencies, the MTF decreases steeply to about 0.4; for the next 35% it is approximately constant, at about 0.4; in the last 30%, the MTF decreases at about the same rate as that of a monolithic mirror. The MTF exhibits a large plateau for the moderate values of the radial spatial frequency, much the same as that for a mirror with a central obscuration (see Figure 11.19). The decrease of performance for intermediate radial spatial frequencies may also be compared to the degradation due to aberrations, illustrated in Figure 11.11. A tradeoff of cost and performance is continually made to determine the largest diameter aperture of a monolithic mirror exhibiting some aberrations versus the challenges of phasing the individual sub-apertures to obtain the optimum performance of a system. Some of these issues are discussed in Hebden [113].

The interest of building diluted apertures has also grown in the Earth remote sensing community. Figure 11.41 shows a configuration of a large mirror, flanked on one side by several smaller ones, with the purpose of using the large mirror for imaging up to its cutoff frequency and the smaller ones for the detection of a specific phenomenon at a given spatial frequency and orientation. For illustrative purposes, we present a potential redesign of this aperture configuration such that a complete coverage of the spatial frequency plane is achieved: sizes of the side mirror apertures are increased and their centers are redistributed in a non-redundant manner, as depicted in Figure 11.42.

Figure 11.43 illustrates the general trend that the increase in the dilution factors results in the decrease of the quality of imaging as measured by the f_R (the maximum spatial frequency for which the MTF is not zero), normalized with respect to the cutoff frequency, f_c.

11.5 Thermal and Solar Telescopes

A heat stop is one of the most important thermal control devices for a large ground-based solar telescope. Liu et al. [114] describe a novel design for the heat stop, incorporating a multi-channel cooling system with high efficiency. Upon the comparison with existing single-channel designs, the performance improvement is achieved in thermal efficiency and thermal transfer coefficient.

The full solar disc telescope is a part of the future solar orbiter of the combined ESA/NASA mission. It will provide photospheric magnetic field vector, line of sight velocity, and continuum in the visible. The thermally-induced refocus is determined with an autonomous image contrast analysis [115]. A refocusing system allows for a lens displacement to find the best focus position under conditions of thermal loads.

Solar telescopes must consider the design of the heat stop as one of the critical issues, in addition to all others taken in consideration in traditional telescopes. Temperature increase will produce "internal seeing" degrading the image quality significantly. Liu et al. [116] describe an integrated analysis based on fluid dynamics to find temperature distribution and geometric optics, resulting in determination of wavefront aberration. The analytical results on Chinese Large Solar Telescope indicate that 50 K is the maximum acceptable temperature increase. This results in internal seeing aberration of 25 nm, RMS.

11.6 Alternate Applications of Telescopes

Telescopes are often used as an optical receiver system in free-space laser communication systems. Large-diameter primary mirrors enhance the photon collection and pointing. Wang et al. [117] propose an easily implementable optical communication system, incorporating an astronomical telescope. The dual use is implemented with a flip-flop mirror. The maximum acquisition time is decreased by a factor of 10, compared to that of the SILEX project of the ESA.

Acknowledgment

Special appreciation is expressed to Gonzalo Paez for preparing a number of illustrations in the first edition of this chapter.

References

1. Strojnik M, Peez G, Scholl MK. Understanding human visual system and its impact on designs of intelligent instruments, Invited, in Honoring John Caulfield, SPIE Proc. 8833, paper 3; 2013. doi: 10.1117/12.2025720.
2. Hopkins RE. Military standardization handbook, *Optical Design*, Government Printing Office, Washington, DC, 1962. Chapters 10, 14, 15.

3. Kingslake R. *Optical System Design*, New York, NY: Academic Press, 1983.

4. Malacara D, Malacara Z. *Handbook of Lens Design*, New York, NY: Marcel Dekker, 1994.

5. Fisher R, Tadic-Galeb B, Yoder PR. *Optical System Design*, 2nd Edition, Bellingham, WA: SPIE Press, 2008.

6. Smith W. *Modern Lens Design*, 1st Edition, New York, NY: McGraw-Hill Book Company, 1994.

7. Goodman JW. *Introduction to Fourier Optics*, New York, NY: McGraw-Hill, 1968.

8. Born M, Wolf E. *Principles of Optics*, New York, NY: Macmillan, 1959.

9. Strojnik and Scholl, 2016.

10. Scholl MS. Experimental demonstration of a star field identification algorithm, *Optics Letters*, **18**(3), 412–404; 1993a.

11. Scholl MS. Design parameters for a two-mirror telescope for stray-light sensitive infrared applications, *Infrared Physics & Technology*, **37**, 251–257; 1996a.

12. Scholl MS. Apodization effects due to the size of a secondary mirror in a reflecting, on-axis telescope for detection of extra-solar planets, in *Infrared Remote Sensing* M. S. Scholl, ed., *Proc. SPIE*, **2019**, 407–412; 1993b.

13. Lin W-C, Chang S-T, Chang S-H, et al. Alignment and assembly process for primary mirror subsystem of a spaceborne telescope, *Optical Engineering*, **54**, 115109; 2015.

14. Gautam S, Gupta A, Singh GS. Optical design of off-axis Cassegrain telescope using freeform surface at the secondary mirror, *Optical Engineering*, **54** (2), 025113; 2015.

15. Smith WJ. *Modern Optical Engineering. The Design of the Optical Systems*, New York, NY: McGraw-Hill, 1990.

16. Scholl MS, Wang Y, Randolph JE, et al. Site certification imaging sensor for mars exploration, *Optical Engineering* **30**(5), 590–597; 1991.

17. Smith W. *Modern Lens Design*, 2nd Edition, New York, NY: McGraw-Hill Book Company, 2004.

18. Arianpour A, Schuster GM, Tremblay EJ, et al. Wearable telescopic contact lens, *Applied Optics*, **54**, 7195–7204; 2015.

19. Schuster GM, Arianpour A, Cookson S, et al. Wink-controlled polarization-switched telescopic contact lenses, *Applied Optics*, **54**, 9597–9605; 2015.

20. Scholl MS. Autonomous star field identification using intelligent CCD-based cameras, *Optical Engineering*, **33**(1), 134–139; 1994a.

21. Rosin S. Ritchey-Chretien corrector system, *Applied Optics*, **5**(4), 475–676; 1966.

22. Rosete-Aguilar M, Maxwell J. Design of achromatic corrector plates for schmidt cameras and related optical systems, *Journal of Modern Optics*, **40**(7), 1395–1410; 1993.

23. Wilson R. The history of optical theory of reflecting telescopes and implications for future projects; Computer lens design workshop, C. Londoño and R. E. Fischer, (eds.), *Proc. SPIE*, **2871**, 784–786; 1995.

24. Wynne CG. Ritchey-Chretien telescopes and extended field systems, *Astrophysical Journal*, **152**, 675–694; 1968.

25. Blakley R. Cesarian telescope optical system, *Optical Engineering*, **35**(11), 3338–3341; 1996.

26. Puryayev DT. Afocal two-mirror system, *Optical Engineering*, **32**(6), 1325–1329; 1993.

27. Wetherell WB. All reflecting afocal telescopes, in *Reflective Optics* C. Londoño and R. E. Fischer, (eds.), *Proc. SPIE*, **751**, 126–134; 1987.

28. Scholl MS. Experimental verification of the star field identification algorithm in the observatory environment, *Optical Engineering*, **34**(2), 384–390; 1996b.

29. Dials MA, Gille JC, Barnett JJ, et al. A description of the High Resolution Dynamics Limb Sounder (HIRDLS) instrument, in *Infrared Spaceborne Remote Sensing VI*, M. S. Strojnik, and B. F. Andresen, eds., *Proc. SPIE*, **3437**, 84–91; 1998.

30. Tacca M, Sorrentino F, Buy C, et al. Tuning of a high magnification compact parabolic telescope for centimeter-scale laser beams, *Applied Optics*, **55**, 1275–1283; 2016.

31. Peez G, Strojnik Scholl M. Recursive relations for ray-tracing through three-dimensional reflective confocal prolate spheroids, *Revista Mexicana de Fisica*, **43**(6), 875–886; 1997.

32. Cook LG. The last three-mirror anstigmat, in *Lens Design* W. J. Smith, ed., Bellingham, WA: SPIE Optical Engineering Press, pp. 310–324; 1992.

33. Paul M. Sistems correcteurs pour reflecteurs astronomiques, *Revue D'Optique*, **5**, 169; 1935.

34. Rumsey NJ. A compact three-reflection astronomical camera, *Optical Instruments and Techniques*, J. H. Dickson, ed., Newcastle-upon-Tyne: Oriel Press, p. 514; 1969.

35. Eisenberg S, Pearson ET. Two-mirror three-surface telescope, in *Reflective Optics*, C. Londoño and R. E. Fischer, eds., *Proc. SPIE*, **751**, 126–134; 1987.

36. Offner A. Unit power imaging catoptric anastigmat, U.S. Patent #3,748,015, July 24, 1973.

37. Korsch D. *Reflective Optics* New York, NY: Academic Press, 1991.

38. Courtes G, Cruvellier P, Detaille M, et al. Some new optical designs for the ultraviolet bidimensional detection of astronomical objects, in *Progress in Optics XX* E. Wolf, ed., New York, NY: North-Holland, 1983.

39. Scholl MS, Peez G. Push-broom reconnaissance camera with time expansion for a (Martian) landing – Site certification, *Optical Engineering*, **36**(2), 566–573; 1997a.

40. Roddier C, Roddier F. A combined approach to HST wave-front distortion analysis, in *Space Optics* Vol. **19** of OSA 1991 Technical Digest Series (Optical Society of America, Washington, D.C., 1991), paper MB5-1, pp. 25–27.

41. Fienup JR. Phase-retrieval algorithms for a complicated optical system, *Applied Optics*, **32**, 1737–1746; 1993.

42. Fienup JR. HST aberrations and alignment determined by the phase retrieval algorithms, in *Space Optics* Vol. **19** of OSA 1991 Technical Digest Series (Optical Society of America, Washington, D.C., 1991), paper MB3-1, pp. 19–21.

43. Niemack DM. Designs for a large-aperture telescope to map the CMB 10X faster, *Applied Optics*, **55**; 2016.

44. Buisset C, Deboos A, Lépine T, et al. Design and performance estimate of a focal reducer for the 2.3 m Thai National Telescope, *Optics Express*, **24**, 1416–1430; 2016.

45. Scholl MS. Stray light issues for background-limited far-infrared telescope operation, *Optical Engineering*, **33**(3), 681–684; 1994b.

46. Beckers MM, Ulich BL, Williams JT. Performance of the Multiple Mirror Telescope (MMT): I. MMT – The first of the advanced technology telescopes, in *Advanced Technology Optical Telescopes* G. Burbidge and L. D. Barr, (eds.), *Proc. SPIE*, **332**, 2–8; 1982.

47. Woolf NJ, McCarthy DW, Angel JRP. Image shrinking in sub-arcsecond seeing at the MMT and 2.3 m Telescopes (MMT): VIII. MMT as an optical-infrared interferometer and phased array, in *Advanced Technology Optical Telescopes*, G. Burbidge and L. D. Barr, eds., *Proc. SPIE*, **332**, 50–56; 1982a.

48. Tarenghi M. European Southern Observatory (ESO) 3.5 m in new technology telescope, in *Advanced Technology Optical Telescopes II* L. D. Barr, ed., *Proc. SPIE*, **629**, 213–220; 1986.

49. Fried DL. Diffusion analysis for the propagation of mutual coherence, *Journal of the Optical Society of America*, **58**, 961–969; 1968.

50. Merkle F, Freischlad K, Reischman H-L. Deformable mirror with combined piezoelectric and electrostatic actuators, in *Advanced Technology Optical Telescopes* G. Burbidge and L. D. Barr, eds., *Proc. SPIE*, **332**, 260–268; 1982.

51. Nian P, Wen-Li M. Performance analysis of the retractable dome for the Chinese Large Telescope, *Optics Express*, **23**, 25376–25404; 2015.

52. Scholl MS, Peez G. Using the y, y-bar diagram to control stray light noise in IR systems, *Infrared Physics & Technology*, **38**, 25–30; 1997b.

53. Strojnik M, Scholl, MK. Extrasolar planet observatory on the far side of the Moon, *Journal of Applied Remote Sensing*, **8**(1), 084982; 2014.

54. Wu, Z, Bai, H, Cui, X. Curvature wavefront sensing based on a single defocused image and intensity compensation, *Applied Optics*, **55**(10), 2791–2799; 2016.

55. Page NA, McGuire JP Jr. Design of wide-field planetary camera 2 for Hubble space telescope, in *Space Optics*, Vol. 19 of OSA 1991 Technical Digest Series (Optical Society of America, Washington, D.C., 1991), paper MC4-1, pp. 38–40.

56. Scholl MS, Lawrence GN. Adaptive optics for in-orbit aberration correction – feasibility study, *Applied Optics*, **34**(31), 7295–7301; 1995.

57. Rausch P, Verpoort S, Wittrock U. Unimorph deformable mirror for space telescopes: Design and manufacturing, *Optics Express*, **23**; 2015.

58. Marino J, Wöger F. Feasibility study of a layer-oriented wavefront sensor for solar telescopes, *Applied Optics*, **53**, 685–693; 2014.

59. Moore KT, Lawrence GN. Zonal model of an adaptive mirror, *Applied Optics*, **29**, 4622–4628; 1990.

60. Wang H, Yang S. Modeling and analysis of the thermal effects of a circular bimorph piezoelectric actuator, *Applied Optics*, **55**, 873–878; 2016.

61. Hardy JH. Active optics – Don't build a telescope without it, in *Advanced Technology Optical Telescopes* G. Burbidge and L. D. Barr, eds., *Proc. SPIE*, **332**, 252–259; 1982.

62. Beckers JM, Roddier FJ, Eisenhardt PR, et al. National Optical Astronomy Observatories (NOAO) Infrared adaptive optics program I: General description, in *Advanced Technology Optical Telescopes III* L. D. Barr, ed., *Proc. SPIE*, **629**, 290–297; 1986.

63. Scholl MS, Lawrence GN. Diffraction modeling of a space relay experiment, *Optical Engineering*, **29**(3), 271–278; 1990.

64. Massie NA. Experiments with high bandwidth segmented mirrors, in *Advanced Technology Optical Telescopes*, G. Burbidge and L. D. Barr, eds., *Proc. SPIE*, **332**, 377–381; 1982.

65. Roddier F. The effects of atmospheric turbulence in optical astronomy, in *Progress in Optics XIX*, E. Wolf, ed., New York, NY: North-Holland, 1981.

66. Lutomirski RF, Yura HT. Aperture averaging factor for a fluctuating light signal, *Journal of the Optical Society of America*, **59**, 1247–1248; 1969.

67. Ruiz E, Sohn E, Salas L, et al. Common-pull, multiple-push, vacuum-activated telescope mirror cell, *Applied Optics*, **53**, 7979–7984; 2014.

68. Fugate RQ. Laser beacon adaptive optics – Boom or bust, in *Current Trend in Optics*, Vol. 19, C. Dainty, ed., New York, NY: Academic Press, pp. 289–304; 1994.

69. Poglitsch A. Far-infrared astronomy from airborne and spaceborne platforms (KAO and FIRST), in *Infrared Spaceborne Remote Sensing II*, M. S. Scholl, ed., *Proc. SPIE*, **2268**, 251–262; 1994.

70. Suzuki M, Kuze A, Tanii J, et al. Feasibility study on solar occultation with a compct FTIR, in *Infrared Spaceborne Remote Sensing V*, M. S. Scholl and B. F. Andresen, eds., *Proc. SPIE*, **3122**, 2–15; 1997.

71. Fienup JR, Marron JC, Schulz TJ, et al. Hubble space telescope characterized by using phase retrieval algorithms, *Applied Optics*, **32**, 1747–1768; 1993.

72. Burrows C, Krist J. Phase retrieval analysis of pre- and post-repair hubble space telescope images, *Applied Optics*, **34**(22), 491–496; 1995.

73. Kessler MF. Science with the infrared space observatory, in *Infrared Spaceborne Remote Sensing*, M. S. Scholl, ed., *Proc. SPIE*, **2019**, 3–8; 1993.

74. Cesarsky CJ, Bonnal JF, Boulade O, et al. Development of ISOCAM, the camera of the infrared space observatory, in *Infrared Spaceborne Remote Sensing*, M. S. Scholl, ed., *Proc. SPIE*, **2019**, 36–47; 1993.

75. Clegg PC. ISO long-wavelength spectrometer, in *Infrared Spaceborne Remote Sensing*, M. S. Scholl, ed., *Proc. SPIE*, **2019**, 15–23; 1993.

76. Graauw Th, Beintema D, Luinge W, et al. The ISO short-wavelength spectrometer, in *Infrared Spaceborne Remote Sensing*, M. S. Scholl, ed., *Proc. SPIE*, **2019**, 24–27; 1993.

77. Cohen R, Mast T, Nelson J. Performance of the W. M. Keck telescope active mirror control system, in *Advanced Technology Optical Telescopes V*, L. M. Stepp, ed., *Proc. SPIE*, **2199**, 105–116; 1994.

78. Lemke D, Garzon F, Gemund HP, et al. ISOPHOT – Far-infrared imaging, polarimetry and spectrophotometry on ISO, in *Infrared Spaceborne Remote Sensing*, M. S. Scholl, ed., *Proc. SPIE*, **2019**, 28–33; 1993.

79. Scholl MS. Image-plane incidence for a baffled infrared telescope, *Infrared Physics & Technology*, **38**, 87–92; 1997.

80. Zhu Y, Zhang X, Liu T, et al. Internal and external stray radiation suppression for LWIR catadioptric telescope using non-sequential ray tracing, *Infrared Physics & Technology*, **71**, 163–170; 2015.

81. Padin S. Inexpensive mount for a large millimeter-wavelength telescope, *Applied Optics*, **53**, 4431–4439; 2014.

82. Hardy T, Cain S, Jeon J, et al. Improving space domain awareness through unequal-cost multiple hypothesis testing in the space surveillance telescope, *Applied Optics*, **54**(17), 5481–5494; 2015.

83. Beckers JM, Williams JT. Performance of the Multiple Mirror Telescope (MMT): III. MMT – Seeing experiments with the MMT, in *Advanced Technology Optical Telescopes*, G. Burbidge and L. D. Barr, (eds.), *Proc. SPIE*, **332**, 16–23; 1982.

84. Kemp JC, Esplin RW. Sensitivity model for the wide-field infrared explorer mission, in *Infrared Spaceborne Remote Sensing III*, M. S. Scholl and B. F. Andresen, eds., *Proc. SPIE*, **2553**, 26–37; 1995.

85. McCarthy DW, Strittmatter PA, Hege EK, et al. Performance of the Multiple Mirror Telescope (MMT): VIII. MMT as an optical-infrared interferometer and phased array, in *Advanced Technology Optical Telescopes*, G. Burbidge and L. D. Barr, eds., *Proc. SPIE*, **332**, 57–64; 1982.

86. Chanan G, Troy M, Sirko E. Phase discontinuity sensing: A method for phasing segmented mirrors in the infrared, *Applied Optics*, **38**(4), 704–713; 1999.

87. Chanan G, Ohara C, Troy M. Phasing the mirror segments of the keck telescopes II: the narrow-band phasing algorithm, *Applied Optics*, **39**(25), 4706–4714; 2000.

88. Nelson JE, Gilingham P. An overview of the performance of the W. M. Keck observatory, in *Advanced Technology Optical Telescopes V* L. M. Stepp, ed., *Proc. SPIE*, **2199**, 82–93; 1994.

89. Mast TS, Nelson JE, Welch WJ. Effects of primary mirror segmentation on telescope image quality, in *Advanced Technology Optical Telescopes* G. Burbidge and L. D. Barr, eds., *Proc. SPIE*, **332**, 123–134; 1982.

90. Cole TW. Quasi-optical techniques of radio astronomy, in *Progress in Optics XV* E. Wolf, ed., New York, NY: North-Holland, 1977.

91. Beckers JM. Field of view considerations for telescope arrays, in *Advanced Technology Optical Telescopes III*, L. D. Barr, ed., *Proc. SPIE*, **629**, 255–260; 1986.

92. Scholl MS. Recursive exact ray trace equations through the foci of the tilted off-axis confocal prolate spheroids, *Journal of Modern Optics*, **43**(8), 1583–1588; 1996c.

93. Scholl MS. Signal detection by an extra-solar-system planet detected by a rotating rotationally-shearing interferometer, *Journal of the Optical Society of America A*, **13**(7), 1584–1592; 1996d.

94. Strojnik Scholl M, Peez G. Cancellation of star light generated by a nearby star-planet system upon detection with a rotationally-shearing interferometer, *Infrared Physics & Technology*, **40**, 357–365; 1999.

95. Gillespie A, Matsunaga T, Rokugawa S, et al. Temperature and emissivity separation from Advanced Spaceborne Thermal Emission and Reflection Radiometer (ASTER) images, in *Infrared Spaceborne Remote Sensing IV*, M. S. Scholl and B. F. Andresen, eds., *Proc. SPIE*, **2817**, 83–94; 1996.

96. Woolf NJ, Angel JRP, Antebi J, et al. Scaling the Multiple Mirror Telescope (MMT) to 15 meters – Similarities and differences, in *Advanced Technology Optical Telescopes*, G. Burbidge and L. D. Barr, eds., *Proc. SPIE*, **332**, 79–88; 1982b.

97. Angel JRP, Hill JM. Manufacture of large glass honeycomb mirrors, in *Advanced Technology Optical Telescopes* G. Burbidge and L. D. Barr, eds., *Proc. SPIE*, **332**, 298–306; 1982.

98. Anderson D, Parks RE, Hansen QM, et al. Gravity deflections of light-weighted mirrors, in *Advanced Technology Optical Telescopes*, G. Burbidge and L. D. Barr, eds., *Proc. SPIE*, **332**, 424–435; 1982.

99. Bely PY. A ten-meter optical telescope in space, in *Advanced Technology Optical Telescopes III*, L. D. Barr, ed., *Proc. SPIE*, **629**, 188–195; 1986.

100. Enard D. The ESO very large telescope project, in *Advanced Technology Optical Telescopes III*, L. D. Barr, ed., *Proc. SPIE*, **629**, 221–226; 1986.

101. Kodaira K, Isobe S. Progress report on the technical study of the Japanese Telescope Project, in *Advanced Technology Optical Telescopes III*, L. D. Barr, ed., *Proc. SPIE*, **629**, 234–238; 1986.

102. Pearson E, Stepp L, Wong W-Y, et al. Planning the National New Technology Telescope (NNT): III. Primary optics – Tests on a 1.8 m borosilicate glass honeycomb mirror, in *Advanced Technology Optical Telescopes III*, L. D. Barr, ed., *Proc. SPIE*, **69**, 91–101; 1986.

103. Siegmund WA, Mannery EJ, Radochia J, et al. Design of the apache point observatory 3.5 m telescope II. Deformation analysis of the primary mirror, in *Advanced Technology Optical Telescopes III*, L. D. Barr, ed., *Proc. SPIE*, **629**, 377–389; 1986.

104. Mountain M, Kurz R, Oschmann J. The GEMINI 8 m telescope project, in *Advanced Technology Optical Telescopes V* L. M. Stepp, ed., *Proc. SPIE*, **2199**, 41–55; 1994.

105. Petrovsky GT, Tolstoy MN, Ljubarsky SV, et al. A 2.7 meter diameter silicon carbide primary mirror for the SOFIA telescope, in *Advanced Technology Optical Telescopes, V* L. M. Stepp, ed., *Proc. SPIE* **2199**, 263–270; 1994.

106. Olbert BJ, Angel RP, Hill JM, et al. Casting 6.5 meter mirrors for the MMT conversion and magellan, in *Advanced Technology Optical Telescopes V*, L. M. Stepp, ed., *Proc. SPIE*, **2199**, 144–155; 1994.

107. Miglietta L, Gray P, Gallieni W, et al. The final design of the large binocular telescope mirror cells for the ORM, in *Optical Telescopes of Today and Tomorrow*, A. Ardeberg, ed., *Proc. SPIE*, **2871**, 301–313; 1996.

108. Espinosa JM, Alvarez Martin P. Gran Telescopio Canarias: A 10 m Telescope for the ORM, in *Optical Telescopes of Today and Tomorrow*, A. Ardeberg, ed., *Proc. SPIE*, **2871**, 69–73; 1996.

109. Meimon S, Petit C, Fusco T. Optimized calibration strategy for high-order adaptive optics systems in closed-loop: The slope-oriented Hadamard actuation, *Optics Express*, **19**(21), 27134–27144; 2015.

110. Ferreira C, Lopez JL, Navarro R, et al. Zernike-like systems in polygons and polygonal facets, *Applied Optics*, **54**, 6575–6583; 2015.

111. Simar JF, Stockman Y, Surdej J. Single-wavelength coarse phasing in segmented telescopes, *Applied Optics*, **54**(5), 1118–1123; 2015.

112. Smith GM. II status report, in *Optical Telescopes of Today and Tomorrow*, A. Ardeberg, (ed.), *Proc. SPIE*, **2871**, 10–14; 1996.

113. Hebden JC, Hege EK, Beckers JM. Use of coherent MMT for diffraction limited imaging, in *Advanced Technology Optical Telescopes III* L. D. Barr, ed., *Proc. SPIE*, **629**, 42–49; 1986.

114. Liu Y, Gu N, Rao C. Quantitative evaluation on internal seeing induced by heat-stop of solar telescope, *Optics Express*, **27**(15), 19980–19995; 2015a.

115. Silva-Lopez M, Garranzo-Garcia D, Sanchez A, et al. *Optical Engineering*, **54**(8), 084104; 2015.

116. Liu Y, Gu N, Rao C, et al. Heat-stop structure design with high cooling efficiency for large ground-based solar telescope, *Applied Optics*, **54**, 6441–6447; 2015b.

117. Wang J, Lv J, Zhao G, et al. Free-space laser communication system with rapid acquisition based on astronomical telescopes, *Optics Express*, **23**, 20655–20667; 2015.

12

Microscopes

Daniel Malacara-
Doblado and
Alejandro
Téllez-Quiñones

12.1 Introduction

A microscope (from the Ancient Greek: μικρός, mikrós, "small" and σκοπεῖν, skopeîn, "to look" or "see") is an instrument used to see objects that are too small for the naked eye. The science of investigating small objects using such an instrument is called microscopy. Microscopic means invisible to the eye unless aided by a microscope.

    There are many types of microscopes. The most common (and the first to be invented) is the optical microscope, which uses light to image the sample. Other major types of microscopes are the electron microscope (both the transmission electron microscope and the scanning electron microscope), the ultramicroscope, and the various types of scanning probe microscope.

12.2 Single Magnifier

The most common and traditional use of a lens is as a magnifier. The stop of the system is the pupil of the observing eye. Since with good illumination this pupil is small (3–4 mm or even smaller), the aberrations that increase with the diameter of the pupil, such as spherical aberrations, do not present any problem. On the other hand, the field of view is not small; hence, field curvature, astigmatism, and distortion are the most important aberrations to be corrected.

The angular magnification M is defined as the ratio of the angular diameter β of the virtual image observed through the magnifier to the angular diameter α of the object as observed from a distance of 250 mm (defined as the minimum observing distance for young adults). If the lens is placed at a distance from the observed object so that the virtual image is at distance l' from the eye, the magnification M is given by

$$M = \frac{\beta}{\alpha} = \frac{250}{(l'+d)}\left(\frac{l'}{f}+1\right), \tag{12.1}$$

where d is the distance from the lens to the eye, f is the focal length, and all distances are in millimeters.

The maximum magnification, obtained when the virtual image is as close as 250 mm in front of the observing eye, and the lens is close to the eye ($d = 0$) is

$$M = \frac{250}{f}+1. \tag{12.2}$$

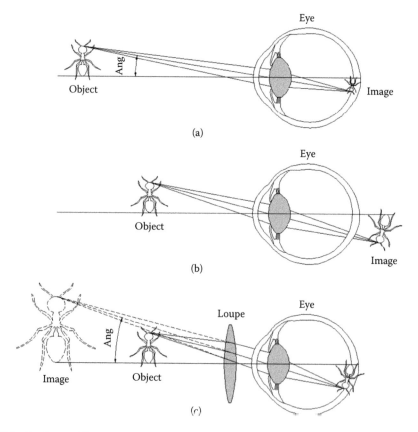

FIGURE 12.1   Single magnifier.

If the virtual image is placed at infinity to avoid the need for eye accommodation, the magnification becomes independent of the distance d and has a value

$$M = \frac{250}{f}. \tag{12.3}$$

Here, we have to remember that eye accommodation occurs when the eye lens (crystalline) modifies its curvature to focus close objects.

We can thus see that for small focal lengths f, the magnification is nearly the same for all lens' positions with respect to the eye and the observed object, as long as the virtual image is not closer than 250 mm from the eye.

It has been shown that if the magnifier is a single plane-convex lens, the optimum orientation to produce the best possible image with the minimum aberrations is

1. With the plane on the eye's side if the lens is closer to the eye than to the object, as shown in Figure 12.1a
2. With the plane on the side of the object if the lens is closer to the object than to the eye, as shown in Figure 12.1b

The disadvantage of the plane-convex magnifier is that if one does not know these rules, one may use it with the wrong lens orientation. A safer magnifier configuration would be a double convex lens, but then the image is not the best. For this reason, most high-quality magnifiers are symmetrical. To reduce the aberrations and have the best possible image, more complicated designs can be used, as illustrated in Figure 12.1c.

12.3 Compound Microscope

A compound microscope uses a very short focal length objective lens to form a greatly enlarged image. This image is then viewed with a short focal length eyepiece used as a simple magnifier. The image should be formed at infinity to minimize eyestrain (Figure 12.2).

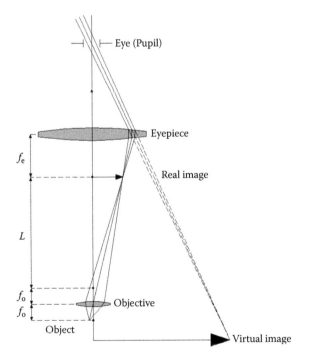

FIGURE 12.2 Compound microscope.

The general assumption is that the length of the tube L is large compared to either f_o or f_e so that the following relationships:

$$m_o = -\frac{L}{f_o} \tag{12.4}$$

where m_o is the objective magnification, L is the length of the tube and f_o is the objective focal distance.

$$m_\alpha = \frac{25}{f_e} \tag{12.5}$$

where m_α is the angular magnification and f_e is the eyepiece magnification.

12.4 Dark Field Microscope

Dark field microscopy is used to enhance the contrast in unstained and transparent specimens. A special sized disc prevents unscattered light from entering the objectives and produces the almost black background. Only scattered light from the tissues or cells of the specimens will be seen as bright objects against the dark background. Despite the simplicity of the system, the quality of the images are quite impressive. Because of the low light levels, the microscopes are equipped with strong illumination systems (Figure 12.3).

12.4.1 When to Use a Dark Field Microscope

Dark field microscopes are used in a number of different ways to view a variety of specimens that are hard to see in a light field unit. Live bacteria, for example, are best viewed with this type of microscope, as these organisms are very transparent when unstained.

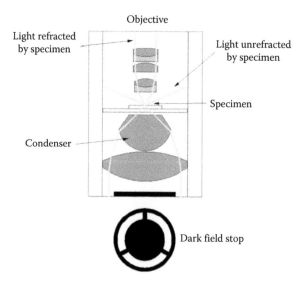

FIGURE 12.3   Dark field microscope diagram.

There are a multitude of other ways to use dark field illumination, often when the specimen is clear or translucent. Some examples are

- Living or lightly stained transparent specimens
- Single-celled organisms
- Live blood samples
- Aquatic environment samples (from seawater to pond water)
- Living bacteria
- Hay or soil samples
- Pollen samples
- Certain molecules, such as caffeine crystals

Dark field microscopy makes many invisible specimens appear visible. Most of the time the specimens invisible to bright field illumination are living, so you can see how important it is to bring them into view.

12.4.2 Advantages and Disadvantages

No one system is perfect and dark field microscopy may or may not appeal to you depending on your needs.

Some advantages of using a dark field microscope are

Extremely simple to use
Inexpensive to set up
Very effective in showing the details of live and unstained samples

Some of the disadvantages are

Limited colors (certain colors will appear, but they are less accurate and most images will be just black and white)
Images can be difficult to interpret to those unfamiliar with dark field microscopy

Although surface details can be very apparent, the internal details of a specimen often do not stand out as much with a dark field setup.

Below there is an example of dark field illumination of lens tissue paper (Figure 12.4).

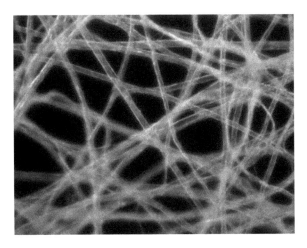

FIGURE 12.4 Image produced by dark field microscope.

12.5 Polarizing Microscope

It is a clear field microscope to which filters that modify the light are added. It is also called a petrographic or metallurgical microscope for its initial use in the study of minerals; however, its application has extended to the field of biology, medicine, chemistry and many other disciplines. This microscopic technique may employ both transmitted light and incident light.

Compared with other contrast enhancement techniques, the use of polarized light is the most effective in the study of samples rich in birefringent materials, since it improves incomparably the quality of the image. Light from a standard source of illumination vibrates and propagates in all directions, but when passing through a polarizing filter, the waves and their electric field oscillate all in the same plane. The polarizer is a device that only lets light pass by vibrating in a certain plane called the polarization axis.

When studying the behavior of light when traversing a specimen, the anisotropic materials have different indices of refraction in relation to the direction of the light beam. On the contrary, the isotropic materials have a constant index of refraction. When a ray of light strikes the surface of a transparent anisotropic material, the phenomenon of double refraction or birefringence occurs; this means that two different refracted rays are produced that vibrate in different planes that propagate with different velocities inside the material.

This microscope facilitates the investigation of the optical properties of the specimens and is ideal to observe and photograph those elements, which are visible thanks to the anisotropy, hence its use in crystallography. However, it is also used to study the birefringent character of many anisotropic cell structures. This microscope is shown in Figure 12.5.

12.5.1 Applications of the Polarized Light Microscope

Some applications of the polarized light microscope are
1. Identify intracellular (cytoskeletal) and extracellular (amyloid, asbestos, collagen, uratic crystals, and others of exogenous origin) or crystalline or fibrous substances.
2. Identify and estimate mineral components quantitatively.

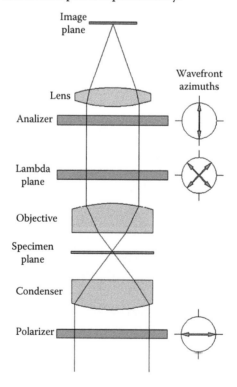

FIGURE 12.5 Polarizing microscope diagram.

12.6 Objective Classes

Objectives are categorized into performance classes on the basis of their transmission, chromatic correction, planarity, and others. From the point of view of the International Organization of Standardization (ISO), there are three groups of objective classes differing in quality of chromatic correction (Table 12.1).

12.6.1 Depth Field

Magnifying glasses enlarge images by small degrees, blurring the image around the edges as lens magnification increases. Microscopes combine two lenses, requiring adjustments in depth of field and magnification to observe small objects clearly.

Microscopes contain two lenses, positioned at opposite ends of a cylinder. The cylinder is connected to an eyepiece, a base and a lighted slide shelf. Placing slides beyond the focal distance of the first lens produces an inverted image approximately 16 cm behind the lens, which is re-magnified by the second lens. Magnification increases as focal length for each lens is shortened (Table 12.2).

$$\sigma_{ob} = \frac{nl}{2NA^2} \tag{12.6}$$

Most compound microscopes employ the Deutsche Industrie Norm, or DIN, standard. The DIN standard has a 160 mm distance from the objective flange to the eyepiece flange as shown in Figure 12.6.

12.7 Confocal Microscopy

A *confocal configuration* can provide, among other several benefits in optical microscopy, the possibility of recovering three-dimensional images of thick samples, such as biological tissue. As described by Wilson [1], the devices that consider this configuration operate well in the bright field and fluorescence modes. A conventional microscope can be considered as a parallel system, where the goal of the system is to recover the images of the entire object field (the complete sample). This goal can be difficult to achieve

TABLE 12.1 Microscope Objective Classes

Objective Class	Field Flatness	Absolute Value(s) of Focus	Application
Achromats	up to 23 mm	Between red and blue wavelength (2 colors) is ≤ 2x depth of field	For standard applications in the visual spectral range
Semi-Apochromats	up to 25 mm	For red wavelength and blue to green wavelength (3 colors) are ≤ 2.5x depth of field	For applications in the visual spectral range with higher specifications
Apochromats	up to 25 mm	For red wavelength and blue to green wavelength (3 colors) are ≤ 1x depth of field	For applications with highest specifications in the visual range and beyond

TABLE 12.2 Microscope Objectives Main Characteristics

Magnification	Numerical Aperture	Depth of Field (mm)
4x	0.10	50
10x	0.25	7.7
20x	0.40	2.9
40x	0.65	0.9
60x	0.85	0.36
100x	0.95	0.17

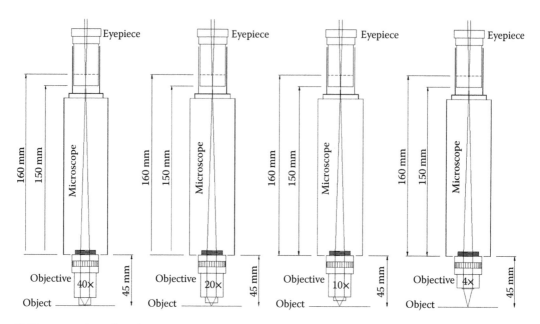

FIGURE 12.6 Deutshe industrie norm (DIN).

with the optical components of a basic system of microscopy; however, the difficulty is minimized, if the image formation is made point by point. This form of image formation is known as the *scanning process*. The scanning process can be carried out successfully, especially when adapting a confocal configuration in the scanning microscopy system. Two of the most basic confocal arrangements are shown in Figure 12.7; Figure 12.7a shows confocal microscope in transmission mode, and Figure 12.7b shows confocal microscope in reflection mode.

In Figure 12.7, the term *focal plane* must be considered as the plane to be focused with respect to both lenses, objective and collector, but as independent systems. In this sense, focal plane is not necessarily the plane where the paraxial focus of the objective lens is located. The same argument applies for the collector lens.

There are several configurations of confocal microscopes based on different types of scanning techniques. All these varieties were invented by different individuals in different parts of the world. By 1961, the scientist Marvin Minsky [2,3] patented a confocal microscope where the specimen was mechanically scanned with respect to the light source. In Minsky's patent two modes of confocal arrangements quite similar to those in Figure 12.7 are described. Mojmír Petráň designed a confocal microscope based on the rotation of a Nipkow disk for scanning the light source on the specimen [2,4]. Similar techniques like this have been used by other authors, such as Xiao Guoqing and Gordon Kino [2,5]. Other methods for the scanning process have been proposed, like the use of a two-sided mirror, according to Svishchev [2,6].

One of the main advantages of adapting a confocal configuration in a microscopy system can be exemplified when discussing microscopy by fluorescence. Among the major drawbacks that had the non-confocal fluorescence microscopy was the depth discrimination or optical sectioning capability [7]. When confocal arrangements were not available, the common solution to this problem was to make observations of very thin specimens, such as cells in tissue culture monolayers or thin smears of cells for pathology [2]. This drawback limited the use of the light microscope for the observation of thick specimens, as human skin *in vivo*, live embryos, brain imaging and studies of hard tissue like teeth and bone. Similar problems occurred in specimens observed by reflection configurations (see Figure 12.7b). However, the adaptation of a confocal system can deal with the problem of depth discrimination, as depicted in Figure 12.8. As described by Müller [7], the exicitation of the specimen is induced by the laser illumination system and a fraction of the fluorescence emitted by the fluorophores in the specimen are

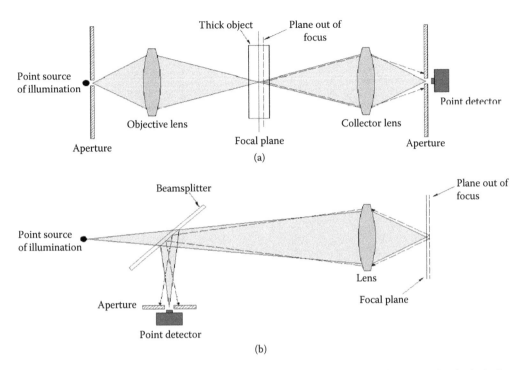

FIGURE 12.7 Two basic confocal arrangements; (a) transmission mode, and (b) reflection mode. The dashed rays in these configurations correspond to sample points of a thick object that are not imaged by the system. In (a), if the collector and the objective lenses, respectively, have the same focal length, then both lenses can be located at the same distance with respect to the focal plane.

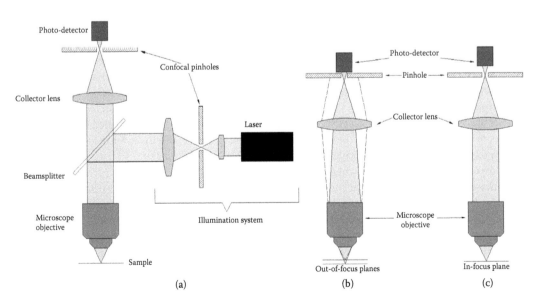

FIGURE 12.8 (a) Depiction of a confocal fluorescence microscope. (b) When the sample is out of focus, it cannot be imaged at the detection pinhole in front of a photo-detector. (c) Once the sample is in focus, the image is formed.

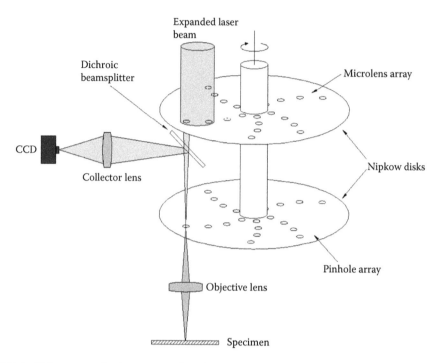

FIGURE 12.9 Yokogawa confocal microscope by reflection.

collected by the microscope objective and the collector lens, which permits the imaging at the detection pinhole (see Figure 12.7a).

There are two types of scanning processes, according to Minsky [2]; scanning the specimen, and scanning the beam of light. In the first type, the optics is kept fixed and the sample is moved for illuminating its portions, point by point, or section by section. Here we are understanding a section, as a trace of the sample, which is formed by a certain number of points. In the second type, the specimen is fixed, and the light is guided for imaging each of the sample points. The use of Nipkow disks in the microscopy system is an example of the second type. To have an idea of the mechanical swept of the specimen points, with this kind of technology, we can consider the Yokogawa confocal microscope by reflection, which is shown in Figure 12.9 [2]. An expanded laser beam illuminates the upper disk, which is conformed by an array of microlenses. All these microlenses have their paraxial focus at the lower disk, which is conformed by an array of pinholes. Once the light passes through the pinholes, different points of the sample are illuminated by the objective lens, and the back projection of the reflected light forms the images of all the sample points at the CCD. The tandem rotation of both disks makes the scanning of many points of the specimen in a wide field fashion possible.

As we can see in Figures 12.7 through 12.9, the term *confocal*, in its most simple representation, can be understood as an optical system conformed by two lenses for imaging the object (the specimen): an *objective lens*, which illuminates the object, and a *collector lens*, which forms the image of the object in a detection plane. In the case of Figure 12.8b, the objective lens and the collector lens coincide. In the case of Figure 12.8a, an object point of the sample induces its corresponding image at the pinhole plane in the illumination system, whereas the collector lens forms the image of this object point at the pinhole plane in front of the photo-detector. All these configurations are referred as confocal, because both lenses (or systems of lenses), objective and collector, are focused on the same point of the complete specimen, that is why all the apertures (or pinholes) in all these setups are generally called as confocal apertures. So, to understand what a confocal arrangement is, first, we are going to describe some mathematical points about the image formation with this setup. For such description, Fresnel and Fraunhofer approximations will be required.

12.7.1 Propagation of the Complex Amplitude through Circular and Annular Apertures

An interesting identity is given by

$$\int_0^x w J_0(w)\,dw = x J_1(x),\tag{12.7}$$

where J_0 and J_1 are first kind Bessel functions of zero and one order, respectively. Equation 12.7 is consequence of the definitions for J_0 and J_1 in power series. These two functions are solutions of the well-known Bessel differential equations [8].

Another important identity about the Bessel function $J_0(x)$ is

$$J_0(x) = \frac{1}{2\pi} \int_0^{2\pi} e^{-ix\cos(t)}\,dt,\tag{12.8}$$

where i is the imaginary unity. Equation 12.8 is easy to prove by integrating the term $e^{ix\cos(t)}$ with respect to t, and considering the complex Maclaurin expansion of this term. Such integration implies

$$\int_0^{2\pi} e^{ix\cos(t)}\,dt = 2\pi J_0(x).\tag{12.9}$$

However, the Bessel function $J_0(x)$ is an even function, which is $J_0(x) = J_0(-x)$, then Equation 12.8 can be obtained from Equation 12.9. Equations 12.7 and 12.8 can be used for some important descriptions about the Fresnel and Fraunhofer diffraction in the propagation of complex optical amplitude functions (propagation of electromagnetic waves in the visible spectrum).

As is known, an electromagnetic field can be expressed by a scalar field amplitude. The propagation of the electromagnetic field can be calculated from one plane to another by considering the Fraunhofer approximation [9]. If $U(x, y)$ denotes the scalar field amplitude in a cartesian xy-plane, and $U(u, v)$ is the same scalar field in a uv-plane, both separated at a fixed distance z, then

$$U(u,v) = g(h,z)F\left[\frac{u}{\lambda z}, \frac{v}{\lambda z}\right],\tag{12.10}$$

where $g(h,z) = \left[\exp(ikz)\exp\left(i\left(\frac{k}{2z}\right)h^2\right)\right] / [i\lambda z]$, and

$$F[u,v] = \int_{-\infty}^{+\infty}\int_{-\infty}^{+\infty} U(x,y)\exp\big(-i2\pi(ux+vy)\big)\,dx\,dy.\tag{12.11}$$

Here, $U(u,v)$ is the Fraunhofer approximation applied to $U(x,y)$, $F[u,v]$ is the Fourier transform of $U(x,y)$, λ is the wavelength, k is the wavenumber, and $h^2 = u^2 + v^2$. If the vector position (u,v) is assumed as a parallel vector to the u–axis, then the dot product $(x,y)\cdot(u,v)$ is such that

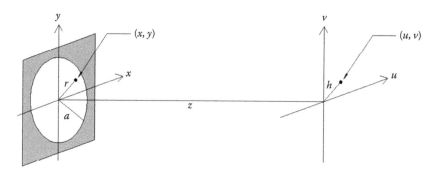

FIGURE 12.10 Schematic sketch for the propagation of a monochromatic wave from the *xy*-plane to the *uv*-plane through a circular aperture with radius *a*. A monochromatic light source is assumed behind the aperture.

$(x,y)\cdot(u,v)=ux+vy=rh\cos(t)$, where t is the angle formed by the vector position (x,y) with the positive part of the *x*-axis, and $r^2=x^2+y^2$ (see Figure 12.1). Now, if we consider the circular aperture with radius *a*, as shown at Figure 12.10, then

$$U(x,y)=U(r)=\begin{cases} 1 & r<a \\ 0 & r\geq a \end{cases},$$
(12.12)

abusing the notation. By means of the change to polar coordinates $x=r\cos(t)$, $y=r\sin(t)$, and by considering Equations 12.7 and 12.8, and Equations 12.10 through 12.12, the Fourier transform term in Equation 12.1 is rewritten as

$$F\left[\frac{u}{\lambda z},\frac{v}{\lambda z}\right]=2\pi\int_0^a hJ_0(2\pi hr/(\lambda z))dr.$$
(12.13)

If the change of variable $w=(2\pi hr/(\lambda z))$, and Equation 12.7, are taken into account in the previous Fourier transform, then the intensity $I(h)=|U(h)|^2=|U(u,v)|^2$, induced by Equation 12.10, will be such that

$$I(h)=\left(\frac{\pi a^2}{\lambda z}\right)^2\left[\frac{2J_1\left(\frac{2\pi ha}{\lambda z}\right)}{\frac{2\pi ha}{\lambda z}}\right]^2,$$
(12.14)

due to the radial symmetry. Equation 12.14 is valid in the context of the Fraunhofer approximation, where the hypothesis $ka/2\ll z$ is required [9]. Equation 12.14 is exactly the same observed for the corresponding intensity in the standard literature [10].

In the case of an annular aperture with outer radius *a* and inner radius γa, the calculation of the scalar field amplitude in the *uv*-plane can be obtained by following similar arguments.

By assuming the radial symmetry of the scalar field, the amplitude for an annular aperture takes the form

$$U(h)=\exp\left(\frac{ikh^2}{2z}\right)\frac{\exp(ikz)}{i\lambda z}2\pi\varepsilon aJ_0(2\pi ha/(\lambda z)),$$
(12.15)

where $1 - \gamma = \varepsilon$, and ε is a small positive value (as described* in [10]). Equation 12.15 leaves us to obtain an intensity that is equal to

$$I(h) = \left(\frac{2\pi\varepsilon a}{\lambda z} \right)^2 \left[J_0 \left(2\pi h a / (\lambda z) \right) \right]^2.$$

(12.16)

12.7.2 Propagation of the Complex Amplitude through Positive Thin Lenses; Circular and Annular Pupils

If we consider a positive thin lens with a circular pupil function $P(x, y)$ (a *circular thin lens*) in an xy-plane, when the lens is illuminated by a plane wave, the amplitude just after the lens is given by

$$U(x, y) = P(x, y) \exp\left[-\frac{ik}{2f} (x^2 + y^2) \right],$$

(12.17)

where f is the focal length of the lens [9]. The Fresnel approximation applied to $U(x, y)$, by considering the distance $z = f$, is

$$U(u, v) = g(h, f) \int_{-\infty}^{+\infty} \int_{-\infty}^{+\infty} U(x, y) \exp\left[\frac{ik}{2f} (x^2 + y^2) \right] \exp\left[-\frac{i2\pi}{\lambda f} (xu + yv) \right] dx dy,$$

(12.18)

where g function is given as in Equation 12.10. Equation 12.18 induces to obtain

$$U(h) = -i \exp(ikf) \exp\left(\frac{ikh^2}{2f} \right) \frac{\pi a^2}{\lambda f} \left[\frac{2J_1 \left(2\pi h a / (\lambda f) \right)}{2\pi h a / (\lambda f)} \right],$$

(12.19)

when considering Equations 12.7, 12.17 and 12.18 and a circular aperture $P(x, y) = P(r)$ with radius a, which implies radial symmetry for the complex amplitude at every plane $(U(x, y) = U(r), U(u, v) = U(h))$. This time, the Fresnel approximation (near field) is taking place instead of the Fraunhofer approximation (far field), because we are considering a finite distance $z = f$, which does not necessarily satisfy $ku / 2 \ll z$. Since $k = 2\pi / \lambda$, Equation 12.19 can be rewritten in terms of the Fresnel number $N = \pi a^2 / (\lambda f)$ and the special coordinate $\rho = 2\pi h a / (\lambda f)$, which is

$$U(\rho) = -iN \exp(ikf) \exp\left(\frac{i\rho^2}{4N} \right) \left[\frac{2J_1(\rho)}{\rho} \right].$$

(12.20)

Thus, the intensity observed in the uv-plane is proportional to the square of the term in the brackets of Equation 12.20 (Figure 12.11).

For a pupil function with an annulus of fractional thickness ε (*annular thin lens*), the corresponding amplitude in the focal plane is

$$U(\rho) = -2iN\varepsilon \exp(ikf) \exp\left(\frac{i\rho^2}{4N} \right) J_0(\rho),$$

(12.21)

---

* However, in reference [10], the sign of the arguments for the exponential terms in Equation 13.15 are negative, because in that reference, the Fourier transform is defined with a positive sign for the argument of the exponential kernel.

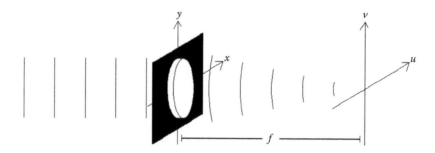

FIGURE 12.11 Propagation of a monochromatic plane wavefront through a circular lens (positive lens with circular aperture). The parallel lines before the *xy*-plane represent planar wavefronts, and the curved lines after this plane represent spherical wavefronts. The intensity observed in the *uv*-plane at the focal length of the lens is given by the square modulus of Equation 12.20.

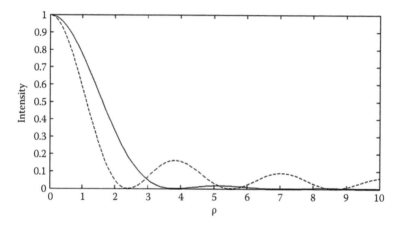

FIGURE 12.12 Intensity distribution in the focal plane of a circular lens (continuous line) and an annular lens (dashed line).

as described in [10]. When N is large, the condition $N \gg \rho^2 / 4$ is satisfied, then the quadratic phase terms in Equations 12.20 and 12.21 are negligible. Thus, the intensity of the circular lens is proportional to $\left[2J_1(\rho)/\rho \right]^2$, while the intensity of the annular lens is proportional to $\left[J_0(\rho) \right]^2$. The functions $\left[2J_1(\rho)/\rho \right]^2$ and $\left[J_0(\rho) \right]^2$ are plotted in Figure 12.12.

12.7.3 Image Formation for Positive Thin Lenses; Circular and Annular Pupils

As in the previous section, the calculation of the complex field amplitude in a *uv*-plane can be obtained from the propagation of a monochromatic spherical wavefront (instead of a plane wavefront). This spherical wavefront can be considered as the resultant complex field produced by a punctual source of illumination (a single punctual object, see Figure 12.12). If the spherical wavefront produced by this punctual object is propagated through a *positive circular thin lens*, the wavefront is changed to another spherical wavefront whose curvature center is located at the image distance of this object. If the image point of this object is observed in a *uv*-plane, located at the corresponding image distance, it can be seen that the image point is not a point. Instead of a point, we have an Airy pattern [11], specifically, an intensity pattern.

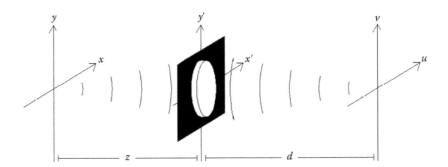

FIGURE 12.13 Propagation of a monochromatic spherical wavefront through a positive circular thin lens. The relation between the distances z and d is given by Equation 12.22.

Let us consider the illustration in Figure 12.13. In this figure, the single object is located at the center of an xy-plane. The object, considered as a punctual source of illumination, produces a monochromatic spherical wavefront that propagates through a positive circular thin lens with focal length f. The lens is located at an $x'y'$-plane whose distance from the xy-plane is z. In the paraxial approximation, the image distance d of the single object is calculated by means of

$$\frac{1}{z} + \frac{1}{d} = \frac{1}{f}. \tag{12.22}$$

The uv-plane is separated from the $x'y'$-plane at the distance d. Then, by using the Fresnel approximation, the complex amplitude just after the lens in the $x'y'$-plane is

$$U(x',y') = \left[P(x',y')\exp\left[\frac{-ik}{2f}(x'^2 + y'^2)\right]\right] \frac{\exp(ikz)}{i\lambda z} \int_{-\infty}^{+\infty}\int_{-\infty}^{+\infty} U(x,y)\exp\left[\frac{ik}{2z}\left((x'-x)^2 + (y'-y)^2\right)\right]dxdy, \tag{12.23}$$

where P is the pupil function, and U is the amplitude produced by the punctual source. If the amplitude U is a simple transmittance function, this function can be considered as $U(x,y) = \delta(x)\delta(y)$, where the symbol δ represents a Dirac delta. From the properties of the Dirac delta, Equation 12.23 implies that

$$U(h) = \left[\frac{\exp(ik(d+z))}{-\lambda^2 zd}\right]\exp\left[\frac{ikh^2}{2d}\right]\int_0^{+\infty} P(r)J_0\left(\frac{2\pi hr}{\lambda d}\right)2\pi rdr, \tag{12.24}$$

by considering Equation 12.22. Here $h^2 = u^2 + v^2$ and $r^2 = x'^2 + y'^2$. Equation 12.24 is achieved from the Fresnel approximation applied to calculate the complex amplitude in the uv-plane, when considering radial symmetry for P and U. Finally, for a pupil function P with radius a, Equation 12.24 implies

$$U(h) = \left[\frac{\exp(ik(d+z))}{-\lambda^2 zd}\right]\exp\left[\frac{ikh^2}{2d}\right]\left[\frac{\pi a^2}{\lambda d}\right]\left[\frac{2J_1(2\pi ha/(\lambda d))}{2\pi ha/(\lambda d)}\right]. \tag{12.25}$$

Again, if the term $\rho = 2\pi ha/(\lambda d)$ is introduced, Equation 12.25 suggests that the intensity observed in the uv-plane is proportional to $\left[2J_1(\rho)/\rho\right]^2$, which profile is observed in Figure 12.3. In a similar manner, it can be concluded that the intensity observed for the configuration in Figure 12.12, with a *positive annular thin lens*, is proportional to $\left[J_0(\rho)\right]^2$, where a in the ρ coordinate is the radius of the outer ring of the annular pupil with a small width ε.

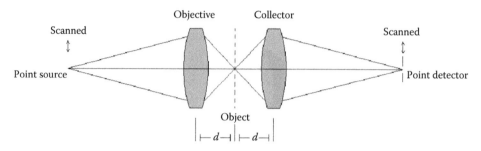

FIGURE 12.14 Confocal scanning microscope.

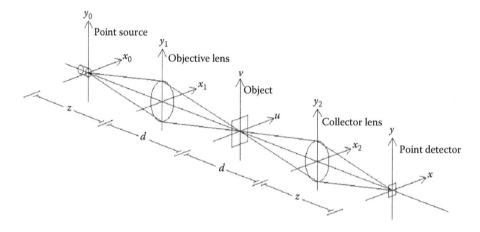

FIGURE 12.15 Image formation in a confocal microscope.

12.7.4 Image Formation in Confocal Scanning Microscopy

As described in [10], there are two basic types of arrangements for a scanning microscope: the simple scanning microscope, and the confocal scanning microscope. Since the confocal scanning microscopy has better optical properties than the simple scanning microscopy, we are going to discuss only the basic confocal arrangement corresponding to this type (see Figure 12.14).

A confocal scanning microscope is formed by two lenses, both separated at the same distance d from the object to be scanned. This configuration is called confocal arrangement, because both lenses are focused on the same point of the object. A point source of monochromatic light is used to illuminate the object by means of a positive lens called objective. The light rays pass through the object until reaching a second positive lens called collector. Finally, the collector lens projects the rays to a point detector. The image of the object is formed point by point (scanning process).

The image formation of a confocal microscope can be approximated by using Fresnel integrals, as in the previous sections. Normally, in the confocal configuration, both lenses have the same focal length, and a detailed description of this image formation is described by Wilson and Sheppard [10]. Here, we are going to describe this image formation by assuming certain particular conditions.

Let us consider the confocal arrangement shown in Figure 12.15. So, by using the Fresnel integral and the pupil-lens function, the complex field produced just after the objective lens, given by a monochromatic point source at the origin of an $x_0 y_0$-plane, is

$$U(x_1, y_1) = \frac{e^{ikz}}{i\lambda z} P_1(x_1, y_1) \exp\left[-\frac{ikr_1^2}{2f}\right] \iint U(x_0, y_0) \exp\left[\frac{ik}{2z}\left((x_1 - x_0)^2 + (y_1 - y_0)^2\right)\right] dx_0 \, dy_0, \qquad (12.26)$$

where, to simplify notation, the symbol $\iint$ will refer the double integral over the complete two-dimensional space $\mathbb{R}^2$. In Equation 12.26, we have taken into account the term $r_1^2 = x_1^2 + y_1^2$ and the term f as the focal length of the objective lens with a pupil function P_1. The $x_1 y_1$-plane is the place where the objective lens is located at a distance z from the point source with wavelength λ, and wavenumber k, respectively. If the complex field in the $x_0 y_0$-plane is a simple transmittance function given by $U(x_0, y_0) = \delta(x_0)\delta(y_0)$, then, by considering Equation 12.22 and the Fresnel approximation, Equation 12.26 implies that

$$U(u,v) = \frac{e^{ik(d+z)}\pi a^2}{i^2 \lambda^3 d^2 z} \exp\left[\frac{ik}{2d}h^2\right]\left[\frac{2J_1\left(\dfrac{2\pi ha}{\lambda d}\right)}{\dfrac{2\pi ha}{\lambda d}}\right]T(u,v), \tag{12.27}$$

where $h^2 = u^2 + v^2$. Equation 12.27 is obtained, when an aperture of radius a is assumed as the pupil function, and T denotes the transmittance function associated to the object. From this last equation, the scalar field just after the $x_2 y_2$-plane is calculated as

$$U(x_2, y_2) = \frac{e^{ik(2d+z)}\pi a^2}{i^3 \lambda^4 d^3 z} \exp\left[\frac{ik}{2d}r_2^2\right]P_2(x_2, y_2)\exp\left[-\frac{ik}{2f}r_2^2\right] \times$$
$$\iint \exp\left[\frac{ikh^2}{d}\right]\left[\frac{2J_1\left(\dfrac{2\pi ha}{\lambda d}\right)}{\dfrac{2\pi ha}{\lambda d}}\right]T(u,v)\exp\left[-\frac{ik}{d}(x_2 u + y_2 v)\right]du\,dv \tag{12.28}$$

by considering a collector lens in the $x_2 y_2$-plane with the same focal length of the objective lens, a pupil function P_2, and the term $r_2^2 = x_2^2 + y_2^2$ (see Figure 12.6). If the transmittance function of the object is $T(u,v) = \delta(u - x_s)\delta(v - y_s)$, where (x_s, y_s) is a particular fixed position in the uv-plane, then Equation 12.28 reduces to

$$U(x_2, y_2) = \frac{e^{ik(2d+z)}\pi a^2}{i^3 \lambda^4 d^3 z} \exp\left[-\frac{ik}{2z}r_2^2\right]\left[\frac{2J_1\left(\dfrac{2\pi h_s a}{\lambda d}\right)}{\dfrac{2\pi h_s a}{\lambda d}}\right]P_2(x_2, y_2). \tag{12.29}$$

where $h_s^2 = x_s^2 + y_s^2$. Equation 12.29 is obtained due to the properties of the two-dimensional Dirac delta, and the paraxial Equation 12.22 for the collector lens. Moreover, this previous equation is consequence of considering $x_s \ll d$ and $y_s \ll d$ (this fact can be assumed when $(x_s, y_s) \approx (0,0)$), therefore, the exponential terms $\exp\left[\dfrac{ikh_s^2}{d}\right]$, $\exp\left[-\dfrac{ik}{d}(x_2 x_s + y_2 y_s)\right]$, are closer to the unity. Again, from the Fresnel approximation applied to Equation 12.29, the incident complex amplitude in the xy-plane of the point detector is

$$U(r) = \frac{e^{i4kd}\pi^2 a^4}{i^4 \lambda^6 d^6} \exp\left[\frac{ik}{2d}r^2\right]\left[\frac{2J_1\left(\dfrac{2\pi h_s a}{\lambda d}\right)}{\dfrac{2\pi h_s a}{\lambda d}}\right]\left[\frac{2J_1\left(\dfrac{2\pi ra}{\lambda d}\right)}{\dfrac{2\pi ra}{\lambda d}}\right], \tag{12.30}$$

where $r^2 = x^2 + y^2$. Equation 12.30 is induced when P_2 is another circular aperture with radius a, and considering the fact that, when $z = 2f$, then $d = 2f$ also, and that implies $d = z$. Moreover, when $d = z$, the objective lens and the collector lens have the same common unitary magnification M, that means $M = d / z = 1$. If Equation 12.30 is evaluated in the position (x, y) of a point detector, in such a way that (x, y) is the corresponding image of the position (x_s, y_s), given by the collector lens, then $x = -Mx_s = -x_s$ and $y = -My_s = -y_s$, and that implies $r = h_s$. So, when $d = z = 2f$, Equation 12.30 can be rewritten in terms of the constant value r and the constant distance d, which is

$$U(r) = \frac{e^{i4kd}\pi^2 a^4}{i^4\lambda^6 d^6} \exp\left[\frac{ik}{2d}r^2\right]\left[\frac{2J_1\left(\dfrac{2\pi ra}{\lambda d}\right)}{\dfrac{2\pi ra}{\lambda d}}\right]. \tag{12.31}$$

From Equation 12.31, the image of the point (x_s, y_s) can be described. The point (x_s, y_s) is a particular scanned position of the total object plane (the uv-plane) and U can be considered as a function of the r distance, measured from the origin of the xy-plane (the detector plane). The function U can be evaluated at any other arbitrary point (x_s, y_s), closer to the origin of the uv-plane, to be scanned. Equation 12.31 describes the image formation for the confocal microscope, and allows us to conclude that the intensity observed in

the incident xy-plane satisfies $I(\rho) = |U(\rho)|^2 \propto \left[\dfrac{2J_1(\rho)}{\rho}\right]^4$, where $\rho = 2\pi ra / (\lambda d) = \pi ra / (\lambda f) \approx \pi r NA / \lambda$.

Of course, this proportional relation is valid when the two lenses of the confocal arrangement have the same pupil function of radius a, and the same focal length f; in other words, the confocal microscope is formed by two circular lenses with common numerical aperture $NA \approx na / f$ (assuming imaging in air, refractive index $n = 1$). A similar result for a confocal microscope with two annular lenses (in such case, a is the radius of the outer ring of the annular pupil) is discussed in reference [10], where the intensity observed in the detector is proportional to $J_0(\rho)^4$. Furthermore, in this reference, it is explained that the resultant intensity

is proportional to $\left[\dfrac{2J_1(\rho)}{\rho}\right]^2 J_0(\rho)^2 = \left[\dfrac{2J_0(\rho)J_1(\rho)}{\rho}\right]^2$, when the confocal arrangement has one circular and

one annular lens of equal radius a. Although in this last case, the coordinate ρ in the factor $\left[\dfrac{2J_1(\rho)}{\rho}\right]^2$ is in

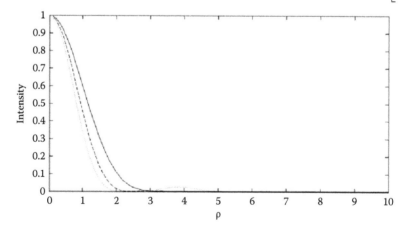

FIGURE 12.16 Intensities for confocal microscopes with different pairs of lenses. The continuous line is for circular-objective and circular-collector. The dashed line is for circular-objective and annular-collector, which is also equivalent for annular-objective and circular-collector. The dotted line is for annular-objective and annular-collector.

terms of a constant a, which is different in meaning to the constant a of the coordinate ρ in the factor $J_0(\rho)^2$; both coordinates are equal, in terms of the common numerical aperture NA, for both lenses. The resultant intensities for two annular lenses and combined forms will not be described here, however, the reader can take as an exercise the deduction of these cases. The intensities for confocal microscopes with different pairs of lenses (objective and collector) are shown in Figure 12.16.

12.8 Phase Contrast Microscopy

According to Ernst Abbe, when an object is illuminated by a coherent light source, the object can be interpreted as a diffraction grating [8]. If this object has only one spatial frequency, like a diffraction grating, light will be diffracted in many beams with different diffraction orders. Therefore, a microscope objective can form the image of this object, by superposing all these beams in the image plane, as illustrated at Figure 12.16. In this sense, the resultant image is an interference pattern given by the diffracted beams. If the lens that represents the microscope objective in Figure 12.17 is too small with respect to the frequency of the grating, the image will not be formed by only the contribution of the zero order beam. The details of the object will be imaged, only if the lens is sufficiently large for passing at least the beams with order one. However, if the lens has the minimal diameter for receiving the beams with order one, this will not be sufficient to get an image with good quality.

This image formation is quite similar to what happens when considering different orders for the Fourier expansions of functions. For instance, if the high quality image of the grating looks like the square signal at Figure 12.18a, then the image of a lens that passes the beams up to the order one will be such as that shown in Figure 12.18b. For a lens that forms an image with the beams up to the second order, the resultant image would be similar to Figure 12.18c.

An application of Abbe's theory is the microscope by phase contrast, invented by Frits Zernike (Nobel prize in Physics, 1953). If the refractive index of a transparent object, the sample, is different from the media that surrounds it, a phase contrast microscope permits the observation of this object, without the need of coloring the sample [8]. An ordinary microscope cannot form the image of the transparent object because the diffracted beams are such that their irradiance (or intensity) are kept constant.

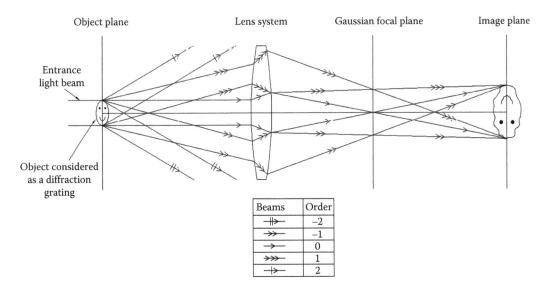

FIGURE 12.17 Abbe's theory for microscopy. In this case, the object is considered as a diffraction grating, and many diffracted beams, with different orders, contribute to the image formation. The illustration exemplifies the case of a lens system, which forms the image with the first order beams.

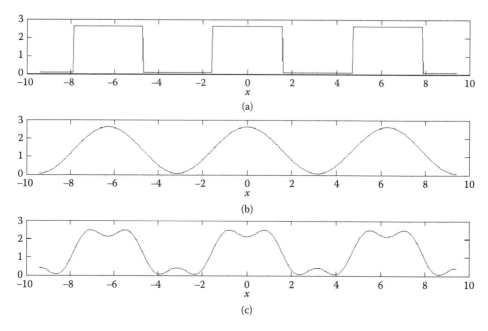

FIGURE 12.18 (a) Image amplitude of an object. (b) Amplitude formed with the first order. (c) Amplitude formed with the first and the second order.

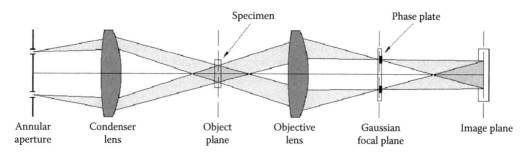

FIGURE 12.19 Simple example of phase contrast microscope.

We can imagine that the complex amplitude of one point of the image is the resultant sum of the non-diffracted beam (the zero order) and one diffracted beam (any order different from zero). So, for two different points of the image, we have two different values of complex amplitudes, with different phase terms, but the same magnitudes. Since the contrast of the image between two different points is related to variations in magnitude of the complex amplitude, the application of a phase shift of 90 degrees to the zero order beam permits these two points to contrast. This phase displacement can be introduced by adding a phase plate in the focal plane of the objective lens, as depicted in the phase contrast microscope, exemplified at Figure 12.18.

In Figure 12.19 we have a phase plate, which is a thin plate of glass with a dielectric material ring that induces the desired phase shift. In this case, the system is illuminated by an annular aperture in such a way that the zero order beam passes through the dielectric annulus. The advantage of using this annular apertures has been explained previously.

Because the recovery of images of thin samples for biological applications is one of the most common uses of optical microscopes, we can think of the possibility of considering confocal configurations for a phase contrast microscope. These samples can be considered so thin in such a way that the absorption is negligible, and only phase shifts to a transilluminating light beam are induced [12]. In this description, we need to understand the geometry of transillumination imaging, which consists of a 4f system. Thus, we are going to describe the 4f system, where a useful function called *coherent spread function* (CSF) plays an important role in the comprehension of the phase contrast microscopy.

12.8.1 Basic 4f System

Let us consider first the scheme in Figure 12.8a of a thin lens with focal length $f > 0$. If $U(\hat{\eta})$ is the complex amplitude at the lens plane ($z = s_0 > 0$), and $U(\hat{p}_0)$ is the complex amplitude at the object plane ($z = 0$), by considering the Fresnel formula, we consider that

$$U(\hat{\eta}) = -\frac{ik}{s_0} e^{i2\pi k s_0} e^{i\pi k \frac{\eta^2}{s_0}} \int U(\hat{p}_0) e^{i\pi \frac{k}{s_0} p_0^2} \exp\left[-i2\pi \frac{k}{s_0} \hat{p}_0 \cdot \hat{\eta} \right] d^2 \hat{p}_0, \tag{12.32}$$

where $k = 1/\lambda$, $\hat{p}_0 = (x_0, y_0)$, $\hat{\eta} = (x, y)$, $p_0 = \| \hat{p}_0 \|$ (Euclidean norm), $\eta = \| \hat{\eta} \|$, $\hat{p}_0 \cdot \hat{\eta} = x_0 x + y_0 y$ (inner product), $d^2 \hat{p}_0 = dx_0 dy_0$, and $\int = \int_{-\infty}^{+\infty}\int_{-\infty}^{+\infty}$ (simplified notation). In a similar way, the complex amplitude in the observation plane ($z = s_0 + s_1$, where $s_1 > 0$) is give by

$$U(\hat{p}_1) = -\frac{ik}{s_1} e^{i2\pi k s_1} e^{i\pi k \frac{p_1^2}{s_1}} \int t(\hat{\eta}) U(\hat{\eta}) e^{i\pi \frac{k}{s_1} \eta^2} \exp\left[-i2\pi \frac{k}{s_1} \hat{\eta} \cdot \hat{p}_1 \right] d^2 \hat{\eta}, \tag{12.33}$$

where $t(\hat{\eta}) = \exp\left[-i\pi \frac{k}{f} \eta^2 \right]$ is the transmittance of the lens, $\hat{p}_1 = (x_1, y_1)$, $p_1 = \| \hat{p}_1 \|$, and $d^2 \hat{\eta} = dxdy$. When substituting Equation 12.32 into Equation 12.33, it is obtained that

$$U(\hat{p}_1) = \frac{-k^2}{f s_1} \exp\left[i2\pi k(f + s_1) \right] \exp\left[i\pi \frac{k}{s_1} p_1^2 \right] \int U(\hat{p}_0) e^{i\pi \frac{k}{f} p_0^2}$$
$$\left(\int \exp\left[i\pi k \frac{\eta^2}{s_1} \right] \exp\left[-i2\pi k \hat{\eta} \cdot \left(\frac{\hat{p}_0}{f} + \frac{\hat{p}_1}{s_1} \right) \right] d^2 \hat{\eta} \right) d^2 \hat{p}_0, \tag{12.34}$$

which is the consequence of using the Fubini's theorem (interchanging the order of the integrals) [13], and considering $s_0 = f$.

However, there is a useful identity that can be applied to the integral with respect to $d^2 \hat{\eta}$ in Equation 12.34. This identity is

$$\int \exp\left[i\beta \eta^2 \right] \exp\left[i\gamma \hat{m} \cdot \hat{p} \right] d^2 \hat{\eta} = i\frac{\pi}{\beta} \exp\left[-i\frac{\gamma^2}{4\beta} p^2 \right], \tag{12.35}$$

where $\beta \neq 0$ and γ are arbitrary scalars. By considering Equation 12.35, it is inferred that

$$U(\hat{p}_1) = -\frac{ik}{f} \exp\left[i4\pi k f \right] \int U(\hat{p}_0) \exp\left[-i2\pi \frac{k}{f} \hat{p}_0 \cdot \hat{p}_1 \right] d^2 \hat{p}_0, \tag{12.36}$$

when considering $s_1 = f$. Equation 12.36 is basically a Fourier transform. Now, let us consider the scheme in Figure 12.19b, a 4f system with two thin lenses of focal lengths $f_0 > 0$, and $f_1 > 0$, respectively. Here the 2D coordinates for the object plane are $\hat{p}_0 = (x_0, y_0)$, for the pupil plane, $\hat{\xi} = (x_*, y_*)$, and for the observation plane, $\hat{p}_1 = (x_1, y_1)$. Thus by using the result in Equation 12.36, from the object plane to the pupil plane (first lens), we have that

$$U(\hat{\xi}) = -\frac{ik}{f_0} \exp\left[i4\pi k f_0\right] \int U(\hat{p}_0) \exp\left[-i2\pi \frac{k}{f_0} \hat{p}_0 \cdot \hat{\xi}\right] d^2 \hat{p}_0. \tag{12.37}$$

When considering again the Fresnel propagation from the pupil plane to the observation plane (second lens), we obtain that

$$U(\hat{p}_1) = -\frac{k^2}{f_0 f_1} \exp\left[i4\pi k(f_0 + f_1)\right] Term(\hat{p}_1), \tag{12.38}$$

where $Term(\hat{p}_1) = \int U(\hat{p}_0) \left(\int P(\hat{\xi}) \exp\left[-i2\pi k\hat{\xi} \cdot \left(\frac{\hat{p}_0}{f_0} + \frac{\hat{p}_1}{f_1}\right)\right] d^2\hat{\xi}\right) d^2 \hat{p}_0$, by considering the Fubini's theorem one more time. Here $P(\hat{\xi})$ is the pupil function. Equation 12.38 represents the complex amplitude observed in a 4f configuration like the one in Figure 12.20b.

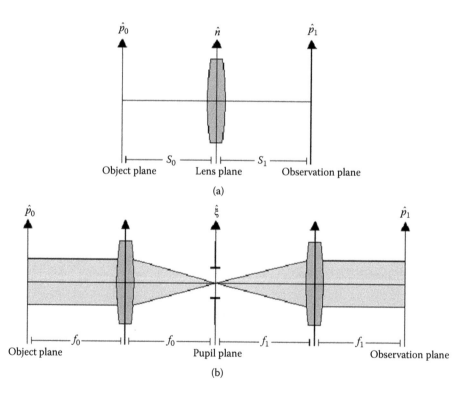

FIGURE 12.20 (a) Single lens system to illustrate the Fresnel propagation from an object plane to an observation plane. (b) System conformed by two lenses and one pupil (a 4f system). In this case, a monochromatic plane wave illuminates the object plane.

12.8.2 The Role of the Coherent Spread Function in Transillumination Microscopy

A compact form of representing Equation 12.38 is given by introducing the definition of

$$CSF(\hat{\eta}) = \left(\frac{k}{f_0}\right)^2 \int P(\hat{\xi}) \exp\left[i2\pi \frac{k}{f_0} \hat{\xi} \cdot \hat{\eta}\right] d^2\hat{\xi},$$ which in connection with the Fourier transform of the

pupil function, permit us to obtain that

$$U(\hat{p}_1) = \frac{1}{M} \exp\left[i4\pi k(f_0 + f_1)\right] \int U(\hat{p}_0) CSF\left(\frac{1}{M}\hat{p}_1 - \hat{p}_0\right) d^2\hat{p}_0. \tag{12.39}$$

Equation 12.39 is the complex amplitude in the observation screen, for a 4f configuration, where $M = -f_1/f_0$ is the magnification of the system. Expression *CSF* defines *the coherent spread function* (CSF), while Equation 12.39 denotes the relation between the image and the object fields, $U(\hat{p}_1)$ and $U(\hat{p}_0)$, respectively [12]. This relation establishes that the image field is a convolution between the object field and the CSF function. When assuming $f_1 = f_0 = f$, we have $M = -1$, then Equation 12.39 is written as

$$U(\hat{p}_1) = - \exp\left[i8\pi kf\right] \int U(\hat{p}_0) CSF(-\hat{p}_1 - \hat{p}_0) d^2\hat{p}_0. \tag{12.40}$$

An interesting property of the CSF function, considering $f_0 = f$, is that $\int CSF(-\hat{p}_1 - \hat{p}_0) d^2\hat{p}_0 = P(0)$.

So, we can assume that the object field is a single transmittance function given by $U(\hat{p}_0) = U_{in} t(\hat{p}_0)$, where U_{in} is the amplitude of a plane wave illumination field, and $t(\hat{p}_0) = e^{i\phi(\hat{p}_0)} = \cos\left[\phi(\hat{p}_0)\right] + i \sin\left[\phi(\hat{p}_0)\right]$.

When considering very thin samples, we can take into account the approximation $U(\hat{p}_0) \approx U_{in}\left[1 + i\,\phi(\hat{p}_0)\right]$.

Here, function $\phi(\hat{p}_0)$ has been considered as a phase term close to zero, then, we can substitute the previous approximation in Equation 12.40 and obtain that

$$U(\hat{p}_1) = -U_{in} \exp\left[i8\pi kf\right]\left(P(0) + i\int \phi(\hat{p}_0) CSF(-\hat{p}_1 - \hat{p}_0) d^2\hat{p}_0\right), \tag{12.41}$$

from the property of the CSF mentioned above.

The intensity observed in the image plane, induced by Equation 12.41, is such that

$$I(\hat{p}_1) = I_{in}\left[|P(0)|^2 - 2Im\left(P^*(0)\int \phi(\hat{p}_0) CSF(-\hat{p}_1 - \hat{p}_0) d^2\hat{p}_0\right)\right], \tag{12.42}$$

by dropping the second order phase term, and considering $I_{in} = |U_{in}|^2$, where *Im* denotes the imaginary part.

Equation 12.42 describes the intensity observed in the transillumination microscopy (see Figure 12.21), and also gives an explanation of the difficulties when trying to recover information of the phase objects. As described by Mertz [12], $\phi(\hat{p}_0)$ is essentially real, and if $P(0)$ and $CSF(\hat{p})$ are also real, as in the case of a standard, well-aligned microscope, the second term in Equation 12.42 must vanish. This implies that, to the first order in $\phi(\hat{p}_0)$, the intensity at the image plane cannot contain information about the phase object.

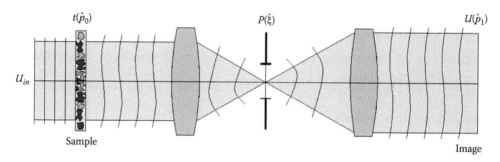

FIGURE 12.21   Transillumination imaging geometry.

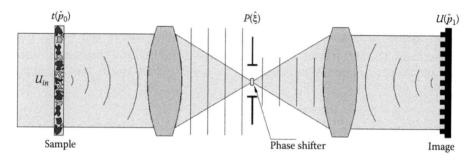

FIGURE 12.22   Zernike phase-contrast configuration.

This problem is also observed either in the case when considering spatially coherent illumination, or in the case with a more realistic illumination geometry, a spatially incoherent illumination. Many solutions to counteract this difficulty have been proposed, but in this description we are going to describe only two of them for a spatially coherent illumination: the Zernike phase-contrast microscopy, and the Schlieren microscopy.

12.8.3 Zernike Phase-Contrast Microscopy with Coherent Illumination

When a simple plane wave is used as the illumination source or the numerical aperture of the illumination optics is smaller than that of the detection optics, the illumination is coherent. For such case, the expression in Equation 12.42 is valid for the first order in $\phi(\hat{p}_0)$. If the term $P^*(0)CSF(-\hat{p}_1 - \hat{p}_0)$ in Equation 12.42 was imaginary, then the information of the first-order phase term would be recovered from the data $I(\hat{p}_1)$. In order to make this possible, a phase shifter, often called a phase mask, can be inserted in the aperture pupil (see Figure 12.22). The introduction of such phase shifter in the pupil plane of the transillumination configuration defines what is called the *Zernike phase-contrast microscopy*.

An example of pupil function, which represents the addition of a given phase shifter, can be

$$
P\left(\hat{\xi}\right) =
\begin{cases}
i & \xi \leq \varepsilon \\
1 & \varepsilon < \xi \leq r \\
0 & \xi > r
\end{cases}
\tag{12.43}
$$

where $= \left(\xi\right)$, r is the radius of the pupil, and ε is a very small value. The pupil function in Equation (E76) induces a phase shift of $\pi/2$ within a tiny region of radius ε. Since ε is very small, the CSF function remains almost completely real because the ratio of imaginary to real contributions in CSF is of order $\varepsilon^2 / r^2 \approx 0$. This implies that $P^*(0)CSF(-\hat{p}_1 - \hat{p}_0)$ becomes dominantly imaginary, which from Equation 12.42, allows us to recover the first-order phase term. However, as described by Mertz [12], Zernike phase-constrast microscopy requires specialized optics, and other difficulties can arise when choosing a standar lamp as the illumination source. When using a standar lamp, it is required to make its illumination spatially coherent; therefore, the illumintion aperture must be quite small, which implies a very low lamp power transmission. To deal with this other problem, annular apertures are more preferable than pinhole ones, but using them requires matched annular phase masks in the conjugate detection aperture.

12.8.4 Schlieren Microscopy

As in the one-dimensional case, the concepts of even function and odd function can be extended to more dimensions. With minor difficulties, for functions from $\mathbb{R}^2$ to $\mathbb{R}$, it can be demonstrated that the Fourier transform of an even function, depending on space coordinates, is an even function, depending on frequency coordinates. The same happens with odd functions; the Fourier transform of an odd function produces another odd function. Additionally, when the two-dimensional integral over $\mathbb{R}^2$ exists for a given function, it can be demonstrated that such integral is zero for odd functions. Thus, from these considerations, we can conclude that, if the pupil function P is real and odd, then its Fourier transform $\wp$ is an odd function such that

$$\wp(\hat{\eta}) = i \wp_{imag}(\hat{\eta}), \tag{12.44}$$

where $\wp_{imag}$ is the imaginary part of $\wp$. In other words, the real part of $\wp$ is zero. On the other hand, when substituting $f_0 = f_1 = f$, and $\hat{\eta} = -\hat{p}_0 - \hat{p}_1$ in the valid relation $\left(\dfrac{f_0}{k}\right)^2 CSF\left(-\dfrac{f_0}{f_1}\hat{p}_1 - \hat{p}_0\right) = \wp\left(\dfrac{k\hat{p}_0}{f_0} + \dfrac{k\hat{p}_1}{f_1}\right)$, we obtain that $CSF(\hat{\eta}) = \left(\dfrac{k}{f}\right)^2 \wp\left(-\dfrac{k}{f}\hat{\eta}\right)$.

This last implies

$$CSF_{imag}(\hat{\eta}) = \left(\frac{k}{f}\right)^2 \wp_{imag}\left(-\frac{k}{f}\hat{\eta}\right). \tag{12.45}$$

So, by evaluating Equation 12.44 at $(-k/f)\hat{\eta}$, multiplying it by $(k/f)^2$, and considering Equation 12.45, we obtain

$$CSF(\hat{\eta}) = iCSF_{imag}(\hat{\eta}). \tag{12.46}$$

In the *Schlieren microscopy*, the strategy proposed to recover information of the first-order phase term consists of the addition of an amplitude mask (instead of a phase mask) in the pupil plane. This mask is chosen as an odd-real pupil function such that $P(0) = 1$. For this kind of mask and the result in Equation 12.46, the intensity given by Equation 12.42 reduces to

$$I(\hat{p}_1) = I_{in}\left[1 - 2\int \phi(\hat{p}_0)CSF_{imag}(-\hat{p}_1 - \hat{p}_0)d^2\hat{p}_0\right], \tag{12.47}$$

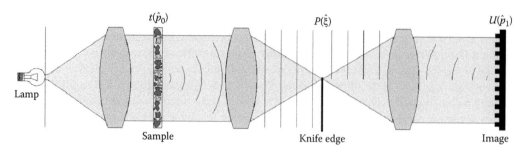

$t(\hat{p}_0)$ $P(\hat{\xi})$ $U(\hat{p}_1)$

Lamp

Sample Knife edge Image

FIGURE 12.23 Schlieren microscopy configuration.

which makes the phase imaging possible. The aforementioned mask can be induced by introducing a knife edge in the pupil plane, as can be seen in Figure 12.23.

When samples are thin, as in the case of individual cells or cell monolayers, the problem of poor performance in illumination does not present great difficulties. However, in the case of thick samples, such as tissue sections, the illumination problem is considerably major. In this last situation, the geometry of an incoherent illumination is preferable to study the image formation by phase contrast microscopy. A concise study on this geometry is provided by Mertz [12], where techniques, such as the *oblique field microscopy* or the *differential interference contrast microscopy* are useful for imaging thick samples.

12.9 Non-Optical Microscopes

The microscopes previously studied are considered as *optical microscopes* or *light microscopes*, in the sense that they use light (photons) to generate the image of the object or sample. There are also other types of microscopes, which can be classified as *non-optical microscopes*, because they do not use light for imaging. Among the non-optical microscopes, we can consider the so-called transmission electron microscopes (TEM), the scanning electron microscopes (SEM), and the scanning transmission electron microscopes (STEM). These instruments use electrons instead of photons for the imaging process [14,15]. Moreover, among the non-optical microscopes, which can be considered as topographers, there are the scanning tunneling microscope (STM) and the atomic force microscope (AFM).

In microscopy by an electron bombardment, the sample or object is considered as a point source of radiation, or more precisely, as a collection of point sources [14]. Each of these sources transmits information on the composition and structure of the sample. In electron microscopy, the sample is irradiated successively at various points and the radiation emitted by each point can be used to construct the complete image of the object. Any mode of radiation emitted by the object can be considered for the process of image formation.

12.9.1 Conventional Transmission Electron Microscope

A basic configuration of a simple transmission electron microscope is sketched in Figure 12.24. For very thin samples, the transmission electron microscope allows the image formation by the electrons passing through the sample. Since the wavelength induced by accelerated electrons is much less than the visible wavelengths, the amplification power of a TEM is higher than that of a light microscope. To understand this aspect, remember the *de Broglie equation*

$$\lambda = \frac{h}{p} = \frac{h}{mv},$$

(12.48)

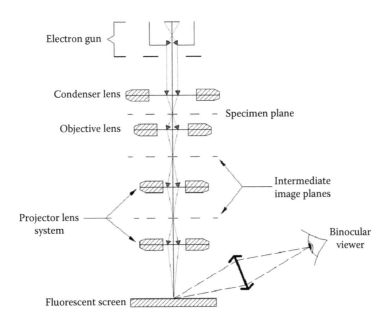

FIGURE 12.24 Conventional transmission electron microscope.

where h is the Planck's constant, p is the electron momentum, m is the electron mass, and v is its velocity. From this equation, and ignoring relativistic changes, the electron wavelength, measured in nanometers, is obtained as

$$\lambda = \frac{1.227}{\sqrt{V}}, \qquad (12.49)$$

where V is the electron potential, measured in volts [14]. Thus, for a typical operating potential of 100 kilovolts, the electron wavelength is 3.8×10^{-3} nanometers.

A conventional TEM instrument is conformed by some basic components: an electron gun, a vacuum system, magnetic (or electron) lenses, and a photographic plate or a fluorescent screen where the specimen image is projected. A vacuum pump is used to maintain the microscope column free of contamination by residual materials interacting with the electron beam. The electron gun provides an intense focused spot of illumination. Electrons can be ejected from a cathode into a vacuum chamber by thermionic emission. For such case, the gun is called *thermionic electron gun*. Additionally, the release of electrons from unheated metal surfaces across a high potential, can occur whenever a sharp metal point is exposed to a potential gradient. This physical phenomena induces an alternative configuration for an electron gun, the so-called *field emission gun* [14,15].

In comparison with the optical microscopes, which use conventional lenses, the TEM instruments use magnetic lenses. Magnetic lenses are employed to generate electric fields that can focus the electron beam on the sample (width reduction of the incident electron beam tip). Once the electron beam has passed through the sample, the information of the image is focused by a projector lens system on a fluorescent screen or a photographic plate. As described by Slayter and Slayter [14], for direct sample observation, the viewing screen consists of a coating of crystallites that fluoresce visible light in response to the electron bombardment; however, for a full detailed viewing of the sample objects, photographic images are required.

One of the most impressive details to point out is that with this kind of microscopes, very high magnification levels, at the order of 1,000,000x, can be achieved. Of course, extreme magnifications are not usually of practical use. As an example, for biological specimens, it is sufficiently enough to have an instrument with 40,000x magnification. In the case of rough objects, composed of heavy atoms, the structures can be recorded at about 80,000x. However, the quest of extreme resolutions, especially in the field of experimental physics, remains the landmark today.

12.9.2 Scanning Electron Microscope

In this kind of microscope, a very fine probe is formed by the radiation from the source, an electron gun. The probe is focused by an electron lens (the objective lens) located at a plane that is close to the specimen. A magnetic deflection device in front of the electron lens, scans the electron probe in a *raster pattern* across the specimen. This process is kept in synchrony with an electron beam of a cathode-ray tube (CRT) [15]. The device is capable of regularly moving the electron beam along an *xy* coordinate system. At speeds of tens of milliseconds, the probe sweeps along successive horizontal lines, which are spaced at a fixed distance.

It is important to remark that both components, the objective lens, and the condenser lens, are electron lenses. An electron lens is essentially a coil, which consists of an axial magnetic field with rotational symmetry. Scanning circuits, in connection with the CRT, permits through a beam deflection coil, the scanning of the probe along the sample. For samples like opaque surfaces, the probe electrons of the SEM are reflected by the sample, such as in the case of light reflection from an irregular surface. In this case, the image intensities are related to roughnesses of the surface topography. The reflected electrons are collected by a detector and finally, the output signal is plotted in a computer (see Figure 12.25b).

For a transparent sample, the use of a STEM (see Figure 12.25c), implies that the output signal was transmitted through the specimen. In this case, the images obtained are similar to those observed in the simple transmission electron microscope as shown in Figure 12.25a; however, the basic difference between STEMs and TEMs, is not only about the scanning process, but is also about the benefits of an immediate processing of the output signal from its separate portions.

In the semiconductor industry, the SEM applications have been of great help, even in the biological field. The SEMs permit us to observe the topography of certain transparent biological samples but with some limitations. Normally, in these cases, the microscopic elements, such as living cells, are usually covered by some kind of metal, allowing the possibility of a better electron reflection, but paying the cost of some dead cells. With the TEMs, the electron bombardment may also affect the sample because

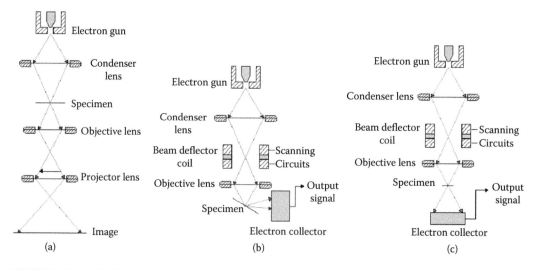

FIGURE 12.25 (a) Transmission electron microscope (TEM). (b) Scanning electron microscope (SEM). (c) Scanning transmission electron microscope (STEM).

the full energy of the radiated beam is absorbed within the specimen. However, since the atomic resolution is not a forced requirement in biological applications, the damage that can make an electron beam to a TEM sample is not often a problem, with the exception of some particular cases. In comparison to the light microscopes, the SEMs do not allow observing, in real time, the cellular interaction; however, this limitation has not stopped the increasing of the SEM applications in the biological field, due to the modern techniques for cellular preparation as fixation, dehydration, infiltration and embedment, sectioning, and staining [14].

12.9.3 Scanning Tunneling Microscope and Atomic Force Microscope

When considering an electrically conducting or semiconducting surface, a metal tip can be used to determine its topography by sensing a tunneling current between the tip and the surface to be measured [16]. For detecting the tunneling current, a voltage must be applied between the tip and the surface. This is the basic operating principle used in the scanning tunneling microscope. The probe tip, controlled by piezoelectric transducers (PZTs) can be moved in three orthogonal directions, x, y, z, allowing the possiblity of the scanning of the surface topography at atomic resolutions (see Figure 12.26a).

The tunneling current can be sensed when the metal tip is less than 1 nm of distance with respect to the sample, and this sensed current, permits us to recover the contour of the sample surface, which is translated in a sample image. However, the STM can measure the sample topography by two modes: the constant-current mode, and the constant-height mode. In a constant-current mode, the STM scans the surface by varying the tip height and finding the points where the tunneling current is constant, whereas in a constant-height mode, the tip is kept constant along the surface while the tunneling current is sensed for each point. For a good image resolution, the geometry of the tip is an important factor. The STM tips can be made from tungsten wire by computer-controlled etching, or by mechanical cutting (preferable for atomic imaging) in the case of tungsten or platinum/iridium wires [16].

Although the main constraint of the STMs is the need of a conductive surface, this technology offers certain advantages over electron microscopes. With the STMs, it is possible to achieve, approximately, a lateral resolution of about 0.2 nm, and a horizontal resolution (normal to the surface) of 0.01 nm. Besides measuring the topography of very small objects, as metal plates, the electrical behavior of metal samples can be measured, allowing the physical structure of the object to correlate with an electrical function.

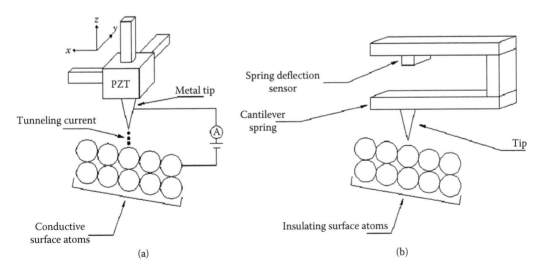

FIGURE 12.26 (a) Scanning tunneling microscope (STM). (b) Atomic force microscope (ATM).

Also, by using probes with low voltage instead of electron beams of high energy, the STM instruments favor to a smaller destruction of the sample [14].

Similarly to the STMs, with an AFM, a probe tip scans the topography of a very small sample, which is not necessarily electrical conductive as shown in Figure 12.26b. The scanning process consists in following contours that represent the force required to maintain a spring-mounted tip in exact contact with the test surface. The expression "exact contact" is not exaggerated because the tip reaches the sample like what happens with a record player. In an AFM instrument, the tip must be fine and sharp, but its tracking force must be sufficiently weak, to keep an adequate contact with the sample surface without damaging it. An AFM probe can be made of a small diamond fragment, which is attached to a weak cantilever spring. Usually in the range of 10^{-9} to 10^{-6} newtons, small repulsive tracking forces between the tip and the surface are recorded by a spring deflection sensor. The spring deflection sensor measures minute deflections of the cantilever, typically about 1 newton per minute [14].

The AFM is basically an extension, or a modification, of the STM with a variation with respect to the form of contour measuring: constant tunnel currents for conductive test surfaces (STM) versus force required to maintain a probe tip in exact contact with conductive (as well as insulating) test surfaces (AFM).

References

1. Wilson T. *Confocal Microscopy*, Cambridge, MA: Academic Press, 1990.
2. Masters BR. *Confocal Microscopy and Multiphoton Excitation Microscopy, The Genesis of Live Cell Imaging*, Bellingham, WA: SPIE Press, 2006.
3. Minsky M. *Microscopy Apparatus*, U.S. Patent 3013467, 1961.
4. Petràn M, Hadravsy M, Egger MD, Galambos R. Tandem-scanning reflected-light microscope, *J. Opt. Soc. Am.*, **58**, 661–664; 1968.
5. Xiao GQ, Kino GS. A real-time confocal scanning optical microscope, *Scanning Imaging Technology*, T. Wilson and L. Balk (Eds.), *Proc. SPIE*, **809**, 107–113; 1987.
6. Svishchev GM. Microscope for the study of transparent light-scattering objects in incident light, *Opt. Spectrosc.*, **26**, 171–172; 1969.
7. Müller M. *Introduction to Confocal Fluorescence Microscopy*, Bellingham, WA: SPIE Press, 2006.
8. Zill DG, Cullen MR. *Differential Equations with Boundary-Value Problems*, 7th Ed., Belmont, CA: Brooks & Cole / Cengage Learning, 2009.
9. Goodman JW. *Introduction to Fourier Optics*, New York, NY: McGraw-Hill, 1968.
10. Wilson T, Sheppard C. *Theory and Practice of Scanning Optical Microscopy*, Cambridge, MA: Academic Press, 1984.
11. Ghatak AK, Thyagarajan K. *Contemporary Optics*, New York, NY: Plenum, 1978.
12. Mertz J. *Introduction to Optical Microscopy*, Greenwood Village, CO: Roberts and Company Publishers, 2010.
13. Marsden JE, Tromba AJ. *Vectorial Calculus*, Upper Saddle River, NJ: Prentice Hall, 1998.
14. Slayter EM, Slayter HS. *Light and Electron Microscopy*, Cambridge: Cambridge University Press, 1992.
15. Reimer L. *Scanning Electron Microscopy*, Berlin, Heidelberg: Springer-Verlag, 1985.
16. Malacara D. *Optical Shop Testing*, Hoboken, NJ: John Wiley & Sons, 2007.

13

Spectrometers

William Wolfe

13.1 Introduction

A spectrum is a representation of a phenomenon in terms of its frequency of occurrence. One example is sunrise. At first thought we might conclude that the sun rises every 24 hours, so that the spectrum would be one placed at a frequency of 1/24 cycle per hour. More careful consideration, of course, tells us that, as the seasons pass and at my latitude, the sun rises anywhere from about once every 12 hours to once every 16 hours. Therefore, for a period of a year, the spectrum would be ones starting at a frequency of 1 at 0.0625 to 0.0833 cycles per hour. There might also be several twos in there near the solstices when the sun has about the same period for several days. In the arctic, there would be a bunch of zeros in the summer and winter when the sun never sets or never rises. Another example, which is more or less constant for all of us, is that we receive from one to 20 advertisements in the mail every day, except on Sundays and official holidays. So this rate would vary from one to 20 per day and have a few, blessed zeros once in a while. In mathematics, such a relationship is often called a spectral density, and the cumulative curve is called the spectrum, that is, about one to 20 at the end of the first week, then more at the end of two weeks and still more by the end of the year. In this chapter, the spectral density will be called the spectrum for brevity. We will not have a need for the cumulative spectrum.

A spectrometer is any device that measures a spectrum. In the example of the advertisements, you, the counter of the ads, would be the spectrometer. In an optical spectrometer, the emission, absorption, or fluorescence spectrum of a material is measured. Spectrometers come in many different forms, and most of them are discussed in this chapter. Their function can be based on any physical phenomenon that varies with optical wavelength or frequency.

Probably the earliest spectrum observed by man was the rainbow. Other observations were surely glories and perhaps the colors seen from oil slicks and other thin films. The first spectrometer may have been "constructed" by Seneca during the first century AD, who observed the colors that came from the edge of a piece of glass were similar to those of the rainbow. More likely it was the prism of Newton. He observed prismatic colors and, indeed, differentiated among different substances by their different *refrangibility* [1].

The first observation of spectral lines was made by Thomas Melville in 1752 [1]. He used a prism and small aperture to observe the different spectra produced by sal ammoniac, alum potash, and other substances. Wollaston (1766–1828) observed the spectrum of the blue base of a candle flame and William Swan saw what are now called the Swan bands in 1856. The first use of the spectrometer as a wave analyzer must belong to Fraunhofer. While measuring the refractive indices of various glasses in about 1815 (independent of the works cited above), he observed the sodium doublet. He then proceeded to observe the Fraunhofer lines of the sun, first with a slit and prism and then with a slit and a diffraction grating. This work was followed in 1835 by Herschel and many other luminaries of physics: Talbot, Foucault, Daguerre, Becquerel, Draper, Angstrom, Bohr, Kirchhoff, Bunsen, Stewart …. The critical step of identifying the molecular and atomic structure was the accomplishment of Bunsen and Kirchhoff, although others were involved, including Miller, Stokes, and Kelvin.

Spectroscopy was critical in most of the advances in atomic and molecular physics. The Bohr atom was defined and delineated by relating spectroscopic measurements to the motion of electrons [2]. The very advent of quantum physics was due to the exact representation of the spectrum of blackbody radiation [3]. Sommerfeld applied the theory of relativity to the motion of the perihelion of an electron in its orbit [4], and Paschen and Meissner confirmed his theoretical predictions with spectroscopic measurements of the fine structure [5].

Prism spectrometers were improved little by little by the introduction of better sources, better detectors, and better optics. Grating spectrometers had these improvements, but were also improved in very substantial ways by the improvements in grating manufacture. These advances were made by Wood, Harrison, Strong, and others [6,7]. Gratings were ruled by very precise mechanical devices. A major advance was the introduction of interferometric control. More modern devices now use holographic and replicated gratings [8].

Spectroscopy is used today in many different applications. It is a standard technique for the identification of chemicals. There are even extensive catalogs of the spectra of many elements, molecules, and compounds to assist in this process. The exhaust gases of our vehicles are tested spectroscopically for environmental purposes, as are many smoke-stack emissions. The medical chemical laboratory uses colorimetry, a simple type of spectrometer for blood, urine, and other fluids analysis. Colorimetry, another simple form of spectroscopy, is used extensively in the garment and paper industries. (There are different types of colorimetry—the first is a transmission measurement at a single color; the second is a determination of chromaticity.)

13.2 Spectrometer Descriptions

Spectrometers are described in terms of how well they resolve lines, how sensitive they are, whether there is a "jumble" of lines, as well as their different geometric and path configurations.

13.2.1 Spectral Lines

The concept of a *spectral line* probably arose when Fraunhofer used a slit and a prism. Where there was a local spectral region of higher intensity, it looked like a line, because it was the image of a slit. The plot of a spectrum as a function of the frequency can be low and flat until a local region of higher intensity occurs. If this is narrow enough, it looks like a vertical line. *Line* is probably a nicer term than *spike* or *pip*, which could have been used. There is a story that has wonderful memories for me. When the animals were being named, the question arose: "Why did you call that a hippopotamus?" The answer was: "Because it looked like a hippopotamus." That applies here.

13.2.2 Spectral Variables

The spectra that are measured by spectrometers are described in terms of their *frequency* and their *wavelength* as independent variables and by *emission, absorption, fluorescence,* and *transmission* as dependent variables. Sometimes these are given as arbitrary output values and as digital counts.

The spectral variables include v, the frequency in cycles per second or Hertz; σ, the wavenumber in cycles per centimeter; k, the radian wavenumber equal to $2\pi\sigma$ in radians per centimeter; and the wavelength λ, given variously in nanometers, micrometers, or angstroms. The wavenumber is also cited as kaysers, and angstroms are no longer on the "approved list" for a wavelength measure. Many authors use $\tilde{v}$, but this is awkward in typography and will not be used here. In summary and in equation form, the spectral variables are related by

$$\sigma = \frac{k}{2\pi} = \frac{1}{\lambda} = \frac{v}{c}.$$
(13.1)

All of these have the same units, usually cm and cm/s. σ is almost always given in reciprocal centimeters and λ in micrometers. In this case

$$\sigma = \frac{10000}{\lambda}.$$
(13.2)

13.2.3 Resolution and Resolving Power

Resolution is a measure of the fineness with which the width of a spectral line can be measured. One measure of this is the full width of the measured line at half the maximum value, the FWHM. This can be given in any of the spectral variables: $d\lambda$, $d\sigma$, dk, and so on. Resolution can also be stipulated as a fraction of the wavelength: $d\lambda/\lambda$, $d\sigma/\sigma$, dk/k, and so on. This is slightly awkward in that lower values are generally better: that is, 0.01 is a higher resolution than 0.1! It is finer. *Resolving power* (RP) is defined as resolution that avoids this little complication. It is the reciprocal of fractional resolution: $\lambda/d\lambda$, $\sigma/d\sigma$, and so on. This is the same definition as the quality factor Q of an electrical circuit. Thus, both RP and

Q are used for the resolving power of a spectrometer, and some authors use 5 and call it resolvance. The fractional resolution and resolving power are equal no matter what the spectral variable:

$$Q = \text{RP} = \frac{\sigma}{d\sigma} = \frac{\lambda}{d\lambda} = \frac{k}{dk} = \frac{\nu}{d\nu} = \cdots. \tag{13.3}$$

However, the resolution $d\sigma$ is not equal to the resolution in wavelength, but, for σ in reciprocal centimeters and λ in micrometers, one finds

$$|d\lambda| = \left| \frac{\sigma}{\lambda} d\sigma \right| = \left| \frac{10000}{\lambda^2} d\sigma \right|. \tag{13.4}$$

Sometimes the base band is specified; it can be the full width at 1% of maximum or the full width to the first zeros of the spectral line. The shape factor is the ratio of the base band to the half width (FWHM).

13.2.4 Free Spectral Range and Finesse

In multiple-beam interference, which includes both Fabry–Perot interferometers and diffraction gratings, there can be an overlapping of orders. The *free spectral range* is the spectral interval between such overlaps. The *finesse* is the ratio of the free spectral range to the spectral slit width.

13.2.5 Throughput and Sensitivity

Whereas resolution is one important characteristic of a spectrometer, defining how well the spectral lines can be determined, throughput, is part of the determination of the sensitivity of the spectrometer. The signal-to-noise ratio of a spectrometer can be written as

$$\text{SNR} = \frac{D^* L_\lambda d\lambda Z}{\sqrt{A_d B}}, \tag{13.5}$$

where D^* is the specific detectivity [9], L_λ is the spectral radiance, $d\lambda$ is the spectral bandwidth, A_d is the detector area, B is the electronic bandwidth, and Z is the throughput, which is defined by the following expression

$$Z = \frac{\tau A_e \cos\theta_f A_f \cos\theta_i}{f^2} = \tau A_f \cos\theta_i \Omega' = \tau A_f \cos\theta_i \frac{\pi}{4F^2} = \tau A_f \pi \text{NA}^2, \tag{13.6}$$

where τ is the transmission of the spectrometer, A_e is the area of the entrance pupil, A_f is the area of the field stop, Ω' is the projected solid angle, f is the focal length, F is the optical speed in terms of the F-number, and NA is the numerical aperture. These will be evaluated in later sections for specific instruments.

The specific detectivity is a function of the detector used, and the radiance is a function of the source used. The throughput is a property of the optical system and is invariant throughout the optics.

In classic treatments [10–12], the product of the resolving power and the throughput has been used as a figure of merit. While this is important and a reasonable figure, it is not the whole story. From the expression for the SNR, it is easy to see that the square root of the detector area and the noise bandwidth are important considerations. An extended figure of merit that incorporates the bandwidth and detector area will be used later to compare these different instruments.

13.3 Filter Spectrometers

In some ways these are the simplest of all spectrometers. There are, however, many different types of filters and filter designs. A filter spectrometer in concept is a device that consists of a number of filters, each of which passes a small portion of the spectrum. Filters are described by their transmission characteristics, such as *narrow band, wide band, cut-on,* and *cut-off.* These are, of course, general descriptions because *narrow* and *wide* are relative terms. Care must be exercised in reading about cut-on and cut-off filters. Usually a *cut-on* filter has zero or very low transmission at short wavelengths and then is transparent at longer wavelengths. The *cut-off* is the opposite. Sometimes, however, authors use the term *cut-on* to mean a filter that has zero or low transmission at short *frequencies.* The former nomenclature will be used in this chapter.

13.3.1 Filter Wheels

It is easy to imagine the simple filter-wheel spectrometer as an optical system with a detector behind a large wheel that has a set of filters in it, as shown schematically in Figure 13.1. The filter wheel is moved from position to position, thereby allowing different portions of the spectrum to fall on the detector. The resolution and resolving power depend on the nature of the filter.

Filters can be classified as absorption filters, interference filters, Lyot filters, Christiansen filters, reststrahlen filters, and acousto-optical tunable filters.

Absorption filters [13] operate on the principle that absorption in any material is a function of wavelength. These filters are manufactured by a number of vendors, including Kodak, Corning, Zeiss, and Schott. [14]

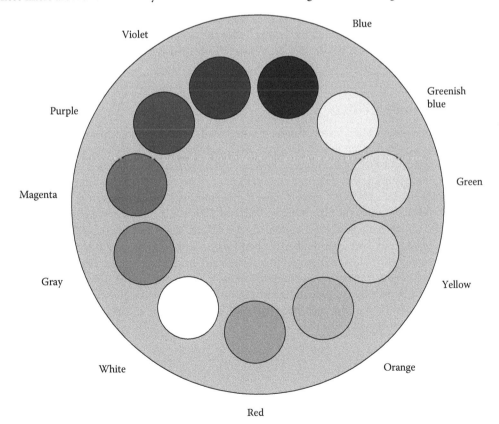

FIGURE 13.1 A filter wheel.

Their spectral bands are not regular (well behaved – neither symmetric nor of equal resolution in d or d_ or resolving power Q) because they are dependent upon the dyes that are usually used in a glass or plastic substrate. Semiconductors make fine cut-on filters. They absorb radiation of wavelengths short enough that the photons are energetic enough to cause an electronic transition. Then, at the critical wavelenght, where the photons are no longer more energetic than the bandgap, there is a sharp transition to transmission.

These can often be used in conjunction with interference filters as shortwave blockers. The transmissions of special crystals and glasses are tabulated in several publications. [15–17]

13.3.2 Christiansen Filters

These filters are based on the fact that scattering of light is a function of wavelength. Solid particles are immersed and spread throughout a liquid. Light is scattered by the particles at every wavelength except the one for which the refractive index of the particles is the same as that of the liquid. Liquids are used because their refractive indices change faster than solids; some combinations of solids can certainly be used if the particles can be properly distributed in the volume. McAlister [18] made a set of Christiansen filters [19] with borosilicate glass spheres in different concentrations of carbon disulfide in benzene. They had center wavelengths from about 450 nm to 700 nm with a resolving power of about 10.

13.3.3 Reststrahlen Filters

Reststrahlen or residual ray filters are based on the variation of refractive indices in the wavelength regime of an absorption band. In this region, the reflectance varies from about 5 or 10% to almost 100%. By generating several reflections from such a material in this region, the high-reflection portion of the spectrum is passed through the system, while the low reflectivities are attenuated. Both the Christiansen and reststrahlen filters are cumbersome and little used as commercial spectrometers. They can be required for some parts of the spectrum, especially the infrared.

13.3.4 Lyot-Öhman Filters

Also called a polarization filter, filters of this type are based on the wavelength variation of the rotation of the plane of polarization by certain materials, quartz being one of the best known. They were developed independently by Lyot in France in 1933 and by Öhman in 1938 in Sweden. A linear polarizer precedes a rotation plate; another polarizer precedes a plate of twice the thickness, and so on, as shown in Figure 13.2. If the rotation is 90 degrees, then the second plate does not pass the polarized light. To achieve 90-degree rotation, the plate must be "half wave"; that is, it must have an optical path difference that is one half wavelength (in the medium), or $\Delta = nd\lambda / 2$. Then, wavelengths of all odd multiples of this will be blocked, and all even multiples, or full waves will be transmitted. The situation is the same for the second plate and its polarizers, but it is twice as thick. It therefore has twice the frequency of maxima. The successive transmissions and resultant are shown in Figure 13.3. The passband is a function of the wavelength because the path difference is a function of the refractive index and therefore of wavelength. Filters of this type have been made with half-widths of about 0.1 nm in the visible ($Q \approx 5000$) and peak transmissions of about 70% [20].

13.3.5 Interference Filters

These are perhaps the most popular and useful filters in the arsenal [21]. They are composed of thin layers of transparent material in various combinations of different thicknesses and refractive indices. Of course the refractive index is a function of the wavelength, so the interference is a function of the wavelength. An analysis of the transmission of a single-layer filter can provide insight into the more

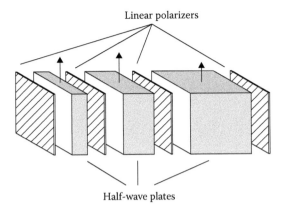

Linear polarizers

Half-wave plates

FIGURE 13.2 The Lyot–Öhman filter.

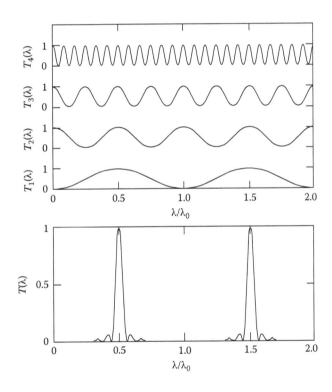

FIGURE 13.3 The filter actions: top, the interference of the components; bottom, the result.

complex designs. The expression for its transmission always includes the cosine of the incidence angle, a fact that can be critical to its use as a spectrometer. Figure 13.4 shows the transmission of the filter as a function of wavelength; Figure 13.5 shows the filter transmission as a function of wavenumber and Figure 13.6 shows a typical transmission as a function of angle. These together illustrate why most theoretical treatment interference filters are done on a frequency basis; that is, the interference filter is a multiple-beam system with a finite free spectral range and transmission is a function of angle of incidence. The last fact implies that the passband can be altered by tipping the filter, and that the passband is broadened when a convergent beam is incident on the filter.

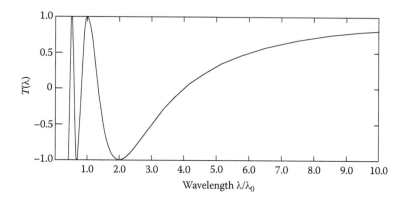

FIGURE 13.4 Filter transmission vs. wavelength.

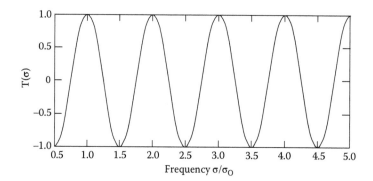

FIGURE 13.5 Filter transmission vs. frequency.

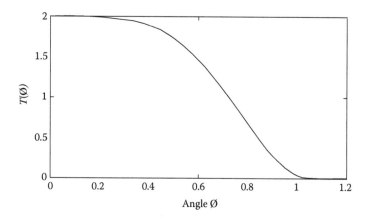

FIGURE 13.6 Filter transmission vs. angle.

Most bandpass interference filters are based on the Fabry–Perot interferometer in which the mirrors and the spacer are all thin films—with differing refractive indices and thicknesses. Typically, one material is the high index and one other material is low index, and the thicknesses are all QWOT—quarter-wave optical thickness. These filters are also subject to the angle changes discussed above, although some special designs can reduce the effect. Details of the Fabry–Perot interferometer, which apply in large measure to the Fabry–Perot filter, are discussed later.

13.3.6 Circular Variable Filters

Interference filters of the types described above can be made in a special way. The different layers can be deposited according to the prescription on a circular substrate—as the substrate is rotated at a variable rate. Therefore, each portion of the circle has layers of slightly different thickness. There is a circular gradient of layer thickness around the filter. Consequently, each different circumferential position of the filter has a different wavelength of maximum transmission. It is a circularly variable filter or CVF. These can be made to order, with different gradients, different resolving powers, and different spectral regions, but there are constraints on them. These filters are used in spectrometers of relatively low resolving power by rotating them in front of the optical system. A slit is used right behind the filter, and the optics then image the slit on the detector. As the filter rotates, different wavelength bands are incident on the detector.

An Optical Coating Laboratories Inc. (OCLI) filter can serve as an example. It covers the region from 2.5 to 14.1 mm in three 86-degree arcs of radii 1.45 and 1.75 in. for bands from 2.5 to 4.3, 4.3 to 7.7, and 7.7 to 14.1 μm. The gradients are therefore 0.75, 1.42, and 2.67 μm per inch, or a slit 0.1 inch (0.25 cm) wide has a resolution of 0.075, 0.142, and 0.267 μm in each of the bands and the resolving powers are about 45. The resolution could be made finer with a finer slit, but the flux is reduced proportionally.

The throughput of a CVF spectrometer will include the slit area, the system focal length, and the detector area, as well as the filter transmission. The throughput will be determined by this aperture area, the size of the detector, and the focal ratio. For the 0.1-inch slit and the wheel cited above, an F/1 lens, and a 25×75 μm detector, the throughput is 10^{-6} in^2 sr or 6.25×10^{-6} cm^2 sr. (I chose a small infrared detector or a big visible one.)

Since this is a plane, parallel plate in front of the optical system, narcissus (internal reflections) may be a problem [22].

13.3.7 Linear Variable Filters

Gradient filters can also be made such that the gradient is linear; therefore, these are called linear variable filters or LVFs. These filters are generally meant for use with detector arrays. They are placed in close proximity with the arrays, thereby ensuring that each pixel in the array senses one component of the spectrum. This close proximity can also cause a certain amount of scattering that may give rise to cross-talk. The spectral range is 2:1 (an octave – do, re, mi, fa, so la, ti, do; do) with resolving power of 10–100 and peak transmission of about 70%. The infrared range of 1–20 μm is covered. The throughput can be larger, since a slit is not necessary.

With both CVFs and LVFs, the manufacturer should be contacted and can surely make them to order with approximately these properties.

13.4 Prism Spectrometers

Prism spectrometers are probably the oldest type of spectrometer. First used by Fraunhofer to determine refractive indices of various materials for better designs of telescopes, they are still in use today and still provide good performance for some applications.

13.4.1 Throughput

The "standard" configuration of a prism spectrometer is shown in Figure 13.7. Light enters a slit with dimensions $h \times w$, and is collimated by the first lens. A collimated beam is incident on the first face of the prism A_{prism}, where it is dispersed in a set of collimated beams that cover a range of angles corresponding to a range of wavelengths. These beams are collected by the camera lens and focused onto the exit slit. The light that passes through the exit slit is then focused onto an appropriate detector.

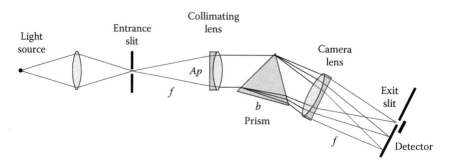

FIGURE 13.7 Prism spectrometer layout.

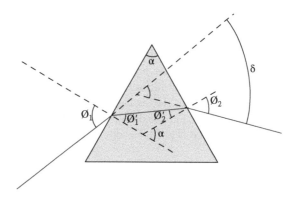

FIGURE 13.8 Ray geometry.

The throughput is determined by the area of the slit (which is normal to the optical axis and is therefore equal to the projected area) and the projected area of the prism and the square of the focal length of the lens. The throughput is constant throughout. Since the prism is the critical element, the collimating lens is slightly larger, and the prism area is the aperture stop of the system. The camera lens must be a little larger than the collimating lens because of the dispersion. In a well-designed spectrometer, the entrance and exit slits are equal in size, and either or both are the field stop. Thus, the throughput is

$$Z = \frac{A_{\text{slit}} A_{\text{prism}} \cos\theta}{f^2} = \frac{hw A_{\text{prism}} \cos\theta}{f^2}. \tag{13.7}$$

13.4.2 Prism Deviation

The collimated beams that pass through the prism can be represented nicely by rays. The geometry is shown in Figure 13.8. The deviation δ is easily seen to be

$$\delta = \theta_1 - \theta_1' + \theta_2 - \theta_2' = \theta_1 + \theta_2 - (\theta_1' + \theta_2') = \theta_1 + \theta - \alpha. \tag{13.8}$$

Minimum deviation can be found by taking the derivative and setting it to zero, as usual:

$$d\delta = d\theta_1 + d\theta_2 = 0. \tag{13.9}$$

The final expression is valid since two of the legs of the triangle formed by the normals and the ray in the prism are perpendicular to each other, so the third angle of the prism is α. Clearly, minimum deviation is obtained when the two angles are equal, but opposite in sign. The minimum deviation angle is $\delta = 2\theta_1 - \alpha$, and $\theta = \alpha/2$. This information can be used to obtain an expression used for finding the refractive index of such a prism. At the first surface, Snell's law is

$$n_1 \sin\theta_1 = n_2 \sin\theta_2, \tag{13.10}$$

$$\frac{n_2}{n_1} = \frac{\sin\theta_1}{\sin\theta_2} = \frac{\sin\dfrac{\delta+\alpha}{2}}{\sin\dfrac{\alpha}{2}}. \tag{13.11}$$

The angular magnification can be found by a reapplication of Snell's law. At the first surface, the relationship is

$$n_1 \sin\theta_1 = n_2 \sin\theta_1'. \tag{13.12}$$

Differentiation yields

$$n\cos\theta_1 d\theta_1 = n'\cos\theta_1' d\theta_1'. \tag{13.13}$$

The same applies to the second surface, with 2s as subscripts. Division and a little algebra yield

$$\frac{d\theta_2}{d\theta_1} = \frac{\cos\theta_1 \cos\theta_2'}{\cos\theta_1' \cos\theta_2}. \tag{13.14}$$

From Equation 13.9, at minimum deviation,

$$\frac{d\theta_2}{d\theta_1} = -1. \tag{13.15}$$

So, when the deviation is at a minimum, $\theta_1 = \theta_2$ and $\theta_1' = \theta_2'$. In this condition, there is complete symmetry with respect to incident and exiting beams, and the beam is parallel to the base (of an isometric prism with base normal to the prism angle bisector). We note in passing that Equation 13.7 provides the expression for the magnification generated by the prisms, which is one for minimum deviation.

13.4.3 Dispersion

One of the classical techniques for measuring the refractive index of a prism was introduced by Fraunhofer. It can be written

$$n = \frac{n_2}{n_1} = \frac{\sin\theta_1}{\sin\theta_1'} = \frac{\sin(\alpha+\delta)/2}{\sin(\alpha/2)}, \tag{13.16}$$

which was obtained from the relationships above at minimum deviation. Then,

$$\frac{d\delta}{dn} = \frac{2\sin\alpha/2}{\cos(\alpha+\delta)/2} = \frac{2\sin\alpha/2}{\sqrt{1-n^2 \sin^2\dfrac{(\alpha+\delta)}{2}}} = \frac{2\sin\alpha/2}{\cos\theta_1}. \tag{13.17}$$

This expression can be used to find the angular dispersion of a prism of relative refractive index n, and the linear dispersion is $f d\delta$, where f is the focal length. The resolving power of the prism is given by

$$Q = \frac{\lambda}{d\lambda} = \frac{\lambda}{d\delta} \frac{d\delta}{d\lambda} = a \frac{d\delta}{d\lambda} = a \frac{d\delta}{dn} \frac{dn}{d\lambda}, \qquad (13.18)$$

where a is the beam width. Then, substitution of $a = p\cos\theta_1$ and $b = 2p \sin\alpha/2$ gives

$$Q = a \frac{b/p}{a/p} \frac{dn}{d\lambda} = b \frac{dn}{d\lambda}, \qquad (13.19)$$

where p is the length of the side of the (square) prism face and b is the base of the prism, or, more accurately, the maximum length of travel of the beam through the prism. The resolving power is just the "base" times the dispersion of the prism material.

13.4.4 Mounting Arrangements

There are both refractive and reflective arrangements that have been invented or designed for prism spectrometers. A few apply to imaging prism spectrometers. Perhaps the most popular is the Littrow arrangement, which is illustrated in Figure 13.9. Only the essential elements are shown. The light is retroreflected so that there are two passes through the prism. Somewhere in the optics in front of the prism there must be a beamsplitter or pick-off mirror to get to the exit slit. In the Wadsworth arrangement, the mirror is set parallel to the base of the prism, as shown in Figure 13.10, in order to have an undeviated beam.

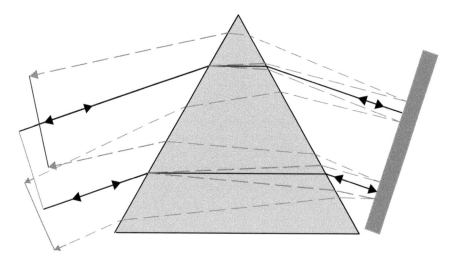

FIGURE 13.9 The Littrow mount.

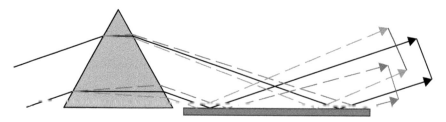

FIGURE 13.10 The Wadsworth mount.

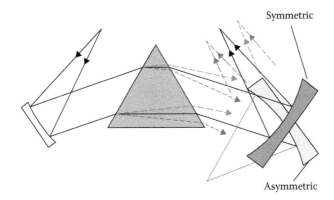

FIGURE 13.11   Symmetric and asymmetric mounts.

Symmetric and asymmetric mirror arrangements are shown in Figure 13.11. The symmetric arrangement is shown as image one, and the other as image two. A little consideration leads to the conclusion that the asymmetric (right-hand) arrangement leads to balancing of off-axis aberrations—because the tilt in one direction helps to offset the aberrations that were introduced by the tilt in the other direction.

13.5 Grating Spectrometers

Grating spectrometers make use of the diffraction of light from a regularly spaced, ruled surface. They disperse the light by a combination of diffraction and interference rather than the refractive index variation with wavelength, as with a prism.

Probably, Joseph Fraunhofer [23] was (again) the first to use diffraction gratings (in 1823). Henry Rowland [24] later built extremely precise ruling engines that could make relatively large, about 10 inch, gratings of high resolving power. Rowland invented the concave grating [25]. Albert Michelson, America's first physics Nobel laureate, developed interferometric techniques for ruling gratings that were used and improved by John Strong and George Harrison [26]. It was R. W. Wood [27] who introduced the blaze, and Strong who ruled on aluminum films that had been coated on glass (for greater stability and precision than speculum). Although Michelson indicated in 1927 the possibility of generating, not ruling, gratings interferometrically, it took the advent of the laser to allow the development of holographic gratings.

There are three types of grating: the rule grating, the holographic grating, and replica gratings made from molds that have been ruled. The latter two are relatively cheap, and are used in most commercial instruments today because they are cheap, reproducible, and do not depend on the transparency of a medium, as does a prism.

13.5.1 Diffraction Theory

It can be shown [28] that the expression for the irradiance pattern from a ruled grating is

$$E = E_0 \operatorname{sinc}^2(\pi w \sin\theta / \lambda) \left[\frac{\sin(N\pi s(\sin\theta_d + \sin\theta_i)/\lambda)}{\sin(\pi s \sin\theta / \lambda)} \right]^2, \tag{13.20}$$

where N is the number of rulings, s is the ruling spacing, θ_i is the angle of incidence, θ_d is the angle of diffraction, and λ is the wavelength of the light. This pattern consists of three terms: a constant E_0 that depends on the system setup; the source, optical speed, and transmission; a single-slit diffraction function and an interference function. The general form is shown in Figure 13.12. If the rulings are sinusoidal rather than rectangular in cross-section, the sinc function is replaced by a sine. (The sinc function is

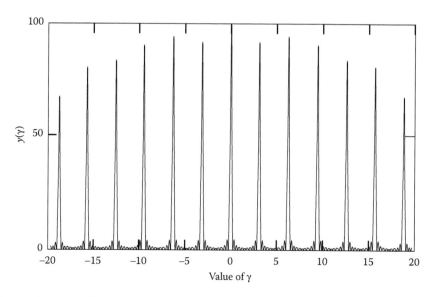

FIGURE 13.12   Multislit diffraction pattern.

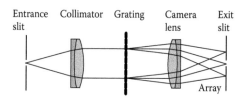

FIGURE 13.13   Transparent grating layout.

given by $\mathrm{sinc}\, x = \sin x / x$.) It can be seen, based on the Fourier transform theory, that this is the Fourier transform of the rectangle, which is the full grating × the comb function that represents grooves [29]. This equation is sometimes written

$$E = E_0 \sin c^2 \beta \left[\frac{\sin N\gamma}{\sin \gamma} \right]^2.$$

(13.21)

In this form, it is emphasized that β is half the phase difference between the edges of a groove, and γ is half the phase difference between rulings. This theory does not take into account the generation of ghosts and other results of imperfect rulings.

13.5.2 Geometric Layout

The layout for a grating spectrometer is almost the same as for a prism. If the grating operates in transmission, then it simply replaces the prism and is straight through, as shown in Figure 13.13. Configurations for plane and concave [30] reflective gratings are shown at the end of this section.

13.5.3 Resolution and Resolving Power

The equation for the position of the peaks may be obtained from the basic expression for the irradiance. The position of the peaks for each wavelength is given by

$$m\lambda = s(\sin\theta_i + \sin\theta_d),$$

(13.22)

where s is the line spacing, often called the grating spacing, θ_i is the angle of incidence and θ_d is the angle of diffraction. The order of interference is m and λ is the wavelength:

$$d\lambda = \frac{s\cos\theta_d d\theta_d}{m}. \tag{13.23}$$

The resolution can be found by differentiation (where only the diffraction angle changes). Better (smaller) spectral resolution is obtained at higher orders, with smaller line spacing and at angles that approach 90 degrees. A grating has constant wavelength resolution as a function of angle in the image plane.

The resolving power is given by

$$Q = \frac{\lambda}{d\lambda} = mN. \tag{13.24}$$

The resolving power is just the order number × the number of lines in the grating. It is not possible to pick an arbitrarily large order or to get an unlimited number of lines. This is true partly because the grating is of finite size, and efficiency generally decreases with order number.

13.5.4 Throughput

The throughput of the grating is identical to that of a prism, as the product of the slit area and projected area of the grating divided by the focal length of the collimating lens. For this reason, it is desirable to have a zero incidence angle when possible. This can be possible with a transmission grating, but is almost never available with a reflection grating. The entrance and exit slits should be identical.

13.5.5 Blazing Rulings

The efficiency may be increased by making the rulings triangular and using the slope to reflect the light to a particular order. The manufacturer will specify the blaze direction and the efficiency, typically about 60%. The rulings are still rectangular in cross-section; they are slanted within the groove.

13.5.6 Grating Operation

The normal operation of the grating is the same as with a prism. The grating is rotated, and wavelength after wavelength passes the slit and its radiation is detected. There is, however, an additional complication. The unwanted orders must be filtered so that they do not taint the measurement.

The requirement to eliminate extraneous orders arises from the basic equation and indicates that lower diffraction orders are preferable. Equation 13.22 has geometry on the right-hand side. On the left-hand side is the product $m\lambda$, the order × the wavelength. So first order and 1 mm, for instance, is the same as second order and 0.5 μm, and third order and 0.33 μm, and so on. It is preferable from one standpoint to operate at low orders, because it makes the separation of overlapping orders easier. The ratio of wavelength is given by

$$\frac{\lambda_2}{\lambda_1} = \frac{m+1}{m}. \tag{13.25}$$

The higher the order, the closer the two wavelengths are. The separation of over-lapping orders is

$$\Delta\lambda = \frac{\lambda}{m}. \tag{13.26}$$

This is the free spectral range.

13.5.7 Mounting Arrangements

Most gratings are reflective and need special mounting arrangements. These include Rowland, Eagle, and Paschen–Runge mountings for a concave grating, the Fastie Ebert, and Czerny–Turner mounts for plane gratings. Clearly, prism mounts, like the Littrow can be used for transparent gratings.

Both systems for use with plane gratings make use of the symmetry principle for the reduction of off-axis aberrations, but there is usually some residual astigmatism. The Czerny–Turner, Figure 13.14, has more flexibility and uses two mirrors, which can be adjusted independently. The Fastie–Ebert, Figure 13.15, uses a larger, single spherical mirror.

The Rowland mounting was, of course, invented by Rowland, himself, the originator of the curved grating. It placed the curved grating and the detector (then a photographic plate) on the ends of a rigid bar, the length of which is equal to the radius of curvature of the grating. The ends of the bar rest in two perpendicular rails, as shown in Figure 13.16. The entrance slit is placed at the intersection of the rails with the jaws perpendicular to the rail that carries the grating. By this arrangement, the slit, the centers of the grating, and the plate are all in a circle, the Rowland circle. This mounting is now obsolete as a result of various technological advances. The Abney mounting, now also obsolete, uses the same

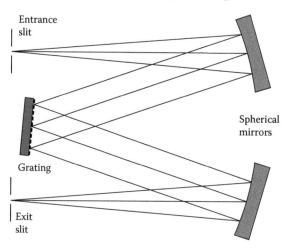

FIGURE 13.14   Czerny–Turner mount.

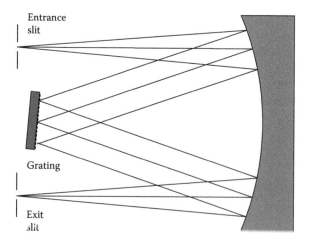

FIGURE 13.15   The Fastie–Ebert mount.

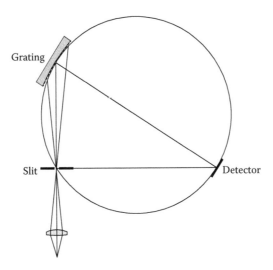

FIGURE 13.16 The Rowland mounting.

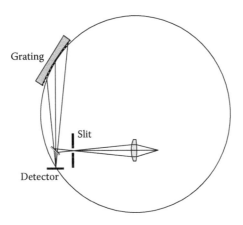

FIGURE 13.17 The Eagle mounting.

geometry, but the bar is rigid, while the slit rotates about a radius arm at the middle of the bar and it rotates to keep the jaws perpendicular to the grating. Although this sounds more complicated, only the light (not heavy) slit arrangement moves.

The Eagle mounting, shown in Figure 13.17, is similar to the prism Littrow mount in that the plate and slits are mounted close together on one end of a rigid bar. The concave grating is at the other end of the bar. The main advantage of the Eagle mount is its compactness. However, there are more adjustments to be made than with either the Abney or Rowland mounting arrangements.

The Wadsworth mounting is shown in Figure 13.18. The most popular modern mounting is the Paschen–Runge mount, shown in Figure 13.19.

13.6 The Fourier-Transform Spectrometer

This very important instrument had its origins in the interferometer introduced by Michelson in 1891 [31] for the examination of high-resolution spectra. It is more often used today in the form of a Twyman–Green interferometer [32], which is essentially a Michelson interferometer used with collimated light. The Twyman–Green was introduced to test the figure and quality of optical components. It has more

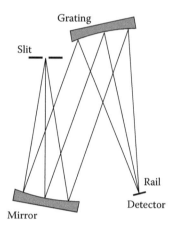

FIGURE 13.18 Wadsworth mounting.

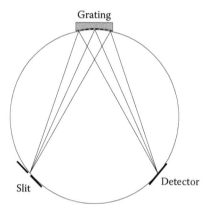

FIGURE 13.19 The Paschen–Runge mount.

recently become the basis of many Fourier-transform spectrometers (FTSs). The use of interferometers as spectral analysis instruments was pioneered by Felgett [33], Jacquinot [34], and Strong [35].

13.6.1 Two-Beam Interference

When two monochromatic beams of light are superimposed, an interference pattern results. If Ψ_1 and Ψ_2 represent the complex electric fields of the two beams that are combined, the expression for the interference pattern in terms of the irradiance E is

$$E = \left\langle (\Psi^2) \right\rangle = \Psi \cdot \Psi^* = \Psi_1^2 + \Psi_2^2 + 2\Psi_1\Psi_2 \cos\left(\frac{2\pi n d \cos\theta}{\lambda}\right). \tag{13.27}$$

The optical quantity that is sensed is the time average of the square of the electric field. The right-hand expression then consists of a dc or constant term plus the interference term.

13.6.2 Interference in the Michelson Interferometer

The Michelson interferometer is shown schematically in Figure 13.20. An extended source illuminates a beamsplitter, where the light is split into two separate beams. The light is returned from each of

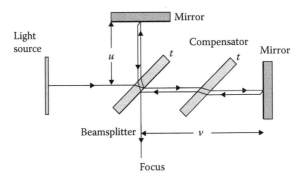

FIGURE 13.20 The Michelson interferometer.

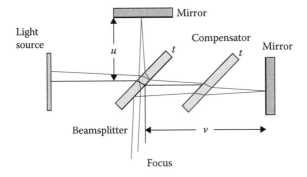

FIGURE 13.21 Off-axis beams in a Michelson.

the mirrors to the beamsplitter, which now acts as a beam combiner. The two beams then form an interference pattern at the focus of the lens [35].

There is a maximum in the pattern whenever the phase term is 0, or a multiple of π. This, in turn, can happen when $n\Delta$, the optical path difference is 0, or when $\cos\theta$ is 0 or a multiple of π. Thus, the pattern is dependent on both the effective separation of the two mirrors d and the angle off-axis θ.

The optical path of the on-axis beam that goes up is $u + 2nt/\cos\theta' - 2t/\cos\theta$, where u is the (central) separation of the top mirror from the beamsplitter, n is the refractive index of the plate, θ' is the refracted angle in the beamsplitter and compensator plate, θ is the incidence angle, and t is their thickness. The path length of the horizontal beam is $2v + 2nt/\cos\theta' - 2t/\cos\theta$, since the beam goes through the beamsplitter twice for this configuration. (I have assumed that the bottom surface has an enhanced reflection while the bottom is coated for low reflection. It could be the other way around. It is troublesome if both surfaces have about the same reflection, because then there is multiple-beam interference generated in the beamsplitter.) This difference in the two path lengths is the reason that many instruments have a so-called compensation plate in one arm. The difference is just $2(u - v)$, but would be a function of refractive index (and therefore wavelength) if the compensator were not there, or if the top surface were the highly reflecting one. An off-axis beam is more complicated, as is shown in Figure 13.21. There are more cosines, but the results are much the same. The differences are that the incidence and refraction angles are not the same as in the first case, and the paths in air are longer by the cosine. The result, with the compensator in place, is that the path difference is $2(u - v)/\cos\theta$. This path difference is part of the argument in the cosinusoidal interference term:

$$E = E_0[1 + \cos(2\pi n\Delta\cos\theta)]. \tag{13.28}$$

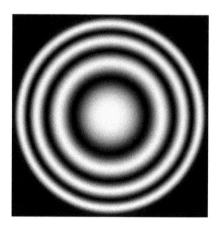

FIGURE 13.22   Plan view of the bull's-eye pattern.

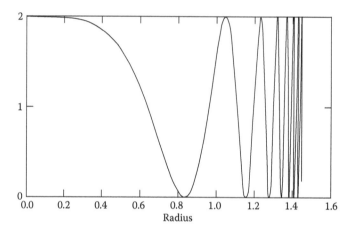

FIGURE 13.23   Sectional view of the bull's-eye pattern.

This gives the so-called bull's-eye pattern of a central ring with annuli that surround it of decreasing width, as shown in Figure 13.22 in plan view and sectionally in Figure 13.23.

13.6.3 The Twyman-Green Interferometer

When the Michelson interferometer is illuminated with monochromatic collimated light, for instance by a lens that has a laser at its front focus, then the bull's-eye pattern disappears, and the pattern is a point. The flux density in this single on-axis spot varies as a function of the separation of the mirrors, the path difference. This is the basis of the Fourier-transform spectrometer.

13.6.4 The Fourier-Transform Spectrometer

If the source of this on-axis interferometer is on-axis, its flux density varies as the separation, in a manner that depends on the optical path difference and the wavelength of the light in the usual way [37]:

$$E + E_0\left[1 + \cos\left(\frac{2\pi\Delta}{\lambda}\right)\right]. \tag{13.29}$$

The optical path difference Δ, as shown above, is twice the difference of the mirror separation from the beamsplitter. If the source is bichromatic, that is, has two beams with two different frequencies and amplitudes, the interference pattern will be the sum of two cosines.

$$E = E_0 \left[\frac{E_1}{E_0} + \cos\left(\frac{2\pi\Delta}{\lambda_1} \right) + \frac{E_2}{E_0} + \cos\left(\frac{2\pi\Delta}{\lambda_2} \right) \right]. \tag{13.30}$$

If there are many, the same is true, but with more components

$$E = \sum_i E_i \cos\left(\frac{2\pi\Delta}{\lambda_i} \right) \Rightarrow \int E(\lambda)\cos\left(\frac{2\pi\Delta}{\lambda} \right) d\lambda \tag{13.31}$$

and the integral represents a continuous summation in the usual calculus sense. The pattern obtained, as the mirror moves, is the sum of a collection of monochromatic interference patterns, each with its own amplitude. This is the interferogram; its Fourier transform is the spectrum. If the interferogram is measured by recording the pattern of a given source as a function of the path difference (translation of one mirror), then the mathematical operation of Fourier transformation will yield the spectrum. We have just seen the essence of Fourier-transform spectroscopy, the so-called FTS technique.

Figures 13.24 and 13.25 further illustrate these relationships. Figure 13.24 shows five cosines plotted as a function of the path difference, which is assumed to be on-axis and is just nd. The wavelengths of these cosines are 10–14 μm in steps of 1 μm. It can be seen how they all "cooperate" in their addition at zero path difference, but as the path difference increases, their maxima and minima separate, and they get "muddled up." As the path difference increases, the interference pattern will decrease, but not monotonically. Figure 13.25 shows the square of the sum of these five functions. It is the interferogram (within a constant and ignoring the constant term).

There is another way to look at this. For every position of the moving mirror, the radiation that reaches the detector is a sum of components of different amplitudes and different frequencies, and for each frequency, there is a different phase difference causing a different state of interference. This is the

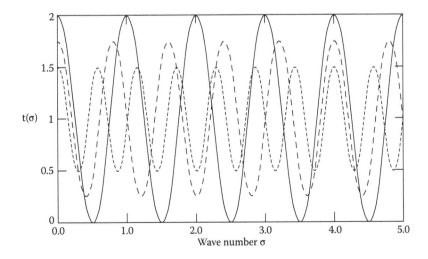

FIGURE 13.24 Waves in an FTS.

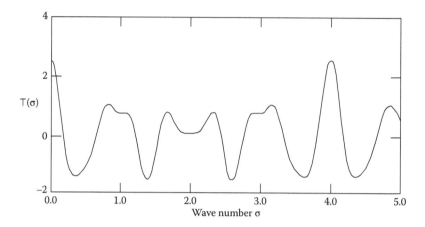

FIGURE 13.25   Superposition of waves in an FTS.

interferogram or interference pattern—a sum of sine waves with different amplitudes. The Fourier transform of this is the spectrum. The interferogram flux density can be written as

$$E(\delta) = E_0\left(1 + \cos 2\pi\sigma\delta\right), \tag{13.32}$$

where $E(\delta)$ is the incidence as a function of the path difference δ. For a nonmonochromatic beam, it is

$$E(\delta) = \int_0^\infty E_0\left(1 + \cos 2\pi\sigma\delta\right)\mathrm{d}\sigma, \tag{13.33}$$

where E_0 is the intensity measured at zero path difference between the two arms and $S(\sigma)$ is the spectrum of the light. Then, ignoring the dc or fixed background term, one has

$$E(\delta) = \int_0^\infty E_0 \cos\left(2\pi\sigma\delta\right)\mathrm{d}\sigma. \tag{13.34}$$

This is in the form of a Fourier cosine transform. The inverse transform provides the spectrum:

$$S(\sigma) = \int_0^\infty E_0 \cos\left(2\pi\sigma\delta\right)\mathrm{d}\delta. \tag{13.35}$$

There is still another way to understand this relationship and process. The motion of the mirror generates the autocorrelation function of the incident light, and the power spectrum is the Fourier transform of the autocorrelation function. Usually the interferogram is recorded by reading the output from the detector as one arm of the interferometer is moved. Then, after the fact, the spectrum is calculated numerically by computer techniques.

The resolution can be found by integrating Equation 13.22 from zero to the full extent of the path difference δ_{max}. The result is a sinc of $2\pi\sigma\delta_{max}$. The first zero occurs when this argument equals π, and that gives the following condition:

$$\Delta\sigma = \frac{1}{2\delta_{max}} = \frac{\lambda^2}{20,000\delta_{max}}. \tag{13.36}$$

By the Shannon sampling theorem, one needs two samples for every cycle. The resolving power, of course, will be given by

$$Q = \frac{\sigma}{d\sigma} = \frac{\lambda}{d\lambda} = \frac{\sigma}{2\delta} = \frac{5000}{\lambda\delta}. \tag{13.37}$$

The relations for wavelength are obtained from those for frequency by the fact that $\lambda/d\lambda = \sigma/d\sigma$ and, because the wavelength is in μm and the wavenumber is in cm^{-1}, $\lambda = 10{,}000/\sigma$. The resolution in the frequency domain is independent of the frequency, but the resolution in wavelength terms is dependent on the wavelength.

13.6.5 Throughput and Sensitivity

The Michelson (or Twyman–Green) has both a throughput and a multiplex advantage. The detector is exposed to all the radiation in the total spectral band during the entire scan, whereas with the grating and prism spectrometers only the part of the spectral region within the resolution cell is incident on the detector. The sensitivity calculations can be complicated. The following is a fairly simple argument that shows this advantage for the FTS. The expression for the interference on-axis (assuming the beam flux densities are equal) is the average over a full scan of the moving mirror. This can be written as [38–40]

$$E = \frac{1}{\Delta}\int_\Delta\int_\sigma E(\sigma)[1 + \cos(2\pi\sigma nd)]d\sigma\, d\Delta = \int_\sigma E(\sigma)d\sigma. \tag{13.38}$$

The first term in the integral is a constant and represents the full density in the spectrum. The second term is a cosine that has many cycles in the full-path averaging and goes to zero. Of course, the optical efficiency of the beamsplitter, compensator plate, mirrors, and any foreoptics need to be taken into account.

The throughput is given by the usual expression,

$$Z = \frac{A_o\cos\theta A_d\cos\theta}{f^2} = A_o\cos\theta\Omega = A_o, \tag{13.39}$$

where A_o is the area of the entrance optics, and equal to the area of the focusing optics, A_d is the detector area, f is the focal length, and Ω is the solid angle that the detector subtends at the optics. The optics is on-axis, so the cosine is 1. The area of the detector can be at a maximum, the area of either the central circle of the bull's eye or any of the annuli or some combination. These can have a width that encompasses one full phase. The solid angle in general is given by

$$\Omega = 2\pi(\cos\theta_2 - \cos\theta_1), \tag{13.40}$$

where the θs represent the minimum and maximum angles of the annuli's radii. The central circle has a zero minimum angle. Since the phase is represented by $2\pi\sigma\,\Delta\cos\theta$, $\cos\theta$ can only be as large as $1/\sigma\Delta$, which is the same as $2d\sigma/\sigma = 2/Q$. This leads to the relationship

$$Q\Omega = 4\pi, \tag{13.41}$$

which is one similar to that reported by Jacquinot [10]. This result is valid for the Michelson interferometer and is 2π for the Fabry–Perot instrument (which makes two passes of the spacing to generate interference), which is discussed next.

13.7 Fabry–Perot Spectrometers

The Fabry–Perot (FP) interferometer is a multiple-wave device. It is essentially a plane, parallel plate of material, that can be air or vacuum, between two partially reflecting plates, as shown in Figure 13.26. (The rays are shown for a slightly off-axis point to emphasize the repeated reflections.) It can be in the form of two partially reflecting mirrors with a space between them, a plate of glass or other dielectric with reflective coatings on the faces, or layers of thin films that form a filter. As shown, there is a walk off of the beams after many reflections. When used on-axis, that is not the case. The figure also shows two separate partially reflective mirrors; it could be a single plate with two surfaces that perform the reflection functions. It is shown later that the field covered by an FP is indeed quite small so that walk off is not a problem, and the classic Airy function applies. Note, however, that in configurations that are used relatively far off-axis, the light is attenuated; the summation of beams is not infinite, and modifications must be made.

The function that describes the transmission of the FP is the so-called Airy function [40]:

$$\tau(\sigma) = \frac{\tau_{max}}{1 + R \sin^2 \phi}, \tag{13.42}$$

where the maximum transmission τ_0 is given by

$$\tau_{max} = \frac{\tau_1 \tau_2}{(1 - \rho)^2}, \tag{13.43}$$

with

$$\phi = \pi\sigma\Delta - \frac{\varepsilon_1 + \varepsilon_2}{2} = \pi\sigma \text{ and } \cos\theta - \frac{\varepsilon_1 + \varepsilon_2}{2}. \tag{13.44}$$

The interferometer reflectivity is

$$R = \frac{4\rho}{(1 - \rho)^2} \tag{13.45}$$

and

$$\rho = \sqrt{\rho_1 \rho_2} \tag{13.46}$$

where the ρ's are the reflectivities of the coating measured from the gap, the τ's are the transmittances of the coatings on the plates, the ε's are the phase shifts on reflection, n is the refractive index of the plate, and d is its thickness. The phase shifts serve only to shift maxima and minima; they do not affect resolution, resolving power, free spectral range, finesse, and throughput.

The pattern is shown in Figure 13.27. The line width (FWHM) is given approximately, for small angles, by the following expression:

$$d\sigma = \frac{1 - \rho}{2\sqrt{\rho}} \frac{1}{\pi\Delta}. \tag{13.47}$$

It can be found by setting the transmission expression to $\frac{1}{2}$ and solving for $\sin \phi$. Then, for small angles, the angle is equal to the sine and the FWHM is twice the value found, yielding Equation 13.47.

The free spectral range is the distance in frequency space between the maxima, that is, when $F \sin \phi$ is zero. This occurs at integral multiples of π, for $\phi = m\pi$, or when $\sigma = m/\Delta$. Therefore, the free spectral range, the distance between the peaks, is $1/\Delta$. The resolving power is given by

$$Q = \frac{\sigma}{d\sigma} = \frac{\lambda}{d\lambda} = \sigma\pi\Delta \frac{\sqrt{\rho}}{1 - \rho} = \pi \frac{\Delta}{\lambda} \frac{\sqrt{\rho}}{1 - \rho}. \tag{13.48}$$

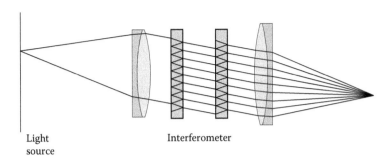

Light
source

Interferometer

FIGURE 13.26 Schematic of a Fabry–Perot interferometer.

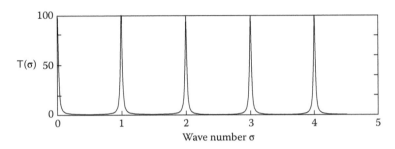

FIGURE 13.27 The Airy function.

It is related to plate separation and the reflectivities, and is proportional to the number of waves in the separation. Assume that the reflectivity is 0.99, that the separation is 5 cm, and that the light is right in the middle of the visible, at 500 nm. Then, the resolving power is

$$Q = \frac{\sigma}{d\sigma} = \frac{\lambda}{d\lambda} = \pi \frac{\Delta}{\lambda} \frac{\sqrt{\rho}}{1-\rho} = 3.14 \frac{0.05}{0.5 \times 10^{-6}} \frac{\sqrt{0.99}}{0.01} = 3.14 \times 10^7 \times 99.5 \approx 3 \times 10^9. \tag{13.49}$$

This extremely high resolving power is the strength of the FP interferometer spectrometer. Most of it, 10^7, comes from the number of waves in the cavity, but having high reflectivity helps (by a factor of about 100) both the Q and the maximum transmission, which increases as the reflectivity of the plates increases. The transmission maximum occurs when the square of the sine is zero and minimum when it is one. The maximum occurs when ϕ is an integral multiple of π, when σnd is a multiple of 1/2, or when nd is an integral number of waves. The separation of the peaks, the free spectral range, is given by

$$\Delta\sigma = \frac{1}{\Delta} = \frac{1}{nd\cos\theta}. \tag{13.50}$$

The free spectral range decreases as the path separation increases (the order number increases), so high resolving power is obtained at the expense of free spectral range. For the example given above, the free spectral range is 0.2 cm^{-1} (the resolution is 2×10^{-12} cm^{-1}). The finesse for collimated light is

$$\frac{\Delta\sigma}{Q} = \frac{1-\rho}{nd\cos\theta\sqrt{\rho m\pi}}. \tag{13.51}$$

The throughput is calculated based on the area of the beam that traverses the plates and the size of the image. The image is a bull's-eye pattern with alternating light and dark annuli, as with the Michelson, and it has the same expression. In fact, the throughput of the Fabry–Perot is the same as that of the

Michelson. Some workers [42] have used curved slits that match an arc of one of the annuli. In that case, the solid angle is $\phi(\cos\theta_2 - \cos\theta_1)$, with the angle ϕ replacing the full circumferential angle 2π.

The expressions given here are for an idealized instrument. Jacquinot provides information on the influence of various imperfections and provides some additional insights into the operation of the performance of Fabry–Perot interferometers.

Fabry–Perot interferometers are research tools, and are not available commercially.

13.8 Laser Spectrometers

Laser spectrometers [43] are different from classical spectrometers in that there is no monochromator; the source is a tunable laser. The line width can be very narrow. These can be used for transmission, reflection, and related measurements, but cannot be used to measure emission spectra. The limitations are strictly those of the bandwidth and spectral range of tunable lasers. Some of the salient characteristics of lasers used in the visible and infrared are power, resolving power, and spectral range.

There are several different types of tunable lasers, all with different characteristics [44]. The most useful type seems to be semiconductor lasers [45], which can span the spectrum from about 350 nm to about 14 μm. This range must be accomplished with different materials, ranging from ZnCdS to HgCdTe, of various mixture ratios. Tuning is accomplished in these semiconductor lasers by varying either the drive current or the pressure. Powers range from microwatts to milliwatts; line widths as small as 10^{-6} cm^{-1} have been observed, but useful widths are somewhat larger. Note that this represents a resolving power of about 10^{10}.

The second most useful type is probably the family of dye lasers [46]. These cover the region from about 350 nm to 1 μm using several different dyes and solvents. Each combination can be tuned several hundred wavenumbers from its spectral peak with a concomitant reduction in power.

Color-center lasers, in which lasers like NdYAG pump certain alkali halides, cover the region from about 1–3 μm, again with several different pump and color-center lasers.

The fourth class is molecular lasers, such as carbon dioxide, carbon monoxide, deuterium fluoride, and carbon disulfide, which are used at relatively high pressures to broaden the vibrational-rotational bands of the gases themselves. These typically have much greater power, but somewhat lower resolving power.

These laser spectrometers have incredibly high resolving powers as a result of the narrowness of the laser lines. They can be used in special situations with relatively limited spectral ranges, with relatively low power and with the specter of mode hopping as a problem to be overcome and controlled.

A summary of the properties of the main tunable lasers that can be used for these spectrometers is given in Table 13.1.

TABLE 13.1 Properties of Spectrometer Lasers

Semiconductor Lasers	Spectral Range	Dye Lasers	Spectral Range	Molecular Lasers	Spectral Range	Color Center Lasers	Spectral Range
ZnCdS	0.3–0.5	Stilbene	0.4–0.42	DF	3.8–4.2	NaF	1.0–1.2
CdSeS	0.5–0.7	Coumarin	0.42–0.55	CO	5.0–8.0	KF	1.2–1.6
GaAlAs	0.6–0.9	Rhodamine	0.55–0.7	CO_2	9.0–12.0	NaCl	1.4–1.8
GaAsP	0.6–0.9	Oxazine	0.7–0.8	CO_2	9.0–12.0	KCl	1.8–1.9
GaAs	0.8–0.9	DOTC	0.75–0.8	–	–	KCl:Na	2.2–2.4
GaAsSb	1.0–1.7	HIDC	0.8–0.9	–	–	KICl:Li	2.3–2.8
InAsP	1.0–3.2	IR-140	0.9–1.0	–	–	RbCl:Li	2.8–3.2
InGaAs	1.0–3.2	–	–	–	–	–	–
InAsSb	3.0–5.5	–	–	–	–	–	–
PbSSe	4.0–8.0	–	–	▦	▦	–	–
HgCdTe	3.2–15	–	–	–	–	–	–
PbSnTe	3.2–15	–	–	–	–	–	–

13.9 Unusual Spectrometers

During the 1950s, largely as a result of the pioneering work of Jacquinot on the importance of throughput in spectrometers and that of Felgett on the value of multiplexing, many innovative designs arose. Some became commercial, while others have died a natural death. The concepts are interesting.

The general treatments have been based on classical spectrometers like the prism and grating instruments described above. Therefore, these newer ones are described as having either a multiplex or a throughput advantage. The spirit of this section is to reveal these interesting concepts for those who might employ one or more of them.

13.9.1 Golay Multislit Spectrometer

Marcel Golay described the technique of using more than one entrance slit and more than one exit slit to gain a throughput and multiplex advantage. His first paper [47] describes what might be called dynamic multislit spectrometry. His example is repeated here; it is for an instrument with six slits. They are modulated by a chopper. Table 13.2 shows how each of the slits is blocked or unblocked by the chopper.

The convention is that one is complete transmission and zero represents no transmission; it is a binary modulation of the slits. During the first half cycle, no radiation that passes through the first entrance slit gets through the first exit slit. During the second half cycle, all of the radiation passes through both. The same is true for the rest of the slits. This provides 100% modulation for corresponding entrance and exit slits. However, during the first half cycle, no light gets through the 1–4, 2–5, and 3–6 slit combinations and one-fourth gets through the others in each half cycle. There is, therefore, strong modulation of the corresponding slits, but no modulation from the "off" combinations.

Static multislit spectroscopy uses some of the same ideas, but establishes two masks, an entrance and an exit mask. Again, using Golay's example [48], assume that there is an optical system with two side-by-side entrance pupils. In each, there is a set of eight slits, and there are corresponding exit pupils and slits. Table 13.3 shows the arrangement.

Light passes through the entire system only when ones are directly above each other in each half. In this arrangement, there are four pairs that allow light through in the first pair of pupils and none in the second. Light of a nearby wavelength is represented by a shift of the entrance pupils. In the first set, only one pair will let light through for a shift of one, and the same is true for the second set. This is also true for shifts of up to six, and no light gets through for shifts of more than six. These, of course, are not

TABLE 13.2 Dynamic Multislit Modulation Scheme

Slit	Entrance Slits		Exit Slits	
	First Half Cycle	Second Half Cycle	First Half Cycle	Second Half Cycle
1	0101	0101	1010	0101
2	0011	0011	1100	0011
3	0110	0110	1001	0110
4	0101	1010	1010	1010
5	0011	1100	1100	1100
6	0110	1001	1001	1001

TABLE 13.3 Modulation Scheme for Static Multislit Spectroscopy

First entrance pupil	11001010
First exit pupil	11001010
Second entrance pupil	10000001
Second exit pupil	01111110

really shifts, but static blockages of unwanted spectra. The scheme can be adapted to more than one row and more slits.

13.9.2 Haddamard Spectrometer

This device bears a very distinct similarity to the Golay devices just described. It consists of a collection of entrance slits and correlated exit slits [49]. Decker [50] describes these as binary masks (100% transmitting and 0% transmitting) arranged in a pseudo-random pattern that does not depend on any prior knowledge. They are an analog of frequency modulation and can be thought of as a set of n equations with n unknowns.

One fairly representative device operated in the 2.5–15 μm region, with a resolution of 3.5 cm^{-1}, from 666–4000 cm^{-1}. It used several gratings and order sorting that will not be described here. The effective throughput was 1.3 mm^2 sr.

13.9.3 Girard Grille

This device is another version of a multislit spectrometer, although one must consider the word *slit* in a very general way. Rather than slits, the entrance and exit planes contain Fresnel zone plates. In other respects, this system operates the same as the Golay static multislit spectrometer. At one specific wavelength, the light passes both grilles without attenuation. At other wavelengths, the dispersed light passes through the first zone plate on-axis, but through the exit plate off-axis. The autocorrelation function of these zone plates is peaked at the reference wavelength (zero displacement) and decays rapidly off-axis. The Girard grille has high throughput, but does not have the multiplex advantage. It was once offered commercially by Huet in Paris, but does not seem to be in existence as a commercial instrument today. The instrument they built was a large (3 m × 2:3 m diameter with 2 m focal length mirrors) laboratory device, which used a vacuum system and PbS, InSb, and extrinsic germanium detectors for operation from 1 to 35 μm [51,52].

13.9.4 Mock Interferometer

Although Mertz described this as a mock interferometer, a more descriptive appellation might be a *rotating multislit spectrometer*. Besides the usual optics, it consists of an entrance aperture, comprising a set of parallel bars, as shown in Figure 13.28, which is a schematic of the arrangement. The exit slit is an image of the entrance slit. As these rotate, a modulation is imposed on the light that is proportional to the distance between the center of the disks. Since the exit slit is an image of the entrance slit in the light dispersed from the prism, the modulation is a function of the frequency of the light. The detector measures a quantity proportional to the integral of the spectral flux × the cosine of the rotational frequency of the disks, that is,

$$V(t) = \Delta \int_{\sigma_1}^{\sigma_2} \Phi(\sigma) \cos^2(2\pi\sigma t + \phi_0) \mathrm{d}r\,, \tag{13.52}$$

where Δ is the responsivity and Φ is the flux on the detector. This system has both the throughput and the multiplex advantage. Although Block engineering offered it commercially for some years, it no longer appears available as a commercial product.

13.9.5 SISAM

The acronym is from the French, *système interferential selection amplitude modulation* (well, it's almost English. Is this the Anglaise that some French despise?). The technique, developed by Connes [53] is

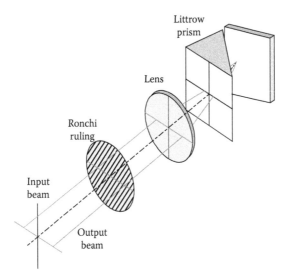

FIGURE 13.28 Mock interferometer.

to use a Michelson interferometer with gratings rather than mirrors. Motion of one of the gratings provides modulation of a given spectral line. The advantage here is throughput. Imagine that a monochromatic beam enters and that the (identical) gratings are set to reflect the beam directly back. Then at zero-path difference, there is constructive interference of that beam. It can be modulated by rotating the compensation plate or by translating one of the gratings. In that way, the radiation of that wavelength is modulated. Radiation of a nearby wavelength will also be modulated, but not as strongly, until the gratings are rotated (in the same direction) to obtain constructive interference for that wavelength.

It can be shown that the resolving power is $2\sigma\Delta$ without apodization and $\sigma\Delta$ with apodization (masking the gratings with a diamond shape). The throughput is the same as for a Michelson; the free spectral range is the same as that for a grating.

13.9.6 PEPSIOSIS

No, this is not some sort of stomach remedy; it stands for purely interferometric high-resolution scanning spectrometer. I confess that I do not understand the acronym. In its simplest form, it is a Fabry–Perot with three etalons. More are described in the pertinent articles. Here it is enough to note that one can eliminate some of the maxima of one etalon by interference with the second etalon and thereby increase the finesse. The arguments and calculations are very similar to those involving the Lyot–Öhman filter [54,55].

13.10 Acousto-Optical Filters

Acousto-optics has provided us with a new method of performing spectral filtering. An acoustic wave can be set up in a crystal. This wave, which is an alternation of rare and dense portions of the medium, provides a diffraction grating in the crystal. The grating spacing can be adjusted by tuning the frequency of the acoustic wave. This is a tunable diffraction grating. It provides the basis for the acousto-optical tunable filter, the AOTF.

There are two types of acousto-optical tunable filters: collinear and non-collinear. In the collinear version, unpolarized light is polarized and propagated through a medium, usually a rectangular cylinder of the proper AO material. An acoustic transducer is attached to the side of the cylinder and by way of a prism, as shown in Figure 13.29, the acoustic waves propagate collinearly with the optical waves.

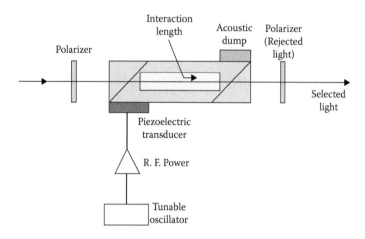

FIGURE 13.29   The collinear AOTF.

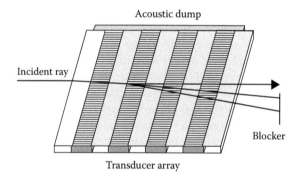

FIGURE 13.30   Non-collinear AOTF.

There is energy coupling so that a new wave with a frequency that is the sum of the acoustic and optical frequencies is generated—as long as the phases match. The acoustic wave is reflected to a beam dump; the output light is passed through an analyzer in order to maintain only the phase-matched light. As a result of the two polarization operations, the maximum transmission is 0.25. In practice, it will also be reduced by the several surfaces and by the efficiency of coupling. The collinear AOTF couples the energy from one polarization state to the other, while the non-collinear system separates the beams in angle, as shown in Figure 13.30.

The center wavelength of the passband λ_0 is given by

$$\lambda_0 = \frac{\upsilon \Delta n}{f},$$
(13.53)

where υ is the acoustic velocity in the material, Δn is the difference between the ordinary and extraordinary refractive indices, and f is the acoustic frequency.

The spectral resolution is given by

$$d\lambda = \frac{0.9\lambda_0^2}{\zeta l \sin^2 \theta_i},$$
(13.54)

where l is the interaction length (the length in the material over which the acoustic waves and the optical waves are superimposed and interact), θ_i is the angle of incidence, and ζ is the dispersion constant, given by

$$\zeta = \Delta n - \lambda_0 \frac{\partial \Delta n}{\partial \lambda_0}, \qquad (13.55)$$

where Δn is the change in refractive index over the band and λ_0 is the center wavelength.

The solid angle of acceptance is given approximately by

$$\Omega = \frac{\pi n^2 \lambda_0}{l \Delta n}. \qquad (13.56)$$

The more accurate representation includes both angles, and is given by

$$d\theta_1 = n \sqrt{\frac{\lambda_0}{d n l F_1}}, \qquad (13.57)$$

$$d\theta_2 = n \sqrt{\frac{\lambda_0}{d n l F_2}}, \qquad (13.58)$$

where

$$F_1 = 2\cos^2 \theta_i - \sin^2 \theta_i$$
$$F_2 = 2\cos^2 \theta_i + \sin^2 \theta_i; \qquad (13.59)$$

where θ_i is again the angle of incidence. The solid angle is obtained by integrating these two differential angles, and when multiplied by the projected area gives the throughput of the system, as usual.

The acoustic power required is given by

$$\tau_0 = \sin^2 \left[\frac{\pi l}{\lambda_0} \sqrt{\frac{M_2 E_a}{2}} \right], \qquad (13.60)$$

where τ_0 is the maximum transmission, E_a is the acoustic power density, and M_2 is the acoustic figure of merit.

In a non-collinear system, the separation of beams is given by

$$\Delta \theta_d = \Delta n \sin 2\theta_0, \qquad (13.61)$$

as illustrated in Figure 13.30.

One of the early AOTFs used in the visible [56] was TeO$_2$, operated from 450 to 700 nm with an acoustic frequency of 100–180 MHZ, a spectral band FWHM of 4 nm and an acoustic power of 0.12 W. Nearly 100% of the incident light is diffracted with an angle of about 6 degrees. The resolution is given by Equation 13.54. The acceptance angles, by Equations 13.57 and 13.58, are 0.016 and 0.015 rad (13.16 and 8.59 degrees and 2.4 msr). The required power is given as 120 mW. Finally, the deviation angle is about 6 degrees.

Westinghouse built an AOTF System [57] for Eglin Air Force Base to measure the spectra of jet aircraft exhaust plumes. It was an imaging spectrometer. The properties are summarized in Table 13.4 [58].

TABLE 13.4   Properties of the Westinghouse AOTF Imaging Spectrometer

Property	Value	Units
Spectral resolution	5	cm^{-1}
Resolving power	400–1000	
Spectral band	2–5	μm
Crystal length, area	3.5, 2:5 × L4	mm
Interaction length	2.6	cm
Acoustic input	0.86	Wcm^{-2} (80% efficiency)
Detector	128 × 128 InSb	$D^* = 4 \times 10^{11}$
Optics	TMA and reimager	
f, F, A	207.4, 3.27, 56:2 × 56:2	mm
SNR[a] vs. 400 K	13 dB @ 2 μm, 31 db @ 3 μm	for 1 frame/s
SNR[a] vs. 400 K	2@ 2 μm, 126 @ 3 μm	for 0.01 frame/s

Source: Bass M, et al. *The Handbook of Optics*, McGraw-Hill, New York, 1995.

[a] The SNR data in db are given directly from the paper, where db = 10 log (ratio). A TMA is a Three-Mirror Anastigmat.

13.11 Spectrometer Configurations

Grating and prism spectrometers come in a variety of configurations: single and double pass, single and double beam, and double monochromator versions.

A classic spectrometer has a source, a monochromator, and a receiver. These configurations are different, according to how the monochromator is arranged. The simplest configuration is a single-pass, single-beam single monochromator, which is shown in Figure 13.31. This configuration uses a source to illuminate the monochromator through the entrance slit, a prism with a Littrow mirror to disperse the light, and an off-axis paraboloidal mirror to focus and collimate the beam. The light of a single wavelength passes through the exit slit to a detector. Although the beam actually passes through the prism twice, it is considered a single-pass instrument. Figure 13.32 shows the double-pass version of this. The use of two passes, provides better resolution at the cost of a second prism and some additional components.

Double-beam spectrometers are used to obtain instrumental corrections for changes in atmospheric transmission, slit size variations, and spectral variations of the source and detector. They are nulling instruments, the most popular of which uses a comb-shaped attenuator to match the signals, and the position of the attenuator is monitored. A pair of choppers is used to switch from beam to beam in the nulling process, as shown in Figure 13.33. With a single-beam instrument, a transmission measurement requires a measurement without a sample and then with, and the ratio-ing of the two. The double-beam does this instrumentally. The two choppers are synchronized so that light passes onto the detector alternately from the sample beam and the reference beam. In this way, everything is the same, except for the comb-shaped attenuator in the reference beam. The position of the comb is monitored, as it always seeks a zero difference between the two beams. Other attenuators can be used, but this one has the advantage of spectral neutrality. The top beam is the sample beam and the bottom beam is the reference beam.

Double-monochromator spectrometers use two monochromators in series to obtain greater immunity from scattered light.

Compound spectrometers may use a prism monochromator as a prefilter for a grating monochromator.

Interferometer spectrometers come in different configurations as well. The two-beam versions most often are in the form of a Michelson interferometer. However, it is critical that the moving mirror does not tip, tilt, or translate. One approach to eliminating tip and tilt is to use a cube corner as the moving mirror; this does not eliminate the possibility of translation. Another approach is the use of cat's-eye mirrors, as shown in Figure 13.34. For Michelsons, in particular, knowledge of the position of the mirror is critical; most systems use a position-sensing laser system to record the position, and this information is used in the Fourier transformation. The transformation can be done accurately even if the velocity of translation is not constant.

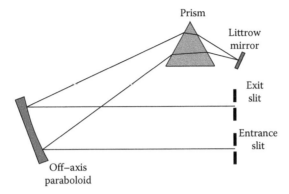

FIGURE 13.31 Single-pass Littrow spectrometer.

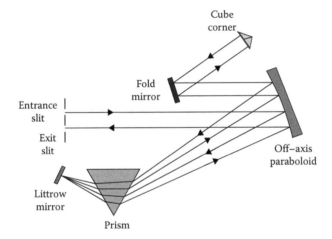

FIGURE 13.32 Double-pass Littrow spectrometer.

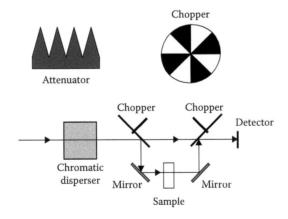

FIGURE 13.33 Schematic of the double-beam principle.

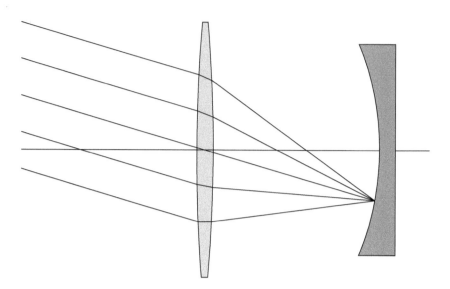

FIGURE 13.34　Cat's-eye reflector.

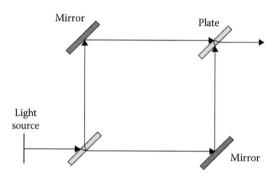

FIGURE 13.35　Mach–Zender interferometer.

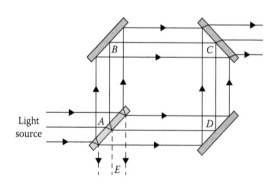

FIGURE 13.36　The Sagnac four-mirror interferometer.

Some have used interferometer configurations other than the Michelson to obtain the advantages of Fourier-transform spectrometers. These include the Mach–Zender, shown in Figure 13.35, and the Sagnac. The Sagnac interferometer is one of a class of interferometers that are of the two-beam variety with beams that traverse the same path in opposite directions. There can be as few as three mirrors and as many as imagination and space will allow. The four-plate Sagnac is shown in Figure 13.36 because it seems to be the easiest to illustrate and explain. The light enters the Sagnac at A where the beam splitter sends it to B above and D to the right. The mirror D sends the light up to C, where it rejoins the beam that went from A to B to C. The two go through the output beamsplitter with a phase difference that is determined by the positions of the mirrors M_1 and M_2. It can be seen that this is not multiple-beam interferometry by considering the lower beam that goes from A to D to C. It is reflected to B and to A, but then it is refracted out of consideration to E. Similarly, the ABCDA beam returns to the source. Similar considerations apply to these counter-rotational interferometers. Some of the history and performance are described by Hariharan, who also gives ample references [59].

There are also several versions of multiple-wave interferometers. These include the double etalon and the use of spherical mirrors [60]. The spherical Fabry–Perot has the same transmission, resolving power, free spectral range, and contrast as the plane version, but the throughput is directly proportional to the resolving power rather than inversely proportional.

13.11.1 Spherical Fabry–Perot

Consider two identical spherical mirrors, the upper parts of which are partially reflecting and the lower parts are completely reflecting, as shown in Figure 13.37. They have centers of curvatures on the opposite mirror, as indicated by C_1 and C_2. A ray, representing a collimated beam, enters the instrument at A and traverses to B, where it is partially reflected. It goes to C on the mirror part, where it is totally reflected towards C and back again to A. There it joins the incident ray with a phase shift that is exactly twice the mirror separation. This is why it has the properties of a Fabry–Perot with twice the etalon separation [58].

The literature also contains descriptions of some designs for maintaining the position of the mirror while it translates [61–65].

Jackson describes a Fabry–Perot with a double etalon to increase the finesse [66].

13.12 Comparisons

There are many similarities among the different types of spectrometers. They are presented here, and a summary of the salient properties is given in Tables 13.5 and 13.6. Table 13.5 lists the different types of spectrometers and the expressions for resolution, $d\sigma$ and $d\lambda$, resolving power Q, free spectral range $\Delta\sigma$, throughput Z, the required number of samples N_s during a spectral scan, and a normalized signal-to-noise ratio, SNR*. For purposes of this summary, the appropriate symbol definitions are reviewed here. The prism base is b, $dn/d\lambda$ is the prism material dispersion, m is the grating order number, N is the number of lines in the grating, Δ is the maximum total path difference, R is the reflection factor in the Fabry–Perot filter; that is, $\sqrt{\rho}/(1-\rho)$, s is the grating spacing, h and w are the linear dimensions of the grating or prism slit, α and β are the angular measures, and the equivalent noise bandwidth B is given by $B = N_s/(2t_s)$.

There have been other such comparisons, and it seems appropriate to compare their results as well. Jacquinot compared gratings, prisms and Fabry–Perots. He compared a grating to a prism that has

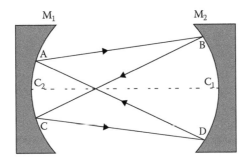

FIGURE 13.37 The spherical Fabry–Perot.

TABLE 13.5   Spectrometer Expressions

Type	$d\sigma$	$d\lambda$	Q	$\Delta\sigma$	Z	N_s	SNR*
One-layer filter	$\dfrac{1}{\pi\Delta}$		$\pi\dfrac{\Delta}{\lambda}$		$\alpha\beta A_{\text{filter}}$	$\Delta\sigma/d\sigma = \Delta\lambda/d\lambda$	$\alpha\beta A_{\text{filter}}\sqrt{\dfrac{\Delta\lambda d\lambda}{A_d}}$
AOTF	$\dfrac{0.9\lambda_0^2}{\zeta l \sin^2\theta_i}$	$\dfrac{\zeta l \sin^2\theta_i}{0.9\lambda_0}$		$\dfrac{\text{AOTF}\pi n^2\lambda}{l\Delta n}$	$\Delta\sigma/d\sigma = \Delta\lambda/d\lambda$	$\alpha\beta A_{\text{AOTF}}\sqrt{\dfrac{\Delta\lambda d\lambda}{A_d}}$	
Prism	$\dfrac{\lambda}{bdn/d\lambda}$	$bdn/d\lambda$	∞		$\alpha\beta A_p\cos\theta$	$\Delta\sigma/d\sigma = \Delta\lambda/d\lambda$	$\alpha\beta A_p\cos\theta\sqrt{\dfrac{\Delta\lambda d\lambda}{A_d}}$
Grating	$\dfrac{s\cos\theta_d d\theta_d}{m}$	$mN=(\Delta/\lambda)(\sin\theta)$	λ_x/m	$\alpha\beta A_p\cos\theta$	$\Delta\sigma/d\sigma = \Delta\lambda/d\lambda$	$\alpha\beta A_g\cos\theta\sqrt{\dfrac{\Delta\lambda d\lambda}{A_d}}$	
FTS	$\dfrac{1}{2\Delta}$		$2\Delta/\lambda$	∞	$\dfrac{4\pi A_{\text{FTS}}}{Q}$	$2\Delta/\lambda_n$	$\dfrac{A_{\text{FTS}}\phi\lambda_{\min}}{\sqrt{A_d}}\sqrt{\dfrac{2}{d\lambda\Delta}}$
FP	$\dfrac{R}{2\pi\Delta}$		$2\pi(\Delta/\lambda)2R$	$1/\Delta$	$\dfrac{2\pi A_{\text{FP}}}{Q}$		$\dfrac{A_{\text{FTS}}\phi\lambda_{\min}}{\sqrt{A_d}}\sqrt{\dfrac{2}{d\lambda\Delta}}$

the same base area as the grating area. The ratio of the flux outputs for the same resolving power was found to be

$$\frac{\Phi_{\text{prism}}}{\Phi_{\text{grating}}} = \frac{\lambda dn/d\lambda}{2\sin\theta} \rightarrow \lambda\frac{dn}{d\lambda}, \tag{13.62}$$

where the blaze angle has been taken as 30 degrees for the final expression. Since the dispersion is very low in the regions where the prism transmission is high, the practical advantage for the grating is about 100.

The comparison of the grating with the Fabry–Perot proceeded along the same lines, whereby

$$\frac{\Phi_{\text{FP}}}{\Phi_{\text{grating}}} = \frac{\tau_{\text{FP}}0.7\pi^2/2}{\tau_{\text{grating}}2\beta\sin\theta} = \frac{3.4}{\beta} \tag{13.63}$$

for equal resolving powers and areas, where the factor of 0.7 is the ratio of the effective resolving power to the theoretical value, β is the angular height of the slit, and θ is again taken as 30 degrees. Values for β typically run from 0.1 to 0.01; the Fabry–Perot is therefore from 34 to 340 times as good as a grating based on these assumptions and criteria. Any other interferometric device that does not have a slit will

TABLE 13.6   Representative Spectrometer Properties

Type	$\Delta\lambda$ (μm)	$\Delta\lambda$ singly (μm)	Q	Aperture size (cm)
Filter	0.3–1000	octave	100	1–20
AOTF	0.5–14	$<\lambda_0$	1000	<1
Prism	0.2–80	10	1000	1–20
Grating	1–1000	octave	10^6	1–40
FTS	1–1000	detector/splitter	10^6	10
FP	1–1000	detector/splitter	10^9	10
Laser	0.3–1.5	$<\lambda_0$	10^{12}	0.5

have an equivalent advantage over the grating and prism instruments. The evaluation criteria are valid, but they are limited. Other considerations may enter, including the size of the free spectral range, ease of scanning, and computation of spectra as well as the size, weight, availability, and convenience of the components and their implementation.

The comparisons change a little when the full sensitivity equation is used as part of the comparison. The equation that determines the signal-to-noise ratio, SNR, is repeated here for convenience:

$$\text{SNR} = \frac{D^* L_\lambda d\lambda Z}{\sqrt{A_d B}}, \tag{13.64}$$

where D^* is the specific detectivity, L_λ is the spectral radiance, $d\lambda$ is the spectral interval, Z is the throughput, A_d is the detector area, and B is the noise bandwidth. The values for the throughput derived above can be used in this expression. For prisms and gratings, the required sample number is $\Delta\lambda/d\lambda$, the total spectrum divided by the resolution. For a Fourier-transform spectrometer, the required sample number is twice the number of minimum wavelengths in the maximum path difference, that is, Δ/λ_{min}, and this can be related to the resolving power, $\sigma_{max}/d\sigma$. The other major difference is that the flux available to the prism and grating is the spectral radiance $L_\lambda \times$ the resolution $d\lambda$, while for the Fourier-transform spectrometer, it is the spectral radiance $\times$ the total spectral interval $\Delta\lambda$ (more precisely, the integral over the band of the spectral radiance). These equations can be reformulated in a variety of ways by the use of the expressions given in Table 13.5, which is a summary of the expressions that have been determined earlier.

The normalized SNR is found from the appropriate expressions for the SNR itself. In terms of the specific detectivity, the SNR is

$$\text{SNR} = \frac{D^* L_\lambda d\lambda Z}{\sqrt{A_d B}} = \frac{D^* L_\sigma d\sigma Z}{\sqrt{A_d B}}. \tag{13.65}$$

Since the detectivity depends on the detector selected, the spectral radiance depends on the source used, and the sample time is a design variable, the normalized SNR takes these out of consideration. The normalized SNR is the SNR per unit specific detectivity, spectral radiance, and the square root of $2t_s$, that is,

$$\text{SNR}^* = \frac{\text{SNR}\sqrt{2t_s}}{D^* L_\sigma} = \frac{\text{SNR}\sqrt{2t_s}}{D^* L_\lambda}. \tag{13.66}$$

It can be expressed in terms of the radiance per unit wavelength or radiance per unit wavenumber as long as the spectral interval is expressed in the proper units.

Perhaps the first and most obvious generality is that grating, Fourier-transform spectrometer, and Fabry–Perot spectrometers, all of which are based on interference and have a resolving power that is proportional to the number of waves in the maximum path difference. The same is true for a Fabry–Perot filter. (This also applies to the MTF of a diffraction-limited imaging system. The cutoff frequency is $1/(\lambda F) = (1/f)(D/\lambda)$, where D is the optics diameter, f is the focal length, and F is the speed.)

Greenler has used the product of throughput and resolving power as a figure of merit for rating spectrometers. Using this lead, I define a figure of merit, FOM, as the normalized SNR $\times$ the resolving power. Then, using the information given above, the FOM is easily generated.

Table 13.6 shows some representative performance characteristics of the different types of spectrometers. Of course, a representation is only a guide and can only be approximate, that is, representative. The spectral range $\Delta\lambda$ may need to be covered by more than one instrument. A second column indicates the range generally covered by one instrument, prism, grating, laser, and so on. The filter spectral range depends on materials, but generally covers the ultraviolet through the long-wave infrared and even the millimeter range. The individual range is usually one octave.

The resolving power is a function of construction; it can be made larger, but usually at the expense of maximum transmission. The AOTF range is covered by different materials. Each of them has a range that is somewhat less than the center wavelength λ_0. The aperture size today is typically less than 1 cm, thereby limiting the throughput. Prisms cover the range from the ultraviolet (with LiF) to the long-wave infrared, but are practically limited at about 80 mm (CsI). The limitation to a single prism is a result of having proper transmission and dispersion over a limited range. The resolving power is a function of the dispersion, typically from 0.001 to 0.01 μm in this range and the size of the prism. The (reflection) grating is limited in its single range by the groove spacing and the blaze efficiency as well as overlapping of orders. The octave is obtained only for use in the first order, where the resolving power is smaller. The Fourier-transform spectrometer has a spectral range that is limited only by the transmission of the beamsplitter and the spectral sensitivity of the detector. Its resolving power is a function of the path difference and the minimum wavelength of the spectral region. The value given is for 1 mm and 1 m, which is somewhat extreme. The Fabry–Perot has the same spectral limitations as the FTS, but it has a higher resolving power as a result of multiple-wave interference. Laser spectrometers, like the AOTF, are limited to an individual spectral range of a little less than the center wavelength. Their resolving power is phenomenal. Size is not so important as the fact that the output is of very high radiance and is very bright.

References

1. Cajori, F., *A History of Physics,* Mishawaka, IN: Dover, 1962.
2. Bohr, N., *Philosophical Magazine,* "On the Constitution of Atoms and Molecules, Part I," **26**, 1–25 (1913).
3. Planck, M., *Physikalische Abhandlungen und Vortrage,* Band III, Braunschweig: Druck und Verlag (1958).
4. Sommerfeld, A., *Atombau und Spektrallinien,* 4th Ed., Braunschweig: Druck und Verlag von Fried, 1921.
5. Millikan, R. A., "The Last Fifteen Years of Physics," *Proceedings of the American Philosophical Society,* **65,** 74–78 (1926).
6. Rowland, H. A., "On the Relative Wave-lengths of the Lines of the Solar Spectrum," *Philosophical Magazine,* **23,** 257–265 (1887);
7. Harrison, G. R., *Journal of the Optical Society of America,* "The Production of Diffraction Gratings I, Development of the Ruling Art," **39,** 413–426 (1949).
8. Wallace, R. J., "Orthochromatic Plates, Preliminary Note On," *The Astrophysical Journal,* **22,** 153 (1905).
9. Dereniak, E. and D. Crowe, *Radiation Detectors,* Hoboken, NY: Wiley, 1984.
10. Jacquinot, P., "Characteristics Common to the New Methods of Interferometric Spectroscopy: The Factor of Merit," *Journal de Physique et le Radium,* **19,** 223 (1958).
11. Greenler, R.,"Interferometry in the Infrared," *Journal of the Optical Society of America;* **45,** 788–791 (1955).
12. Jacquinot, P. "The Luminosity of Spectrometers with Prisms, Gratings or Fabry-Perot Etalons.", *Journal of the Optical Society of America,* **44,** 761–765 (1954).
13. Dobrowolski, J., "Coatings and Filters," in Driscoll, W. and W. Vaughan, eds, *Handbook of Optics,* New York, NY: McGraw-Hill, 1978.
14. Strong, J., *Procedures in Experimental Physics,* Englewood Clifs, NJ: Prentice Hall, 1938.
15. Wolfe, W., "Optical Materials," in Wolfe, W. and G. Zissis, Eds, *The Infrared Handbook,* Environmental Research Institute of Michigan
16. Wolfe, W., Volume 3, Chapter 1 "Optical Materials," in Acetta, J. and Schumaker, D., Eds, *The Intrafred and Electro-Optical Systems Handbook*; Bellingham, WA: SPIE Press, 1993.

17. Tropf, W., Thomas M., and T. Harris, "Properties of Crystals and Glasses," in Bass, M.,. Palmer, D., Van Stryland, E. and W. Wolfe, eds, *Handbook of Optics*, McGraw-Hill, 1995.

18. McAlister, E., "The Christiansen Light Filter: Its Advantages and Limitations," *Smithsonian Miscellaneous Collection*, **93**, 11 (1935).

19. Christiansen, C., "Untersuchungen tiber die optischen Eigenschaften von fein vert- heilten Korpern," *Annalen der Physik und Chemie*, **23**, 298 (1884).

20. Strong, J., *Procedures in Applied Optics*, New York, NY: Dekker, 1988.

21. Dobrowolski, J., "Optical Properties of Films and Coatings," in Bass, M., Palmer, D., Van Stryland, E., and W. Wolfe, Eds, *Handbook of Optics*, New York, NY: McGraw-Hill, 1995.

22. Wolfe, W., *Introduction to Infrared System Design*, Bellingham WA, SPIE Press, 1996.

23. Fraunhofer, J., "Kurzer Bericht von den Refultaten neuerer Verfuche uber dieGefetze des Lichtes un die Theorie derfelben," *Annalen der Physik*, **74**, 337–373 (1823).

24. Rowland, H. A.,"Preliminary Note on the Results Accomplished in the Manufacture and Theory of Grating for Optical Purposes," *Philosophical Magazine*, **13**, 469–474 (1882).

25. Rowland, H. A., "On Concave Gratings for Optical Purposes," *Philosophical Magazine*, 16, 197–210 (1883).

26. Harrison, G. R., The Production of Diffraction Gratings. I. Development of the Ruling Art," *Journal of the Optical Society of America*, **39**, 413–426 (1949).

27. Wood, R. W., *Nature*, "Recent Improvements in Diffraction Gratings and Replicas," **140**, 723–724 (1937).

28. Jenkins, F. A. and H. E. White, *Fundamentals of Optics*, 3rd Ed., New York NY: McGraw-Hill, 1957; and other basic texts.

29. Goodman, J. E., *Introduction to Fourier Optics*, New York, NY: McGraw-Hill, 1968.

30. Wood, R. W., *Physical Optics*, 3rd Ed., New York, NY: Macmillan Co., 1934.

31. Michelson, A. A., "The Relative Motion of the Earth and the Luminiferous Ether," *American Journal of Science*, **22**, 120–129 (1881).

32. Twyman, F. and A. Green, British Patent 103382 (1916).

33. Felgett, P., Thesis, Cambridge MA: University of Cambridge, 1951.

34. Jacquinot, P., and C. Dufour, *Journal of Research of CNRS*, **6**, 91 (1948).

35. Strong, J., *Concepts of Classical Optics*, San Francisco, CA : W. H. Freeman and Co, 1958.

36. Steel, W. H., *Interferometry*, Cambridge, MA: Cambridge University Press, 1967.

37. Bell, R. J., *Introductory Fourier Transform Spectroscopy*, Boston, MA: Academic Press, 1972.

38. Treffers, R. R., *Applied Optics*, ""Signal-to-Noise Ratio in Fourier Spectroscopy." **16**, 3103–3106 (1977).

39. Carli, B. and V. Natale, *Applied Optics*, "Efficiency of Spectrometers," **18**, 3954–3958 (1979).

40. Junttile, M., *Applied Optics*, "Stationary Fourier Transform Spectrometer," **31**, 4106–4112 (1993).

41. Anonymous, "Mil Handbook 141," *Military Standardization Handbook, Optical Design*, Defense Supply Agency, 1962.

42. Hirshberg, J. G., "The Use of Curved Slits for the Placing in Tandem of a Diffraction Grating with a Fabry-Perot Interferometer with Scanning by Pressure Variation," *Journal de physique et la de radium*, **19**, 256 (1958).

43. Hinkley, E. D., K. W. Nill, and F. A. Blum, "Infrared Spectroscopy with Tunable Lasers," in H. Walther, ed., *Laser Spectroscopy of Atoms and Molecules*, Springer-Verlag, 1974.

44. Weber, M. J., ed., *CRC Handbook of Laser Science and Technology*, *Vol. 1* Boca Raton, FL: CRC Press, 1986.

45. Koechner, W., *Solid State Laser Engineering*, Berlin Heidelberg: Springer-Verlag, 1976.

46. Schaefer, F. P., "Principles of Dye Laser Operation," in F. P. Schaefer, ed., *Dye Lasers*, Berlin Heidelberg: Springer-Verlag, 1973.

47. Golay, M.,, "Multi-Slit Spectrometry," *Journal of the Optical Society of America*, **39**, 437–444 (1949).

48. Golay, M., "Static Multislit Spectrometry and its Application to the Panoramic Display of Infrared Spectra," *Journal of the Optical Society of America*, **41**, 468–472 (1951).

49. Harwit, M. and N. Sloane, *Haddamard Transform Optics,* New York, NY:Academic Press, 1979.

50. Decker, J., "Haddamard Transform Spectroscopy," *Industrial Research,* **2,** 61 (1973).

51. Girard, A., "Nouveaux Dispositifs de Spectroscopie a Grande Luminosite," *Optica Acta,* **7,** 81–97 (1960);

52. Girard A.,"Spectrometre a Grilles," *Applied Optics,* **2,** 79–87 (1963).

53. Connes, P., "An Interferometric Spectrometer for Selecting by Amplitude Modulation," *Journal de Physique et la de Radium,* **19,** 215–222 (1960).

54. Mack, J. E., D. P. McNutt, F. L. Roessler, and R. Chabbal, "The Pepsios Purely Interferometric High-Resolution Scanning Spectrometer I, The Pilot Model," *Applied Optics,* **2,** 873–980 (1963).

55. McNutt, D. P., "The Pepsios Purely Interferometric High-Resolution Scanning Spectrometer II. Theory of Spacer Ratios," *Journal of the Optical Society of America,* **55,** 288 (1965).

56. Bass, M., E. Van Stryland, D. Williams, and W. Wolfe, eds, *Handbook of Optics,* McGraw-Hill, New York, 1995, p. 12.30.

57. Taylor, L. H., D. R. Shure, S. A. Wutzke, P. L. Ulerich, G. D. Baldwin, and M. T. Myers, "Infrared Spectroradiometer Design Based on an Acousto-Optical Tunable Filter," *Proc. SPIE,* **2480,** 334–345 (1995).

58. Bass, M., D. Palmer, E. Van Stryland, and W. Wolfe, *Handbook of Optics,* New York, NY: McGraw-Hill, 1995.

59. Hariharan, P., *Applied Optics,* "Sagnac or Michelson-Sagnac Interferometer?," **14,** 2319_1-2321 (1975).

60. Connes, P., "The Spherical Fabry-Perot Etalon," *Le journal de physique et la radium,* **19,** 262 (1958).

61. Chabbal, R. and M. Soulet, "a Device Permitting the Mechanical Displacement of a Fabry-Perot Flat," **19,** 274 (1958).

62. Gorbert, J., "Electromagnetic Sweeping of Fabry-Perot Rings," **19,** 278 (1958).

63. Roig, J., "Thermal Sweeping of Fabry-Perot Rings," **19,** 284 (1958).

64. Dupeyrat, R., "A Study of Electrical Scanning Methods for Recording Interferometers," **19,** 290 (1958).

65. Peck, E. R., "The Method and Results of Wavelength Methods with a Corner Reflector Interferometer, " **19,** 397 (1958).

66. Jackson, D. A., "Precision Measurements with the Fabry-Perot Double Etalon," **19,** 379 (1958).

Appendix 13.1 Glossary

a	beam width
A_d	detector area
A_e	entrance pupil area
A_f	field stop area
A_{slit}	slit area
A_{prism}	prism face area
b	prism base
B	effective noise bandwidth
c	light speed in vacuo
d	distance
$d\lambda$	resolution
$d\sigma$	resolution
D^*	specific detectivity
E	irradiance
f	focal length
F	focal ratio, speed
$\wedge$	finesse
h	slit height

k	radian wavenumber
l	interaction length
L	radiance
L_λ	spectral radiance
m	order number
M_2	acoustic figure of merit
n	refractive index
N	number of rulings
NA	numerical aperture
p	length of prism side
Q	resolving power
R	$4\rho/(1-\rho)^2$
5	responsivity
s	ruling spacing
SNR	signal-to-noise ratio
t	time
υ	velocity
w	slit width
Z	throughput

Greek Symbols

α	prism angle
β	groove width phase
δ	deviation angle
ε	phase shift on reflection
ζ	dispersion constant
θ	general angle
$\Delta\lambda$	wavelength spectral range
Δ	path difference
$\Delta\sigma$	wave number spectral range
γ	half spacing phase
λ	wavelength
ν	frequency
ρ	reflectance
σ	wavenumber
τ	transmittance
Φ	flux
Ψ	wave function
Ω	solid angle
Ω'	projected solid angle

Subscripts and Superscripts

d	diffraction
i	incident
d	detector
0	optics
1	first
2	second

Index

9 780367 872960